D1154777

UNIVERSITY OF WATERLOO

FEB 4 1993

DAVIS CENTRE LIBRARY
GOVERNMENT PUBLICATIONS

Withdrawn
University of Waterloo

Optimization and Artificial Intelligence in Civil and Structural Engineering

Volume I: Optimization in Civil and Structural Engineering

Withdrawn
University of Waterloo

NATO ASI Series

Advanced Science Institutes Series

A Series presenting the results of activities sponsored by the NATO Science Committee, which aims at the dissemination of advanced scientific and technological knowledge, with a view to strengthening links between scientific communities.

The Series is published by an international board of publishers in conjunction with the NATO Scientific Affairs Division

A Life Sciences	Plenum Publishing Corporation
B Physics	London and New York
C Mathematical	Kluwer Academic Publishers
and Physical Sciences	Dordrecht, Boston and London
D Behavioural and Social Sciences	
E Applied Sciences	
F Computer and Systems Sciences	Springer-Verlag
G Ecological Sciences	Berlin, Heidelberg, New York, London,
H Cell Biology	Paris and Tokyo
I Global Environmental Change	

NATO-PCO-DATA BASE

The electronic index to the NATO ASI Series provides full bibliographical references (with keywords and/or abstracts) to more than 30000 contributions from international scientists published in all sections of the NATO ASI Series.
Access to the NATO-PCO-DATA BASE is possible in two ways:

– via online FILE 128 (NATO-PCO-DATA BASE) hosted by ESRIN, Via Galileo Galilei, I-00044 Frascati, Italy.

– via CD-ROM "NATO-PCO-DATA BASE" with user-friendly retrieval software in English, French and German (© WTV GmbH and DATAWARE Technologies Inc. 1989).

The CD-ROM can be ordered through any member of the Board of Publishers or through NATO-PCO, Overijse, Belgium.

Series E: Applied Sciences - Vol. 221

Optimization and Artificial Intelligence in Civil and Structural Engineering

Volume I: Optimization in Civil and Structural Engineering

edited by

B. H. V. Topping

Department of Civil and Offshore Engineering,
Heriot-Watt University,
Edinburgh, U.K.

UNIVERSITY OF WATERLOO
FEB 4 1993
DAVIS CENTRE LIBRARY
GOVERNMENT PUBLICATIONS

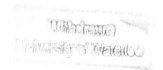

Withdrawn
University of Waterloo

Kluwer Academic Publishers

Dordrecht / Boston / London

Published in cooperation with NATO Scientific Affairs Division

Proceedings of the NATO Advanced Study Institute on
Optimization and Decision Support Systems in Civil Engineering
Edinburgh, U.K.
25 June–6 July, 1989

Library of Congress Cataloging-in-Publication Data

Optimization and artificial intelligence in civil and structural
 engineering / edited by B.H.V. Topping.
      p.   cm. -- (NATO ASI series. Series E, Applied sciences ; vol.
 221)
    Contents: v. 1. Optimization in civil and structural engineering -
 - v. 2. Artificial intelligence in civil and structural engineering.
    ISBN 0-7923-1957-5 (set : acid free paper). -- ISBN 0-7923-1955-9
 (v. 1 : acid free paper). -- ISBN 0-7923-1956-7 (v. 2 : acid free
 paper)
    1. Civil engineering--Mathematical models. 2. Structural
 engineering--Mathematical models. 3. Artificial intelligence.
 I. Topping, B. H. V. II. Series: NATO ASI series. Series E,
 Applied sciences ; no. 221.
 TA153.O68   1992
 624'.01'5118--dc20   92-26741

ISBN 0-7923-1955-9 (Volume I)
ISBN 0-7923-1956-7 (Volume II)
ISBN 0-7923-1957-5 (Set)

Published by Kluwer Academic Publishers,
P.O. Box 17, 3300 AA Dordrecht, The Netherlands.

Kluwer Academic Publishers incorporates the publishing programmes of
D. Reidel, Martinus Nijhoff, Dr W. Junk and MTP Press.

Sold and distributed in the U.S.A. and Canada
by Kluwer Academic Publishers,
101 Philip Drive, Norwell, MA 02061, U.S.A.

In all other countries, sold and distributed
by Kluwer Academic Publishers Group,
P.O. Box 322, 3300 AH Dordrecht, The Netherlands.

Printed on acid-free paper

All Rights Reserved
© 1992 Kluwer Academic Publishers and copyright holders as specified on appropriate pages within
No part of the material protected by this copyright notice may be reproduced or
utilized in any form or by any means, electronic or mechanical, including photo-
copying, recording or by any information storage and retrieval system, without written
permission from the copyright owner.

Printed in the Netherlands

Contents

Part I

vi

DIRECTOR

Dr B H V Topping, Department of Civil Engineering, Heriot-Watt University, Edinburgh, United Kingdom

ORGANISING COMMITTEE

Professor C B Brown, Department of Civil Engineering, University of Washington, Seattle, Washington, United States of America

Professor J S Gero, Department of Architectural Science, Univesity of Sydney, Sydney, Australia

Professor D Grierson, Department of Civil Engineering, University of Waterloo, Ontario, Canada

Professor P W Jowitt, Department of Civil Engineering, Heriot-Watt University, Riccarton, Edinburgh, United Kingdom

Professor A B Templeman, Department of Civil Engineering, The University of Liverpool, Lancashire, England

Dr B H V Topping, Department of Civil Engineering, Heriot-Watt University, Edinburgh, United Kingdom

PRINCIPAL LECTURERS

Professor C B Brown, Department of Civil Engineering, University of Washington, Seattle, Washington, United States of America

Professor D G Elms, Department of Civil Engineering, University of Canterbury, Christchurch, New Zealand

Professor J S Gero, Department of Architectural Science, Univesity of Sydney, Sydney, Australia

Professor L Gründig, Institut für Geodäsie und Photogrammetrie, Technische Universität Berlin, Berlin, West Germany

Professor D Grierson, Department of Civil Engineering, University of Waterloo, Ontario, Canada

Dr M H Houck, Purdue University, West Lafayette, Indiana, United States of America

A T Humphrey, GEC Marconi Research Centre, Great Baddow, Chelmsford, Essex, England

Dr D.G. Jamieson, Thames Water, Reading, Berkshire, England

Professor A Jennings, Department of Civil Engineering, Queeen's University of Belfast, Belfast, United Kingdom

Professor P W Jowitt, Department of Civil Engineering, Heriot-Watt University, Riccarton, Edinburgh, United Kingdom

Professor U Kirsch, Department of Civil Engineering, Technion, Israel Institute of Technology, Technion City, Haifia, Israel

Professor M.L. Maher, Department of Civil Engineering, Carnegie-Mellon University, Pittsburgh, United States of America

Dr R N Palmer, Department of Civil Engineering, University of Washington, Seattle, Washington, United States of America

Professor G I N Rozvany, Department of Civil Engineering, University of Essen, Essen, West Germany

Professor A B Templeman, Department of Civil Engineering, The University of Liverpool, England

Professor G Thierauf, Department of Civil Engineering, University of Essen, Essen, West Germany

Dr B H V Topping, Department of Civil Engineering, Heriot-Watt University, Edinburgh, United Kingdom

Dr G N Vanderplaats, VMA Engineering, Goleta, California, United States of America

PARTICIPANTS

Professor A Adao-da-Fonseca, Universidade do Porto, Gabinete de Estruturas, Faculdade de Engenharia, Porto, Portugal

Dr G Akhras, Department of Civil Engineering, Royal Military College of Canada, Kingston, Ontario, Canada

Professor D Altinbilek, Department of Civil Engineering, Middle East Technical University, Ankara, Turkey

Dr T Arciszewski, Department of Civil Engineering, Wayne State University, Detroit, Michigan, United States of America

A M Azevedo, University of Porto, Faculty of Engineering, Gabinete de Estructures, Porto, Portugal

Dr B W Baetz, Department of Civil Engineering, McMaster University, Hamilton, Ontario, Canada

J Bäuerle, Institut für Geodäsie und Photogrammetrie, Technische Universität Berlin, Berlin, West Germany

S Bitzarakis, Department of Civil Engineering, National Technical University, Athens, Greece

Professor D G Carmichael, School of Civil Engineering, University of New South Wales, Kensington, New South Wales, Australia

Duane Castaneda, Department of Civil Engineering, University of Washington, Seattle, Washington, United Sates of America

S Coelho, LNEC - National Laboratory of Civil Engineering, Lisboa, Portugal

J B Comerford, Department of Civil Engineering, Bristol University, Bristol, England

J W Davidson, Department of Civil Engineering, University of Manitoba, Winnipeg, Manitoba, Canada

I Enevoldsen, University of Aalborg, Aalbourg, Denmark

J H Garcelon, Department of Aerospace Engineering, University of Florida, Gainesville, United States of America

Professor W J Grenney, Department of Civil & Environmental Engineering, Utah State University, Logan, Utah, United States of America

Professor D F Haber, Department of Civil Engineering, University of Idaho, Moscow, Idaho, United Sates of America

Professor P Hajela, Department of Aerospace Engineering, University of Florida, Gainesville, Florida, United States of America

T B Hardy, Department of Civil Engineering, Utah State University, Logan, Utah, United States of America

Professor M I Hoit, Department of Civil Engineering, University of Florida, Gainesville, Florida, United States of America

Dr B Heydecker, Transport Studies Group, University College London, London, England

D Howarth, Department of Civil Engineering, Heriot-Watt University, Riccarton, Edinburgh, United Kingdom

Dr S Hernandez, Department of Mechanical & Environmental Engineering, University of California, Santa Barbara, Santa Barbara, California, United States of America

J Jih, Dept of Aerospace Engineering, University of Florida, Gainesville, Florida,

United States of America

D Kallidromitou, National Technical University of Athens, Hydraulics Laboratory, Athens, Greece,

K Kayvantash, Baumechanik Instatik, Universität Essen, Essen, West Germany

Dr J O Kim, Department of Civil Engineering, University of Liverpool, Brownlow Street, Liverpool, England

Dr V K Koumousis, Department of Civil Engineering, National Technical University of Athens, Athens, Greece

Dr B Kumar, Department of Civil Engineering, Heriot-Watt University, Edinburgh, United Kingdom

Dr J Liu, Department of Civil Engineering, University of Liverpool, Brownlow Street, Liverpool

Professor I A Macleod, Department of Civil Engineering, University of Strathclyde, Glasgow

S Maheepala, Department of Civil Engineering, University of Newcastle upon Tyne, Newcastle upon Tyne

Professor M Malafaya-Baptista, Faculdade de Engenharia, Laboratorio de Hidraulica, Universidade do Porto, Porto, Portugal

Dr I M May, Department of Civil Engineering, University of Bradford, Bradford, England

Profesor P H McDonald, Department of Civil Engineering, North Carolina State University, Raleigh, North Carolina, United States of America

S Meyer, Department of Civil Engineering, Carnegie Mellon University, Pittsburg, United States of America

Dr T C K Molyneux, Department of Civil Engineering, University of Liverpool, Brownlow Street, Liverpool, England

Dr J Oliphant, Department of Civil Engineering, Heriot-Watt University, Edinburgh, United Kingdom

Dr M Oomens,Technische Universiteir, Civiele Techniek, Delft, The Netherlands

Professor G Palassopoulos, Military Academy of Greece, Athens, Greece

J C Rodrigues, Universidade de Coimbra, Departmento de Engenharia Civil, Coimbra, Portugal,

G Schoeppner, Department of Civil Engineering, Ohio State University, Columbus,

Ohio, United States of America

Professor L M C Simoes, Departmento de Engenharia Civil, Universidade de Coimbra, Coimbra, Portugal

A Soeiro, Universidade do Porto, Departmento Engenharia Civil, Porto, Portugal

Dr D G Toll, School of Engineering & Applied Science, University of Durham, Durham, England

Dr F Turkman, Dokuz Eylul University, Department of Civil Engineering, Bornova, Izmir, Turkey

Dr G J Turvey, Engineering Department, Lancaster University, Bailrigg, Lancaster, England

N A Tyler, Transport Studies Group, University College London, London, England

Professor G Ulusoy, Department of Industrial Engineering, Bogazici University, Istanbul, Turkey

P Von Lany, Sir William Halcrow and Partners, Burderop Park, Swindon, Wiltshire, England

Dr S Walker, North West Water Authority, Warrington, England

Dr G A Walters, School of Engineering, University of Exeter, Exeter, England

Professor W E Wolfe, Department of Civil Engineering, Ohio State University, Columbus, Ohio, United States of America

C Xu, Department of Civil Engineering, Heriot-Watt University, Riccarton, Edinburgh, UK

PREFACE

This volume and its companion volume includes the edited versions of the principal lectures and selected papers presented at the NATO Advanced Study Institute on *Optimization and Decision Support Systems in Civil Engineering.*

The Institute was held in the Department of Civil Engineering at Heriot-Watt University, Edinburgh from June 25th to July 6th 1989 and was attended by eighty participants from Universities and Research Institutes around the world. A number of practising civil and structural engineers also attended. The lectures and papers have been divided into two volumes to reflect the dual themes of the Institute namely Optimization and Decision Support Systems in Civil Engineering. Planning for this ASI commenced in late 1986 when Andrew Templeman and I discussed developments in the use of the systems approach in civil engineering. A little later it became clear that much of this approach could be realised through the use of knowledge-based systems and artificial intelligence techniques. Both Don Grierson and John Gero indicated at an early stage how important it would be to include knowledge-based systems within the scope of the Institute.

The title of the Institute could have been: 'Civil Engineering Systems' as this would have reflected the range of systems applications to civil engineering problems considered by the Institute. These volumes therefore reflect the full range of these problems including: structural analysis and design; water resources engineering; geotechnical engineering; transportation and environmental engineering.

The systems approach includes a number of common threads such as: mathematical programming, game theory, utility theory, statistical decision theory, networks, and fuzzy logic. But a most important aspect of the Institute was to examine similar representations of different civil and structural engineering problems and their solution using general systems approaches. This systems approach to civil and structural engineering is well illustrated in the first paper in this volume. The Decision Support aspect of the Institute was reflected by the knowledge-based systems and artificial intelligence approach. Papers discussing many aspects of knowledge-based systems and artificial intelligence in civil and structural engineering are included in the second volume of these proceedings.

I should like to thank all the members of the organising committee for their assistance given so readily both before and during the Institute: Professor C B Brown, Department of Civil Engineering, University of Washington, Seattle, United States of America; Professor J S Gero, Department of Architectural Science, Univesity of Sydney, Australia; Professor D Grierson, Department of Civil Engineering, University of Waterloo, Canada; Professor P W Jowitt, Department of Civil Engineering, Heriot-Watt University, Edinburgh, United Kingdom; and Professor A B Templeman, Department of Civil Engineering, The University of Liverpool, United

Kingdom. The Director and Members of the Organising Committee would like to thank the NATO Scientific Affairs Committe for funding the Institute. Without the support of the Committee this Institute could not have been held.

I should also like to thank Professor A D Edwards, Head of the Department of Civil Engineering and Dean of the Faculty of Engineering at Heriot-Watt University for all the support he provided with this and other projects.

My sincere thanks are also due to the research students of the Department of Civil Engineering at Heriot-Watt University including: David Howarth, Chen Chao Xu, Jim Milne and John Hampshire who helped me considerably during the Institute. I should also like to thank Erik Moncrieff who was responsible for typesetting many of the papers in these proceedings and Asad Khan who so kindly came to my assistance during the final stage of preparation of these proceedings.

Finally I should also like to thank Dr L V da Cunha, Director of the NATO ASI Programme for all his help in the organisation of the Institute.

Barry H V Topping
Department of Civil Engineering
Heriot-Watt University
Edinburgh

The Optimization of Civil Engineering Networks

Andrew B. Templeman
Department of Civil Engineering
University of Liverpool
Liverpool
United Kingdom

Abstract This paper describes similarities in the system mathematical models of water supply networks and structural trusses, showing that both are examples of a general class of non-linear engineering networks. Comparisons are made between the two networks in respect of constitutive equations, methods of analysis, optimum design and reliability-based design. The similarities are valuable at a teaching level in unifying the study of engineering networks. The differences point the way to interesting areas of research.

1 Introduction

One aspect of civil engineering systems which seldom fails to appeal to students is the realization that a relatively small number of mathematical methodologies such as linear programming, dynamic programming and graph theory are capable of solving or shedding light upon an enormous range of different practical problems. Graph and network theory always seem to go down well with civil engineering students. Through engineering design classes they are familiar with handling information which is presented pictorially, and the node-arc representation of a graph comes as no surprise to students who have already met pin-jointed trusses and water supply networks in their structures and hydraulics lectures. However, when such students have been fed with the usual diet of shortest paths, maximum flows, spanning trees and various types of tours. it is somewhat deflating to both lecturer and students to have to acknowedge that none of these standard graph theory methods is of much value in connection with the most obvious civil engineering networks—the pin-jointed truss and the water supply network.

There is, however, a large body of literature and much current research on various aspects of these civil engineering networks. Unfortunately from a systems

1

B. H. V. Topping (ed.),
Optimization and Artificial Intelligence in Civil and Structural Engineering, Volume I, 1–18.
© 1992 *Kluwer Academic Publishers. Printed in the Netherlands.*

viewpoint much of this work is still highly compartmentalized according to discipline. Structures people do structural work and publish in structures journals, rarely interacting with water people who keep fairly ridigly to their own world of research and journals. The purpose of this paper is to try and demonstrate that from a systems viewpoint there is a considerable overlap in the problems and solution techniques used in these separate specialisms. It is a very worthwhile exercise to look over the fence into one's neighbour's garden occasionally and to see what he is doing. It turns out that there is potential for the development of a true systems approach to civil engineering networks which can be as intellectually stimulating for both lecturers and students as the more conventional aspects of systems theory.

This paper therefore examines both structural trusses and water supply networks and shows that to a very large extent they use methods which are often very similar but are occasionally very different. On reflection, the similarities are highly satisfying from a pure systems viewpoint but the differences are equally stimulating from a research viewpoint because they pose the question of why there is a difference. It would have been instructive also to include electrical power distribution networks, gas supply networks and perhaps transportation networks in this compatative study as they are also important practical examples of this class of networks but this must wait for another occasion.

2 Analysis

Previous work by the present author [1] has examined in detail the considerable similarities which exist between pin-jointed trusses and water supply networks. Ref. [1] considered both networks in terms of the definitions of their various quantities, their constitutive equations, their analysis problem formulations and the various solution techniques which are used for solving the problems of analysis. That work is briefly summarised here and shows some striking parallels. Much of the detail is omitted but can be found in Ref. [1].

Fig. 1 shows two general networks of nodes and arcs. A structural engineer can intuitively recognize Fig. 1a as a pin-jointed truss and will be familiar with the notation. A water engineer can similarly recognize Fig. 1b as a water supply network with its specialist notation. In fact the network diagrams are the same and Table 1 list equivalences between the quantities of Fig. 1a and 1b. It is clear that both networks are examples of a single general network model, so striking are the parallels. One significant difference is apparent, however: all the water quantities are scalars but some of the structural quantities are vectors. For the structural truss we need a set of reference axes and an angular parameter θ in order to define the directions of the vector quantities.

The constitutive equations for both the networks in Fig. 1 with M arcs and N nodes can be expressed as follows. For the structural truss:

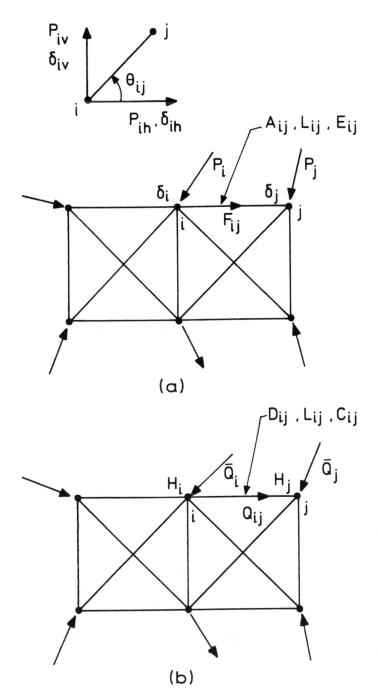

Figure 1. *Non-linear networks: (a) Truss, (b) Water supply.*

Table 1. *Notational equivalences for Fig. 1.*

Structures Notation		Water Notation	
P	applied force	$\overline{Q}$	supply/demand
F	bar force	Q	pipe flow rate
δ	nodal displacement	H	nodal pressure head
Γ	bar deformation	ΔH	head loss
A	bar area	D	pipe diameter
L	bar length	L	pipe length
$1/E$	Elastic modulus	C	friction coefficient

$$\left. \begin{array}{l} \sum_{i=1}^{N} F_{ij} \sin \theta_{ij} - P_{jv} = 0 \\[2ex] \sum_{i=1}^{N} F_{ij} \cos \theta_{ij} - P_{jh} = 0 \end{array} \right\} \qquad j \in N \qquad (1)$$

$$\cos \theta_{ij} \left(\delta_{jh} - \delta_{ih} \right) + \sin \theta_{ij} \left(\delta_{jv} - \delta_{iv} \right) = \Gamma_{ij} \quad ij \in M \qquad (2)$$

$$\Gamma_{ij} = \frac{L_{ij} F_{ij}}{E_{ij} A_{ij}} \qquad\qquad ij \in M \qquad (3)$$

In Eqs. (1) to (3) subscript ij refers to the arc between nodes i and j, and subscripts h and v refer to horizontal and vertical components of a quantity. For the water supply network the constitutive equations are:

$$\sum_{i=1}^{N} Q_{ij} - \overline{Q}_j = 0 \qquad j \in N \qquad (4)$$

$$H_i - H_j = \Delta H_{ij} \qquad ij \in M \qquad (5)$$

$$\Delta H_{ij} = \frac{C_{ij} L_{ij} Q_{ij}^{\beta_1}}{D_{ij}^{\beta_2}} \qquad ij \in M \qquad (6)$$

For the water supply network the exponents β_1 and β_2 in Eq. (6) are approximately 2 and 5 respectively, depending upon the particular empirical form of Eq. (6) used.

Comparing the two sets of constitutive Eqs. (1) to (3) and (4) to (6) it is evident that they are closely similar. Eqs. (1) and (4) are nodal flow continuity equations which contain equivalent quantities and are linear. Eqs. (2) and (5) are also linear and represent loss of nodal potential on an arc. The quantities in these first two linear sets are linked by Eqs. (3) and (6) which for the structural truss are linear, but for the water supply network are highly non-linear. There are therefore close similarities but significant differences.

2.1 Solution Methods

The problem of analysis in respect of both examples consist of solving these equations to find the arc flows (bar forces, F, or pipe flow rates Q) and nodal potentials (nodal displacements, δ, or nodal pressure heads, H), given values for all the other quantities. Ref. [1] examines analysis methods in detail; here some salient features are briefly noted:

1. Both examples have a special case in which the sets of linear equations (1) or (4) are square and may be solved to give unique values for all arc flows. For the structural truss this special case corresponds to a statically determinate truss as shown in Fig. 2a; for the water supply network it corresponds to a branched network as shown in Fig. 2b.

2. For statically indeterminate trusses and looped water networks the analysis is much more difficult to perform. Truss analysis is the easier of the two problems since all equations (1) to (3) are linear but water equation (6) is non-linear.

3. There is no truss analysis method which is similar to the Hardy Cross method [2] for water network analysis. The Hardy Cross method is a simple iterative means of handling the non-linear equation (6) and is unnecessary in structural analysis because the corresponding equation (3) is linear.

4. Some other methods for water supply analysis consist of linearizing the non-linear equation (6), solving the resulting sets of linear equations, relinearizing and iterating to a solution. The ways in which the linearization and solution of the linearized equations are done turn out to be identical to well-known structural analysis methods. The linear theory method of Wood and Charles [3] for water supply analysis is thus identical to the stiffness method (displacement method) of structural analysis when Eq. (6) is inverted to give Q as a function of ΔH and linearized by a Taylor series expansion. The Newton-Raphson method (flow method) of Martin and Peters [4] for water network analysis turns out to be identical to the flexibility method (force method) of structural analysis when Eq. (6) is linearized directly by a Taylor series expansion.

5. In the comparatively rare cases when truss equation (3) is non-linear (usually material non-linearities or large displacements) the methods used are even closer to those used for general water supply analysis. Typically Eq. (3) is iteratively linearized as in the methods of Refs. [3] and [4]. An alternative approach for the analysis of non-linear trusses is to abandon Eqs. (1) to (3) completely and, instead, to perform a direct numerical minimization of the total energy function π of the structure. This is particularly useful for highly

6

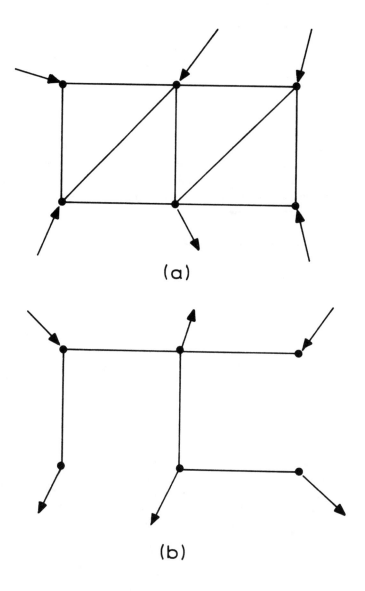

(a)

(b)

Figure 2. *Special cases:* (a) *Statically determinate truss,* (b) *Branched water supply system.*

nonlinear trusses such as cable trusses or cable net structures but is, of course, unnecessarily time-consuming and complicated for the usual linear structure. Nevertheless, for general water supply networks a similar energy minimization approach can be used and has been shown by Kaya and Simon [5] to be as efficient as other methods in common use.

3 Optimum Design

The survey of network analysis methods in the previous section has shown some very close parallels between the methodologies of the two specialisms. The vectorial nature of truss analysis turns out not to be a significant hindrance and the non-linear nature of the water supply problem can be efficiently handled through iterative linearizations. When design methods are examined comparatively a similar set of close parallels between the two specialisms can be seen. Here the term design is assumed to mean optimum design since we are essentially concerned with computational methods and there seems little point in using a computer to produce a less-than-optimum design when it could produce an optimum one.

3.1 Formulation

For design, the constitutive equations (1) to (3) for the truss and (4) to (6) for the water supply network have the arc size quantities, A or D, as unknowns in addition to the arc flows, F or Q, and the nodal potentials, δ or H. Whereas in analysis there were just enough equations to determine values for all the unknowns, in design we still have the same number of equations but have an extra lot of unknown arc sizes. An infinite number of design solutions is possible. Specifying additional sets of minimum or maximum limits upon desirable values of the unknown quantities reduces the range of possible solutions but still does not produce a unique design.

For truss design these constraints limit nodal displacements, bar stresses and bar sizes and are:

$$-\overline{\delta} \leq \delta_i \leq \overline{\delta} \qquad\qquad i \in N \qquad\qquad (7)$$

$$-\overline{\sigma} \leq \frac{F_{ij}}{A_{ij}} \leq \overline{\sigma} \qquad\qquad ij \in M \qquad\qquad (8)$$

$$A_{ij} \geq \overline{A}; \qquad\quad A_{ij} \text{ discrete} \quad ij \in M \qquad\qquad (9)$$

Other constraint forms are possible but (7) to (9) represent the most common ones. For the water supply network design, constraints usually limit nodal pressure heads, pipe flow velocities and pipe sizes and are:

$$H_{\min} \leq H_i \leq H_{\max} \qquad\qquad i \in N \qquad\qquad (10)$$

$$V_{\min} \leq \frac{4Q_{ij}}{\pi D_{ij}^2} \leq V_{\max} \qquad\qquad ij \in M \qquad\qquad (11)$$

$$D_{\min} \leq D_{ij} \leq D_{\max}; \quad D_{ij} \text{ discrete} \qquad ij \in M \qquad\qquad (12)$$

The discreteness requirements upon A and D in (9) and (12) are practical necessities from an engineering viewpoint but are often omitted from optimum design formulations which consequently obtain continuous-valued solutions for A and D.

These constraints are additional to the constitutive equations (1) to (3) or (4) to (6) and they clearly involve equivalent quantities in similar functional forms. As they are inequalities they are insufficient to delineate a unique design solution. An objective function is added, the minimization of which yields the desired optimum design. For trusses the objective function usually used is weight or cost which leads to a linear function in the bar size variables:

$$\text{Minimize}_{A} \ f = \sum_{ij \in M} L_{ij} A_{ij} \qquad\qquad (13)$$

For water supply systems the objective function of cost has the form:

$$\text{Minimize}_{D} \ f = \sum_{ij \in M} L_{ij} D_{ij}^{\beta_3} \qquad\qquad (14)$$

in which the exponent $\beta_3 > 1$. The objective function (13), constitutive equations (1) to (3) and constraints (7) to (9) represent a general truss optimum design problem; Eqs. (14), (4) to (6) and (10) to (12) represent the corresponding water supply network optimum design problem. It can be noted that for both examples the optimum design problem is non-linear.

3.2   Solution Methods

A literature search quickly reveals that there has been a very large amount of work on solution methods for the truss optimization problem. Innumerable different methods exist. Water supply optimization, however, has been less well served by good solution methods. Before examining particular solution methods, one observation may be made which applies generally to both problems. This is that the least cost design for either example under a single pattern of applied loads or supply/demand quantities is the special case noted above of a statically determinate truss or a branched water supply network.

Even if the geometry of a truss is initially specified to be statically indeterminate with many redundant members, in the absence of a minimum size limit on all members the cost optimization process will reduce some member sizes down to zero and effectively remove them from the design leaving a statically determinate truss. If a minimum size limit is specified, the cost optimization process will generate a statically determinate main structure with several minimum size redundant members which play very little part in the load carrying capacity of the truss—ideally they are not really required. A least cost water supply system has similar characteristics: for a single pattern of supply/demand, optimization produces a branched system in the absence of a minimum pipe size limit, or a branched system with several loop-creating minimum size pipes if a minimum size limit is specified. The result is intuitively obvious: it is always cheaper to supply a fixed quantity of water to a node by one pipe rather than by two. Consequently, the optimization process tries to generate a design in which each node is supplied by only one pipe, i.e. a branched network. For the truss a similar argument applies though not quite as strongly: it is never cheaper to transmit force by two members rather than one (though it may be as cheap). In consequence the optimum design will most likely be a determinate truss though an equally cheap (but not cheaper) indeterminate truss may exist. These results may be formally proved. Their main importance lies in the fact that they produce a paradox for water engineers who would clearly like to be able to design least cost supply systems but who also want it to be a looped design. Structural engineers are perhaps slightly less prejudiced against statically determinate structures so an optimized design which turns out to be determinate may be acceptable. We shall return to this paradox later but it should be borne in mind that all the optimum design methods now to be discussed should, if properly implemented, lead to branched or determinate designs under a single dominant loading system.

The optimum design problem for water supply systems is the more difficult of the two to solve and forms the basis of the methodology discussion. Space precludes a full treatment of methods: as with analysis methods, only the main features are noted here.

1. The optimum design problem for even a modest pipe network is large with many variables and many constraints and is highly non-linear throughout. It is also non-convex. Even though it has explicit algebraic forms for all functions it turns out to be extremely expensive in CPU time to solve it by standard non-linear programming methods. Because of this, alternative strategies have been adopted.

2. The method devised by Alperovits and Shamir [6] is both popular and ingenious. It also has the advantage that, unlike other methods, it yields a design which uses only the discrete available pipe sizes. The problem is simplified by specifying an initial design and analysing it to determine pipe flows Q. Pipe head losses ΔH then become functions of pipe size D only. An ingenious transformation of variables then results in an optimum design problem which is entirely linear and can be solved by linear programming. The change of variables is done by changing the fundamental assumption that each pipe has a known length L and unknown diameter D. Instead it is assumed that each

pipe is made up of several segments, each of known diameter (the discrete available sizes) but unknown length ΔL. Extra constraints are added to ensure that for each pipe the sum of the unknown segment lengths is equal to the known total length of the member. Fig. 3a shows this transformation of variables. Putting this into the optimization problem results in a linear programming problem. Solving this LP gives a least cost design for the specified flows. Quindry, Brill and Liebman [7] have shown how these specified flows may then be changed in such a way as to further reduce the cost. The solution proceeds iteratively. In the optimum design generated there are usually relatively few pipes with more than one segement. Such pipes may easily be rounded up to give a single pipe of the diameter of the largest segment.

3. The Alperovits and Shamir method [6] has been successfully used for optimum truss design by Templeman and Yates [8]. Bars which are usually assumed to be of known length but unknown area are here assumed to be made up of several segments, each of known discrete size but unknown segment length, exactly as in the Alperovits and Shamir method and as shown in Fig. 3b. This results in a LP which is easily solved. The method of Ref. [8] turns out to be one of the most effective methods available for discrete optimum truss design and, though the idea has been known to the water supply specialism for some years, it was new to the structures community.

4. An alternative simplification to the water supply optimization problem is afforded by specifying known values of nodal pressure heads calculated from some initial design. After some manipulation of the optimization problem it can be expressed as a linear programming problem which is easily solved. This method was first proposed by Lai and Schaake [9] and is described in Ref. [7] which also outlines how the nodal pressure heads may subsequently be altered to give an iterative solution method which leads to an optimum design.

5. The Lai and Schaake method [9] also has a direct parallel in optimum truss design. McKeown [10,11] was the first to use the idea of specifying a set of nodal displacements and finding by linear programming the least cost structure which has those displacements. In Ref. [11] McKeown describes how the initial set of displacements may be changed so as to further reduce the cost, thus deriving an iterative optimum design procedure. The Lai and Schaake method [9] and McKeown's method [11] are exact parallels but appear to have developed entirely independently.

6. One attempt to apply an optimization method with origins in structural engineering to water supply system optimization may be noted. This has been proposed by Cheng and Ma [12]. For complicated optimum structural design problems where the structural analysis is performed by the finite element

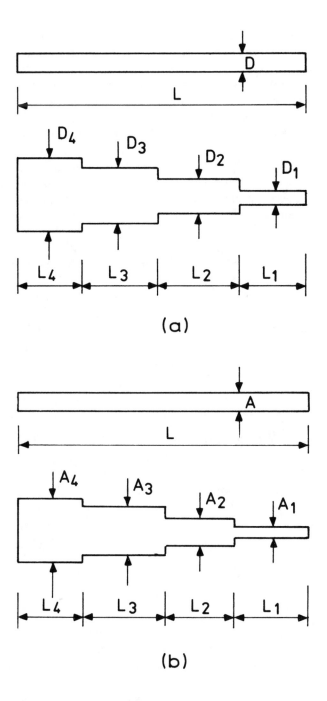

Figure 3. *Prismatic and segmental arcs: (a) pipe, (b) bar.*

method explicit algebraic forms for the constraint functions are not available. The optimization process must therefore be based upon approximate models. An initial design is first analysed in detail to determine numerical values for all the objective and constraint functions. A sensitivity analysis is then performed on the initial design to calculate numerical values for the first and sometimes second derivatives of all the objective and constraint functions with respect to all the design variables (the bar sizes A in the case of a truss). This numerical data is then used to set up an approximating linear programming or quadratic programming problem in place of the actual non-explicit optimization problem. This is solved to give a new, improved design which is then re-analysed. A sensitivity analysis of the new design sets up a second iteration, and iterations continue until a solution is reached. Cheng and Ma [12] have successfully applied this approach to a water supply system. One advantage is that the analysis is done separately from the optimization, so the optimization problem is much smaller and has fewer variables since it no longer contains the unknowns Q or H. Also, LP or QP problems of relatively large size can be solved quickly. Disadvantages are that the iterative optimizations are done on approximate models, and that the sensitivity analysis of each design which is essential to set up the next LP or QP model is very expensive in CPU time. Cheng and Ma show how this can be done analytically. However, in their solved example with a single supply/demand pattern, the design created at the end of the iterations was not essentially a branched system, as it should have been. This suggests that they may not have actually reached an optimum design.

7. There seem still to be opportunities for investigating the many other truss optimum design methods in the context of water supply systems. Sufficient close parallels have already been demonstrated for this to be at least an interesting task and possibly a fruitful one. The fact that for a single supply/demand pattern the least cost design should be branched still presents a major obstacle to the practical acceptance of water supply system optimization, however. It needs to be overcome and the next section addresses some aspects of this problem.

4 Reliability

That water supply networks need to be looped seems inescapable. That designing them to a single dominant supply/demand pattern and for least cost leads to a branched system is also inescapable. Some compromise upon the least cost criterion must, therefore, be made in order to generate practical, looped designs.

If the question is asked, why is a looped system essential? there are usually two answers. First, the existence of loops ensures that nodes can be supplied by several different paths in the event of some failure on a particular path, and second, that loops ensure that water is kept moving and does not have the chance to stagnate as might be the case at the ends of branches. The first of these is one of the great myths of water supply system design. Providing loops merely ensures that possible alternative supply paths are available in the event of some failure. It does not ensure that such alternative paths are actually useable for supply purposes unless they have been specifically hydraulically designed to be so used. Current design procedures do not usually include any alternative path contingency requirements. Whether looped systems can actually supply nodes in the event of some failure is therefore somewhat serendipitous. The second reason, however, has considerable merit and cannot be ignored. The real question, then should not be whether loops are necessary but whether they should be hydraulically designed to have an alternative path capability.

Exactly the same reasoning in respect of alternative paths exists for structural trusses. It is commonly assumed (particularly by students) that statically indeterminate trusses are inherently safer than statically determinate ones. Failure of a member in an indeterminate truss still leaves at least a determinate and rigid structure. However, this ignores the fact that unless the truss has been very carefully designed, the force originally carried by the failed member will be redistributed around the other bars causing overloading and possibly failure of more bars leading to progressive collapse of the truss. Merely providing an indeterminate arrangement of bars is therefore no guarantee of safety.

Structural engineering research first addressed this question of safety and alternative load paths many years ago and 'damage tolerant design' is now widely used in respect of military and aerospace strcutures. A fighter aircraft structure must obviously be capable of sustaining some damage without complete failure. It must be noted, however, that damage tolerance is not the same as resilience (the ability to carry other loads than those designed for) or reliability. These three different concepts are intrinsically distinct but tend to have been lumped together under the general name 'reliability.'

In looking for parallels between structural trusses and water supply systems at this 'reliability' level it should first be recognised that structural reliability theory is concerned with probabilistic events. The reliability of a structural element or system is defined as the complement of the probability of failure of the element or system. Three decades of research has been done in developing methods for the calculation of failure probabilities for structures; methods which are widely used in the practical design of offshore and nuclear structures and are approaching Code of Practice status for general structures. Structural reliability calculation methods are based upon a definition of failure as the probability that load exceeds resistance as shown in Fig. 4. This definition is so firmly rooted in the structural engineering literature that it is almost heretical to point out that it is a definition of very limited value since real structures mostly fail for entirely different reasons than that covered by the definition. Indeed, the present methodology of structural reliability can be seriously criticised at all levels from the philosophical, through the theoretical, to the practical [13], but to examine this further would be a digression.

resistance, R

F ← load, F

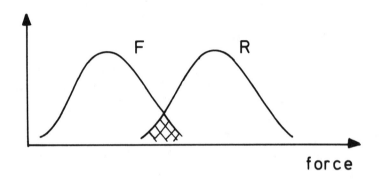

F   R

force

Probability of failure, $p_f = p([R-F] < 0)$

Reliability, R, $= 1 - p_f$

Figure 4. *Structural reliability.*

   The main point to be made here is that in examining water supply system reliability (and here reliability implies methods for the assurance of continuity of supply) most work on structural reliability is irrelevant for comparative purposes. Water supply system design does not yet operate in terms of probability distributions of supply/demand and of the flow carrying capacity of pipes. Probability-based structural reliability methods therefore have no current counterparts in water supply networks.

   The concept of damage tolerant design appears to be the closest comparator between structural and water systems. In respect of structures this has been modelled in two ways. One is to impose the requirement that the truss must be capable of meeting its full performance specification with any one member removed. In terms of optimum design this implies that the initial design must be statically indeterminate, that any analysis and sensitivity analysis must be performed many times for the basic structure with each possible member removed in turn, and that an extra set of constraints is necessary for each of these different sub-structures. Naturally, this leads to a very large optimization problem which is computationally very expensive to solve. Furthermore, the resulting designs turn out to be considerably heavier or costlier than structures optimized without these extra conditions. A second approach retains the requirement that any member should be capable of removal but, in order to reduce weight or cost, accepts a reduced performance specification for the structure in its damaged state.

Translating this to water supply systems is straightforward in concept, but would prove enormously expensive in terms of the computer resources needed to solve the resulting optimum design problem. Also, as with the first approach outlined above it can be forseen that the resulting designs would be considerably more expensive than conventional ones. A simple example demonstrates this. Fig. 5a shows a single supply node A providing a supply flow rate of one unit. Two demand nodes B and C each require flow rates of half a unit. The three node are at the vertices of an equilateral triangle. For simplicity it is assumed that B and C are at the same level and below A so that a pressure head of one unit is available to drive the flows.

The simple solution to this problem is shown in Fig. 5b and consists of pipes AB and AC. Each pipe has to be designed to carry a flow rate of half a unit and may use the total available one unit of pressure head. Actually this is not the cheapest possible design which is that shown in Fig. 5c. The existence of Steiner-type layouts such as this is yet another aspect of water supply system design which is very much unexplored. A looped system is created by pipe BC in Fig. 5d. If this system is designed such that each demand node still receives its half unit of flow even though any one pipe is removed, the system must be designed to carry one unit of flow in each of the pipes AB and AC and half a unit of flow in either direction in pipe BC as shown in Fig. 5d. Note that the full one unit of available head must not be used in pipes AB and AC otherwise there will be no pressure head to drive the flow along pipe BC to the other demand node in the event of either AB or AC being removed. The design of Fig. 5d will obviously have a much larger cost than the two previous designs.

This example shows that the very cheapest design (Fig. 5c) is also the least damage tolerant of all since a failure of the pipe AD will cut off supply to both nodes B and C. The branched design of Fig. 5b is only slightly more expensive and is marginally more damage tolerant since failure of any pipe always leaves the supply intact to one node. The most damage tolerant design of all, Fig. 5d, is also by far the most expensive. If the idea of a reduced performance requirement for the

16

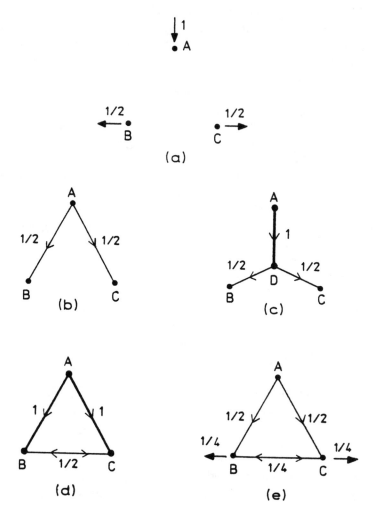

Figure 5. *Damage tolerant water supply networks.*

system in the damaged state could be accepted it would be possible to reduce the cost of the design of Fig. 5d. Fig. 5e shows the flows which must be designed for if it is permissible for demand nodes to receive only half of their demand flow rate when the system is in a damaged condition. This design would be much cheaper than the most damage tolerant design of Fig 5d but more expensive than the design of Fig. 5b. However, the trade-off here is between the advantage of maintaining continuity of supply to all nodes and the disadvantage that only part of the demand can be met.

The biggest obstacle in trying to design cheap and 'reliable' waer supply systems is the lack of any firm criteria for measuring reliability. In the structures area these problems are fairly well sorted out; damage tolerant design is recognized as a desirable thing in certain sectors of structural engineering, probabilistic reliability analysis is thought to be desirable in others. Frustration occurs with the water supply industry because they have not yet decided upon their design criteria for 'reliability.' The only small consolation is that the gas distribution industry (at least in the UK) is at about the same level of thinking as the water industry, and both trail a little way behind the electric power distribution industry which has the requirement that at least two arcs should serve each demand node.

5   Conclusions

The comparative study of different types of civil engineering networks is interesting and instructive. At the analysis level a high degree of similarity between methodologies is found across the range of applications. This similarity extends to optimum design methodologies also but reveals that least cost designs may not necessarily be appropriate engineering design solutions. The reliability and cost optimization aspects of design are inextricably linked. In the water supply industry current design practice is nominally directed towards reliability issues but is actually inadequate in its treatment of system reliability. It is unlikely that further civil engineering systems research will unearth new concepts and methods which will satisfy the water industry's desire for cheap and 'reliable' systems in the absence of clear definitions from the industry of what constitutes a 'reliable' system.

References

[1] Templeman, A. B. and Yates, D. F., (1984), 'Mathematical similarities in engineering network analysis,' *Civil Engineering Systems* **1**, 114–122.

[2] Cross, H., (1936), *Analysis of flow in networks of conduits or conductors*, Bulletin No. 286, Univ. of Illinois Engineering Experimental Station, Urbana, Illinois.

[3] Wood, D. J. and Charles, C. O. A., (1972), 'Hydraulic network analysis using linear theory,' *J. Hydraulics Div., ASCE* **98**, HY7, 1157–1170.

[4] Martin, D. W. and Peters, G., (1963), 'The application of Newton's method to network analysis by digital computer,' *J. Inst. Water Eng.* **17**, 115–129.

[5] Kaya, A. and Simon, A. L., (1974), 'An optimization method to design hydraulic networks,' *Engineering Optimization*, **1**, 2, 71–78.

[6] Alperovits, E. and Shamir, U., (1977), 'Design of optimal water distribution systems,' *Water Resources Research* **13**, 6, 885–900.

[7] Quindry, G. E., Brill, E. D. and Liebman, J. C., (1981), 'Optimization of looped water distribution systems,' *J. Environmental Engineering Div, ASCE* **107**, EE4, 665–679.

[8] Templeman, A. B. and Yates, D. F., (1983), 'A segmental method for the discrete optimum design of structures,' *Engineering Optimization* **6**, 3, 145–155.

[9] Lai, D. and Schaake, J., (1969), 'Linear programming and dynamic programming applications to water distribution network design,' Report 116, Dept of Civil Engineering, MIT, Cambridge, Mass.

[10] McKeown, J. J., (1977), 'Optimal composite structures by deflection-variable programming,' *Computer Methods in Applied Mechanics and Engineering* **12**, pp. 155–179.

[11] McKeown, J. J., (1989), 'The design of optimal trusses via sequences of optimal fixed-displacement structures,' *Engineering Optimization* **14**, 3, 159–178.

[12] Cheng, G. D. and Ma, H. T., (1989), 'The optimal design of water distribution systems,' *Civil Engineering Systems*, **6** 3, 111-121.

[13] Templeman, A. B., (1988), 'Some reservations about reliability-based design,' *New Directions in Structural System Reliability*, D. M. Frangopol (Ed.), University of Colorado Press, Boulder, Colorado pp. 265–275.

Fuzzy Sets Applied to Civil Engineering Systems

Colin B. Brown
Department of Civil Engineering
University of Washington
Seattle, Washington
United States of America

Abstract The inclusion of verbal uncertainty into a civil engineering decision system can be accomplished by the use of fuzzy set theory. The fuzzy set is defined, and then the specific operations of complement, union, intersect, relation and composition are constructed. With this mathematical background the organization of a fuzzy decision logic is demonstrated. Additionally, methodologies for determining fuzzy supports and the use of entropies to obtain combined random and fuzzy probabilities are examined.

1 Introduction

The theory of fuzzy sets was first mentioned at a seminar at Berkeley by Lotfi Zadeh in 1964. Since then it has attracted an increasing band of devotees and a sterling chorus of critics. The literature on the subject has increased exponentially and falls into two obvious classes: fuzzy mathematics as a generalization of set theory; and fuzzy applications. Here we will need both classes in order to provide a reasonable method for employing fuzzy sets in civil engineering decision making.

Essentially the intention of fuzzy sets is to enrich the understanding of uncertainty. Richard Bellman made the comment,

> One of the difficulties we would like to emphasize is that we have all been very badly educated: we have all been brain-washed so that we automatically say uncertainty is probability. This is not true at all. In the universe of uncertainty, there are only little points that can be handled by classical probability theory. In general, classical probability theory is of little value in the study of uncertainty. In practice, we use it because we do not know anything better to do.

B. H. V. Topping (ed.),
Optimization and Artificial Intelligence in Civil and Structural Engineering, Volume I, 19–32.
© 1992 *Kluwer Academic Publishers. Printed in the Netherlands.*

In later years he combined with Zadeh in trying to expand the understanding of uncertainty beyond the constraints of the Kolomogorov axioms of probability. The intention was to introduce a mathematical expression for the 'maybe' of Lukasiewicz logic and thus extend the binary 'truth-false' logic into a middle ground. This allows the incorporation of verbal labels such as 'important,' 'good,' or 'very good' into a logical argument. Clearly these words are not precise. The set of good engineers has a certain vagueness which does not exist in the set of engineers who have practiced over 30 years. The likelihood of an engineer being in either of these sets may be of interest. The sets have nothing to do with each other unless the length of practice is an unfailing measure of good quality practice. The likelihood of an engineer having practiced over 30 years may be appropriately treated by classical probability. However, the set of good engineers is a fuzzy set, and the likelihood of being a member requires a description of a fuzzy set.

In this lecture a tidy but not complete view of fuzzy set application to civil engineering will be provided. In the end it will be necessary to have a method of decision making. We will not quarrel with either the Von-Neumann-Morgenstern decision model, or the seeking of extrema on an objective function subject to constraints, as two methods of making professional decisions. These do require that the fuzzy sets can be combined with probabilities for use in these models. The plan of the lecture is to:

- introduce the idea of fuzzy sets,

- develop elements of a fuzzy calculus,

- construct n-wise fuzzy supports,

- construct mergings between supports of fuzzy sets and probabilities of other, crisp sets.

2 Description

The process of describing fuzzy sets must, when taken to a limit, also describe a crisp or normal set. Then the crisp set can be regarded as a special case of a fuzzy set.

One approach to thinking of a crisp set is to consider a universe of discourse, U, and then apply constraints on U which leave the sub-set A. The same approach can be applied to the description of fuzzy sets. If the variable of the universe of discourse is considered as the abscissa and the support for the value or grade of the variable is the ordinate, then the graph for U contains the same support for all values or grades. If the support is limited to the range $[0,1]$, then all supports in U will be unity. If the values or grades are x_i, $i = 1$ to n, then

$$U \equiv 1 \mid x_1 + 1 \mid x_2 \cdots + 1 \mid x_n \tag{1}$$

where the solidus divides the support from the value or grade of the variable. Constraints on U leave a sub-set A. For instance the constraints that exclude all but x_2 and x_3 results in

$$A \equiv 0 \mid x_1 + 1 \mid x_2 + 1 \mid x_3 + 0 \mid x \cdots + 0 \mid x_n \tag{2}$$

as the crisp set A.

The separation between the set A and the complement, $\overline{A}$, is precise; thus, an event in U must either occur in A or $\overline{A}$. This is appropriate for concrete with strengths of 4 and 5 ksi but not appropriate for good concrete. The universal set of (1) could then refer to concrete strengths with $x_1 = 3$, $x_2 = 4$, $x_3 = 5$, $x_4 = 6$, $x_5 = 7$, $x_6 = 8$ and $x_7 = 9$ ksi. Then $n = 7$ and U is as designated in (1). The sub-set A which excludes all but 4 and 5 ksi concrete is as (2). However, the qualitative statement that the concrete is 'good' may involve incomplete support for 5 and 6 ksi strengths and complete support for 7, 8 and 9 ksi strengths. Thus,

$$A \equiv \text{good} \equiv 0 \mid x_1 + 0 \mid x_2 + 0.3 \mid x_3$$

$$+ 0.7 \mid x_4 + 1 \mid x_5 + 1 \mid x_6 + 1 \mid x_7 \tag{3}$$

This is the fuzzy set for good—a sub-set of the universe of concrete strengths. The separation between A and the complement.

$$\overline{A} \equiv \text{not good} \equiv 1 \mid x_1 + 1 \mid x_2 + 0.7 \mid x_3$$

$$+ 0.3 \mid x_4 + 0 \mid x_5 + 0 \mid x_6 + 0 \mid x_7 \tag{4}$$

is crisp for x_1, x_2, x_5, x_6 and x_7 and is imprecise for x_3 and x_4. The sets A and A are fuzzy sets for good and not good, respectively. In general the fuzzy set is

$$A \equiv \sum_{i=1}^{n} \mu(x_i) \mid x_i; \qquad 0 \leq \mu(x_i) \leq 1 \tag{5}$$

where the x_i refer to a quality variable of the universe of discourse.

These arrangements ensure that the crisp set is a constrained fuzzy set where all the $\mu(x_i)$ are either unity or zero. The continuous form of the discrete set of (5) is

$$A = \int_A \mu(x) \mid x \tag{6}$$

3   Operations

The operations of crisp set theory—union, intersection, relation and composition—must have their counterpart in fuzzy set theory. It is required that the special case of the fuzzy operation results in the crisp operation. The calculus of fuzzy sets is confined to the determination of minima and maxima from amongst candidate supports.

3.1 Single Operations

These consist of the complement of the fuzzy set A, which is illustrated in (4), and union and intersect of sets A and B.

The complement of the fuzzy set A of (5) is

$$\overline{A} \equiv \sum_{i=1}^{n} (1 - \mu(x_i)) \mid x_i; \quad 0 \le \mu(x_i) \le 1 \tag{7}$$

This produces the crisp complement when $\mu(x_i)$ are either zero or unity.

The union of the fuzzy sets A and B where

$$A = \sum_{i=1}^{n} \mu_A(x_i) \mid x_i \tag{8}$$

$$B = \sum_{i=1}^{n} \mu_B(x_i) \mid x_i \tag{9}$$

produces the result A or B which can be interpreted as the maximum support between $\mu_A(x_i)$ and $\mu_B(x_i)$ for all x_i in the n values of (8) and (9). The fuzzy union is designated as $\vee$ and defined as

$$C = A \vee B = \sum_{i=1}^{n} \max\left(\mu_A(x_i), \mu_B(x_i)\right) \mid x_i \tag{10}$$

Conversely, the intersect is the minimum support between $\mu_A(x_i)$ and $\mu_B(x_i)$ for all x_i in the n values of (8) and (9). The fuzzy intersect is designated as $\wedge$ and defined as

$$D = A \wedge B \equiv \sum_{i=1}^{n} \min\left(\mu_A(x_i), \mu_B(x_i)\right) \mid x_i \tag{11}$$

These definitions degenerate to crisp unions and intersects when the $\mu_A(x_i)$ and $\mu_B(x_i)$ are either zero or unity.

Examples of these fuzzy operations are

$$A \equiv \text{good design} \equiv 1 \mid 1 + 0.7 \mid 0.9 + 0.4 \mid 0.8 + 0.1 \mid 0.7 \tag{12}$$

$$B \equiv \text{average construction} \equiv 0.2 \mid 0.7$$
$$+ 0.8 \mid 0.6 + 1 \mid 0.5 + 0.8 \mid 0.4 + 0.2 \mid 0.3 \tag{13}$$

Good design or average construction $\equiv$

$$C \equiv A \vee B \equiv 1 \mid 1 + 0.7 \mid 0.9 + 0.4 \mid 0.8 + 0.2 \mid 0.7 + 0.8 \mid 0.6$$
$$+ 1 \mid 0.5 + 0.8 \mid 0.4 + 0.2 \mid 0.3 \tag{14}$$

Good design and average construction $\equiv$

$$D \equiv A \wedge B \equiv 0.1 \mid 0.7 \tag{15}$$

In these examples the quality of design and construction is measured on a discrete scale from 0 to 1 at intervals of 0.1 with 1 being best and 0 being worst. The qualities of these professional activities are provided as grades over the $[0,1]$ scale. This indicates that the variable of the fuzzy set $A(x_i)$ need not refer to measurable quantities. For instance, in (3) the x_i are concrete strengths, whereas in (12) or (13) they are qualitative grades.

3.2 Fuzzy Relations

The intention of a relation is to express ordered m-tuples at which the variables of m sets are associated. These can be regarded as a m dimensional graph. In fuzzy relations we seek a fuzzy graph with ordinates expressed as the variables of the m fuzzy sets.

In crisp set we have the $m = 2$ sets A and B and we seek the ordered pairs of the elements of A and B in the form of the Cartesian product, $A \times B$. With the use of probability, each ordered pair is associated with a joint probability derived from $p(A)$ and $p(B)$. As an example, if $A = (100, 200, 300)$ and $B = (5, 7)$ where A is the set of bridge spans and B the bid price unit length, then

$$A \times B = \begin{bmatrix} 100,5 & 200,5 & 300,5 \\ 100,7 & 200,7 & 300,7 \end{bmatrix} \tag{16}$$

Clearly, $A \times B \neq B \times A$

In fuzzy sets we can think of the labels

$$\text{unlikely} \equiv A = 1 \mid 0 + 0.9 \mid 0.1 + 0.5 \mid 0.2 + 0.2 \mid 0.3 + 0.1 \mid 0.4 \tag{17}$$

$$\text{little} \equiv B = 1 \mid 0 + 0.8 \mid 0.1 + 0.6 \mid 0.2 + 0.2 \mid 0.3 \tag{18}$$

as

$$A = \sum_{i=1}^{n} \mu_A(x_i) \mid x_i \tag{19}$$

$$B = \sum_{j=1}^{s} \mu_B(x_j) \mid x_j \tag{20}$$

The ordered pairs are the (x_i, y_j) and their support is the minimum of $\mu_A(x_i)$ or $\mu_B(y_j)$. The fuzzy Cartesian product is

$$A \times B = \begin{bmatrix} 0,0 & 0,0.1 & 0,0.2 & 0,0.3 \\ 0.1,0 & 0.1,0.1 & 0.1,0.2 & 0.1,0.3 \\ 0.2,0 & 0.2,0.1 & 0.2,0.2 & 0.2,0.3 \\ 0.3,0 & 0.3,0.1 & 0.3,0.2 & 0.3,0.3 \\ 0.4,0 & 0.4,0.1 & 0.4,0.2 & 0.4,0.3 \end{bmatrix} \tag{21}$$

and the supports for the 20 elements (x_i, y_j) are

$$R = \begin{bmatrix} 1 & 0.8 & 0.6 & 0.2 \\ 0.9 & 0.8 & 0.6 & 0.2 \\ 0.5 & 0.5 & 0.5 & 0.2 \\ 0.2 & 0.2 & 0.2 & 0.2 \\ 0.1 & 0.1 & 0.1 & 0.1 \end{bmatrix} \tag{22}$$

The R is support for AxB and can be written in a tabular form rather than (21) and (22).

A ≡ Unlikely to be solvent	B ≡ little hope of a good job			
	0	0.1	0.2	0.3
0	1	0.8	0.6	0.2
0.1	0.9	0.8	0.6	0.2
0.2	0.5	0.5	0.5	0.2
0.3	0.2	0.2	0.2	0.2
0.4	0.4	0.1	0.1	0.1

These fuzzy supports, $\mu_R(x_i, y_j)$, are a general measure of uncertainty which occurs in the same manner as the joint probabilities, $p(x_i, y_j)$ in the crisp situation.

The formal statement of the fuzzy relation is

$$R = A \times B = \sum_{i=1}^{n} \sum_{j=1}^{s} \mu_R(x_i, y_j) \mid x_i, y_j \tag{23}$$

where

$$\mu_R(x_i, y_j) = \min\left(\mu_A(x_i), \mu_B(y_j)\right) \tag{24}$$

The relations can be considered as laws between the variables or grades of sets A and B. They can reflect a base of general information which can be construed as

If A is x_i and B is y_j, **then** the support is $\mu_R(x_i, y_j)$
else ...

Each of these if-then statements is a relationship, R_i, and the collected statements are the fuzzy union of these R_i.

3.3 Fuzzy Composition

The knowledge base in the form of a fuzzy relation, R, of the previous section considers a sub-set, A, of the universe $U = \sum_i 1 \mid x_i$ and a sub-set, B, of the universe $V = \sum_j 1 \mid y_j$.

Empirical or other information can provide, x a fuzzy sub-set of U, and it is reasonable to ask, in the light of this information and the above relation, what fuzzy subset, y, of V is induced? This is the fuzzy composition.

$$y = x \circ R \tag{25}$$

In practical terms we have a fuzzy relation, R, between A and B and fuzzy information $A^\star$, from which we desire the fuzzy set, $B^\star$ as

$$B^\star = A^\star \circ R \tag{26}$$

with

$$B^\star = \sum_{i=1}^{n} \mu_B(x_i) \mid x_i \tag{27}$$

and

$$\mu_B^\star(x_i) = \max\left\{ \mu_{A^\star} \circ R(x_i, y_j) \right\} \tag{28}$$

and

$$\mu_A^\star \circ_R (x_i, y_j) = \min\left\{ \mu_A{}^\star(y_j), \mu_R(x_i, y_j) \right\} \tag{29}$$

As an example, consider the relationship previously obtained between the contractor's solvency and the contractor's ability to perform a good job. This is a part of a knowledge base and we have information on the solvency of the contractor as the fuzzy set.

$$A^\star = 0.5 \mid 0 + 0.6 \mid 0.1 + 1 \mid 0.2 \tag{30}$$

Then (29) appears as

$$\begin{bmatrix} 0.5 & 0.5 & 0.5 & 0.2 \\ 0.6 & 0.6 & 0.6 & 0.2 \\ 0.5 & 0.5 & 0.5 & 0.2 \\ 0 & 0 & 0 & 0 \\ 0 & 0 & 0 & 0 \end{bmatrix} \tag{31}$$

and from (27) and (28)

$$B^\star = 0.6 \mid 0 + 0.6 \mid 0.1 + 0.6 \mid 0.2 + 0.2 \mid 0.3 \tag{32}$$

which is the expectation of a good job being completed.

3.4 Fuzzy System Analysis

The operations provided allow a fuzzy decision system to be constructed. Consider

(a) a measure of importance, M. where m is the value which is independent of fuzzy information and

$$M = (m-3, m-2, m-1, m, m+1, m+2, m+3) \qquad (33)$$

In the presence of fuzzy information, we wish to find

$$M = \sum_{i=m-3}^{m+3} \mu_i \mid i \qquad (34)$$

where μ_i is the support for the measure i.

(b) factors which effect the measure M. Term these E_j; then the relation

$$R_j = E_j M \qquad (35)$$

where $j = 1$ to f factors and f relationships R_j exist.

(c) The combination of the R_j by a fuzzy union.

$$R = \overset{V}{\underset{j}{}} R_j \qquad (36)$$

These provide general if-then laws between the effects E_j and the measure M for all j.

(d) A state $\overline{E}_j C E_j$ are assessed for a special situation. Additionally, the severity, $\overline{G}_j$, in the special situation are assessed. The intersect

$$\overline{P}_j = \overline{G}_j \wedge E_j \qquad (37)$$

is computed and for all factors

$$P = \overset{V}{\underset{j}{}} P_j \qquad (38)$$

(e) The intersect of P with the general law of (36) relates the special situation information to the measure M to provide M_s. This is provided by the composition

$$M_s = P \circ R \qquad (39)$$

This systematic approach to modifying critical measures by fuzzy information has the characteristics of a simple expert system.

4   Support Construction

A method of constructing the supports $\mu(x_i)$ to x_i has been provided by Saaty. It is based upon the proposition that an attempt to compare the n values of $\mu(x_i)$ for n states x_i ($i = 1$ to n) for a fuzzy modifier is much more difficult than attempting the $n(n-1)/2$ pair wise comparisons between the various x_i. Consider for instance the material strengths $x_1 = 2$, $x_2 = 3$, $x_3 = 4$, $x_4 = 5$ and $x_5 = 6$ ksi. The fuzzy modifier is 'good' and we seek the $\mu(x_i)$ to construct the fuzzy set.

$$\text{Good} \equiv \sum_{i=1}^{6} \mu(x_i) \mid x_i \tag{40}$$

Initially, let us presume that the $\mu(x_i)$ are appropriate. Then the comparison between the states x_i and x_j would be $\mu(x_i) \cdot \mu(x_j)^{-1}$. This quotient is an element a_{ij} in a n x n matrix.

$$A = \begin{bmatrix} a_{11} & a_{12} & \ldots & a_{1n} \\ a_{21} & a_{22} & \ldots & a_{2n} \\ a_{n1} & a_{n2} & \ldots & a_{nn} \end{bmatrix} \tag{41}$$

where $a_{ij} = a_{ji}^{-1}$. The n vector μ also exists where

$$\mu^T = [\mu(x_1), \mu(x_2), \ldots, \mu(x_n)] \tag{42}$$

The product

$$A \cdot \mu = n\mu \tag{43}$$

Now let us consider an application where A is not known but where pairwise comparisons between x_i and x_j, namely, $\bar{a}_{ij}$, have been made to reflect the fuzzy information. We desire the supports $\mu(x_i)$ obtained by solving the system

$$(A - nI)\,\mu = 0 \tag{44}$$

where A is the n x n matrix of $\bar{a}_{ij}$ elements. This system has a non-trivial solution for if μ and only if n is an eigenvalue of A. Only one non-zero eigenvalue exists, $\lambda_{\max}$, when A is consistent; then all $a_{ik} = a_{ij}a_{jk}$ and $\lambda_{\max} = n$. Otherwise $\lambda_{\max} > n$ and the difference, $\lambda_{max} - n$ is a measure of consistency.

Saaty suggested rules for establishing the $\bar{a}_{ij}$:

states i and j are equally important	— 1
state i is weakly more important than j	— 3
state i is strongly more important than j	— 5
state i is clearly more important than j	— 7
state i is absolutely more important than j	— 9

Now reconsider the fuzzy modified 'good' with respect to material strengths, x_i. For instance, neither $x_1 = 2$ ksi nor $x_5 = 6$ ksi make sense with respect to a good material. Therefore, $\bar{a}_{15} = 1$. On the other hand, a good material compared to $x_1 = 2$ ksi, hence $\bar{a}_{13} = 9$. On this basis a matrix of pairwise comparisons

$$A = \begin{bmatrix} 1 & 1/3 & 1/9 & 1/7 & 1 \\ 3 & 1 & 9 & 1 & 1 \\ 9 & 1/9 & 1 & 9 & 9 \\ 7 & 1 & 1/9 & 1 & 1 \\ 1 & 1 & 1/9 & 1 & 1 \end{bmatrix} \tag{45}$$

is constructed. This has obvious inconsistencies and the eigenequation (44) has $\lambda_{\max} = 7.47$ which is much bigger than $n = 5$. This would suggest a reconsideration of A in (45). However, as an example, the solution of (44) with $n = 7.47$ and A as in (45) for provides the fuzzy set.

$$\text{Good} \equiv 0.06 \mid 2 + 0.44 \mid 3 + 0.31 \mid 4 + 0.12 \mid 5 + 0.09 \mid 6 \tag{46}$$

5 Mergings

The construction of fuzzy measures of subjective uncertainties may have to be combined with probabilistic measures of objective uncertainties. Two procedures are described.

5.1 Fuzzification of Probabilities

Consider the fuzzy set M in (33) with elements which measure objective safety as in the probability of failure

$$P_f = 10^{-m} \tag{47}$$

The description of fuzzy system analyses in Section 3.4 leads to a fuzzy set.

$$M_s = \sum_{i=m-n}^{m+n} \mu_i \mid x_i \tag{48}$$

for the description of the exponents of (47) about m. Clearly a high support for $(m \pm n)$ indicates n order of magnitudes change from the objective probability of safety. This is a fuzzification about the objective probability measure, m.

5.2 Entropy Arguments

Considerations of probabilistic uncertainty involve random discrete variables, x_i, with probabilities of occurence, $p_i \cdot (i = 1$ to $n)$. The average uncertainty of the set of events X is the information entropy $H(p_1 \cdot p_n)$. This satisfies the Shannon *desiderata*:

- H is continuous in the various p_i,

- When all p_i are equal to $1/n$, then $H(1/n, \ldots, 1/n) < H(1/m_1, \ldots, 1/m)$ when $m > n$

- When $p_E = p_1 + p_2 + \cdots, p_e$ and $p_N = p_{e+1} + p_{e+2} + \cdots p_n$

then

$$H(p_1, \ldots, p_n) = H(p_E, p_N) + H(p_E) H \left(\frac{P_1}{P_E}, \frac{P_2}{P_E}, \ldots, \frac{P_e}{P_E} \right)$$

$$+ H(p_N) H \left(\frac{P_{e+1}}{P_N}, \ldots, \frac{P_n}{P_N} \right) \tag{49}$$

when

$$H = -C \sum_{i=1}^{n} p_i \ln p_i$$

Jaynes' maximum entropy principle provides for the least biased p_i when (49) is a maximum subject to the information or statistical constraints.

Considerations of fuzzy uncertainty involve discrete variables, x_i, with fuzzy supports, $\mu(x_i)$ $(i = 1$ to $n)$. The fuzzy entropy, $G(\mu(x_1), \ldots, \mu(x_n))$, expresses the average imprecision of X. *Desiderata* for the fuzzy entropy are:

- $G_i = 0$ iff $\mu = 0$ or $\mu = 1$

- G_i is a maximum iff $\mu = 0.5$

- $G_i \geq \overline{G}_i$ where $\begin{cases} \overline{\mu} \geq \mu \text{ for } \mu \geq 0.5 \\ \mu \leq \mu \text{ for } \mu \leq 0.5 \end{cases}$

These lead to

$$G = K \sum_{i=1}^{n} G_i \tag{50}$$

where

$$G_i = -\mu(x_1) \ln \mu(x_i) - (1 - \mu(x_i)) \ln (1 - \mu(x_i)) \tag{51}$$

C in (49) and K in (50) are positive constants.

Considerations of combined probabilistic and fuzzy uncertainty with variables of X of x_i with probabilities $p(x_i)$ and fuzzy supports $\mu(x_i)$ in which $\sum_{i=1}^{n} p_i = 1$ and $0 \le \mu(x_i) \le 1$ lead to a combined measure of average uncertainty and average imprecision with an entropy of $F(p_1, \ldots, p_n; \mu(x_1), \ldots \mu(x_n))$ which fulfills the *desiderata*

- F is continuous in p_i and $\mu(x_i)$

- $F(p_1\mu) = F(p) + F(\mu \mid p)$

- when a $p_i = 1$ then $F(p) = 0$ and $F(p, \mu) = G(\mu(x_i))$

- when all $\mu(x_i) = 0$ or 1 then $F(p, \mu) = H$,

when

$$F(p, \mu) = -\sum_{i=1}^{n} p_i \ln p_i + D \sum_{i=1}^{n} p_i G_i \qquad (52)$$

with D as a constant.

Considerations of combined probabilities, $\bar{p}_i$, in which

$$-\sum_{i=1}^{n} \bar{p}_i \ln \bar{p}_i = F(p, \mu) \qquad (53)$$

reveal a more platykurtic probability distribution, $\bar{p}_i$, than that of p_i, when $G \ne 0$. This $\bar{p}_i$ is the probability that reflects uncertainties due to all p_i and $\mu(x_i)$.

The constant, D, combining the two entropy measure H and G requires a criterion for its enunciation. One possibility is that when all $\mu(x_i) = 0.5$, then all $p_i = 1/n$ in (53). Then

$$D = -1.44270 \quad (\ln 1/n + H) \qquad (54)$$

In principle the $\bar{p}_i$ can be determined from (53) and (54). However, the individual p_i have at least $n!$ values. This may be remedied by splitting the sequence of x_i into $x_i \ge \Theta_A$ and $x_i < \Theta_A$. The probabilities of these states are termed $\bar{p}_A$ and $\bar{p}_B$ and are related by the normality constraint

$$\bar{p}_A + \bar{p}_B = 1 \qquad (55)$$

Such an approach provides two values $\bar{p}_A$ and $\bar{p}_B$. A decision concerning the appropriate probability value requires consideration of the constant D. The value in (54) is associated with $\mu(x_i) = 0.5$ and $p_i = 1/n$. This produces a maximum value of combined entropy, F_{max}. In general $H < F < F_{max}$ and one or more $\mu(x_i) > 0.5$ and the distribution of $\bar{p}_i$ is more kleptokurtic. At $F = H$ all $\mu(x_i) = 1$

and all $p_i = pi$. Therefore, this approach to determining D ensures that $0.5 \leq \mu(x_i) \leq 1$ and p_i to lie inclusively between p_i and $1/n$.

As an example, for $\mu(x_A) = 0.8$ and $\mu(x_B) = 0.9$, then $G_A = 0.50040$ and $G_B = 0.32508$. If $p_A = 0.975$ and $p_B = 0.025$, then $F = 0.52868$. The values of $\bar{p}_A$ and $\bar{p}_B$ are either 0.7786 or 0.02214. However, from the previous remarks, $1/n = 0.5$ and $0.5 \leq \bar{p}_A \leq 0.975$ and $0.025 \leq \bar{p}_B \leq 0.5$. Therefore, $\bar{p}_A = 0.7786$ and $\bar{p}_B = 0.2214$. The conclusion is that the property x_i exceeds or equals Θ_A with a probability of 0.7786 in the light of both objective and subjective information.

6 Conclusion

The intention of this material is to extend the understanding of uncertainty beyond the objectivity of probabilistic information and to include subjectivity in the form of fuzzy information. This has required the development of the meaning of fuzzy sets; the operations for fuzzy combination, relationships and composition; a method for constructing the fuzzy supports; a method for fuzzifying objective probabilities; a method for combining objective and subjective uncertainties to produce a combined entropy and probabilities.

These results provide a basis for civil engineering decision making where both objective and subjective information can be used without the ubiquitous imposition of probabilistic constraints.

Acknowledgment   The National Science Foundation supported this work over a six-year period.

References

Blockley, D. I., (1977), 'Analysis of Structural Failures,' *Proc. Inst. Civ. Engr.*, Pt. 1, **62**, 51–74.

Blockley, D. I., (1979), 'The Calculation of Uncertainty in Civil Engineering,' *Proc. Inst. Civ. Engr.*, Pt. 2, **67**, 313–326.

Brown, C. B., (1979), 'The Fuzzification of Probabilities,' *Proc. Spec. Conf. on Prob. Mech. and Struct. Rel.*, ASCE, Tuscon.

Brown, C. B., (1979), 'A Fuzzy Safety Measure,' *J. Engr. Mech. Div., ASCE* **105**, EM5, 855–872.

Brown, C. B., (1980), 'The Merging of Fuzzy and Crisp Information,' *J. Engr. Mech. Div., ASCE* **106**, EM1, 123–133.

Brown, C. B., (1980), 'Entropy Constructed Probabilities,' *J. Engr. Mech. Div.*, *ASCE* **106**, EM4, 633–640.

Brown, C. B., Johnson, J. L. and Loftus, J. J., (1984), 'Subjective Seismic Safety Assessments,' *J. Struct. Engr., ASCE* **110**, 9, 2212–2233.

Brown, C. B. and Louie, D. H., (1984), 'Uncertainty in Civil Engineering Systems: Probability and Fuzzy Sets,' *Civ. Engr. Syst.* **1**, 282–287.

Brown, C. B. and Yao, J. T. P., (1982), 'Instructions and Opinions: The Use of Fuzzy Subjectivity,' *Architectural Science Rev.* **25**, 3, 70–74.

Brown, C. B. and Yao, J. T. P., (1983), 'Fuzzy Sets and Structural Engineering,' *J. Struct. Engr., ASCE* **109**, 5, 1211–1225.

DeLuca, A. and Termini, S., (1972), 'A Definition of Non-Probabilistic Entropy in the Setting of Fuzzy Sets,' *Inf. and Cont.* **20**, 301–312.

Johnston, D. M. and Palmer, R. N., (1988), 'Application of Fuzzy Decision Making: An Evaluation,' *Civ. Engr. Syst.* **5**, 2, 87–92.

Saaty, T. L., (1977), 'A Scaling Method for Priorities in Hierarchial Structures,' *J. Math. Psych.* **15**, 235–281.

Civil Engineering
Optimizing and Satisficing

Colin B. Brown
Department of Civil Engineering
University of Washington
Seattle, Washington
United States of America

Abstract Decision making can be viewed as providing extrema on a measure of value, the objective function, in the light of practical constraints. This optimization approach may not be appropriate in civil engineering decision making where uncertainties dominate. In its place, the process of satisficing is offered.

1   Introduction

In symbolic form the mathematical program,

$$
\left.
\begin{aligned}
&\min_{\overline{x}} \quad z(\overline{x}) \\
&\text{s.t.} \quad f(\overline{x}) \geq \overline{G} \\
&\qquad\quad \overline{x} \geq \overline{0}
\end{aligned}
\right\}
\tag{1}
$$

provides an objective function Z which depends upon the elements of the vector $\overline{x}$. The search for elements of $\overline{x}$ that minimize Z is constrained by the various inequalities.

The simplest form of (1) is the linear form for both the objective function and the constraints. Then the linear program is

$$
\left.
\begin{aligned}
&\min_{\overline{x}} \quad \{Z = \overline{C}^T \overline{x}\} \\
&\text{s.t.} \quad \overline{A}(\overline{x}) \geq \overline{G} \\
&\qquad\quad \overline{x} \geq \overline{0}
\end{aligned}
\right\}
\tag{2}
$$

B. H. V. Topping (ed.),
Optimization and Artificial Intelligence in Civil and Structural Engineering, Volume I, 33–41.
© 1992 *Kluwer Academic Publishers. Printed in the Netherlands.*

The objective function and the constraints can be fuzzified. For instance the elements of $\overline{G}$, namely G_i, can be replaced by a fuzzy mean with connotation of 'small, but not too small.' The objective function can also be softened. This is reasonable when multiple objectives exist. These can then be dealt with as goals in the form.

$$\overline{C}^T \overline{x} \leq \overline{Z}^1 \tag{3}$$

This approach is still crisp but can be interpreted as fuzzy objectives. The constraints now appear as the inequalities

$$\overline{k}\overline{x} \geq \overline{h} \tag{4}$$

where

$$\overline{k} = \left(-\overline{C}^T \overline{A}\right)^T \tag{5}$$

and

$$\overline{h} = \left(-\overline{Z}^1 \overline{G}\right)^T \tag{6}$$

The crisp form of the i-th constraint is

$$\sum_j k_{ij} x_j \geq h_i \tag{7}$$

which can be softened to

$$\sum_j k_{ij} x_j \geq h_i - S_i \tag{8}$$

where S_i is the slack variable and may have fuzzy form as

$$\mu_i \mid S_i$$

with, for instance,

$$\mu = 1, \qquad \text{if (7) applies}$$
$$\mu = 0, \qquad \text{if (8) does not apply}$$
$$\mu = \sum_j \left(\frac{k_{ij} x_j - h_i + S_i}{S_i}\right), \quad \text{if the form is between (7) and (8)}$$

The optimizing problem of (2) has now been changed to determing, for any vector $\overline{x}$, the maximum of the minimum supports among these fuzzy inequalities. In this way, the minimum supports are the sequence $\{a_i\}$ and the maximum among these is the 'optimum' decision variable.

More formally, the program is

$$\left. \begin{aligned} \max \quad & \{a_i\} \\ \text{s.t.} \quad & a_i \le \sum_j \left(\frac{k_{ij} - h_i + S_i}{S_i} \right); \quad 0 \le a_i \le 1 \\ & \overline{x} \ge 0; \qquad\qquad\qquad i = 1, 2, \ldots \end{aligned} \right\} \qquad (9)$$

In this way, the a_i provides a support for 'optimal' solution, $\overline{x}$, where optimal is interpreted as having the greatest support.

When the objective function is crisp, then we have a mixed program between (2) and (9); namely,

$$\left. \begin{aligned} \min \quad & \left\{ z^1 = \sum_j c_j x_j \right\} \\ \text{s.t.} \quad & a_i \le \sum_j \left(\frac{k_{ij} - h_i + S_i}{S_i} \right); \quad a_i \ge a_L \\ & \overline{x}_j \ge 0, \qquad\qquad\qquad i = 1 \text{ to } m \\ & \qquad\qquad\qquad\qquad\quad j = 1 \text{ to } n \end{aligned} \right\} \qquad (10)$$

In this case, a_L, is the minimum support for barely acceptable design.

These approaches whereby the optimization problems of (1) and (2) are softened to the fuzzy problems of (9) and (10) are developed in the book by Zimmermann. The intention in this paper is to treat this softening as a trend from optimization to satisficing. The reasons and justifications for these forms of decision making are discussed. The terms of the discussion are those of structural engineering, but some application to other areas of civil engineering is possible.

2 Objective Function

The obvious objective function in decision making in capital intensive schemes is the cost of the design and construction. These costs also involve the weighted cost of replacement in the event of failure. The weighting will be a probability of the failure occurring. Additional to this will be the weighted cost of litigation and the present worth costs of maintenance and operation. However, in structural optimization, the weight of the structure is a more usual objective function. This moves the problem away from the world of uncertainty in the future and hence makes such an objective popular.

2.1 Costs

The practical selection of a minimum initial cost structure requires the bidding by competing contractors on a design supplied by the owner's representative. The design meets the intentions of the owner and is the instrument for obtaining cost bids. Such a design must attract enough bidders to ensure a competition on the basis of cost and therefore must not include features which would favor one bidder and exclude others. This process usually guarantees that the successful bid is not one of minimum cost; a design which utilized the particular skills and equipment of that or one of the other contractors would usually result in a lower cost. Collusion in this form may be encouraged in parts of the world, but not in North America. It may lead to the minimum cost structure if, and only if, that structure is included among the competing alternatives.

The design costs are related to the length of time that the process takes. Table 1 shows the design time for each ton weight of structure that falls into the purvey of civil engineers. The design of ships requires about the same time as bridges.

Table 1. *Design time per ton weight of structure.*

Structure	Man Days
Novel offshore structures	2.50
Conventional offshore structures	0.25
Bridges	0.20
Buildings	0.07

However, aircraft need about 10 years of a person's time to design a structural ton and spacecraft claim 500 man years.

As well as the design and construction costs, the insurance premiums must be included in initial costs. These premiums spread the risk costs of failure, litigation and replacement over a broader community. Such an activity is of apparent benefit to engineers and owners but still provides an enduring cost to society.

Expenses of maintenance and operation depend upon future inflation and interest rates. It is not clear that the true least-cost solution will be independent of these future rates.

Such a discussion suggests that:

(a) the process of civil engineering bidding,

(b) the limited design time available,

(c) the hedging on the future,

makes the use of future costs as an objective function somewhat debatable.

2.2 Weight

Certainly in aircraft and spacecraft the use of weight as an objective function to be minimized appears popular and sensible. The design of such structures has requirements on the kinematic performance and the pay load. Of interest is the structure necessary to carry the pay load and fuel to support itself and the thrust units. Essentially, the kinematics (a) and the pay load (m^*) are constraints. The variables are the thrust (F) and the other weights of the structure, fuel and engines (m). Newton's second law applies and

$$F = (m + m^*)\, a \tag{11}$$

The design process involves minimizing the thrust and hence the fuel requirements and hence the vehicle weights. The minimization of m ensures this situation and therefore the use of weight as an objective function has merit. No such argument is valid for civil engineering structures. In fact, examples abound where extra weight reduces the total cost. What appears to be a fruitful approach in a design governed by dynamics is not appropriate in civil engineering where statics provides the physical laws.

3 Constraints

The constraints for cost or weight objective functions are the same; the structure should function and not fall down. Formally, these amount to

(a) stress safety constraints,

(b) displacement constraints,

(c) stability safety constraints,

(d) frequency constraints.

(a) and (c) are to do with safety, (b) is a functional constraint and (d) serves both roles. The constraints (a) and (b) require the completion of structural analysis

$$K \circ \gamma = Q \tag{12}$$

for each loading and each alternative in the optimality path. The constraints (c) and (d) require the solution of eigenvalue problems where

$$(K - \lambda M) \cdot \gamma = 0 \tag{13}$$

Here M is the geometric stiffness for (c) and the mass for (d). Additionally, constraints on available materials and members exist.

The computational activity in any but simple structural problems is extensive in order to cover all of these constraints. The design time is large, but except for aircraft and spacecraft, the time available is small. More usually the design intention is changed to a desired state where each member is fully stressed to the extent of constraint (a) under some loading state. Thus, the optimality criterion of this fully stressed design becomes the surrogate for minimum weight. The process is iterative and the constraints are (b), (c) and (d). These are checked at the fully stressed design. This apes the process of conventional design. Of interest is the work of Spillers, who has shown that in trusses this process leads to a convergence onto a fixed weight which is the minimum weight state. No such convergence properties have been displayed for flexural structures.

The constraints listed have differences in their return periods. The functional constraints, (b) and (d), are exercised on a return period that is small compared to the life of the structure. The safety constraints, (a) and (c), are concerned with gravity loads of the same return period as the functional effect, and extreme lateral loads where the return periods are much the same as the intended structural life. These comments suggest that the constraints could be appropriately reorganized into:

(a) short return period effects with high chances of occurrence,

(b) long return period effects with much smaller chances of occurrence.

Such an approach must be reflected in the uncertainties expressed in a true cost objective function. However, exactness requires precise statements on uncertainty. It well may be the crispness of the exact optimum solution is barmecidal.

4 Comment

In the previous discussion we have been concerned about the introduction of reality into the optimization process. It appears that an extremum on costs requires a realistic approach to civil engineering bidding procedures, as well as considerable information on the future. A change of objective to weight makes sense with respect to vehicles but not to static structures. The constraints that are posed may have the same features of unsureness that distinguish costs and even the usual way of dealing with weight minimization; namely, the design for a fully stressed structure, avoids the main objective of cost reduction. Capping all this is the absence of design time to study truly complex systems.

Constraints and objectives are limited by the human mind; they may, in fact, be an incomplete model of the real world. In all significant practical problems the human mind fails in the modeling of the actual reality; it is this inability to capture and solve real, complex problems that suggested to Simon the necessity of a Principle of Bounded Rationality. In this Principle formal solutions, in which the alternatives, the constraints and the objectives are incomplete, cannot produce global answers. Simon separates the limited rationality employed by man from the global rationality necessary to attain a complete solution to a real problem. The limited rationality is what professionals use when they seek solutions to real problems in a finite time in the face of uncertainty about the world. Such a definition of a professional is fitting; engineers, lawyers, doctors and soldiers would find it applicable. Based upon this Principle, Simon urged decision makers to abandon optimization and instead seek solutions which are acceptable. He termed the process whereby the acceptable, rather than the best of all possible worlds, is sought as satisficing. Certainly in complicated problems of structural design, the conditions for the Principle of Bounded Rationality exist and satisficing would seem to be appropriate.

The incompletness of data and understanding in realistic situations when coupled with short design time for highly complex systems describes the environment facing civil engineering decision makers. The unambiguous extremum of (1) requires equally clear objectives and constraints. These are only possible in the simplest of all systems, and as complexity increases, the effectiveness of precise answers diminishes. Zadeh addressed this problem of complexity in his Principle of Incompatibility which states that:

> ... as the complexity of a system increases, our ability to make precise and yet significant statements about its behavior diminishes until a threshold is reached beyond which precision and significance (or relevance) become almost mutually exclusive characteristics.

He takes this as an apology for the use of fuzzy sets by offering the paraphrase 'The closer one looks at a real world problem, the fuzzier becomes its solution.'

The previous discussion displays a separation between optimizing and satisficing. Optimizing is a procedure based upon strong mathematical foundations which allows precise statements to be made in the objective function, constraints and solution. Satisficing is an expectation with formal trappings which admits of imprecision in the objectives and constraints and robustness in the solution. Applications of optimization in design requires an act of confidence in the procedures which can seldom be supported by the reality of the problem. Applications of satisficing reveal the reality with confidence but are based upon undeveloped procedures. Optimizing is an act of functional and limited rationality; satisficing has the appearance of global rationality and respects the Principle of Bounded Rationality. Certainly an optimization scheme ignores the Principal of Incompatibility; satisficing may respect the Principle.

Satisficing in the form of (10) is one approach to the design problem in civil engineering. It ensures a formal mathematical procedure which leads to results which are of practical interest. A desideratum is that the results should not be critically sensitive to changes in the elements of x. In this way the design solution is robust and in the geometric sense has a broad plateau about the extremum. Such a requirement is important; then the vagaries of the objective functions and constraints, together with the incompleteness of data, are not critical. A false extremum does not lead to an invalid solution.

Other approaches have been suggested in which the optimum solution, $\bar{x}$, is first found and termed the 'best' solution. Then solutions in the admissible set are identified that are 'good' and others which are 'satisfactory.' To these solutions are applied indifference tests to determine classes of solutions which are of equivalent value. Such an approach meets the requirement of robustness and a value indifference between solutions. This approach allows for uncertainty and imprecision to be directly taken into account.

5 Conclusion

An attempt has been made to discuss the characteristics of optimizing and satisficing. Both are mathematical activities and can be formally examined. However, the extent that the formalism can ape reality varies. To guide the realism we have mentioned Simon's Principle of Bounded Reality and Zadeh's Principle of Incompatability, and in the light of these it is suggested that satisficing may embrace the understanding of uncertain data, objective functions and constraints in a more realistic manner than optimizing. In the end, both optimizing and satisficing deal with exactly the same decision problem and available information. The way this material is viewed is the key to the difference. Complete bounds on the same information are provided by J. B. Cabell when he wrote:

> The optimist proclaims we live in the best of all possible worlds; and the pessimist fears this is true.

Certainly the optimizer sees the world through rosy glasses. We do not suggest the other extreme, but do suggest somewhere in between and urge that civil engineer decision-makers become satisficers.

Acknowledgment   The National Science Foundation supported this work.

References

Brown, C. B., (1990), 'Optimizing and Satisficing,' *Structural Safety*, **7**.

Brown, C. B., (1989), 'Structural Safety and Satisficing,' *Proc. ICOSSAR '89*, San Francisco, California.

Brown, C. B., Furuta, H., Shiraishi, N., and Yao, J. T. P., (1986), 'Civil Engineering Applications of Fuzzy Sets,' *Analysis of Fuzzy Information*, Vol. III, J. C. Bezdak (Ed.), CRC Press.

Roy, B., (1971), 'Problems and Methods with Multiple Objective Functions,' *Math. Prog.* **1**.

Simon, H. A. (Ed.), (1957), *Models of Man, Social and Rational: Mathematical Essays on Rational Human Behavior in Social Settings*, Wiley, New York.

Simon, H. A., (1957), *Administrative Behavior*, 2nd Edition, Free Press, New York.

Spillers, W. R. and Farrell, J., (1969), 'On the Analysis of Structural Design,' *J. Math. Anal. and Appl.* **25**, 2.

Turner, B. A., (1978), *Man-made Disasters*, Wykeham Publications (London) Ltd., Crane, Russak and Co., Inc., New York.

Zadeh, L. A., (1973), 'Outline of a New Approach to the Analysis of Complex Systems and Decision Processes,' *IEEE Trans. Syst., Man. and Cyber.*, SMC-3, 1.

Zimmermann, H. J., (1987), *Fuzzy Sets, Decision Making and Expert Systems*, Kluwer Academic Publishers, Boston, Massachusetts.

Network Structuring Algorithms

D. G. Elms
Department of Civil Engineering
University of Canterbury
Christchurch
New Zealand

Abstract For some projects an appropriate optimal design strategy is to minimise functional clashes by dealing first with the most highly interacting design requirements. A graph is formed from a list of requirements and the links between them. Graph theory algorithms are given for reducing the graph to a set of clusters, then reassembling them in hierarchical order as a guide for design. Examples are given for layout generation, and an extension is proposed for general systems.

1 Introduction

To optimise is to achieve what is best. This talk discusses one strategy for achieving it; but first is is worthwhile to consider some ways in which we might talk of 'the best.'

The most obvious way to define 'best' is in terms of a numerical quantification of worth. Maximum profit, measured by money or utility, is an example. So are maximum efficiency or territorial gain, or minimum weight, energy consumption or risk. However, there are many areas in which 'best' cannot be so easily quantified: a hiking route, for instance, could be picked for least effort or maximum view, a commercial strategy for the least likelihood of problems, or a design for greatest functional efficiency. Then there is the area of aesthetics: presumably in art or music the best can be chosen, as winners of competitions are selected in these fields. Admittedly, the choice must be subjective as objective criteria cannot be used. However, I would claim that choices of this nature are not so far removed from engineering practice, because most engineering decisions, most choices of the best, are made subjectively.

This is not at all to say that engineering decisions made with a major subjective component are made arbitrarily. Far from it. Good decisions, however they are made, must be founded on all the information possible within the constraints of the

43

B. H. V. Topping (ed.),
Optimization and Artificial Intelligence in Civil and Structural Engineering, Volume I, 43–60.
© 1992 *Kluwer Academic Publishers. Printed in the Netherlands.*

problem. Rather, what is being said is that subjective decision-making is a normal and legitimate engineering activity; that it exists, that it could be well or poorly done, and that we should properly spend more effort than we have done in the past on understanding the processes and establishing clear criteria and guidelines for good subjective decision-making.

It is my belief that in civil engineering at least, the more complex the system being dealt with, the more appropriate it is for an engineer to become immersed in the system and to act subjectively from within it rather than to stay outside and attempt to use only objective, analytic tools. For dealing with complex systems, the engineer needs a new set of techniques which I shall call 'subjective tools.' These are aids within the process, interactive tools by which the engineer's own capabilities are helped and enhanced, as opposed to objective techniques in which a model is set up on, say, a computer and then run, with no interaction taking place during the process. Finite elements and linear programming are examples of objective tools. The talks I give in this Institute are all concerned with subjective tools.

An underlying implication of the concept of subjective tools is that it is the engineer who is the primary tool, and who is acting on and within the system in the process of achieving the system goal. It is therefore not only a question of engineers learning about a series of techniques. Rather, they themselves have to be tuned, as tools in the process. This is often a matter of acquiring attitudes, of learning sometimes subtle criteria as to how to proceed, and how to know what is good. Suggestions for educating engineers in these directions have been made elsewhere (Elms (1989)).

Returning to the question of 'best,' a serious and interesting question facing engineers is, how do we know what we do is good? There is of course a sense in which we do not have to know what is good; we only have to do what is good. We have to produce a good design, as opposed to mediocre, so-so. The question was tackled by Pirsig (1974), whose answer was that merely, in the end, one has to care about what one is doing. This may be true, but there is more to it than that. There is also the matter of not making an error. In this talk, optimisation is primarily concerned with minimising errors.

In the context of engineering design and planning there are four types of error. The first is straightforward: simply doing something wrong of a trivial nature, such as making a mistake in a calculation, or perhaps making a wrong assumption. The answer to such errors lies in quality control of some sort, such as sytematic checking. The second error type is omission; some effect should have been taken into account, but was missed. Generally this is a matter of using a wrong or inadequate model, particulary if the omission is major. (Major, though, can sometimes be more apparent by hindsight than at the time when initial design decisions have to be made). In the context of complex civil engineering systems, forming an appropriate and gap free conceptual model at the early stages of proposal developments is a difficult task, which could be helped by applying systematic criteria to model development (Elms (1988)). The third type of error is one of poor style, of awkwardness. It is perhaps not strictly an error at all, but it is included here for the sake of completeness, particularly as we can quite reasonably argue that a poor piece of design, which is too complex and not particularly functionally or economically efficient, is wrong and, ideally, should not be done. Such designs can often lead to trouble later, as with the much-criticised cargo doors of the DC10 aircraft. The fourth error type occurs as an aspect of the development of an engineering project over time, and concerns functional clashes at a late and expensive stage of a project. One example is the Sydney Opera House, where the initial design of the main auditorium could not be used for symphony concerts, though the brief required it: the partly constructed auditorium had to be demolished and a new one built beneath the outer structure of the building. Other examples exist in the area of the environment, where the very nature of the process whereby an environmental impact assessment is not done until a project exists on which to do it necessarily leads to trouble if environmental requirements cannot be satisfied. It is this final type of error that is principally addressed by the network structuring discussed below.

The approach relies heavily on a design strategy developed by Christopher Alexander (1964) for the development of the layout of physical facilities. It is based on the observation that people take correct functioning for granted, but what they notice is the things that are wrong. (This is sometimes called the thumbtack syndrome, from the scenario of a comfortable armchair with a drawing pin on its seat: what the sitter notices is the pin, and not at all the comfort). Thus an optimum design in this sense is the one with the fewest things wrong. In the layout of a physical facility, the things seen to be wrong are concerned with poor functionality, which generally derives from conflict between basic design requirements. Alexander thus devised a strategy for minimising functional clashes by first forming a list of design requirements. From this, a graph is constructed by linking related requirements. The graph is then examined to identify the most highly interacting clusters of requirements, which are combined in a hierarchy, forming a tree. Alexander (1964) formulated the problem in terms of information theory. The algorithm was programmed by Daish and Kleem (1969), and its use was explored by Clark and Elms (1976) who found there was much more required for the success of the total process then possession of the computer program alone: they went on to formulate principles for successful use. Later, Elms (1983) developed graph theoretic algorithms for efficient decomposition of a graph into a hierarchy of constituent clusters. These algorithms will be described next, followed by examples of use.

2 Building a Structure into a Tree

2.1 Problem Structure as a Graph

Let us start with an interaction graph consisting of a number of elements, which may be assigned weights, and a series of links between them. The links may be directed, and weighted. An undirected link may be thought of as two directed links of equal weights making connections in opposite directions. If the elements refer to the basic requirements for a design, then the graph represents the intrinsic structure of the problem. We shall therefore call such a graph a 'structure.'

2.2 Generating Clusters

The graph may be expressed as an association matrix A with the off-diagonal terms representing links between elements. The matrix is necessarily square. The integer value a_{ij} of a term of the matrix represents the number of single link paths between the corresponding elements, in this case elements i and j. If the association matrix is multiplied by itself to form

$$A^2 = AA \tag{1}$$

then a term $a_{ij}^{(2)}$ of A^2 represents the number of 2-link paths from i to j (the word 'term' is used rather than 'element' to avoid confusion with elements of the original graph). Similarly a term $a_{ij}^{(3)}$ of

$$A^3 = AA^2 \tag{2}$$

represents the number of 3-link paths from i to j.

Two quantities are now of particular interest. Firstly, the row sums of A^3 are a measure of the relative dominance of the corresponding elements. The normalised row sums approach stablised values as the matrix power (the exponent in A^n) increases (Berge (1962)). Secondly, the diagonal terms of A^3 are a measure of the degree of clustering or interactiveness of the corresponding elements. It is this latter measure that is particularly important as it is the tightness of clustering and its nature that we want to explore in the original problem structure. It represents the degree to which an element is mutually interconnected with pairs of other elements.

The average value of the A^3 diagonal terms is a measure of the degree of clustering of the graph as a whole: we shall call this the 'tightness' of the structure.

This definition of tightness can also be applied to any subgraph of the main structure. It is particularly useful in that it enables us to grow tightly-interconnected clusters from an original nucleus, chosen from the relative ranking of elements on the basis of their A^3 diagonal terms. Once a cluster has been formed and its tightness calculated, whether or not an additional element should be added will depend on whether the tightness of the cluster is increased. The cluster can be made to grow in this way until no other element can be found whose inclusion will increase the tightness. Another cluster can then be initiated and the process continued until the original graph has been entirely broken down into a set of clusters.

Consider, for example, the problem structure given in Table 1. This is a simple structure with unweighted elements and unweighted symmetric links. Figure 1 shows the graph pictorially. Following the process outlined above, the first cluster to be completed is shown in Fig. 2. The complete set of basic clusters is given in Table 2.

2.3   Building the Clusters into a Hierarchy

Having broken down the initial structure into a set of clusters, the next step is to reassemble them in a hierarchy. This may be done by first forming the clusters into a new graph, with the clusters as its elements. We need to know two things: the weights to put on the nodes of the graph, and the weights to put on the links.

The weight assigned to a cluster is simply the sum of the weights of its component elements. The weights of the links between clusters need to be defined with some care as the clusters may overlap. For two clusters P and Q, let the weight of the link from P to Q be the sum of the weighted links passing from all elements of P across the boundary of P to the rest of the elements of Q. For the example of Fig. 3, the links from P Q and from Q to P have weights of 2 and 3 respectively.

We next need a measure of attraction or closeness between clusters. Using an approach analogous to the gravitational law, in which both the size of the clusters and their degree of interlinking combine to give a measure of attraction, we define the closeness of two clusters A_i and A_j with weights c_i and c_j as

$$d_{ij} = c_i c_j \left(c_{ij} + c_{ji} \right)^2 \tag{3}$$

where c_{ij} is the weight of the link between A_i and A_j.

A partitioning process now takes place which begins by initiating a splinter group based on the cluster with the smallest average closeness to all other clusters. To this splinter group is added any cluster closer to the splinter group than to the main group, whose average closeness to the main group is least. The splinter group grows by this process until there is no longer a cluster in the main group closer to the splinter group than to the rest. The splinter group is removed from the graph and the partitioning process continued by applying it to the reduced graph. The process is carried on until no further reduction is possible. The splinter groups can of course be reduced too, where appropriate. Figure 4 shows the tree formed in this way from the graph of Fig. 1, using the clusters listed in Table 2.

48

Table 1. *Diagonal terms of A^3 matrix.*

Element	A^3 diagonal	Rank
1	60	2
2	54	4
3	62	1
4	20	14
5	56	3
6	40	7
7	52	5
8	14	17
9	52	5
10	22	13
11	20	14
12	14	17
13	14	17
14	20	14
15	28	11
16	28	11
17	32	9
18	38	8
19	30	10

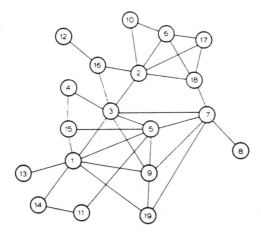

Figure 1. *Pictorial representation of graph.*

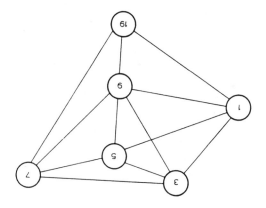

Figure 2. *First cluster.*

Table 2. *Complete set of clusters.*

Cluster	Elements						Tightness
A	3	1	5	9	7	19	41.0
B	2	6	18	17	10		33.2
C	15	4	3	1	5		27.2
D	16	12	2	3			21.5
E	11	14	1	5			20.0
F	7	8					14.0
G	1	13					14.0

3 Layout Generation

So far we have dealt with the mechanics of the process. In practice, using it to generate physical layouts, there is much more to it. The initial graph must be set up with great care, not so much for individual details as for its overall intent and structure. Once a tree has been formed, its subsequent use relies very much on experience and an appropriate attitude. There are certain underlying principles of style and approach.

The first step in the process is to form a list of the attributes the final design should have, and to link them appropriately. The following principles apply:

1. *Generality.* The attributes must be as general as possible and should not presuppose solutions.

2. *Completeness.* No major design requirements should be omitted. This re-

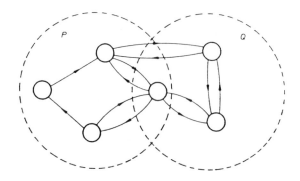

Figure 3. *Links between clusters.*

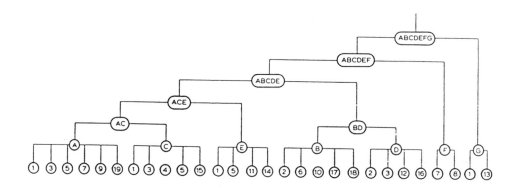

Figure 4. *Complete tree corresponding to Fig. 1.*

quires careful analysis.

3. *Balance.* Elements should be of the same order of magnitude. It does not make sense to include requirements for the size of faucets or the positioning of electrical outlets in a list dealing with entire buildings.

4. *Consistency.* Attribute definitions need to be consistent within a list, and concerning the same things. Chalk and cheese cannot easily be compared and should not be in the same list.

5. *Discrimination.* The function of the elements of the attribute list should be to assist in breaking the design into parts. Some parts should be affected by an attribute and others not. Thus 'lighting' in a building would be a poor attribute as every part of the building and virtually every activity within it would need lighting. It would therefore be linked with every other attribute, and would be no help in structuring the problems. (In fact, it would be a hindrance as it could mislead the designer). Similarly, there is no point in including an attribute if is has no connection to any other except, perhaps, to one, for it, too, would be of no help in structuring the problem. A further point under this heading is that where possible, attribute definitions should not overlap.

6. *Coherence.* This is a difficult principle to explain though it becomes clearer with practice. It is concerned with wholeness, with looking at the totality of the list and its links throughout.

7. *Richness.* The list and its links must have an appropriate degree of richness. Too short a list, with too few links, and the problem becomes trivial. Too much complication and inter-connectedness, and the structure of the problem becomes lost in detail. About 50 attributes is about right.

Table 3 shows the attribute list and links for the design of a structural test laboratory. No differential weights have been assigned in this example. The resulting tree is shown in Fig. 5. Alexander's (1964) original algorithm has been used for this example.

Fortunately the tree has far too much detail in it for the designer to be able to use it as a series of specific instructions. It should be, and can only be, used as a guide, as a series of signposts.

Neither can each cluster and stage be dealt with individually. There are far too many for that. The best approach is to pin the tree on the wall to act as a map to show where progress can best be made.

What is then done is to take suitable clusters at a reasonably low level in the tree and make outline sketches. These are at first rudimentary (Fig. 6), but along with them is developed a graphic vocabulary which can be used to capture a great deal of information in a seemingly-simple diagram. The first elementary sketches are used to outline possible solutions to the most highly interactive functional requirements. Wandering up the tree by combining the ideas of the elementary sketches, intermediate diagrams become more complex (Fig. 8) until eventually a final diagram is developed (Fig. 9). (The sketches shown here are a draughtsman's cleaned-up version of the original rough sketches rapidly drawn by the project architect and the writer).

The final diagram is not meant to represent a real layout. Rather, its symbols, labels and proximities are a representation of the information an architect should

Table 3. *Structural test laboratory—attributes and links.*

#	Attribute	Links
1	Test floor	2, 3, 4, 6, 7, 8, 9, 17, 20, 22, 27, 33, 38, 40, 42, 46
2	Large test manufacture	1, 2, 3, 4, 5, 6, 7, 20, 22, 33, 42
3	Large test storage	1, 2, 3, 4, 6, 7, 33, 42
4	Crane	1, 2, 3, 4, 6, 7, 20, 27
5	Access to concrete mixer	2, 3, 6, 7
6	Truck access	1, 2, 3, 4, 6, 7, 10, 11
7	Flexibility	1, 6, 7
8	Data logger, instrumentation room	1, 12, 16, 17, 23, 25, 35, 41, 45, 47, 49
9	Test machines	1, 6, 11, 20, 28, 39, 44, 49
10	Site testing store	6, 11
11	Caravan park	6, 10
12	Model test area	8, 14, 15, 16, 17, 23, 34
13	Equipment servicing, strain gauging, workshop etc—clean	10, 14, 15, 17, 23, 24, 34, 45, 47
14	Storage for instruments, cable, gauges	12, 17, 23, 42
15	Small workshop area—fitting bench—less clean	2, 12, 23, 34
16	Stable thermal environment—model test	8, 10, 12, 13
17	Reasonably stable environment	1, 8, 9, 14, 28
18	Offices—technician	19, 24, 29, 31, 32, 36, 45, 47, 49
19	Offices—professional	18, 29, 31, 32, 36, 21, 23, 24, 37, 47, 49
20	Noisiness	1, 2, 4, 9
21	Quietness	2, 18, 19, 41, 47, 49
22	Dirtiness	2, 9, 15
23	Cleanliness	9, 12, 13, 14, 43, 47, 49
24	Direct access to administration	13, 18, 19, 42
25	Access to building science from test m/c's	9, 19, 26
26	Test m/c's available to other sections	9, 25
27	Rubbish storage	1, 6
28	Shaking table	4, 17, 20, 22, 47
29	View between labs and offices	18, 19, 20, 47, 49
30	Minimal pedestrian movement	49
31	Natural lighting	18, 19, 46, 49
32	Natural ventilation	18, 19, 22, 42, 49
33	Workbench—heavy duty	1, 2, 13
34	Site testing labs	10, 12, 13, 14, 22, 24, 48
35	Epoxy materials and inflammable liquids store	12, 18, 19, 31
36	Ablutions central in labs	18, 19, 21, 23, 25, 26, 34, 44
37	Office space near computer facilities	19, 48
38	Test m/c's near strong floor	1, 9, 17, 26, 29, 30
39	Sound insulation for test m/c pumps	9, 20, 43, 45, 49
40	Sound insulation for strong floor	1, 20, 29, 30
41	Filing room	18, 21, 23
42	Ability to wheel equipment between labs, and workshop	1, 2, 3, 5, 9, 10, 14, 15, 24, 25, 26, 34
43.	Display area	23, 24, 48, 42, 2, 3, 5, 9
44	Electric hoist for large testing m/c	6, 9, 42, 47, 48
45	Notice boards (for programme, whereabouts)	18, 19, 21, 23, 24, 31, 36
46	Ability to photograph tests	1, 12, 31
47	Good relations between staff	18, 19, 21, 23, 26, 29, 32, 36, 43, 45, 47, 49
48	Good communications with other sections	24, 26, 29, 34, 37, 43, 44, 45, 48
49	Good staff morale	18, 19, 23, 24, 29, 30, 32, 36, 47, 48

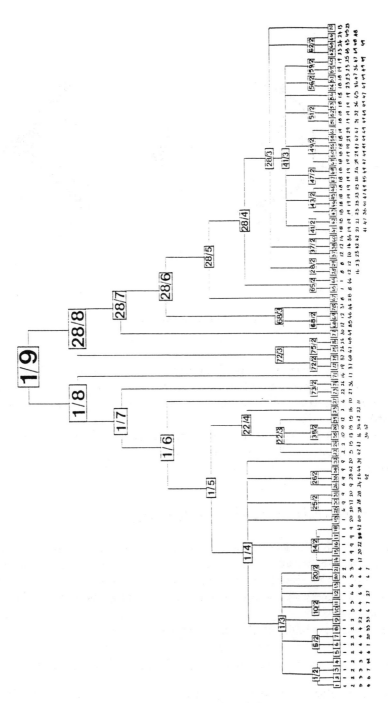

Figure 5. *Tree for structural test laboratory.*

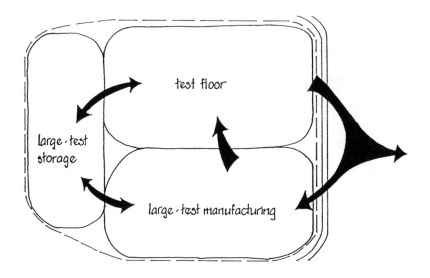

Figure 6. *Early sketch—1/2 on tree.*

know about the requirements of a project which could then be used as part of a brief. Supplemented by text, area schedules and so on, it would then be the basis for proceeding to conventional sketch plans and working drawings (Elms and Clark (1976)).

Figures 10 and 11 give other final diagrams developed in this way for the New Zealand Ministry of Works, for other sections of a major engineering laboratory complex.

4   Towards a General Design Aid

The next step is to generalise the process so that it is more universally applicable and not restricted to the generation of physical layouts. Work is at present underway in this direction, with application aimed at the difficult problem of the scoping and preliminary design of engineering projects involving interaction with the environment.

In design problems of this type, the engineer needs:

(a) to learn about the problem and understand all relevant aspects

(b) to develop an appropriate model which is complete, with no gaps, at a consistent level

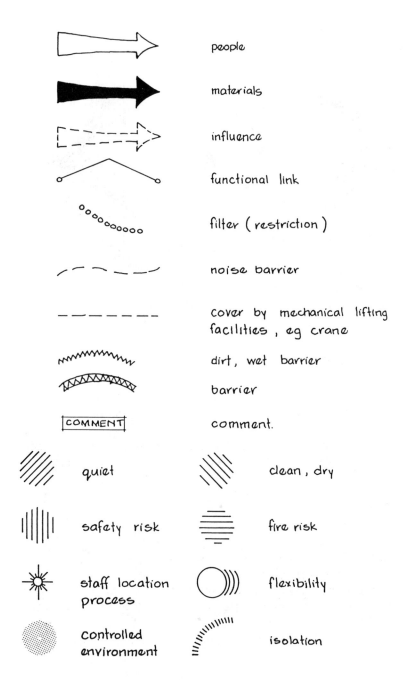

Figure 7. *Graphic vocabulary.*

56

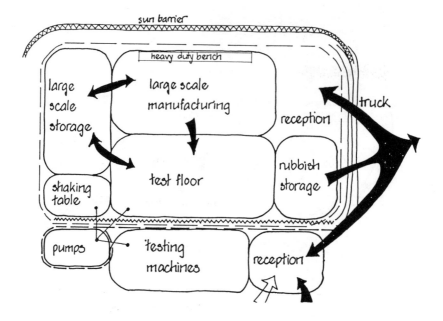

Figure 8. *Intermediate diagram—1/4 on tree.*

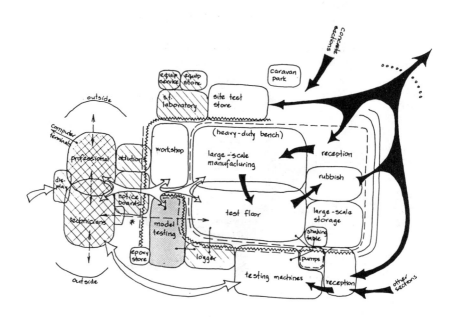

Figure 9. *Final diagram, structural test laboratory.*

57

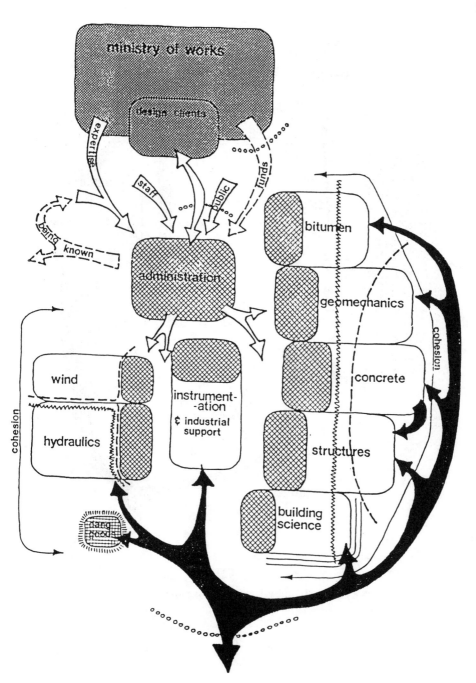

Figure 10. *Laboratory complex—overall layout.*

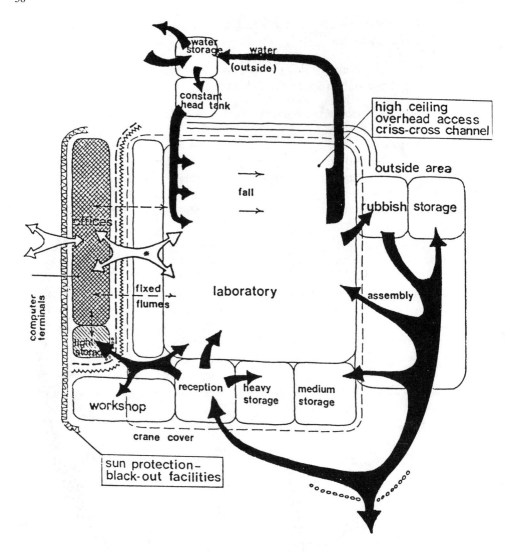

Figure 11. *Hydraulic laboratory.*

(c) to achieve an optimum solution which is, in Alexander's terms, a result with no functional clashes

(d) to achieve a preliminary design with no major errors, which needs a quality assurance approach of systematic checking

(e) to achieve as much as possible an elegant, or quality, solution

The hierarchical clustering approach described above will help with (c) and (e). Items (a), (b) and to some extent (d) can be dealt with by a systematic procedure for developing the structure of the problem, i.e. the list and its links. The development of such a procedure will need a significant effort, and current proposal aims to construct a pseudo-intelligent interactive computer program to carry out the task.

A further development needs to take place with regard to the final product of the process. The diagrams for Figs. 8–10 are appropriate for physical layouts, but a means of summarising the information required for more general design problems is not as easily achieved, particularly as the process requires the design solution to be synthesised as a step-by-step process. The advantage of a diagram or pattern is that it can illustrate required interactions over two-dimensional space in a way that could hardly be achieved by a linear (print, lists etc) form of information display. Any two-dimensional diagram will have to employ a number of different symbols in order to contain sufficient richness of information: a simple bubble diagram would be insufficient. Perhaps Alexander (1977) has once again been here before us, with his work on pattern languages.

References

Alexander, C. J., (1964), *Notes on the Synthesis of Form*, Harvard University Press, Cambridge, Mass., USA.

Alexander, C. J., (1977), *A Pattern Language: Towns, Buildings, Construction*, Oxford University Press.

Berge, C., (1962), *The Theory of Graphs and its Application*, Methven, London.

Clark, S. J. and Elms, D. G., (1976), *The HIDECS Process*, Ministry of Works and Development, Wellington, New Zealand.

Daish, J. and Kleem, J., (1969), *HOPAD—House Parts for Designers*, Ministry of Works, Wellington, New Zealand.

Elms, D. G. and Clark, S. J., (1976), *Briefing with Graphics*, Ministry of Works and Development, Wellington, New Zealand.

Elms, D. G., (1983), 'From a structure to a tree,' *Civil Engineering Systems* **1**, 95–106.

Elms, D. G., (1985), 'Steps beyond technique—education for professional attitude,' *Civil Engineering Systems* **2**, 55–59.

Elms, D. G., (1989), 'Wisdom engineering,' *Proc. World Conf. on Engineering Education for Advancing Technology*, Sydney, Australia, 574–578.

Pirsig, R. M., (1974), *Zen and the Art of Motorcycle Maintenance*, Bantam Books, New York, USA.

Risk Balancing Approaches

D. G. Elms
Department of Civil Engineering
University of Canterbury
Christchurch
New Zealand

Abstract An optimisation procedure is described which minimises the risk in a system
in which various failure modes can occur, by optimally allocating expenditures in a set of
investment modes and subject to an overall cost constraint. An example is given, which
illustrates that the model is transparent.

1 Introduction

One of the more urgent problems in the field of civil engineering systems is that
methods have outstripped methodology: that we know much less about general
strategies and approaches to problems that we do about specific techniques. This
is hardly surprising as the enterprise of research is constrained both by tacit assumptions as to what constitutes 'legitimate' research and by practical constraints
such as the requirement to tailor problems into forms appropriate for graduate students. Research involving the development of specific techniques fits more easily
into the mould than work on methodology, which tends to be vaguer and more
open-ended.

Therefore, although the lectures I am giving at this Institute deal with specific
techniques, I am unashamedly using them to talk about matters of methodology.

In my last talk (Network Structuring Algorithms) I referred to the idea of
subjective tools. In this lecture, I want to emphasise the role of transparent models.

Transparent models are the opposite of black box or opaque models. Subjective
tools, with their high degree of user interaction, often need to use transparent
models as the transparency is a necessary part of the interaction. Transparency
usually means that the inner workings of the model can be seen and understood
at all stages. Note the two parts to this: 'understood' is as necessary as 'seen.'
Some simulation models have been highly opaque because of their complexity even
though all their workings have been open to view.

B. H. V. Topping (ed.),
Optimization and Artificial Intelligence in Civil and Structural Engineering, Volume I, 61–70.
© 1992 *Kluwer Academic Publishers. Printed in the Netherlands.*

In the area of risk (which is the subject of this talk), fault trees are an ideal example of transparent models when probability values are included in the logic diagrams. The user can easily see both the logical structure of the problem and also the relative contributions of different events to the failure probability of the whole.

However, in what follows I shall deal with a different sort of transparency, which I shall call 'data transparency.' In many practical situations, and especially in the risk area, much of the available data is of poor quality. Besides reasonably sound data based on statistics or on known frequencies (such as train timetables), use has to be made of comparative likelihoods or even of anecdotal information. In this sort of situation the user of a model must be able to check that the input data is self-consistent and seems right. A transparent model in this sense must therefore use variables which can be readily understood by common sense and compared with experience. This sort of data transparency is a central feature of the risk balancing models described below.

2   Risk Balancing Models

2.1   General Models

In the risk management of complex systems a central problem is this: given a certain budgeted expenditure for risk reduction, how can that expenditure be best distributed to achieve the greatest overall reduction in risk? We shall call this 'risk balancing.' Clearly it is an example of the classical marginal analysis problem (de Neufville and Stafford (1971)), though Wiggins (1972) developed the concept of balanced risk using a narrower definition.

Risk has two parts; the probability of an event happening, and its consequences. Numerically, they are multiplied. The total risk involved in a situation is the sum of the risks associated with each risky event, which is thus equal to the expected value of the consequences. The level of risk can be decreased by reducing either the probability or the consequences, or both. We assume this can be effected by the expenditure of money. We now follow and expand upon an earlier development (Elms (1979)).

Let us assume the total sum C to be spent on reducing the risk within a system is allocated to m different areas of expenditure. We shall call them 'investment modes,' and denote the amount to be spent in mode j by a_j. Let us suppose the first k modes reduce the cost consequences of failure $(0 < k < m)$ while the remaining $(m - k)$ modes are aimed at reducing the probability of failure. Clearly

$$\sum_{j=1}^{m} a_j = C \qquad (1)$$

Let there be n different types of failure, each with a cost consequence c_j and a probability of occurrence p_i. The expected cost of failure; that is, the risk, will be

$$R = \sum_{i=1}^{m} c_i p_i$$

$$= \sum_{i=1}^{m} c_i(a_i, \ldots, a_k) p_i(a_{k+1}, \ldots, a_m) \tag{2}$$

To optimise the risk, R must be minimised. We form the Lagrangian

$$L = \sum_{i=1}^{m} c_i p_i - \lambda \left(\sum_{j=1}^{m} a_j - C \right) \tag{3}$$

The condition for a minimum is the set of m equations.

$$\sum_{i=1}^{n} \frac{\partial c_i}{\partial a_j} p_i - \lambda = 0 \qquad j = 1, \ldots, k \tag{4}$$

$$\sum_{i=1}^{n} c_i \frac{\partial p_i}{\partial a_j} - \lambda = 0 \qquad j = k+1, \ldots, m \tag{5}$$

with one equation corresponding to each investment mode.

We have taken the general case of expenditure being made to reduce both components of risk. A common situation, in structural safety for example, is that investment is made only to reduce the probability of failure and not in the direction of lessening the consequences (the consequences of structural collapse cannot easily be reduced, for instance). In such a case, the problem simplifies, $k = 0$ and Eq. 4 is eliminated. The complementary simplification where expenditure is only aimed at reducing consequences is rarer and will not be considered.

We now have two tasks: to express the cost consequences and failure probabilities in terms of parameters which have a clear physical meaning (referred to earlier as 'data transparency'), and to obtain an explicit solution for the risk investments a_j.

Let us deal first with reductions in failure probability. Fig. 1 gives a simple form to the problem: it assumes the effect of an expenditure a_j on a failure probability with initial value p_i^0 is to reduce it in a negative-exponential manner, with the reduction being limited to a maximum of k_{ij}. The sensitivity of the reduction, or the rate at which the limit is approached, is given by the slope b_{ij} of the curve at zero expenditure. We shall call b_{ij} the 'standard investment coefficient.' Unlike the initial probability p_i^0 and the limiting failure reduction k_{ij}, b_{ij} is not dimensionless but is in the same cost units as a_j. It may be in interpreted as the expenditure necessary to reduce the failure probability by roughly $2/3$ of the possible maximum reduction.

Figure 1 shows the failure probability reduction in mode i due to expenditure in investment mode j alone. The equation of the curve is

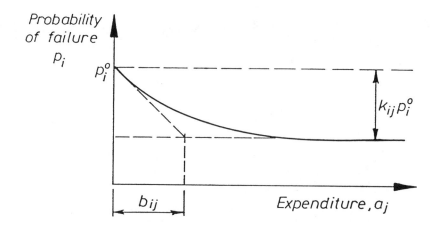

Figure 1. *Reduction in failure probability p_i due to expenditure a_j.*

$$p_i = p_i^0 \left(1 - k_{ij} \left(1 - e^{-a_j/b_{ij}}\right)\right) \tag{6}$$

The effect of expenditure in all investment modes can be expressed by a simple summation to give

$$p_i = p_i^0 \left(1 - \sum_{j=1}^{m} k_{ij} \left(1 - e^{-a_j/b_{ij}}\right)\right) \tag{7}$$

This has the advantage that linearisation leading to an explicit expression for the expenditure a_j is relatively straightforward, as will be seen.

An alternative multiplicative formulation could also be used:

$$p_i = p_i^0 \prod_{j=1}^{m} \left(1 - k_{ij} \left(1 - e^{-a_j/b_{ij}}\right)\right) \tag{8}$$

Here, k_{ij} has a somewhat different meaning from Eq. (7). There is also an implication that the underlying system has parallel components, whereas Eq. (7) relates to a series system in which failure results from failure of one or other of a set of components or potential failure modes. For these reasons, and also because an explicit formulation in terms of the expenditures a_j is somewhat complex, Eq. (8) will not be pursued further, and development will be confined to the model of Eq. (7).

Bounds must be placed on the limiting failure reduction k_{ij}. It is necessary that

$$0 \leq \sum_{j=1}^{m} k_{ij} \leq 1.0 \tag{9}$$

and it is generally to be expected (though logic does not strictly demand it) that

$$0 \le k_{ij} \le 1.0 \tag{10}$$

Taking the derivative of the expression in Eq. (7), we find that the summation disappears and

$$\frac{\partial p_i}{\partial a_j} = -\frac{p_i^0 k_{ij}}{b_{ij}} e^{-a_j/b_{ij}} \tag{11}$$

so that Eq. (5) becomes

$$\sum_{i=1}^{n} \left(\frac{c_i}{b_{ij}} \right) k_{ij} p_i^0 e^{-a_j/b_{ij}} = -\lambda \qquad (j = k+1, \ldots, m) \tag{12}$$

The next step is to linearise the expression in order to obtain an explicit solution for a_j. A first-order Taylor's series approximation at a point a_j^A for

$$y = \left(\frac{c_i}{b_{ij}} \right) k_{ij} p_i^0 e^{-a_j/b_{ij}} \tag{13}$$

will be of the form

$$y = A_{ij} - B_{ij} a_j \tag{14}$$

where

$$A_{ij} = r_{ij} \left(1 + \frac{a_j^A}{b_{ij}} \right) \tag{15}$$

$$B_{ij} = \frac{r_{ij}}{b_{ij}} \tag{16}$$

and where

$$r_{ij} = \left(\frac{c_i}{b_{ij}} \right) k_{ij} p_i^0 e^{-a_j^A/b_{ij}} \tag{17}$$

Equation (12) then becomes

$$\sum_{i=1}^{n} (A_{ij} - B_{ij} a_j) = -\lambda \qquad (j = k+1, \ldots, m) \tag{18}$$

for which a_j can be written as

$$a_j = \frac{\lambda + \sum_{i=1}^{n} A_{ij}}{\sum_{i=1}^{n} B_{ij}} \qquad (j = k+1, \ldots, m) \tag{19}$$

The constant λ can be evaluated using Eq. (1) which is simply the condition that all the expenditure a_j must add up to the total cost C. From this,

$$\lambda = \frac{C - \sum_{j=k+1}^{n}\left(\sum_{i=1}^{n} A_{ij} / \sum_{i=1}^{n} B_{ij}\right)}{\sum_{j=k+1}^{n}\left(1/\sum_{i=1}^{n} B_{ij}\right)} \tag{20}$$

Equation (19) together with Eq. (15), (16), (17) and (20) give the required explicit expressions for the modal expenditures. The solution requires iteration if the initial results are different from the assumed modal expenditures a_j^A, using the results as a new linerisation point.

So far it has been assumed that expenditure is made only to reduce the probabilities of failure. In the more general case, funds could also be directed to reducing the consequences c_i of failure, and Eq. (4) would apply as well as Eq. (5). In such a case the formulation could be expanded to include consequence reductions by assuming precisely the same form as Eq. (6) and applying a similar linearisation. The solution would still be in the form of Eq. (19), but Eq. (15)–(17) would be somewhat modified.

2.2 Expansion Path—Appropriate Overall Expenditure

In some cases, negative values of a_j may be obtained. This could arise when the baseline probabilities p_i^0 are achieved with a finite expenditure assumed in various investment modes. It could be that the approach is being used to appraise an existing situation, which is taken as the baseline. A negative value of a_j simply means that the existing expenditure in that mode is too great.

It follows from this approach that it would be perfectly legitimate for C to be zero, in which case the problem would be one of a redistribution of existing resources. There could even be situations in which it would be reasonable for C to be negative.

Indeed, an appropriate approach would be to compute the modal expenditures for different values of C, and then calculate for each the resulting risk level R, obtained from Eq. (2) and (7). Plotting R against C gives an expansion path from which the appropriate overall expenditure to achieve a desired level of risk can be obtained.

2.3 Example

As an example, consider the design of the fire prevention measures in a building. The total cost C to be allocated to fire protection is 30 \$/m². The three failure modes to be considered are

1. Smoke spread blocks exits, with consequent loss of life

2. Structural collapse occurs after exit of occupants

3. There is excessive fire, smoke and water damage

Three investment modes are considered. They are

1. Degree of compartmentalisation

2. Degree of control of fuel load

3. Insulation of structural members

The following parameter values are used: they are not based on actual data though they have been chosen with some care to ensure the realism of the example.

$$
k_{ij} = \begin{bmatrix} 0.5 & 0.2 & 0 \\ 0 & 0 & 1.0 \\ 0.7 & 0.1 & 0 \end{bmatrix}
\begin{array}{c} \text{investment} \\ \leftarrow \text{mode} \rightarrow \end{array}
$$

(failure mode indicated by vertical arrow)

$$
b_{ij} = \begin{bmatrix} 30 & 20 & \infty \\ \infty & \infty & 1.0 \\ 20 & 80 & \infty \end{bmatrix} \quad \$/\text{m}^2
$$

(failure mode indicated by vertical arrow)

$$
p_i = \begin{bmatrix} 0.00095 \\ 0.0001 \\ 0.01 \end{bmatrix} \quad
c_i = \begin{bmatrix} 1500 \\ 400 \\ 100 \end{bmatrix} \$/\text{m}^2 \quad
a_i^A = \begin{bmatrix} 30 \\ 20 \\ 1.0 \end{bmatrix} \$/\text{m}^2
$$

The costs are based on a total building cost of 300 $/\text{m}^2$. Loss of life causes the failure cost to exceed this figure significantly.

The initial expenditure values a_i^A should really sum to C. However, this is not strictly necessary (as with the figures above) as iteration is inevitable and will rapidly mend matters.

Carrying out the calculations we find that $\lambda = -0.0164$ and that

$$a^1 = 30.3 \ \$/\text{m}^2 \tag{21}$$

$$a^2 = -1.2 \ \$/\text{m}^2 \tag{22}$$

$$a^3 = 0.9 \ \$/\text{m}^2 \tag{23}$$

The result means that there should be a fairly high investment in compartmental-isation and in structural insulation (for which the unit cost is relatively low), but that little effort is worthwhile in controlling the fuel load. Thus the initial assumption of 20 \$/m² for the fuel load was poor, though the other two were close. Indeed, even the existing expenditure on fuel reduction is too high by some 1.20 \$/m².

Using Eqs. (2) and (7) the level of risk is calculated to be $R = 1.46$ \$/m². If the failure probabilities are taken to be annual probabilities, then the risk figure can be related to insurance costs, though the insurance company overheads and profit will increase the figure considerably. It is interesting to note that the figure obtained is in the right range.

2.4 Comments and Assumptions

A number of comments can be made, both on the example and on the theory.

2.4.1 Failure Consequences

In the example, the failure consequences c_i were expressed in terms of monetary units, in dollars. Any other units can be used, however, such as loss of life, or utility. If the consequences were quantified in terms of lives lost, for instance, the problem formulation would be unchanged except that λ would no longer be dimensionless but would have units of lives lost/monetary units. The modal expenditure would still be in monetary units.

If the time frame of a problem is other than the annual situation assumed for the example given above, then failure consequences should be discounted to present value, as should other than initial expenditures.

2.4.2 Data transparency

A major reason for including the example given above was that it should illustrate the point made at the beginning about data-transparent models. Consider the initial data for the example. Even to those not familiar with fire issues, the values chosen for k_{ij} must seem reasonable. Investment in compartmentalisation (first column) can at most (says the data) decrease the probability of life loss by 50%, though it will have a greater effect (70%) on fire, smoke and water damage, mainly through restricting the spread of fire. However, compartmentalisation will have no effect on structural collapse as even a confined fire can seriously weaken unprotected structural elements. On the other hand, insulation of structural members (third column) will have no effect on smoke spread or fire damage, but it could wholly prevent structural collapse. The values of b_{ij} are the costs to reduce the failure probability by 2/3 of the maximum reduction. Note the use of infinity for terms corresponding to zero values of k_{ij}. Obviously, no value of b_{ij} should be zero. This would imply instantaneous maximum reduction with zero expenditure; and in any case the problem formulation would lead to division by zero. Apart from this, the b_{ij} values can be seen to be reasonable guesses. If the user is unsure of the data, the formulation given makes it easy to estimate reasonable bounds for the b_{ij} values, and a sensitivity study can be carried out.

This leads to another point: the intended means of use of the technique. It is assumed that the most fruitful use of risk balancing will be at the early stages of an engineering project where data is not known accurately. The idea is that the technique will give order-of-magnitude indications of relative levels of expenditure. It is not really intended for use as a tool for detailed analysis, and therefore the numerous assumptions made in the derivation are justified. The later talk, 'Consistent Crudeness in System Construction,' expands on this point.

2.4.3 Discontinuous expenditures In many if not most practical situations, expenditure cannot be varied continuously, but only in discrete amounts. However, given the above intention that the technique should be used only for order-of-magnitude investigations, the approximation of continuity is justified.

2.4.4 Constraints Risk problems often have bounding constraints. For reasons of safety, politics or otherwise, failure probabilities might have to be less than specified values. Alternatively, expenditure in some investment mode might be limited. Such constraints can be dealt with by examining the results. If constraints are not satisfied, the solution has to be brought back to the boundary and fixed there: a probability, for instance, can be fixed instead of using Eq. (7). The expansion path could be significantly modified by constraints.

2.4.5 Linearity The formulation assumes the effect of investment is cumulative and linear. Thus the result of expenditure in two investment modes simultaneously is seen as the sum of the effects due to the same investment in each mode separately. This assumption is not always justified, but it is retained on the ground that any other assumption would lead to great complications, and that in any case the analysis is intended for relatively crude use, as mentioned above.

2.4.6 Correlated failure modes In practice, failure modes are often correlated in some way, perhaps due to a common cause. Alternatively, one mode might dominate another and prevent its occurrence. This is difficult territory, and can only be dealt with on a case-by-case basis. Notice, though, that risk formulation of Eq. (2) requires a set of probabilities of mutually exclusive events.

2.4.7 Repairability If the problem is formulated in terms of a design lifetime, then more than one occurrence of the same failure mode may take place. If a failure happens, costs will vary depending on whether the system is repairable or not.

2.4.8 Simplification A major simplification occurs if the investment and failure modes correspond to each other and an expenditure affects only one failure mode.

The formulation and an example are given elsewhere (Elms (1979)). Other simplified uses have involved code design factors (Elms (1980)) and optimum expenditure distribution for earthquake hazard mitigation (Elms, Berrill, Darwin (1981)).

3 Conclusions

Optimisation in the risk area is infrequent. It is often hard enough to carry out a risk assessment or audit, without going further and trying to modify the risk regime systematically other than stand-alone changes to high risk elements identified in a system. However, with the current growth in importance of risk assessment and risk management in the area of civil engineering, it is becoming more important to adopt systematic methods. This paper has adapted the techniques of marginal analysis to the problem of risk management. Its use has proved to be reasonably simple. More importantly, it has been deliberately formulated as a transparent model, especially with regard to data transparency, so that it can be used with some confidence in an areas where data quality is quite notoriously poor.

References

de Neufville, R. and Stafford, J. H., (1971), *System Analysis for Engineers and Managers*, McGraw-Hill, New York.

Elms, D. G., (1979), 'Risk balancing for code formulation and design,' *Proc. Third International Conference on the Application of Statistics and Probability in Soil and Structural Engineering*, Sydney, Australia, pp. 701–713.

Elms, D. G., (1980), 'Rational derivation of risk factors,' *Proc. Seventh Australasian Conference on the Mechanics of Structures and Materials*, Perth, pp. 149–153.

Elms, D. G., Berrill, J. B. and Darwin, D. J., (1981), 'Appropriate distribution of resources for optimum risk reduction,' in *Large Earthquakes in New Zealand: Anticipation, Precaution, Reconstruction*, Misc. Series No. 5, Royal Society of New Zealand, Wellington, pp. 69–75.

Wiggins, J. H., (1972), 'The balanced risk concept: new approach to earthquake building codes,' *Civil Engineering, ASCE*, August, pp. 55–59.

Consistent Crudeness in System Construction

D. G. Elms
Department of Civil Engineering
University of Canterbury
Christchurch
New Zealand

Abstract   The principle of consistent crudeness is a guide for optimal development of complex system models. It requires consistency between the quality-levels of input information items, modified by sensitivity, and the precision of the model. Examples of application are given.

1   Introduction

It all began during a conversation at the University of British Columbia with that great iconoclast and generator of good ideas, Borg Madsen. We were talking of codes and of the struggles we had both had in their development. It was a time when limit states or LRFD codes were coming into fashion and first-order second-moment methods were being used for formulation and calibration. The more extreme advocates toured the international circuit spreading the new doctrine. They would have had us believe a new era of rationality had begun, a sort of code format millenium. And yet when in New Zealand we tried to follow our distinguished colleagues overseas and develop our own rational codes, we found we could not. The ideas were fine, but the practicality was not. The principles could be followed in some areas—gravity load conditions, for instance, though that was difficult enough—but in most cases, either whole areas of data did not exist, or the necessary design or computational models had to be absurdly simple. Earthquake loading cases, which dominate structural design in many countries, could only be dealt with on a crude static-loading single-member basis, whereas the structural action was complex, dynamic and involved the action of an entire structure. Then there were the effects of human error in its various forms which as Colin Brown showed (1970) completely dominate structural failure but which could only be taken into account by, with a desperate leap of faith, putting arbitrarily higher load factors on those effects which could be calculated. The whole system

71

B. H. V. Topping (ed.),
Optimization and Artificial Intelligence in Civil and Structural Engineering, Volume I, 71–85.
© 1992 *Kluwer Academic Publishers. Printed in the Netherlands.*

seemed glaringly inconsistent, thought Madsen and I, with enormous effort being applied in some areas and equally enormous gaps remaining untouched in others. We could not decide which of two results was the worse: the great waste of money and effort in developing detail unjustified by the necessary crudeness of the result, or the way in which many designers and researchers were being led to believe that because of the technical sophistication of a part of the effort, the resulting code requirements were solidly based on a sound foundation of rationality. What was needed was some guiding principle for an optimal and consistent expenditure of effort; a principle of consistent crudeness, insofar as the results seemed dominated by the crudest parts, with unbalanced sophistication having little effect.

The idea, and the name, seemed to stick. Borg was soon referring in public to the principle of consistent crudeness (Madsen (1984)). Later, I decided to formalise the thought and explore it a little using fuzzy set theory, which was presented to a fuzzy set workshop at Purdue (Elms (1985)). In what follows, I shall review the Purdue concept and develop it further.

2 The Principle Of Consistent Crudeness

2.1 General Model

The name 'Principle of Consistent Crudeness' implies that to some extent the choice of level of detail in any part of an engineering procedure must to some extent be governed by the crudest part of that procedure. We first set out to show that this is so, though with a more precise formulation.

The issue is basically a matter of the cost consequences of obtaining more precise information, or of using a more elaborate and exact model. The cost of greater precision is generally not measurable in monetary units: more likely it is a question of effort or time expended. Nevertheless in an overall sense we are talking of obtaining value for money. Value systems must therefore come into the formulation. As the value of information can usually be described rather crudely in linguistic terms such as 'good,' 'sound' or 'pretty crude,' we shall be using fuzzy set theory.

An appropriate general model for dealing with the information processing aspects of engineering procedures is shown in Fig. 1. We assume that a series of items of input information are processed by a model or a transformation rule and are combined in some way to form output information. This is a fundamental building block, as it were, and will be the basis of the subsequent discussion. In practice most engineering procedures are far more complex and will be built up as shown in Fig. 2 from a number of basic blocks.

The model seems general enough to take into account virtually all engineering procedures, whether for direct or inverse problems, deterministic or stochastic situations, descriptive or prescriptive information sets, or even those where chaos

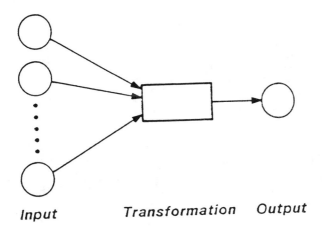

Figure 1. *Basic information processing model.*

theory applies. Admittedly it is a gross simplification. The input information could be anecdotal, the output could be multifaceted, or an engineering drawing. Nevertheless, we will try to stay at a consistent level of approximation.

Figure 1 is intended to represent what an engineer normally does. For our purposes, however, we must associate the basic model with additional information on the precision of the data and its transformation. Thus in Fig. 3, the input information has had added to it a fuzzy measure of its quality, shown cross-hatched in the diagram. Similarly the transformation model, the central box in the diagram, has associated with it a fuzzy measure of its precision. Finally, the output information also has a quality measure associated with it.

Technically, the problem is firstly to devise an appropriate fuzzy description of the quality of an item of information or the precision of a model, and secondly to find a means by which such descriptions can be carried through the transformation process of the model and related to its numerical procedures so as to obtain in the end a fuzzy description of the quality of the output.

2.2 Quality

The quality of an item of information can be described by using a scale from 1 to 5 with supports in the range $0 < \mu \leq 1.0$ as follows, where '+' indicates union:

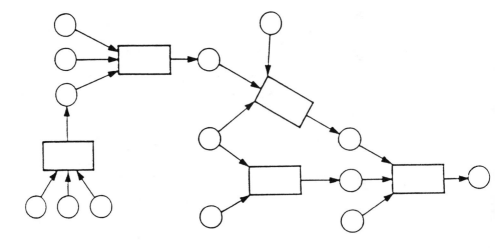

Figure 2. *Extended model.*

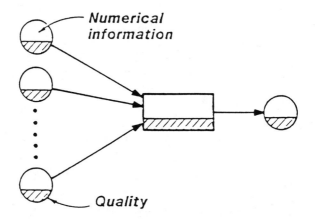

Figure 3. *Basic model including quality information.*

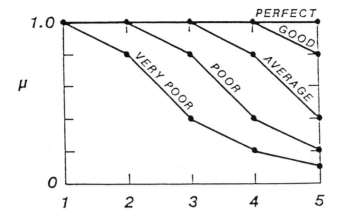

Figure 4. *Degrees of quality.*

perfect:	1.0|1	+	1.0|2	+	1.0|3	+	1.0|4	+	1.0|5
good:	1.0|1	+	1.0|2	+	1.0|3	+	1.0|4	+	0.8|5
average:	1.0|1	+	1.0|2	+	1.0|3	+	0.8|4	+	0.4|5
poor:	1.0|1	+	1.0|2	+	0.8|3	+	0.4|4	+	0.2|5
very poor	1.0|1	+	0.8|2	+	0.4|3	+	0.2|4	+	0.1|5

The scheme is shown graphically in Fig. 4. It reflects the idea that perfect information, or a perfectly appropriate model, will cause no degradation of information, but neither will it increase its quality. On the other hand, a less than perfect item of information has the power to decrease the quality of the result. This is the reason for the asymmetry of the diagram. In a sense, the degree of quality is shown by the area above the line in Fig. 4.

2.3   Sensitivity

A connection between the fuzzy descriptions of quality and the numerical procedure of the underlying model can be made through the sensitivity of the model; that is, the sensitivity of the output to changes in the input. Sensitivity can be expressed in different ways depending on the nature of the problem. We will restrict discussion to one definition and express sensitivity thus: if the numerical model is described by

$$y = f(x_1, \ldots, x_n) \tag{1}$$

where $x_1, \ldots, x_n$ are the items of input information, then the sensitivity of the output y to changes in x_i is given by

$$\gamma_i = \frac{x_i}{y} \frac{\partial y}{\partial x_i} \tag{2}$$

The quantity γ_i gives, as it were, the percentage increase in y due to a one percent increase in x_1, so it is a proportionate rather than an absolute measure of sensitivity. Sensitivity can be converted to qualitative linguistic terms using a fuzzy definitional algorithm (Zadeh (1973)):

1. IF $\gamma_i < 0.01$ THEN y is *insensitive* to x_i ELSE

2. IF $0.01 < \gamma_i < 0.05$ THEN y is *not very sensitive* to x_i ELSE

3. IF $0.05 < \gamma_i < 0.1$ THEN y is *somewhat sensitive* to x_i ELSE

4. IF $0.1 < \gamma_i < 0.3$ THEN y is *quite sensitive* to x_1 ELSE

5. IF $0.3 < \gamma_i < 0.7$ THEN y is *sensitive* to x_i ELSE

6. y is *very sensitive* to x_i

This is of course somewhat arbitrary, but generally seems to cover the required range.

Equation (1) assumes the numerical model can be expressed analytically. Where this is not possible, as when the model is discrete or an algorithm, Eq. (2) will no longer be appropriate. However, equivalent definitions of sensitivity can easily be made.

2.4   Quality of the Model

Next comes the question of assessing the contribution of the numerical model itself to the quality of the output information. It is not entirely correct to look at the situation only in information-processing terms, as a matter of degrading the quality of the information being processed. The appropriateness of a model depends on a number of factors, only one of which is the degree to which it degrades information by, as it were, introducing noise. More important is the relationship of the model to the reality it purports to represent, in the context of the intended use of its output. Some models might be crude. Others might be complex and precise but might solve the wrong problem because of, say, the need for simplified boundary conditions. It is difficult to sum all this up in one word. 'Precision,' 'aptness' and 'reliability' all come to mind, but none seems wholly appropriate. To avoid complication, then, we will simply once again use the word 'quality' with the same meaning and categories as were applied, in Fig. 4, to items of input information. This has an advantage when the various fuzzy measures have to be combined, for as the following section shows, the need for a fuzzy relation is avoided.

A further advantage of using the 'quality' definition to describe model precision is that it implies, in the nature of its formulation, that a model can only degrade information; it cannot improve it.

2.5 Combination

Having defined input quality, sensitivity, and model quality, the next step is to produce rules for combining them to form an estimate of output quality. Consider first the linking of an input quality with the corresponding model sensitivity. This could be done by using a composition. However a more appropriate way is to regard the sensitivity of the model as a modifier of the input quality. Following an earlier approach (Elms (1984)) we will use Zadeh's 'concentration' operation in which a Support μ is acted on by an exponent α to produce a modified support μ^{α}. The effect is to modify the support spread of the fuzzy set. In the present case we shall choose the values of α given in Table 1 to define the effect of sensitivity in modifying the input quality.

Table 1. *Sensitivity exponents.*

Sensitivity	α
insensitive	0.2
not very sensitive	0.5
somewhat sensitive	1.0
quite sensitive	1.5
sensitive	2.0
very sensitive	4.0

For example, if an item of input information is of *poor quality* and the model is *quite sensitive* to it, then the modified quality will be

$$1.01|1 + 1.0|2 + 0.72|3 + 0.25|4 + 0.93|5$$

Note that the poorer the quality of the data, the more it is affected by the sensitivity of the model. If the input data is perfect, then the contribution to the quality of the output will also be perfect, no matter what the sensitivity.

The values chosen are somewhat arbitrary, but not entirely so. Firstly, they provide a reasonably even spread on either side of unity, which relates to the intuitive ideas that some average or normal sensitivity should produce no modification

The fuzzy intersection of these is

$$1.0|1 + 0.96|2 + 0.83|3 + 0.64|4 + 0.16|5 \qquad (5)$$

which is the required result, apart from consideration of the quality of the model. The result is shown graphically in Fig. 5.

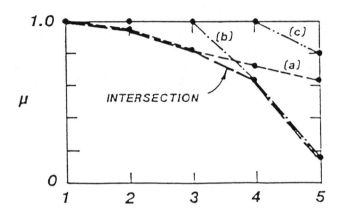

Figure 5. *Example result.*

The precision (or quality) of the model will, or can, further modify the output information. As mentioned previously, it cannot improve the result; the output quality can only be degraded. For instance in the present example, if the model were of *average* quality, then it would not affect the result; expression (5) above would be unchanged. However, if the model quality was poor then the supports of 3 and 4 would be modified.

We can now begin to see the effect of the principle of consistent crudeness. Figure 5 shows that input (b) essentially governs the quality of the output. Input (a) governs the supports for 2 and 3, but they are not very different from unity. As for (c), its quality is unnecessarily high. It can be seen from the diagram that even if the quality of (c) were changed from *good* to *average*, so lowering the (c) line, it would not meet the intersection envelope and would have no effect on the result. Likewise, an improvement in the quality of (a) would lift the (a) line. If (a) is changed from *very poor* to *poor*, though, it can easily be seen that there would be very little change to the final result: the supports for 2 and 3 would be increased slightly, but that is all.

In this case, therefore, the Principle of Consistent Crudeness could be stated thus:

of the input quality, and that only sensitivities greater or less than the norm should produce change. Secondly it must be admitted that the values have been chosen because when applied to examples they produce results which seem believable, and of an expected order of magnitude.

Let us consider for the moment that the model quality is perfect, so that output quality depends merely on the sensitivity-modified input qualities. They must be combined in some way. The appropriate operation for combining one quality and another is to form the intersection of the associated fuzzy sets. Two forms of fuzzy intersection exist, both converging for crisp sets on the conventional definition of intersection. The first is to find the minimum of the supports for each value of the fuzzy sets concerned, while the second is to form the products of the supports. Trials have shown that in the present formulation, both types of intersection give similar results. The simpler form, the minimum, will therefore be chosen.

Once the modified input qualities have been combined by forming their intersection, the model quality can be introduced. This effect, too, may be thought of as being added in some way to the other results, so that again, the appropriate operator is intersection. The intersection of the model quality with the combined and modified input qualities is found, and a fuzzy measure of output quality is produced.

2.6 Example

As an example, suppose we have three items of input data, (a), (b) and (c), with values x_1, x_2 and x_3 combined to form an output y by the relationship (or model)

$$y = 20 + 0.1x_1x_2 + 5x_2 + 0.5x_3^2 \tag{3}$$

If for the state under consideration the inputs have values of 1, 4 and 2 respectively, then $y = 42.4$ and the subsequent sensitivities are given by the following table:

Table 2. *Example sensitivities and exponents.*

Input	i	x	$\gamma = \frac{x_i}{y}\frac{\partial y}{\partial x_i}$	Sensitivity	a_i
(a)	1	1	0.009	insensitive	0.2
(b)	2	4	0.481	sensitive	2.0
(c)	3	2	0.094	somewhat sensitive	1.0

Let the quality of the three inputs be assessed as *very poor*, *average* and *good* respectively. The sensitivity-modified qualities become

(a) $1.0|1 + 0.96|2 + 0.83|3 + 0.72|4 + 0.63|5$
(b) $1.0|1 + 1.0|2 + 1.0|3 + 0.64|4 + 0.16|5$ $\qquad$ (4)
(c) $1.0|1 + 1.0|2 + 1.0|3 + 1.0|4 + 0.8|5$

80

*The sensitivity-modified quality of any item of input information should
not be made significantly better than that of either the item with the
lowest quality, or the model.*

The result was corroborated (Elms (1985)) by considering a parallel development using variances.

3 Consistent Crudeness in System Construction

In practical situations there is little point in applying the principle in detailed
fuzzy set terms. Though it could be done, there would be a danger of violating
the principle itself. Its appropriate use would normally be in broad terms only.
Nevertheless, it should be used with care. If applied unthinkingly (or worse), it
could be used as a licence for making quite unjustified approximations. It is very
much a subjective tool, as discussed earlier at this Institute, in which responsibility
for use must fall squarely on the shoulders of the user.

Two problems can be addressed using the principle. The first is the effort
needed to improve a data set to achieve consistency and an optimal quality of
result. The second is the development of consistently simplified models in complex
system situations. The first is easier, the second probably more important. The
examples that follow illustrate both approaches.

3.1 Risk Balancing

The normal and proper use of the principle should be as a general guide, bearing
in mind the interaction between quality and sensitivity. It is seldom sensible to
apply the principle in detail. However, in this section we shall do so simply to
show the possibility, but more importantly to illustrate some practical details and
implications.

Let us return to the example given earlier in the lecture 'Risk Balancing Approaches.' Recollect that the investment mode expenditures were expressed as

$$a_j = \frac{\lambda + \sum_{i=1}^n A_{ij}}{\sum_{i=1}^n B_{ij}} \tag{6}$$

where

$$A_{ij} = \frac{c_i p_i^\circ k_{ij}}{b_{ij}}(1 + \frac{a_j^A}{b_{ij}})e^{-a_j^A/b_{ij}}$$

$$B_{ij} = \frac{c_i p_i^\circ k_{ij}}{b_{ij}^2}e^{-a_j^A/b_{ij}}$$

Using Eq. (2), the sensitivity of the output a_j to c_i, p_i° or k_{ij} is

$$\gamma_{ij} = \frac{b_{ij} B_{ij}}{a_j \sum_{i=1}^{n} B_{ij}} \tag{7}$$

and its sensitivity to b_{ij} is

$$\gamma_{ij} = \frac{(\frac{A_{ij}}{a_j - B_{ij}})(\frac{a_j}{b_{ij}} - 1)}{\sum_{i=1}^{n} B_{ij}} \tag{8}$$

The sensitivities of a_j to changes in k_{ij} and b_{ij}, written in parentheses alongside the original data, become

$$k_{ij}(\gamma_{ij}) = \begin{pmatrix} 0.5(.424) & 0.2(-16.3) & 0 \\ 0 & 0 & 1.0(1.1) \\ 0.7(.378) & 0.1(-1.6) & 0 \end{pmatrix} \tag{9}$$

$$b_{ij}(\gamma_{ij}) = \begin{pmatrix} 30(.0043) & 20(17.3) & - \\ - & - & 1.0(-.11) \\ 20(.194) & 80(1.39) & - \end{pmatrix} \tag{10}$$

while the sensitivities to p_i and c_i are given in Table 3. Recall that the final expenditures were $a_1 = \$30.3/m^2$, $a_2 = -\$1.2/m_2$ and $a_3 = \$0.9/m^2$. The sensitivity values show a wide variation. The values of γ_{12} and γ_{32} for both k_{ij} and b_{ij} are very high, especially γ_{12}. This is in part due to the definition of sensitivity used in Eq. (2), as a relative percentage of the output: a_2 turns out to be a smaller number, particularly in comparison to b_{12} and b_{32}, so that a relatively small variation in output in absolute terms is seen as a significant percentage variation. This indicates that care must be taken in the definition of sensitivity, and that Eq. (2) may not always be the best description. However, we will carry on.

Table 3. *Sensitivities for risk balancing example.*

i	p_i	c_i	γ_{i1}	γ_{i2}	γ_{i3}
1	.00095	1500	.424	−16.3	−
2	.0001	400	−	−	1.1
3	.01	100	.378	−1.6	−

As to the quality of the input data, we can assume the initial probability figures p_i^0 are obtained from baseline statistics and are quite good. The cost consequences can be obtained from insurance data, and they too are good. However, the maximum-reduction figures k_{ij} are estimated from experience and are average, and some of the b_{ij} figures are unreliable. Call them poor.

Following the same process as the previous example, the sensitivity-modified qualities of the input are calculated for output a_1 and appear as the continuous lines in Fig. 6. The fuzzy intersection is of course the fuller line representing b_{31}, which dominates the result. It indicates the quality of the result is not at all good, being, by comparison with Fig. 5, somewhat worse that 'poor.' The other two outputs could be shown in separate diagrams, or they could be added to Fig. 6, with the fuzzy intersection of the whole representing the overall output quality. However, we will start with a_1.

One way of looking at the result is to say that the precision (quality) of the risk balancing model need be no better than poor, which justifies the assumptions discussed in the earlier talk. In fact, I believe the quality of the model is better than that, with an average degree of precision.

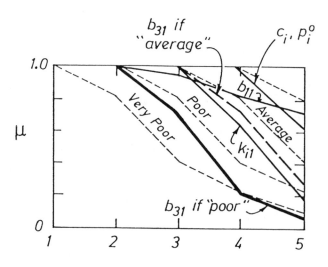

Figure 6. *Risk balancing example.*

Another approach is to say that Fig. 6 indicates that to improve the quality of the result to, say, average, which would bring it into line with the quality of the model, the sensitivity-modified quality of input item b_{31} would have to be improved. This cannot be done by changing the sensitivity, which is inherent in the model. It can only be achieved by changing the quality of b_{31}, which was originally classified as poor. Let us suppose that by extra effort, supplementary investigations or whatever, the quality of b_{31} is raised to average. This raises the result to the dashed line of Fig. 6. The overall quality is no longer dominated by b_{31} but by k_{i1}, and the fuzzy intersection is somewhat lower that 'average.' A small amount of further work on improving the quality of k_{i1} would make the quality of the whole reach a level of 'average.' Consistency would be achieved between crudeness of model and data.

3.2 Calibration Model for Earthquake Loading Codes

The foregoing was an artificial and simple exercise designed to help the reader get a better grasp of the ideas, and show that the principle of consistent crudeness both could and should only be used in a rough way. In contrast, the next example describes a situation in which the principle was used in earnest, as it were, as a helpful guide in the development of a consistently simplified model for use in a complex situation.

The issue was the need to assess the probability of failure due to earthquakes of buildings designed to proposed building code provisions. It is a difficult task. To begin with, an earthquake has very variable ground motion characteristics and it is difficult to estimate the frequency of earthquake occurrence. As for the action of a structure in an earthquake, its response depends on 3-dimensional nonlinear dynamic behaviour which is complex because of

- irregular geometry

- uncertain mass characteristics

- uncertain material behaviour, which is nonlinear and time-varying

- effects of secondary structure and foundation behaviour

- simultaneous action of the entire structural system

To make things worse, there is the difficulty of specifying a clear failure criterion.

Most of the above effects can be taken into account by a 3-dimensional nonlinear time history model. To get a failure probability, a Monte Carlo approach could be used; but even restricting variability to only a few major variables, the computer time requirement would be very large indeed, and in most cases prohibitive.

Yet no structural code could be called rational unless the range of failure probabilities implied by its provisions were known. In practice this would require a great many failure probability determinations for a range of typical designs and earthquake types. For gravity and wind loads the problem was relatively easy and was simplified by the use of the so-called first-order second-moment (FOSM) approach, applied to the behaviour of single structural members. What was needed for the earthquake case was a simple technique allowing many probability determinations to be made with relative ease, preferably using a FOSM approach. It was in achieving this that the principle of consistent crudeness was invoked a number of times as a guiding criterion.

84

One single and major overall simplification, such as assuming that earthquakes could be represented by a static lateral load, could not be used. The problem was too complex for that, and it was necessary to show that the results from any model developed were reasonably close to reality.

The approach adopted involved the use of a number of interrelated simplified analyses whose interaction and degree of approximation were carefully gauged at each stage by the use of sensitivity studies and by comparison with full-scale (as it were) computer analyses, the largest of which took some 30 hours central processing unit time on a Vax 750 computer. The work was restricted to reinforced concrete frame structures.

The final simplified model used a performance function (for FOSM calculations) based on the maximum cumulative inelastic damage energy in a storey. Without going into detail, for each multi-storey building analysed it involved

- an elastic modal analysis of the structure

- a set of single-degree of freedom inelastic dynamic analyses corresponding to each elastic mode

- a frequency-domain random vibration analysis of the structure represented as a simplified multi-degree of freedom system

Clearly, the 'simplified model' is still very complex. Still it is simple enough for its use to be feasible and it has been used, at the time of writing, on four different structures. However, its complexity is such that it could not have been developed in a balanced and consistent way without the guidance of the principle of consistent crudeness. As Ogawa (1988), who developed the model, said,

> It is not easy to apply the principle [in an exact form] to real problems. In practice, engineering judgement must generally be used bearing the principle in mind.

This indeed sums up the appropriate use of the principle.

References

Brown, C. B., (1970), 'A fuzzy safety measure,' *Jour. Engineering Mechanics Div., ASCE* **105**, pp. 855–872.

Elms, D. G., (1984), 'Use of fuzzy sets in developing code risk factors,' *Civil Engineering Systems* **1**, 178–184.

Elms, D. G., (1985), 'The principle of consistent crudeness,' in C. B. Brown, J-L. Chameau, R. Palmer and J. T. P. Yao (Eds.), *Proceedings of NSF Workshop on Civil Engineering Applications of Fuzzy Sets*, Purdue Univ. Sch. of Civil Engineering, pp. 35–44.

Madsen, B., (1984), 'A design code for contemporary timber engineering and its implications for international timber trade,' *Proc. Pacific Timber Engineering Conference*, Auckland, New Zealand, pp. 950–971.

Ogawa, S., (1988), 'A simplified procedure for assessing failure probabilities of reinforced concrete frame buildings under earthquake loading,' Civil Engineering Research Report 88-11, University of Canterbury, New Zealand.

Zadeh, L. A., (1973), 'Outline of a new approach to the analysis of complex systems and decision processes,' *IEE Trans. on Systems, Man and Cybernetics*, SMC-3, 1, 28–44.

Entropy and Civil Engineering Optimization

Andrew B. Templeman
Department of Civil Engineering
University of Liverpool
Liverpool
United Kingdom

Abstract   Recent research is surveyed which has demonstrated the value of the Maximum Entropy Principle in many aspects of civil engineering analysis as a means of logically infering missing data. The numerical optimization process is characterized as one of infering a solution from partial information. New strategies for numerical optimization based upon entropy maximization are introduced.

1   Introduction

In recent years entropy has emerged as an important and powerful concept in a wide variety of different fields. Entropy is most commonly known in the physics and engineering communities in connection with the second law of thermodynamics—the entropy law—which states that entropy, or amount of disorder, in any closed conservative thermodynamic system tends to a maximum. The consequences and ramifications of this law are enormous and fundamental, against which the uses and interpretations of entropy described in this paper are truly puny. Entropy governs the untire universe from the Big Bang to the dim future and is inescapable. What recent research in many fields other than physics has shown is that entropy can be found influencing events there also. Even in the area of optimization and decision support systems in civil engineering, entropy has a place and a part to play which is still relatively unexplored. The purpose of this paper is to describe some of the early explorative uses of entropy in this field.

B. H. V. Topping (ed.),
Optimization and Artificial Intelligence in Civil and Structural Engineering, Volume I, 87–105.
© 1992 *Kluwer Academic Publishers. Printed in the Netherlands.*

2 Thermodynamic Entropy

Before looking at entropy in an engineering optimization context it is useful to remind ourselves of the origins of thermodynamic entropy. Classical thermodynamics in which the entropy concept originated is concerned only with the macroscopic states of the matter, i.e. with experimentally observable properties such as temperature, pressure, volume, etc. Clausius defined entropy in this non-probabilistic context as a function of these macroscopic quantities. Classical entropy, as defined by Clausius, does not concern itself with what happens on a microscopic scale, and so does not concern us here.

Boltzmann was the first to investigate the microscopic states of a thermodynamic system in the hope that they would shed light upon macroscopic behaviour. When a small number of macroscopic quantities such as temperature, pressure, etc of a system are known, the macroscopic thermodynamic state of the system is also known. A description of this kind clearly leaves open many possibilities regarding the detailed states of the system at a microscopic level. A single macroscopic thermodynamic state may be reliazable by very many different micro-states, all of which combine to give the single measurable macro-state. Boltzmann investigated such micro-states and defined entropy in a new way such that the macroscopic maximum entropy state corresponded to a thermodynamic configuration which could be realized by the maximum number of different micro-states. He noticed that the entropy of a system can be considered as a measure of the disorder in the system and that in a system having many degrees of freedom, the number measuring the degree of disorder also measured the uncertainty in a probabilistic sense about particular micro-states. For a system with N micro-states, if p_i, $i = 1$, ..., N is the probability of occurrence of micro-state i, Boltzmann defined the entropy of the system as:

$$S = -K \sum_{i=1}^{N} p_i \ln p_i \tag{1}$$

in which S is the entropy and K is a positive constant called the Boltzmann constant. Boltzmann was the first to emphasize the probabilistic meaning of entropy and the probabilistic nature of thermodynamic micro-states from which stemmed the discoveries associated with what is now known as statistical thermodynamics.

3 Informational Entropy

A fundamental step in using entropy in new contexts unrelated to thermodynamics was provided by Shannon [1] who realised that entropy could be used to measure other types of disorder than that of thermodynamic micro-states. Shannon was interested in information theory, particularly in the ways in which information can be conveyed via a message. This led him to examine probability distributions in a very general sense and he wanted some way of measuring the levels of uncertainty in different distributions. For example, consider tossing two coins, one of which is perfectly fair and the other of which is loaded or biased. For the fair coin it is obvious that the probability of obtaining a head (H) on the next toss is 0.5 and that this is also the probability of obtaining a tail (T). Suppose that the biased coin is loaded such that the probability of tossing a head on the next throw is 0.7 and that of tossing a tail is therefore 0.3. The distributions

$$p(H) = 0.5; \qquad p(T) = 0.5 \qquad \text{for the fair coin}$$

and

$$p(H) = 0.7; \qquad p(T) = 0.3 \qquad \text{for the biased coin}$$

clearly contain different amounts of uncertainty. The uncertainty about the outcome of the next toss is much greater for the fair coin than for the biased coin which is far more likely to give a head than a tail. But how can this uncertainty be quantified? Is there some algebraic function which measures the amount of uncertainty in any probabilistic distribution in terms of the individual probabilities?

Shannon extended this thought-experiment further by imagining a fair three-faced coin (with the third face labelled Z). For such a coin it is clear that

$$p(H) = \frac{1}{3}; \qquad p(T) = \frac{1}{3}; \qquad p(Z) = \frac{1}{3}$$

For the fair two-faced coin we have a one-in-two change of being right but for the fair three-faced coin we have only a one-in-three chance of being right. Thus the amount of uncertainty associated with a fair three-faced coin must be greater than that associated with a fair two-faced coin.

From these types of simple examples and others Shannon was able to devise a set of criteria which any measure of uncertainty may satisfy. He then tried to find an algebraic form which would satisfy his criteria and discovered that there was only one formula which fitted: the amount of uncertainty in any discrete probability distribution in which p_i, $i = 1, \ldots, N$, is the probability of event i is proportional to

$$-\sum_{i=1}^{N} p_i \ln p_i \qquad (2)$$

Eq. (2) is identical to the entropy relationship, Eq. (1), if the constant of proportionality is taken as the Boltzmann constant, K. Thus Shannon showed that entropy, which measures the amount of disorder in a thermodynamic system, also measures the amount of uncertainty in any probability distribution.

Shannon's entropy function (2) has several useful properties. For any fixed N, Eq. (2) is a continuous and symmetric concave function with respect to all its arguments and has a (global) maximum when all the probabilities are equal ($p_i = 1/N$, $i = 1, ..., N$). It has the value zero if any one of the p_i is equal to unity. (It is axiomatic that the N probabilities p_i are collectively exhaustive and mutually exclusive, thus their sum must be unity).

Shannon's great contribution, from which the present paper stems, was to recognise that entropy was not a purely thermodynamic quantity. It was possible to use entropy to measure purely abstract entities such as information and uncertainty in a general probabilistic sense. The next key step in opening up new applications for the Shannon measure of uncertainty, which we have now identified as entropy, was taken by Jaynes [2]. Until Ref. [2] appeared, Shannon entropy was simply a very useful measure of the amount of uncertainty in a probability distribution, permitting different distributions to be compared quantitatively in respect of their total uncertainties. Jaynes suggested that the Shannon entropy measure could be used in a reverse sense to generate or infer a probability distribution which would have a maximum entropy. Since entropy is a measure of uncertainty, a maximum entropy distribution must have maximum uncertainty, must be maximally non-committal and must contain minimum bias. Jaynes's work [2] therefore extended Shannon's entropy function from being merely a measure of uncertainty to being a means of making inferences about probability distributions. In Ref. [2] Jaynes states:

> In making inference on the basis of partial information we must use that probability distribution which has maximum entropy subject to whatever is known. This is the only unbiased assumption we can make; to use any other would amount to an arbitrary assumption of information which by hypothesis we do not have.

The implication of the above statement is that if we have any partial information about some random process we should not simply look at the partial information and arbitrarily choose some probability distribution which appears to fit; we should choose that probability distribution which maximizes the Shannon entropy measure subject to the partial information which we have. The probability distribution which results from this constrained maximization process will then be one which introduces minimum bias into the probability estimation process. This is very significant for civil engineers, who are very familiar with situations in which only partial or incomplete information is available and who usually respond by the use of what is euphemistically called 'engineering judgement' to fill in the missing information so that progress can be made. Essentially this judgement consists of making an educated guess which may or may not be a good one and may or may not introduce considerable bias into all further calculations.

Jaynes's work [2] which is now referred to as the Maximum Entropy Principle has shown, in theory, how to avoid the introduction of bias into problem solving with incomplete information. Much recent research has been devoted to translating that theory into useable methodology in the civil engineering area and this paper will shortly describe some applications. First, however, we shall look a little more closely at the Maximum Entropy Principle.

4 The Maximum Entropy Principle

Consider a random process which can be described by a discrete random variable x which may take several discrete values x_i, $i = 1, \ldots, N$. Define p_i, $i = 1, \ldots, N$ to be the probability that x has the value x_i, $i = 1, \ldots, N$. i.e. $p_i = p(x = x_i)$. Jaynes's Maximum Entropy Principle casts the problem of determining the discrete probabilities p into the form of an optimization problem. The Shannon entropy function (2) (with some arbitrary coefficient of proportionality which we can take as K) must be maximized over variables p_i, $i = 1, \ldots, N$, subject to any available information we have about the random process. Precisely what constitutes this information, what can logically be inferred and how it can be encoded in constraint form clearly depends upon the particular random process under examination. However, one constraint is axiomatic: the normality condition,

$$\sum_{i=1}^{N} p_i = 1 \tag{3}$$

Maximization of Eq. (2) subject only to Eq. (3) is a simple calculation which yields uniform probabilities, $p_i = 1/N$, $i = N, \ldots, N$ and an entropy value of $S = K \ln(N)$. This result simply says that if we have no information other than the normality of the probabilities then we must assign all probabilities the same value. There is no justification for any other assignment of values. This corresponds to Laplace's 'Principle of Insufficient Reason' which has never had the formal status of a physical principle (but is nevertheless very useful) and can now be seen to be a consequence of the Maximum Entropy Principle.

Suppose now that some information is available about the random process in the form of M expectation functions of the form,

$$\sum_{i=1}^{N} p_i f_{ji}(x) = E[f_j] \qquad j = 1, \ldots, M \tag{4}$$

for which values of $f_{ji}(x)$ and $E[f_j]$, $j = 1, \ldots, M$; $i = 1, \ldots, N$ are known. These expectation functions may typically be such things as the mean, standard deviation and perhaps other statistical moments of data collected from the random process under examination. (It is assumed that $M < N - 1$ otherwise Eqs. (3) and (4) would themselves be sufficient to determine the unknown probabilities uniquely and would consist of complete rather than partial information).

The required probabilities must now maximize Eq. (2) subject to Eqs. (3) and (4). This problem is stated below for future reference and is named problem $\mathcal{E}$.

4.1 Problem $\mathcal{E}$

$$\underset{p_i,\, i=1,\,\ldots,\,N}{\text{Maximize}} \qquad S = -K \sum_{i=1}^{N} p_i \ln(p_i)$$

Subject to: $\qquad \sum_{i=1}^{N} p_i = 1$

$$\sum_{i=1}^{N} p_i f_{ji}(x) = E\left[f_j\right] \qquad j = 1, \ldots, M$$

$$p_i \geq 0 \qquad\qquad i = 1, \ldots, N$$

The solution of problem $\mathcal{E}$ can be found by examining the stationarity of its Lagrangean and is:

$$p_i = \frac{\exp\left(\sum_{j=1}^{M} \beta_j f_{ji}(x)/K\right)}{\sum_{i=1}^{N} \exp\left(\sum_{j=1}^{M} \beta_j f_{ji}(x)/K\right)} \qquad i = 1, \ldots, N \tag{5}$$

in which β_j, $j = 1, \ldots, M$ are the Lagrange multipliers associated with the expected value constraints in problem $\mathcal{E}$.

In order to determine the probabilities p_i, $i = 1, \ldots, N$ from Eq. (5) it is therefore necessary first to determine values for the Lagrange multipliers β_j, $j = 1, \ldots, M$. This can be done by substituting result (5) back into the expected value constraints in problem $\mathcal{E}$ and solving the M non-linear equations for the M Lagrange multipliers. This is awkward and tedious. A better approach has been given by Templeman and Li [3] who realised that problem $\mathcal{E}$ is a convex programming problem since it consists of maximizing a strictly concave and continuous function subject to linear constraints. Since problem $\mathcal{E}$ is convex it should have a dual form which may be easier to solve directly than problem $\mathcal{E}$. This turned out to be the case. The dual form of problem $\mathcal{E}$ is here called problem $\mathcal{DE}$ and is:

4.2 Problem $\mathcal{DE}$

$$\underset{\beta_j,\, j=1,\,\ldots,\,M}{\text{Minimize}} \qquad D = -K\left(\sum_{j=1}^{M} \beta_j E\left[f_j\right] + \ln\left(\sum_{i=1}^{N} \exp\left(\sum_{j=1}^{M} \beta_j f_{ji}(x)\right)\right)\right)$$

Problem $\mathcal{DE}$ is an unconstrained minimization over the space of the Lagrange multipliers β_j, $j = 1, \ldots, M$. Since we do not usually have more than a few expectation constraints M is small and problem $\mathcal{DE}$ can be solved numerically using any standard non-linear unconstrained optimizer to yield values for the M Lagrange multipliers. These can then be substituted into Eq. (5) to give the maximum entropy probabilities.

This concludes our formal examination of the Maximum Entropy Principle in mathematical terms. The above treatment has been directed towards the use of the Maximum Entropy Principle in connection with discrete probability distributions since this provides the easier demonstration. Where a random process is continuous the Maximum Entropy Principle applies equally well. In general, all the above theory remains unaltered but integral expressions over the continuous domain replace summations over the discrete probabilities and the Shannon entropy function includes a weighting element. The continuous Maximum Entropy Principle is not described more fully here since most subsequent applications are discrete. We now return to an examination of its use in various civil engineering applications.

5 Inference using the Maximum Entropy Principle

The most obvious and straightforward application of the Maximum Entropy Principle in civil engineering is that of fitting probability distributions to available probabilistic data. As noted earlier, it has become almost customary in many instances to plot the data histograms, inspect the shape of the envelope and then make some assumption about the shape to the effect that it is approximated by some well-known distribution which is used in place of the actual data in subsequent calculations. In many instances this procedure is entirely valid, typically when previous research and data analysis has firmly established that a particular distribution describes a particular random process very well. Indeed, many aspects of civil engineering use specially-developed probability distributions which have stood the test of time and are entirely appropriate. However, there are also many instances in the literature of probability distributions being selected on an ad hoc basis to represent a random process. The phrase 'assuming that the random variables are normally distributed ...' occurs frequently in the literature without any justification other than that the subsequent calculations would be far more difficult without this assumption. According to Jaynes this sort of assumption is 'an arbitrary assumption of information' which introduces extrinsic bias into the subsequent calculations and clouds the validity of any subsequent results.

The rigorous alternative to choosing some arbitrary probability distribution to represent data from some random process is to use the Maximum Entropy Principle to generate a least biased probability distribution, Eq. (5), to fit the data. One example of this is afforded by the calculation of structural reliability. Fig. 1a shows an axial force bar, the resistance properties of which are represented by a random variable R, loaded by an axial force, the magnitude of which is also represented by a random variable F. Fig. 1b shows, on a force scale, typical normalized distributions of R and F. The probability of failure, p_f of the bar is then defined as

$$p_f = p\left([R - F] < 0\right) \tag{6}$$

On Fig. 1b the failure region coresponds to the overlapping area of the tails of the probability distributions of R and F. To evaluate p , therefore, we need to evaluate the probability that $[R - F]$ is less than zero. A common way of doing this is to assume that $[R - F]$ can be represented by a normal (Gaussian) distribution whose mean and variance can be easily calculated from the means and variances of R and F. The assumption that $[R - F]$ is normally distributed leads to a relatively simple method of calculating p_f, using tabulated values of the standard normal distribution. This approach to calculating the probability of failure of a structural member (and hence calculating the reliability of the member as the complement of p_f) is now very widely known and accepted. It already forms a key element in the next generation of structural design codes of practice which will probably be reliability-based.

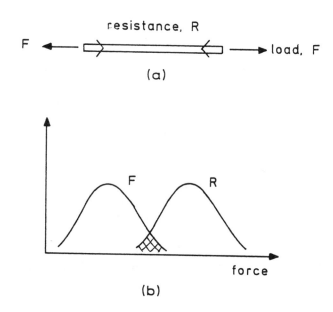

Figure 1. *Axial force bar with probabilistic load and resistance.*

However, the validity of the value of p calculated by the above method is entirely dependent upon the assumption that $[R - F]$ is normally distributed. There is no reason to expect that loads F will always be normally distributed; indeed for many types of structural loading they are not. Similarly there is no reason to expect or assume that R will be normally distributed for all types of materials. Consequently the assumption that $[R - F]$ is normally distributed represents 'an arbitrary assumption of information which by hypothesis we do not have.' The question then arises, how accurate will p be if $[R - F]$ is not normally distributed? Basu and Templeman [4] have used the Maximum Entropy Principle to investigate this question.

Given data collected from typical load and materials tests the first four statistical moments of the data may be calculated. From these moments the first four statistical moments of $[R - F]$ can be determined and these values form the expected values in the right-hand sides of the expectation constraints in the continuous version of problem $\mathcal{E}$. The least biased probability distribution of $[R - F]$ can then be generated by using the Maximum Entropy Principle. The probability of failure, p_f, may then be evaluated from Eq. (6) using the maximum entropy distribution of $[R - F]$ and compared with the value calculated by assuming that $[R - F]$ is normally distributed. Ref. [4] gives numerous examples which show the two calculated values of p_f differing by several hundred percent. This shows fairly conclusively that the 'arbitrary assumption of information' is not merely a hair-splitting rigorous exactitude; it can lead to grossly erroneous estimates of structural reliability.

There are many other examples of the use of the Maximum Entropy Principle to estimate least biased probability distributions from statistical data which could have been presented here. Munro and Jowitt [5] have used it to estimate order probabilities in the ready-mixed concrete industry, Guiasu [6] has used it in queueing systems. The book by Levine and Tribus [7] gives other examples from widely different fields.

The above use of the Maximum Entropy Principle to infer least biased probability distributions is the most obvious use which first springs to mind. However, civil engineers exist in a world which is dominated by incomplete information. Where it is a probability distribution which is missing we can use the Maximum Entropy Principle to infer one, but such cases are fairly rare. Can the Maximum Entropy Principle be used to generate solutions to wider, more general problems where the available information is not complete and is not directly concerned with probability distributions either? The following example shows that the applicability of the Maximum Entropy Principle does indeed extend to wider problems.

5.1 Water Supply Network Analysis

Fig. 2 shows a looped water supply network with 7 nodes and 8 pipes. The data for the example (but not the method) is taken from Alperovits and Shamir [8]. The network has a single source supplying 1120 m^3/hr and 6 demands as shown. Suppose that we need to estimate the flow rates Q in the individual pipes of the network given only the data shown in Fig. 2. Note that Fig. 2 shows the connectivity of the system and the flow direction in each pipe but does not give any data for the individual pipes.

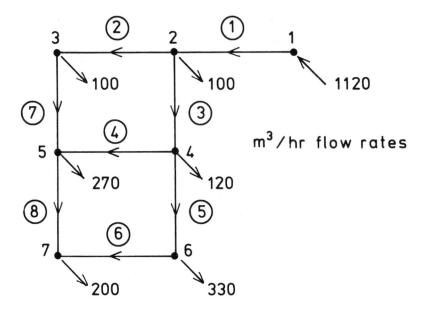

Figure 2. *Looped water supply network.*

If the network were branched we would have no difficulty in calculating the flow rates in the pipes of the branched system. We could simply work backwards along each branch accumulating demand quantities as flow rates until we reach the single source. The flow rates in all pipes could easily be found. However, the network is not branched; it is a looped system with two loops. In order to calculate the flow rates in the pipes of a looped system accurately we need additional information to enable us to calculate the head losses around the two loops and so determine the flow rates. The conventional computer-oriented pipe network analysis methods [9,10,11] are useless in this situation as they require data on pipe lengths, diameters and friction coefficients which we do not have. We could, of course, gather this information but this may not necessarily be an easy task. Can we infer some sort of 'most likely' flow rates based only upon the incomplete information which we possess?

Fig. 3 shows a modified version of the network of Fig. 2. It is obvious that the flow rate in pipe number 1 of Fig. 2 must be the same as the total supply rate, 1120 m³/hr, thus this pipe may be omitted and node 2 is now considered as the supply node for the system of 7 pipes and 6 nodes. All input and output flow rates have also been scaled by a factor of 1020 m³/hr (1120 minus the demand of 100 at node 2). We need to calculate values for the scaled flow rates p_i, $i = 2, \ldots, 7$.

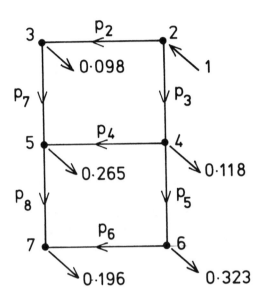

Figure 3. *Simplified and scaled network.*

Mass flow rate equilibrium at nodes 2 to 6 then gives:

$$p_2 + p_3 = 1 \tag{7a}$$

$$p_2 \qquad\qquad - p_7 = 0.098 \tag{7b}$$

$$p_3 - p_4 - p_5 = 0.118 \tag{7c}$$

$$p_4 \qquad + p_7 - p_8 = 0.265 \tag{7d}$$

$$p_5 - p_6 = 0.323 \tag{7e}$$

i.e. five equations in seven unknowns. Equilibrium at node 7 does not provide an independent equation as the supply and demands are in balance.

The p_i, $i = 2, \ldots, 7$ may be interpreted as the probability that a particle of water in the supply passes along a particular pipe. Eqs. (7) then express the known information which relates these probabilities together. If we maximize the Shannon entropy function, Eq. (2), of these probabilities subject to Eqs. (7) this will give us a least biased assignment of probabilities and we shall consequently have calculated the least biased assignment of pipe flow rates in the system. Eqs. (7) may be reduced to the form

$$p_2 = 0.363 - p_4 + p_8 \tag{8a}$$

$$p_3 = 0.637 + p_4 - p_8 \tag{8b}$$

$$p_4 = \qquad p_4 \tag{8c}$$

$$p_5 = 0.519 \qquad - p_8 \tag{8d}$$

$$p_6 = 0.196 \qquad - p_8 \tag{8e}$$

$$p_7 = 0.265 - p_4 + p_8 \tag{8f}$$

$$p_8 = \qquad p_8 \tag{8g}$$

Substituting these relationships into the Shannon entropy function (2) and calculating its unconstrained maximum over the two variables p_4 and p_8 gives a maximum entropy value of 2.0869 and the following assignment of probabilities:

$$p_2 = 0.358 \quad p_3 = 0.642 \quad p_4 = 0.146 \quad p_5 = 0.378$$
$$p_6 = 0.055 \quad p_7 = 0.260 \quad p_8 = 0.141$$

Reapplying the scaling factor gives the required least biased estimates of the pipe flow rates in m^3/hr in the network of Fig. 2 as

$$Q_1 = 1120 \quad Q_2 = 365 \quad Q_3 = 655 \quad Q_4 = 149$$
$$Q_5 = 386 \quad Q_6 = 56 \quad Q_7 = 265 \quad Q_8 = 144$$

There is, of course, no guarantee that these calculated flow rates correspond to actual flows in the network; nor is there any way of checking whether or not they are 'correct.' In the absence of information about lengths, sizes and friction coefficients for individual pipes these calculated values simply represent the least biased estimates that we can make of the pipe flow rates. Even if we had this extra information it would be pointless to check the accuracy of the above estimates; the extra information should logically be used to calculate more precise values of pipe flows.

It may be noted here that the above estimates of least biased pipe flow rates are for the flow directions shown in Fig. 2. If we had not possessed that information we would have been faced with a more difficult problem. Jaynes's requirement that no arbitrary assumptions of information should be made would have prevented us from simply assuming some flow directions. In order to satisfy Jaynes's requirement to the letter in the case where flow directions are unknown we should perform the above analysis for every possible set of flow directions. Each analysis would give a set of least biased flow rates and a corresponding maximum value of the Shannon entropy function. We should then select as our least biased estimates that set of values which gives the highest value of maximum entropy, i.e. that set which contains the maximum uncertainty.

The above example has shown how the entropy measure of uncertainty and the Maximum Entropy Principle can be used in a very practical civil engineering application to infer least biased results using incomplete data. The example did not directly involve probabilities but was capable of being interpreted, somewhat deviously, in a probabilistic light. Many other similar examples exist. For instance, the problem of estimating the axial forces in the bars of a statically indeterminate pin-jointed truss in the absence of data on the cross-sectional areas and material properties of the individual bars can be solved in an analogous fashion to the pipe network problem described above. It seems highly likely that what can be done for a statically indeterminate truss could also be done for any statically indeterminate structure including frames and finite element discretized continua. This is an untouched area of research. For large engineering systems such as pipe networks, trusses or finite elements the possibility of being able to infer 'most likely' performance estimates from partial information seems worth exploring further. For such systems, complete and accurate data is rarely available and often the calculated performance depends upon 'engineering judgement' in supplying guesses at values for data items in order to be able to use existing computer programs. The possibility which entropy-based inference holds out is that in the future it might not be necessary to have to make quite so many arbitrary guesses in order to obtain sensible performance estimates.

6 Entropy in Optimization Processes

So far, this paper has examined the Shannon entropy measure and Jaynes's Maximum Entropy Principle in the context of civil engineering inference under uncertainty. We have seen how entropy can be used to deduce desired results when only limited information is available. Optimization is itself a deductive process. Given some implicit or explicit function f of variables x_i, $i = 1, \ldots, N$, and some constraint functions, the process of locating a minimum value of f commences with no numerical information whatsoever. An initial point is then chosen and information is calculated about the objective and constraint functions, typically their numerical values and gradients at the design point. This numerical information is then used in some deterministic mathematical programming algorithm to infer where the next trial point should be placed so as to get closer to the constrained optimum of the problem. The new trial generates more information from which another point is inferred and eventually the solution is reached by this process of gathering better and better information and using it in an inference-based algorithm.

Almost all such optimization algorithms use some form of geometrical inference to generate a sequence of improving trial points. The functions in the problem are interpreted as geometrical hypersurfaces with contours, slopes and gradients. Actually, we never have sufficient information to be able to plot the geometry, except for the simplest of problems, but it is convenient to imagine that these hypergeometrical shapes exist because it helps us to visualise what a numerical search algorithm is doing and to think about different search strategies and develop new solution algorithms based upon our knowledge of geometry. We pretend, for convenience, that everything in the problem before us has a totally deterministic geometric representation. It is important to realise that the geometrical interpretations placed upon optimization processes are only interpretations. There is nothing sacrosanct about them. It is perfectly reasonable to discard this deterministic, geometric interpretation of optimization and to attempt to develop new optimization algorithms based upon totally non-geometrical, non-deterministic concepts.

Given that conventional optimization methods use geometrical methods to infer a sequence of improving points, the work now to be outlined stemmed from asking the question whether other methods of inference could not also be used for the same task. In particular, instead of using the currently available incomplete numerical information about the problem in a geometrical inference process, could not the same information be used in some sort of entropic inference process? The answer turned out to be, yes. Refs. [12–17] describe in detail some of the work of Templeman and Li which is continuing. Here, only the main ideas are outlined.

6.1 Constrained Non-linear Programming

Ref. [12] considers the general inequality constrained non-linear programming problem:

$$
\begin{aligned}
&\underset{\mathbf{x} \equiv x_i,\, i=1,\dots,N}{\text{Minimize}} && f(\mathbf{x}) \\
&\text{Subject to:} && g_j(\mathbf{x}) \le 0 && j = 1, \dots, M
\end{aligned}
\tag{9}
$$

and develops an interpretation of the problem in probabilistic terms which then uses entropy to infer a sequence of improving solutions. Problem (9) has an equivalent surrogate form:

$$
\begin{aligned}
&\underset{x_i,\, i=1,\dots,N}{\text{Minimize}} && f(\mathbf{x}) \\
&\text{Subject to:} && \sum_{j=1}^{M} \lambda_j g_j(\mathbf{x}) = 0
\end{aligned}
\tag{10}
$$

in which λ_j, $j = 1$, ..., M are non-negative surrogate multipliers which satisfy a normality condition,

$$\sum_{j=1}^{M} \lambda_j = 1 \tag{11}$$

These surrogate multipliers are in many respects similar to normalized Lagrange multipliers. It can be shown that optimum values exist for the vector of surrogate multipliers λ^* such that the optimum vector x^* which solves problem (10) with λ^* also solves problem (9). The difficulty arises in that the optimum values of the surrogate multipliers are unknown and must be found.

In Ref. [12] the surrogate multipliers are interpreted as probabilities which must satisfy the normality condition (11) and the expected value condition which forms the single constraint of problem (10). Values of the surrogate multipliers can then be inferred by the Maximum Entropy Principle. This leads to a two-phase solution procedure for problem (9) via problem (10) which works as follows. First, it is assumed that all M constraints have an equal probability of being active at the problem solution, thus the surrogate multipliers are assigned values of $\lambda_j^0 = 1/M$, $j = 1$, ..., M. This corresponds to Laplace's 'Principle of Insufficient Reason' and also to the maximum entropy solution when no information other than normality is available. Problem (10) is then solved over variable vector x with these values of λ^0 to give x^0. Values of all the constraint functions are then evaluated at x^0 which provides information than can be used to infer improved values of the surrogate multipliers λ^1 by solving the maximum entropy problem:

Maximize
$\lambda_j^1, j=1,...,M$
$\qquad S = -K \sum_{j=1}^{M} \lambda_j^1 \ln \lambda_j^1$

Subject to:
$\qquad \sum_{j=1}^{M} \lambda_j^1 = 1 \tag{12}$

$$\sum_{j=1}^{M} \lambda_j^1 g_j(x^0) = \epsilon$$

ϵ is an error term reflecting the fact that constraint function values $g(x^0)$ have been used in place of $g(x^1)$ which are not yet available. ϵ should be small, positive and decrease towards zero as iterations proceed. Values of λ^1 which solve the above maximum entropy problem are given by a result similar to Eq. (5):

$$\lambda_j^1 = \frac{\exp\left(\mu g_j(x^0)/K\right)}{\sum_{k=1}^{M} \exp\left(\mu g_k(x^0)/K\right)} \qquad j = 1, ..., M \tag{13}$$

in which μ is the Lagrange multiplier for the expected value constraint in problem (12). Since ϵ is not uniquely known and K is any positive constant, μ/K may be considered as a control parameter in result (13). For ϵ to display the desired convergence characteristics μ/K must be positive and increase towards infinity with successive iterations. Eq. (13) therefore gives new estimates for values of the surrogate multipliers λ^1 which replace the previous estimates λ^0. Solving problem (10) again over variable vector $\mathbf{x}$ starts another cycle of iterations which converge to the solution of problem (9) at $\mathbf{x}^\star$.

The essence of the above method consists of using maximum entropy to make least biased estimates of the optimum values of the surrogate multipliers. Each estimate leads to a new problem in the space of the x-variables and generates new information upon which to base an improved estimate of the optimum surrogate multipliers. Ref. [12] then extends the above method by combining the two phases into a single phase which consists of solving the single unconstrained problem (14) over variable vector $\mathbf{x}$, i.e.

$$\text{Minimize}_{\mathbf{x}} \quad f(\mathbf{x}) + \left(\frac{1}{\rho}\right) \ln \sum_{j=1}^{M} \exp\left(\rho\mu g_j(\mathbf{x})\right) \tag{14}$$

with $\rho\mu$ taking an increasing positive sequence of values tending towards infinity. The resulting values of the variable vector $\mathbf{x}^\star$ are the solution of problem (9).

For some time the validity or otherwise of the above entropy-based methods rested upon whether it was valid to interpret the surrogate multipliers as probabilities. Many test problems were solved quickly and efficiently with the methods and they were applied to the optimum sizing of truss structures in Ref. [13] giving smooth and rapid convergence. The proof of the validity of all the assumptions and of the methods was eventually obtained and is given in Ref. [14]. It turned out to be very simple, requiring only the use of Cauchy's inequality (the arithmetic-geometric mean inequality). In consequence, all the results can now be proved entirely deterministically and without recourse to probabilistic interpretations. Nevertheless, without the probabilistic interpretation and the use of maximum entropy the results would not have been initially derived. Also, the fact that maximum entropy can be used to develop probabilistic solution algorithms for deterministic mathematical programming problems opens the way for further uses of maximum entropy.

6.2   Multi-objective and Minimax Optimization

Following on from the above work on scalar optimization, Ref. [15] examined the role of maximum entropy in vector and minimax optimization. If $\mathbf{X}$ is some constrained feasible region of variables $\mathbf{x} \equiv x_i$, $i = 1, \ldots, N$ and $\mathbf{G}$ is a vector of real-valued goal functions $G_j(\mathbf{x})$, $j = 1, \ldots, M$ the multi-objective (or multi-criteria or vector) optimization problem can be stated as:

$$\text{Minimize}_{\mathbf{x}\in\mathbf{X}} \quad \{G_1(\mathbf{x}), G_2(\mathbf{x}), \ldots, G_M(\mathbf{x})\} \tag{15}$$

The minimax optimization problem can also be defined as:

$$\begin{array}{c} \text{Maximize} \\ \text{Minimize} \\ x \in X \end{array} \qquad \langle G_1(\mathbf{x}), G_2(\mathbf{x}), \ldots, G_M(\mathbf{x}) \rangle \qquad (16)$$

It is well-known that a Pareto-optimum solution of the multi-objective problem (15) can be found by solving the scalar optimization problem:

$$\operatorname*{Minimize}_{x \in X} \quad \sum_{j=1}^{M} \lambda_j G_j(\mathbf{x}) \qquad (17)$$

in which λ_j, $j = 1, \ldots, M$ are specified non-negative multipliers satisfying a normality condition such as Eq. (11). In the context of this paper these non-negative normalized multipliers clearly have a probabilistic interpretation and invite further exploration by entropy maximization.

In Ref. [15] it is shown that maximum entropy provides close links between multi-criteria optimization (problem (15)), minimax optimization (problem 16)) and the scalar Pareto problem (17). Specifically, it is shown that the minimax problem (16) can be solved by minimizing a continuous, differentiable scalar optimization problem. Thus:

$$\operatorname*{Min}_{x \in X} \operatorname*{Max}_{j \in M} \langle G_j(\mathbf{x}) \rangle = \operatorname*{Min}_{x \in X} \left(\frac{1}{\rho}\right) \ln \left(\sum_{j=1}^{M} \exp\left(\rho G_j(\mathbf{x})\right)\right) \qquad (18)$$

Furthermore (18) is an upper bound to an entropy-penalised version of problem (17). Thus:

$$\operatorname*{Min}_{x \in X} \operatorname*{Max}_{j \in M} \langle G_j(\mathbf{x}) \rangle \geq \operatorname*{Min}_{x \in X} \left(\sum_{j=1}^{M} \lambda_j G_j(\mathbf{x}) - \left(\frac{1}{\rho}\right) \sum_{j=1}^{M} \lambda_j \ln \lambda_j\right) \qquad (19)$$

with equality when the right-hand side is maximized over the multipliers λ. In both relationships (18) and (19) ρ should have a positive and increasing value towards infinity.

From a maximum entropy viewpoint results (18) and (19) are particularly satisfying since they show quite clearly that maximizing the entropy of the Pareto multipliers in a multi-criteria optimization problem generates the solution of the minimax optimization problem. Maximum entropy therefore forms the link between multi-criteria and minimax optimization. Also, the scalar optimization problem in (18) provides a convenient and effective means of solving minimax optimization problems. Ref. [15] outlines how this scalar optimization problem could be used to find shape-optimal designs of finite element systems which are governed by the need to minimize the maximum stress occuring anywhere. Ref. [16] describes the use of this scalar optimization approach to solving minimax optimization problems. There it was used to determine the optimum distribution of pretensioning forces in a highly non-linear cable net structure. The minimax element of the formulation required that criteria representing maximum forces in all cable elements, displacements of nodes of the cable net, and the magnitudes of the pretensioning forces themselves all be simultaneously minimized.

Current work is concerned with developing algorithms for the above entropy-based methods and using them on a variety of practical applications. Results (18) and (19) have been explored in respect of their ability to generate sets of Pareto-optimal solutions. A further investigation is currently being made into solving unconstrained optimization problems by entropy-based methods. A final reference should be quoted for anyone wishing to read more of the background to the use of entropy in solving optimization problems. This is the Ph.D. thesis of Li [17].

7 Conclusions

This paper has attempted to present an introduction to the use of entropy for measuring uncertainty and to the use of the Maximum Entropy Principle for making inferences in the presence of incomplete information. Finally it has described early research into the use of informational entropy as a means of solving a variety of optimization problems. This has been a lot of material to cover in a single short paper and much of the detail has been omitted from necessity. However, it is hoped that some of the flavour of the work is evident even if the full taste cannot be appreciated. Perhaps the most satisfying aspects of working with entropy are that (a) there are not many other people doing it, and (b) it is still an under-researched field with what currently appears to be excellent potential for the future. Above all, it is interesting work to do and this provides a considerable attraction in an increasingly complicated academic life.

References

[1] Shannon, C. E., (1948), 'A mathematical theory of communication,' *Bell System Technical Jnl.* **27**, 3, 379–428.

[2] Jaynes, E. T., (1957), 'Information theory and statistical mechanics,' *The Physical Review* **106**, 620–630 and **108**, 171–190.

[3] Templeman, A. B. and Li, X. S., (1985), 'Entropy duals,' *Engineering Optimization* **9**, 2, 107–119.

[4] Basu, P. C. and Templeman, A. B., (1985), 'Structural reliability and its sensitivity,' *Civil Engineering Systems* **2**, 1, 3–11.

[5] Munro, J. and Jowitt, P. W., (1978), 'Decision analysis in the ready-mixed concrete industry,' *Proc. Inst of Civil Engrs.*, Pt. 2, **65**, 41–52.

[6] Guiasu, S., (1986), 'Maximum entropy condition in queueing theory,' *Jnl. of the Operational Research Society* **37**, 3, 293-301.

[7] Levine, R. D. and Tribus, M., (Eds.) (1979), *The Maximum Entropy Formalism*, MIT Press.

[8] Alperovits, E. and Shamir, U., (1977), 'Design of optimal water distribution systems,' *Water Resources Research* **13**, 6, 885–900.

[9] Cross, H., (1938), *Analysis of flow in networks of conduits or conductors*, Bulletin No. 286, Univ. of Illinois Engineering Experiment Station, Urbana, Illinois.

[10] Martin, D. W. and Peters, G., (1963), 'The application of Newton's method to network analysis by digital computer,' *Jnl. of the Institution of Water Engineers* **17**, 115–129.

[11] Wood, D. J. and Charles, C. O. A., (1972), 'Hydraulic network analysis using linear theory,' *Jnl. of the Hydraulics Division, ASCE* **98**, HY7, 1157–1170.

[12] Templeman, A. B. and Li, X. S., (1987), 'A maximum entropy approach to constrained non-linear programming,' *Engineering Optimization* **12**, 3, 191–205.

[13] Li, X. S. and Templeman, A. B., (1988), 'Entropy-based optimum sizing of trusses,' *Civil Engineering Systems* **5**, 3, 121–128.

[14] Templeman, A. B. and Li, X. S., (1989), 'Maximum entropy and constrained optimization,' In *Maximum Entropy and Bayesian Methods*, J. Skilling (Ed.), Kluwer.

[15] Templeman, A. B., (1989), 'Entropy-based minimax applications in shape-optimal design,' In *Lecture Notes in Engineering, No. 42: Discretization Methods and Structural Optimization—Procedures and Applications*, H. A. Eschenauer and G. Thierauf (Eds.), Springer-Verlag, 335–342.

[16] Simoes, L. M. C. and Templeman, A. B., (1989), 'Entropy-based synthesis of pretensioned cable net structures,' *Engineering Optimization* **15**, 2, 121-140.

[17] Li, X. S., (1987), 'Entropy and Optimization,' Ph.D. thesis, Univ. of Liverpool.

An Assessment of Current Non-Linear Programming Algorithms for Structural Design, Part I: Basic Algorithms

G. N. Vanderplaats

Vanderplaats, Miura & Associates, Inc.
Goleta, California
United States of America

Abstract Numerical optimization methods offer an efficient design methodology for engineering design. The concepts are not particularly difficult, it is only that they are unfamiliar to most practicing engineers. The purpose here is to outline the general methods of numerical optimization and to identify their basic features. Several basic optimization strategies will be described to demonstrate just how optimization can solve the engineering design problem.

Relative to other applications, structural optimization is a special case that has received considerable attention in the past 30 years. Presently, most major finite element analysis program suppliers are adding optimization to their codes. In Part II of this paper, some recent approximation methods are outlined which help improve efficiency for structural optimization.

1   Introduction

Numerical optimization methods provide a uniquely general and versatile tool for design automation. Research and applications to structural design has been extensive and today these methods are finding their way into engineering offices. The methods that form the basis of most modern optimization were developed roughly 30 years ago, and the first application to nonlinear structural design was in 1960 [1]. Much of the research in structural design in the past 15 years has been devoted to creating methods that are efficient for design problems where the analysis is expensive. This has resulted in various approximation methods that allow a high degree of efficiency while maintaining the essential features of the original problem.

Here, we will first define the general design task in terms of optimization. We will briefly discuss several common algorithms for solving this general problem and will identify their basic features. Some of the special techniques that have been

B. H. V. Topping (ed.),
Optimization and Artificial Intelligence in Civil and Structural Engineering, Volume I, 107–125.
© 1992 *Kluwer Academic Publishers. Printed in the Netherlands.*

devised to make structural optimization efficient are discussed in Part II. Several design examples are offered to demonstrate the power of optimization as a design tool. References are offered for more thorough study.

2 Basic Concepts of Optimization

Mathematical programming (the formal name for numerical optimization) provides a very general framework for scarce resource allocation, and the basic algorithms originate in the operations research community. Engineering applications include chemical process design, aerodynamic optimization, nonlinear control system design, mechanical component design, multidiscipline system design, and a variety of others. Because the statement of the numerical optimization problem is so close to the traditional statement of engineering design problems, the design tasks to which it can be applied are inexhaustible.

In the most general sense, numerical optimization solves the nonlinear, constrained problem, Find the set of design variables, X_i, $i = 1$, N contained in vector $\underline{X}$, that will

$$\text{Minimize} \quad F(\underline{X}) \tag{1}$$

$$\text{Subject to;} \quad g_j(\underline{X}) \leq 0 \qquad j = 1, M \tag{2}$$

$$h_k(\underline{X}) = 0 \qquad k = 1, L \tag{3}$$

$$X_1^l \leq X_i \leq X_1^u \qquad i = 1, N \tag{4}$$

Equation 1 defines the objective function which depends on the values of the design variables, $\underline{X}$. Equations 2 and 3 are inequality and equality constraints respectively, and Eq. 4 defines the region of search for the minimum. This provides limits on the individual design variables. The bounds defined by Eq. 4 are referred to as side constraints. A clear understanding of the generality of this formulation makes the breadth of problems that can be addressed apparent. However, there are some important limitations to the present technology. First, it is assumed that the objective and constraint functions be continuous and smooth (continuously differentiable). Experience has shown this to be a more theoretical than practical requirement and this restriction is routinely violated in engineering design. A second requirement is that the design variables contained in $\underline{X}$ be continuous. In other words, we are not free to chose structural sections from a table. Also, we cannot treat the number of plies in a composite panel as a design variable, instead treating this as a continuous variable and rounding the result to an integer value. It is not that methods do not exist for dealing with discrete values of the variables. It is just that available methods lack the needed efficiency for widespread application to real engineering design. Finally, even though there is no theoretical limit to the number of design variables contained in $\underline{X}$, if we use optimization as a 'black box' where we simply couple an analysis program to an optimization program, the number of design variables that can be considered is limited to the order of fifty. Again, there are many exceptions to this, but it is still a conservative general rule. Also, this is not too great a restriction when we recognize that using graphical methods would limit us to only a few design variables.

The general problem description given above is remarkably close to what we are accustomed to in engineering design. For example, assume we wish to determine the dimensions of a structural member that must satisfy a variety of design conditions. We normally wish to minimize weight, so the objective function, $F(\underline{X})$, is just the weight of the structure, which is a function of the sizing variables. However, we also must consider constraints on stresses, deflections, buckling and perhaps dynamic response limits. Assuming we model the structure as an assemblage of finite elements, we can calculate the stresses in the elements under each of the prescribed load conditions. Then a typical stress limit may be stated as

$$\sigma^l \le \sigma_{ijk} \le \sigma^u \tag{5}$$

where i = element number, j = stress component, k = load condition. The compression and tensile stress limits are σ^l and σ^u, respectively (if we use a von Mises stress criterion, only σ^u would be used). While Eq. 5 may initially appear to be different from the general optimization statement, it is easily converted to two equations of the form of Eq. 2 as

$$g^1(\underline{X}) = \frac{\sigma^l - \sigma_{ijk}}{|\sigma^l|} \le 0 \tag{6}$$

$$g^2(\underline{X}) = \frac{\sigma_{ijk} - \sigma^u}{|\sigma^u|} \le 0 \tag{7}$$

Thus, the formal statement of the optimization task is essentially identical to the usual statement of the structural design task. The denominator of Eqs. 6 and 7 represents a normalization factor. This is important since it places each constraint in an equal basis. For example, if the value of a stress constraint is -0.1 and the value of a displacement constraint is -0.1, this indicates that each constraint is within 10% of it's allowable value. Without normalization, if a stress limit is $20,000$, it would only be active (within 0.1) if it's value was $19,999.9$. This accuracy is probably impossible to achieve on a digital computer. Also, it is not meaningful since loads, material properties, and other physical parameters are not known to this accuracy.

It is often assumed that for optimization to be used, the functional relationships must be explicit. However, this is categorically untrue. It is only necessary to be able to evaluate the objective and constraint functions for proposed values of the design variables, $\underline{X}$. Normally, this is done by a computer program, but we are not limited to this. For example, in [2], jet engine compressor vane settings are determined by passing information between the optimizer and an engine test rig, where the objective and constraint functions are evaluated experimentally. Using optimization, the efficiency was improved compared to the previous trail and error method. Furthermore, the number of experimental data points was reduced by 40% in a very costly study.

Using optimization as a design tool has several advantages; We can consider large numbers of variables relative to traditional methods. In a new design environment such as with composites, we do not have a great deal of experience to guide us and so optimization often gives unexpected results which can greatly enhance the final product. One of the most powerful uses of optimization is to make early design tradeoffs using simplified models. Here we can compare optimum designs instead of just comparing point designs. Furthermore we can obtain the optimal sensitivity with respect to design parameters and use this to guide the decision making process.

On the other hand, optimization has some disadvantages to be aware of; The quality of the result is only as good as the underlying analysis. Thus, if we ignore or forget an important constraint, optimization will take advantage of that, leading to a meaningless if not dangerous design. There is a danger that by optimizing we will reduce the hidden factors of safety that now exist. In this context, we should re-think our use of optimization, using it as only a design tool and not as a sole means to an end product.

However, assuming we agree that optimization is useful, it is also important to know how these algorithms solve our design problems. In the following sections we briefly outline several common algorithms to provide some insight into the numerical techniques used. For brevity as well as clarity, we will avoid some of the more modern methods that tend to be a bit esoteric, and will limit our discussion to a few basic concepts.

2.1 The Optimization Process

Most optimization algorithms do just what good designers do. They seek to find a perturbation to an existing design which will lead to an improvement. Thus, we seek a new design which is the old design plus a change so

$$\underline{X}^{\text{new}} = \underline{X}^{\text{old}} + \delta \underline{X} \tag{8}$$

Optimization algorithms use much the same formula, except it becomes a two step process. Here we update the design by the relationship

$$\underline{X}^q = \underline{X}^{q-1} + \alpha \underline{S}^q \tag{9}$$

where $\alpha \underline{S}^q$ in Eq. 9 is equivalent to $\delta \underline{X}$ in Eq. 8. Here q is the iteration, or design cycle, number. The engineer must provide an initial design, X^0, but it need not be feasible (it may not satisfy the inequalities of Eq. 2). Optimization will then determine a 'Search Direction,' $\underline{S}^q$ that will improve the design. If the design is initially feasible, the search direction will reduce the objective function without violating the constaints. If the initial design is infeasible, the search direction will point toward the feasible region, even at the expense of increasing the objective function.

The next question is how far can we move in direction $\underline{S}^q$ before we must find a new search direction. This is called the 'One-Dimensional Search' since we are just seeking the value of the scalar parameter, α, to improve the design as much as possible. If the design is feasible and we are reducing the objective function, we seek the value of α that will reduce $F(\underline{X})$ as much as possible without driving any $g_j(\underline{X})$ positive or violating any bound on the components of $\underline{X}$. If the design is initially infeasible, we seek the value of α that will overcome the constraint violations if possible, or will otherwise drive the design as near to the feasible region as possible. Note that this is precisely what a design engineer does under the same conditions. The difference is that optimization does it without the need to study many pages of computer output.

There are a wide variety of algorithms for determining the search direction, $\underline{S}$, as well as for finding the value of α [3]. Determining α is conceptually a simple task. For example, we may pick several values of α and calculate the objective and constraint functions. Then we fit a polynomial curve to each function and determine the value that will minimize $F(\underline{X})$ or drive $g_j(\underline{X})$ to zero. Since we picked a search direction that will improve the design, we need only consider positive values of α. The minimum positive value of α from among all of these curve fits is the one we want. Other methods for finding α are called by such names as Golden Section, Bisection, and Fibonacci search, as examples.

2.2 Unconstrained Minimization

Unconstrained minimization problems are defined by Eq. 1 only, where Eqs. 2–4 are omitted. Most engineering problems are not of this form, and so we will only offer some basic concepts here. The motivation for considering unconstrained minimization methods is that they provide a basis for the constrained problem, and that they can sometimes be used indirectly to solve constrained problems. Also, some engineering analysis problems can be posed as unconstrained minimization tasks, an example being nonlinear structural analysis. Finally, unconstrained minimization is familiar to us from calculus, where the minimum or maximum of a function is known to be a point where the gradient of the function vanishes. Perhaps the oldest, best known, and worst unconstrained minimization algorithm is known as the Steepest Descent Method. Here, the search direction, $\underline{S}^q$, is calculated as the negative of the gradient of the objective function;

$$\underline{S}^q = -\underline{\nabla}F(\underline{X}^{q-1}) \tag{10}$$

where

$$\underline{\nabla} F(\underline{X}^{q-1}) = \left\{ \begin{array}{c} \frac{\partial F(\underline{X}^{q-1})}{\partial X_1} \\ \frac{\partial F(\underline{X}^{q-1})}{\partial X_2} \\ \vdots \\ \frac{\partial F(\underline{X}^{q-1})}{\partial X_N} \end{array} \right\} \tag{11}$$

This search direction is now used in Eq. 9 to update the design. It is surprising how often this method is used today and presented as an 'advanced' algorithm. However, it must be stated that, this method is one of the worst available and should never be used. The steepest descent method will seldom converge reliably to a solution in even the simplest of nonlinear minimization tasks. Furthermore, it is almost trivial to modify this method to make it efficient, although this simple modification is still not considered to be the best algorithm available. The conjugate gradient method [4] represents a simple modification to the steepest descent method, but provides dramatic improvements in optimization efficiency. Here, we still use the steepest descent direction on the first iteration ($q = 1$). On subsequent iterations, we use a conjugate direction defined by

$$\underline{S}^q = -\underline{\nabla} F(\underline{X}^{q-1}) + \beta \underline{S}^{q-1} \tag{12}$$

where

$$\beta = \frac{\left| \underline{\nabla} F(\underline{X}^{q-1}) \right|}{\left| \underline{\nabla} F(\underline{X}^{q-2}) \right|} \tag{13}$$

Equations 12 and 13 have a simple physical interpretation. If we were making progress on the last iteration, it makes sense to move partly in that direction (as defined by β), while including the gradient information at the present design point. Other, more powerful methods, are known as Variable Metric Methods and go by such names as DFP and BFGS methods. Each such method attempts to create second order information using the first order (gradient) information, and can be shown to converge in N or fewer iterations for a quadratic problem. On the other hand, the Steepest Descent Method cannot be guaranteed to converge under any circumstances.

2.3 Constrained Minimization

The vast majority of engineering design tasks are constrained problems of the form of Eq. 1–4. Just as for unconstrained problems, there are perhaps as many methods available as there are researchers in the field. Here, we will discuss four methods which provide a basic understanding of the concepts. The first method converts the constrained problem to a sequence of unconstrained problems. The second two methods have been shown to be powerful tools for engineering design, even though each method is considered to be 'poor' by theoretical standards. The last method has been developed more recently than the others and is considered to be a theoretically 'good' method. First, we consider the simplest case of what are referred to as Sequential Unconstrained Minimization Techniques or SUMT. The basic concept is to convert the original problem to an unconstrained problem that can be solved by unconstrained minimization methods. The most direct way to do this is to create a 'penalty' function that will increase the objective for any constraint violation. Thus, we define a Pseudo-objective function of the form

$$F'(\underline{X}) = F(\underline{X}) + R\left(\sum_{1}^{M} \max\{0, g_j(\underline{X})\}^2 + \sum_{1}^{L} (h_k(\underline{X}))^2\right) \tag{14}$$

The parameter, R, is referred to as a penalty parameter. Initially, R is taken as a relatively small number and $F'(\underline{X})$ is minimized as an unconstrained function. Then R is increased, typically by a factor of 10 and the process is repeated. The process is terminated when no improvement in $F(X)$ is found and all constraints are satisfied within a small tolerance. The reason that R cannot just be set to a very large value from the start is that this would create a high degree of nonlinearity, making it impossible to solve the unconstrained problem. This method is called the Exterior Penalty Function Method since it penalizes the objective only if a constraint is violated. This method has the advantages of simplicity and reliability, but the disadvantages that it approaches the optimum from the infeasible region and requires more function evaluations than competing methods. Another implicit advantage is that by coming from the infeasible region, there is an improved likelihood of finding the best optimum for cases where relative minima exist. A variety of other such methods are available and some of them (referred to as interior penalty function methods) approach the optimum from inside the feasible region [5,6]. Also, the Augmented Lagrange Multiplier Method [7] is considered to be a more modern sequential unconstrained minimization method.

The second method we consider is the Method of Feasible Directions [8]. This is referred to as a direct method since if deals with the constraints directly in the optimization process. The basic algorithm is for inequality constraints, although equality constraints may be included in a variety of ways if necessary. Assuming we begin in the feasible design region (there are no active or violated constraints. That is all $g_j(\underline{X}) < 0$), we begin with a steepest descent search direction. If it takes more than one iteration to encounter a constraint, we use a conjugate direction or variable metric method on subsequent iterations, until a constraint is encountered. Once having encountered a constraint boundary, we must find a 'Usable-Feasible' direction, where a usable direction is one that reduces the objective and a feasible direction is one that either follows or moves inside of a constraint boundary. This requires solving the following sub-problem:

Find the components of $\underline{S}^q$ and β that will

$$\text{Minimize} \qquad \underline{\nabla} F\left(\underline{X}^{q-1}\right) \cdot \underline{S}^q \tag{15}$$

$$\text{Subject to;} \qquad \underline{\nabla} g_j\left(\underline{X}^{q-1}\right) \cdot \underline{S}^q \leq 0 \qquad j \in J \tag{16}$$

$$\underline{S}^q \cdot \underline{S}^q \leq 1 \tag{17}$$

where J is the set of active constraints $\left(\text{all } g_j(\underline{X}^{q-1}) \leq \epsilon\text{, where } \epsilon \text{ is a small negative}\right.$ number, say -0.003). Equation 15 attempts to reduce the objective function as much as possible, while Eq. 16 prevents the search direction from pointing into the infeasible region. The purpose of Eq. 17 is to prevent an unbounded solution to this sub-problem. Methods for solving this problem are beyond the scope of this discussion. However, the resulting search direction will reduce the objective function without violating constraints. Similar methods will actually move away from the presently active constraints. This method, which 'follows' constraints requires modifications to the one-dimensional search to bring the design back to the constraint boundaries. This is only one of a class of algorithms based on the feasible directions concept, but serves to define the general method.

The third method we consider is the Sequential Linear Programming (SLP) method. Sequential Linear Programming, also known as Kelley's Cutting Plane Method [9], was developed in the early 1960's. This method is considered unattractive by theoreticians, but has proven to be quite powerful and efficient for engineering design. The basic concept is that we first linearize the objective and constraint functions and then solve this linearized problem by the optimizer of our choice (the obvious choice is the standard linear programming method, but we need not restrict ourselves to this. The Method of Feasible Directions will work very well). Thus, we create the following approximation;

$$\tilde{F}(\underline{\delta X}) = F(\underline{X}^{q-1}) + \underline{\nabla} F(\underline{X}^{q-1}) \cdot \underline{\delta X} \tag{18}$$

$$\tilde{g}_j(\underline{\delta X}) = g_j(\underline{X}^{q-1}) + \underline{\nabla} g_j(\underline{X}^{q-1}) \cdot \underline{\delta X} \tag{19}$$

where

$$\underline{\delta X} = \underline{X}^q - \underline{X}^{q-1} \tag{20}$$

We now solve the approximate optimization problem;

$$\text{Minimize} \qquad \tilde{F}\left(\underline{\delta X}\right) \tag{21}$$

$$\text{Subject to;} \qquad \tilde{g}_j\left(\underline{\delta X}\right) \leq 0 \qquad j \in J \tag{22}$$

$$\delta X_i^L \leq \delta X_i \leq \delta X_i^U \qquad i = 1, N \tag{23}$$

Here, the set of active constraints, J, includes all potentially active constraints (say all $g_j\left(\underline{X}^{q-1}\right) > -0.3$). It is not necessary to approximate all constraints, which may number in the thousands. Usually we can limit the set J to about $3N$.

The move limits, δX_i^L and δX_i^U are needed to prevent unlimited moves as well as limiting the design changes to the region of applicability of the approximation. Once the approximate optimum is found, the problem is re-linearized and the process is repeated until it converges to the optimum. During the process, the move limits are reduced to insure convergence.

Finally, we consider a relatively recent method called Sequential Quadratic Programming [10,11]. While this is a relatively complicated method, it has been found to be quite powerful if reasonable care is taken in formulating the optimization problem. The basic concept is to find a search direction, $\underline{S}$, that will minimize a quadratic approximation to the Lagrangian function subject to linear approximations to the constraints. That is, Find the components of $\underline{S}$ that will

$$\text{Minimize} \qquad Q(\underline{S}) = F(\underline{X}^0) + \underline{\nabla}F(\underline{X}^0) \cdot \underline{S} + \tfrac{1}{2}\underline{S}^T \mathbf{B}\underline{S} \qquad (24)$$

$$\text{Subject to;} \qquad g_j(\underline{X}^0) + \underline{\nabla}g_j(\underline{X}^0) \cdot \underline{S} \leq 0 \qquad j = 1,\, M \qquad (25)$$

This is a quadratic programming problem which is solved for the components of S. A variety of modifications are available to insure this sub-problem has a feasible and bounded solution. The matrix, $\mathbf{B}$, is initially the identity matrix, but as the optimization proceeds, $\mathbf{B}$ is updated using gradient information to approximate the Hessian of the Lagrangian function.

Once the search direction is found, a one-dimensional search is performed using an exterior penalty formulation which includes the Lagrange multipliers from the process of finding $\underline{S}$.

As noted above, this is a relatively complicated algorithm and has been found by experience to be somewhat sensitive to the care with which the overall optimization problem is formulated as well as being sensitive to several internal parameters. However, for those classes of problems where it works, it has been found to be particularly powerful.

In considering the variety of algorithms available, there are no reliable rules to determine which method is best. However, experience shows that the most important thing is to use an algorithm that provides acceptable results on the average problem of interest. The more complicated algorithms are usually considered best by the theoreticians, but also are found to be less reliable for problems that are not carefully formulated. On the other hand, algorithms like the feasible directions method the sequential linear programming are considered 'poor' by the theoreticians, but usually perform reliably in a practical design environment.

3 Using Optimization as a 'Black Box'

Assuming we wish to use optimization in the design environment, the issue now becomes one of implementation. To begin with, it is not necessary for every design group to create its own optimization program. Just as with finite element analysis, there is a growing number of optimization software products becoming commercially available. The real issue is how to get these codes used in an already overworked design department. This is primarily a management and training issue.

The key is a very minor amount of standardization and centralization. By writing engineering analysis software in a consistent format, it can be easily coupled with an optimizer and optimization can be done with very little effort beyond that needed for a single analysis. Now, by proposing standardization and centralization we suggest neither an additional level of bureaucracy nor a removal of software coding from the engineering office. On the contrary, engineers should be allowed and encouraged to code their own special purpose analysis software. Of course, they shouldn't code what they can buy any more than they should design a bolt that is available from stock. Also, they should be not burdened with a long list of rules on structured programming, documentation and the like.

In order to write software that can be readily coupled to optimization, only three simple rules are needed;

1. Write the analysis as a subroutine with a very simple main calling program. This subroutine can call any number of additional subroutines.

2. Separate the analysis into *Input*, *Execution* and *Output*. The optimizer will require that execution be performed repeatedly for different combinations of the design variables.

3. Store every parameter that may be a design variable, objective or constraint function in a single labeled COMMON BLOCK.

Following these rules is good programming practice and will not be offensive except to the most individualistic of engineers/programmers. Yet by following these rules, over 90% of all engineering analysis programs where optimization may be useful can be coupled to an optimizer in a matter of minutes. The main program will be replaced by an optimization control program such as COPES [12] and the input data will define the optimization problem. Furthermore, once the combined program is working, the choice of design variables, objective and constraint functions, and constraint bounds can be changed by only changing a few lines of data. Finally, the issue of optimization is only marginally considered in writing the analysis code. It is only the data to the combined program that makes it a design optimization task.

The motivation for general purpose optimization programs such as COPES is to make it so easy to try optimization that there is little excuse not to. More than that, once the engineer/programmer is accustomed to writing design oriented codes instead of standard analysis codes, a library of design programs evolves which can be repeatedly used as the design process goes forward or as new design problems are addressed.

4 Impact of Advanced Computer Architectures

In seeking efficiency enhancements for optimization algorithms, it is important to remember where the computational cost is greatest and attack that area first. In the case of linear programming, where the objective and constraint functions are linear, the solution procedure bears a strong resemblance to solving simultaneous equations iteratively. Therefore, for this case, vectorization and judicious use of parallel computations are both attractive. However, here we are interested in the nonlinear optimization task where the design will be repetitively updated using Eq. 9. Also, because we are interested primarily in engineering design, the analysis task will require roughly 99 percent of the computational effort for optimization. Therefore, it is logical to address those parts of the algorithm that relate to the number of analyses needed, since this is a direct measure of the cost of optimization.

Most modern nonlinear optimization algorithms require gradient information to determine the search direction, $\underline{S}$. However, with the notable exception of structural analysis, very few large scale analysis programs can provide the gradient of the calculated responses with respect to the design variables. Therefore, this information must be calculated by finite difference methods. Thus, the gradient of some function, $F(\underline{X})$, is calculated from

$$\nabla F(\underline{X}) = \left\{ \begin{array}{c} \frac{F(\underline{X}+\delta X_1)}{\delta X_1} \\[2mm] \frac{F(\underline{X}+\delta X_2)}{\delta X_2} \\[2mm] \vdots \\[2mm] \frac{F(\underline{X}+\delta X_N)}{\delta X_N} \end{array} \right\} \tag{26}$$

where $F(\underline{X})$ may be the objective function or any constraint. Since these functions are calculated by the analysis, it is usual to calculate all functions needed so that one entry of each needed gradient vector is calculated for each perturbation of a design variable. Because this requires one complete analyses for each design variable perturbation, it is efficient to perturb as many variables simultaneously as possible, sending each modified design vector to a different processor for the analysis. Thus, we can simultaneously calculate as many entries in the gradient vectors as there are parallel processors available. This same argument applies if the analyses are sent to separate single processor computers, so most of the discussion here that refers to parallel processing is equally applicable to distributed computing.

In a typical optimization, it is common to require at least ten search directions used in Eq. 9. Thus, for gradient calculations alone, $10N$ analyses are needed, where N is the number of design variables. For a given number of parallel processors, the 'turn-around' time for gradient computations is reduced accordingly. The second opportunity for efficiency improvements is contained in Eq. 9, where we wish to find the value of α that will minimize the objective function subject to the constraints. This requires that we choose several candidate values of α and evaluate the functions corresponding to that $\underline{X}$. Normally, we use only a few candidate values of α and then use polynomial interpolation to estimate the best value. We then analyze this proposed design and accept it as the result of this design iteration.

Given the ability to simultaneously evaluate designs for several proposed values of α, we again have a direct reduction in overall computational times. The only difference between what is done here and what is done in finite difference gradient calculations is that now all components of $\underline{X}$ are changed, instead of just individual components. Also, we can now refine the solution by repeating the process. For example, we may first estimate α based on four proposed values (assuming four processors are available), followed by polynomial interpolation. We can then repeat the process using this proposed design as well as three additional nearby values. Finally, we may analyze the new proposed design as well as three nearby designs and pick the best from among them. This will improve the precision of this 'one-dimensional search' process, with the additional advantage that fewer search directions may be required.

When considering cost savings from parallel processing, it is important to understand that the primary saving is in the real time taken to solve a problem and the ability to solve design tasks that can only be considered if the computational power is available. In other words, four parallel processors are not as cheap as one serial processor. They just do the job faster. Also, in using parallel calculations in optimization, we limit our ability to use these processors during the function evaluation (analysis) phase unless we can use parallelism in a multilevel mode, either by distributing the analysis to computers with parallel processors or by virtue of multilevel parallelism on a single computer.

Typically, optimization requires about $10N + 40$ function evaluations to achieve a practical solution, where the $10N$ function evaluations are used for gradient computations and the 40 are used in the one-dimensional searches. This assumes a total of ten iterations. Assuming five processors are available, the time savings possible through parallel computations for these two components is shown in Fig. 1. Again, this assumes that the parallelism will not be used in the analysis. In general, if the machine architecture can be used to reduce the analysis times by a factor equal to the number of processors available, than it should be used for that purpose with optimization done in the usual manner.

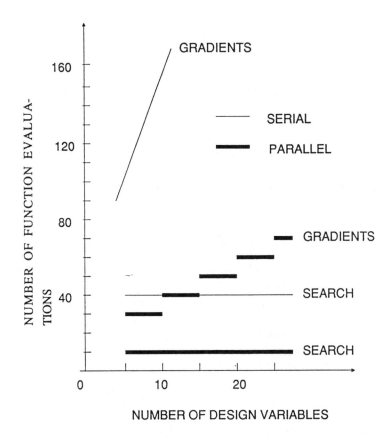

Figure 1. *Possible time savings.*

To summarize, those portions of the optimization task that require large scale analysis are good candidates for efficiency improvements through parallel computation. The internal process of determining the search direction, S, from the gradient information is usually a minor computational task compared to the analysis task. Therefore, the gradient computations and one-dimensional search process, which benefit most by performing analyses in parallel, offer the most direct opportunities for parallel computations.

Table 1. *Panel loads.*

Load Condition	N_x	N_y	N_{xy}
1	20,000.0	0.0	0.0
2	15,000.0	−15,000.0	5,000.0
3	−15,000.0	10,000.0	10,000.0
4	0.0	0.0	20,000.0

5 Examples

Here two optimization examples are offered to indicate the ease with optimization may be used as a 'black box' as well as the computational cost of doing so. Part II provides additional examples using recent approximation techniques to improve design efficiency.

5.1 Case 1: Design of a Simple Composite Panel

Design of a composite membrane panel is complicated, relative to isotropic structures, because of the increased number of design variables and relative complication of the constraints. Therefore, this is a natural candidate for optimization and has been addressed some time ago [13]. Consider the design of a simple panel shown in Fig. 2. The panel is assumed to be balanced (not a necessary assumption, but used here for simplicity) and symmetric about the midplane so that no bending is introduced by inplane loads. The panel must support multiple inplane loads, N_{xk}, $N_{y,k}$, $N_{xy,k}$ where k represents the loading condition. A typical set of loading conditions may be as given in Table 1.

The panel is a typical graphite epoxy with $(0, +45, -45, 90)$ fiber orientation. It is desired to design the panel for minimum thickness subject to strain limits in each ply. The ply longitudinal strain was required to be between +0.00857, the transverse compressive limit was taken as −0.0176 and the transverse tensile limit was taken as 0.00471. The shear strain limit was taken as 0.0184. Since the panel is required to be balanced, the thickness of the +45 and −45 degree plies are required to be equal. The optimum thicknesses are 0.16, 0.18, 0.18, 0.06 for the 0, +45, −45 and 90 degree plies, respectively. The controlling constraints are the transverse tensile allowable strain in ply 4 under loading condition 2, ply 1 under condition 3 and ply 3 under condition 4.

While this is perhaps the simplest composite design case, the optimum solution is neither obvious nor easily found without using numerical search methods. More importantly, this problem is easily solved on a personal computer in about one minute. Finally, if the materials, loads, or strain limits are changed, it is trivial

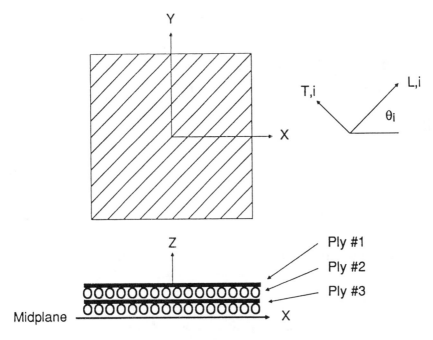

Figure 2. *Simple composite panel.*

to re-solve the problem for the new requirements. Also, more complicated failure criteria, as well as constraints on the panel stiffness can be included without increasing the computational effort.

5.2 Case 2: Construction Management

A contractor is considering two gravel pits from which he may purchase material to supply a project [14]. The unit cost to load and deliver the material to the project site is $5.00/yd^3 from pit 1 and $7.00/yd^3 from pit 2. He must deliver a minimum of 10,000 yd^3 to the site. The mix that he delivers must consist of at least 50 percent sand, no more than 60 percent gravel, nor more than 8 percent silt. (Note that there is some redundancy in the requirements.) The material at pit 1 consists of 30 percent sand and 70 percent gravel. The material at pit 2 consists of 60 percent sand, 30 percent gravel, and 10 percent silt. Determine how much material should be taken from each pit. Since the gravel from pit 1 does not contain the minimum amount of sand to meet project requirements, the contractor may not utilize the cheaper material exclusively. He must mix the material from pits 1 and 2 to produce the required proportions. We define the decision variables to be

$$X_1 = \text{amount of material taken from pit 1 (in cubic yards)}$$

$$X_2 = \text{amount of material taken from pit 2 (in cubic yards)}$$

The objective function is

Minimize $\quad C = 5X_1 + 7X_2$

Let $X_1 + X_2$ equal the total amount of standard mix delivered to the project site. The contractor must deliver at least 10,000 cubic yards, thus the delivery constraint is

$$X_1 + X_2 \geq 10,000$$

The mixture must contain at least 50 percent sand. The contractor may obtain the desired amount of sand by combining the materials from each pit.

$$0.3X_1 + 0.6X_2 \leq 0.5(X_1 + X_2)$$

The products $0.3X_1$ and $0.6X_2$ are the amounts of sand taken from pits 1 and 2, respectively. The term $0.5(X_1 + X_2)$ is the amount of sand in the mix. Similarly, the constraint on the amount of gravel to be delivered is

$$0.7X_1 + 0.3X_2 \leq 0.6(X_1 + X_2)$$

Finally, the constraint equation for silt is

$$0.1X_2 \leq 0.08(X_1 + X_2)$$

The minimum cost model may be written as

Minimize $\quad C = 5X_1 + 7X_2$

Subject to: $\quad X_1 + X_2 \geq 10,000$ $\qquad$ (delivery)

$$0.3X_1 + 0.6X_2 \geq 0.5(X_1 + X_2) \qquad \text{(sand)}$$

$$0.7X_1 + 0.3X_2 \leq 0.6(X_1 + X_2) \qquad \text{(gravel)}$$

$$0.1X_2 \leq 0.08(X_1 + X_2) \qquad \text{(silt)}$$

$$X_1, X_2 \geq 0$$

or in standard form,

Minimize $\quad F(X_1, X_2) = 5X_1 + 7X_2$

Subject to: $10,000 - X_1 - X_2 \le 0$

$2X_1 - X_2 \le 0$

$X_1 - 3X_2 \le 0$

$X_2 - 4X_1 \le 0$

$X_1, X_2 \ge 0$

This is a linear problem, although nonlinear functions can be used just as easily. The problem was solved in a few minutes using a personal computer, where most of the time was spent inputting the data. While it is not a structural design example, it serves well to demonstrate the generality of optimization for solution of everyday problems. The optimization solution is

6 Summary

The purpose here has been to present the basic concepts of numerical optimization methods. This technology has been around for a long time but is only now being widely recognized as a valid and efficient design tool. The development of these methods has matured to the point that they are relatively easy to use in modern engineering design. Optimization by itself doesn't save design time or money. There are fundamental natural laws that state that we will use all of the time and funds available for design. What it does is provide us with a tool to reach a high quality design much faster, allowing us to investigate a wide variety of alternatives. The final result is not a product that costs less in terms of design time and money, but a product that is superior. I suggest that this is what design is about. What is important is that it is better than what the competition can produce in the same time frame. This is an argument for the use of formal optimization methods that, if true, is compelling. There is now sufficient evidence to indicate that it is indeed true and so it can be safely stated that optimization is a design tool whose time has come!

Parameter	Initial value	Optimum
X_1	3,000.0	3,333.2
X_2	3,000.0	6,666.9
Objective	36,000.0	63,333.3
Maximum constraint	0.40	0.00

124

References

[1] Schmit, L. A., (1960), 'Structural Design by Systematic Synthesis,' *Proc. 2nd Conference on Electronic Computation*, American Society of Civil Engineers, New York, pp. 1249–1263.

[2] Garberoglio, J. E. and Song, J. O. and Boudreaux, W. L., (1982), 'Optimization of Compressor Vane and Bleed Settings,' ASME Paper No. 82-GT-81, *Proc. 27th International Gas Turbine Conference and Exhibition*, London, April 18–22.

[3] Vanderplaats, G. N., (1984), *Numerical Optimization Techniques for Engineering Design; with Applications*, McGraw-Hill.

[4] Fletcher, R. and Reeves, C. M., (1964), 'Function Minimization by Conjugate Gradients,' *British Computer Journal* **7**, 2, 149–154.

[5] Fiacco, A. V. and McCormick, G. P., (1968), *Nonlinear Programming: Sequential Unconstrained Minimization Techniques*, John Wiley and Sons, New York.

[6] Cassis, J. H. and Schmit, L. A., (1976), 'On Implementation of the Extended Interior Penalty Function,' *Int. J. Num. Methods for Engineering* **10**, 1, 3–23.

[7] Pierre, D. A. and Lowe, M. J., (1975), *Mathematical Programming via Augmented Lagrangians*, Applied Mathematics and Computation Series, Addison-Wesley, Reading, Mass.

[8] Vanderplaats, G. N., (1984), 'An Efficient Feasible Directions Algorithm for Design Synthesis,' *AIAA Journal* **22**, 11, (November).

[9] Kelley, J. E., (1960), 'The Cutting Plane Method for Solving Convex Programs,' *J. SIAM*, pp. 703–713.

[10] Powell, M. J. D., (1978), 'Algorithms for Nonlinear Constraints that Use Lagrangian Functions,' *Math. Prog.* **14**, 2, 224–248.

[11] Vanderplaats, G. N. and Sugimoto, H., (1985), 'Application of Variable Metric Methods to Structural Synthesis,' *Engineering Computations* **2**, 2, (June).

[12] Vanderplaats, G. N., (1988), 'COPES/DOT—A FORTRAN Program for Engineering Synthesis Using the EDO Design Optimization Tools,' Engineering Design Optimization, Inc., Santa Barbara, California, (February).

[13] Schmit, L. A. and Farshi, B., (1973), 'Optimum Laminate Design for Strength and Stiffness," *Int. Journal for Numerical Methods in Engineering* **7**, 4, 519–536.

[14] Ossenbruggen, P. J., (1984), *Systems Analysis for Civil Engineers*, John Wiley and Sons, New York.

An Assessment of Current Non-Linear Programming Algorithms for Structural Design, Part II: Some Recent Approximation Methods

G. N. Vanderplaats

Vanderplaats, Miura & Associates, Inc.
Goleta, California
United States of America

Abstract Approximation methods for efficient structural optimization have been the subject of active research for nearly fifteen years. The basic concept is to use the finite element analysis to create a high quality approximation to the original problem so that the expense of detailed analyses can be avoided during the actual optimization. Here, some recent approximation methods for application to structural optimization are presented which are applicable to stress and frequency constrained problems. The design variables may include shape as well as member sizing parameters.

1 Introduction

The purpose here is to present some recent research results in the area of efficient structural optimization. It is assumed that the underlying analysis is based on the finite element method. Examples are offered to demonstrate the methods.

2 Structural Optimization

Consider the general methodology of structural optimization (often called structural synthesis). Virtually all modern structural optimization is based on the finite element displacement method, where for linearly elastic analysis we have the familiar set of simultaneous equations to be solved;

$$\mathbf{K}\,\underline{U} = \underline{P} \tag{1}$$

B. H. V. Topping (ed.),
Optimization and Artificial Intelligence in Civil and Structural Engineering, Volume I, 127–141.
© 1992 *Kluwer Academic Publishers. Printed in the Netherlands.*

where $\mathbf{K}$ is the master stiffness matrix, $\underline{U}$ is the vector of joint displacements and $\underline{P}$ is the vector of joint loads. The load and displacement vectors usually contain multiple columns, representing separate loading conditions. Equation 1 is of course only the most restricted case and, in general, may be expanded to include mass, dynamic and aeroelastic terms. Also geometric and material nonlinearities as well as time dependent loads may be included. Thus, it is clear why we chose to work with such a standard. While the discussion here is limited, it should be remembered that the state of the art is well beyond this in terms of the classes of problems that can be solved. Effective structural optimization requires that the gradients (sensitivities) of the structural responses in terms of the design variables be calculated. Since member stresses are based on the displacements calculated in Eq. 1, once displacement sensitivities are available, sensitivity of stresses is calculated at the element level. It is interesting to note that, prior to 1965, all gradients were calculated by finite difference. This is simply because no one had observed that a simple chain rule differentiation of Eq. 1 yields the required information;

$$\frac{\partial \mathbf{K}}{\partial X_i} \underline{U} + \mathbf{K} \frac{\partial \underline{U}}{\partial X_i} = \frac{\partial \underline{P}}{\partial X_i} \tag{2}$$

from which

$$\frac{\partial \underline{U}}{\partial X_i} = \mathbf{K}^{-1} \left(\frac{\partial \underline{P}}{\partial X_i} - \frac{\partial \mathbf{K}}{\partial X_i} \underline{U} \right) \tag{3}$$

Noting that $\mathbf{K}^{-1}$ is already available in decomposed form, the solution of Eq. 3 requires only the information in the brackets and then solution of the original equations for a new set of 'loads.' Since this first observation, analytic or semi-analytic (where $\frac{\partial \mathbf{K}}{\partial X_i}$ is calculated as a finite difference subproblem) gradients have been calculated for system buckling, frequency, dynamic response, aeroelastic response, and nonlinear time dependent response as examples [1,2].

Efficient structural optimization does not require that the finite element analysis and gradient computations be done whenever the optimization program requests them. What we actually do is perform a detailed analysis at the beginning, together with an evaluation of all of the constraint functions. The constraints are then sorted the only those that are critical or near critical are retained for the current design step. Typically, this requires only retaining about $3N$–$5N$ constraints, where N is the number of design variables. Gradients are then calculated for the objective and this retained set of constraints. This information is then used to create a 'high quality approximation' to the original problem.

For example, the simplest approximation would be a Taylor series expansion of the form;

$$F(\underline{X}) \approx F(\underline{X}^0) + \underline{\nabla} F(\underline{X}^0) \cdot (\underline{X} - \underline{X}^0) \tag{4}$$

where $F(\underline{X})$ is any objective or constraint function we wish to approximate and $\underline{\nabla}$ is the gradient operator. If we create such an approximation for the objective function and all retained constraints, we could then solve this with optimization without continually calling the finite element analysis code for function and gradient information. After an approximate optimization is complete, we would repeat the process until it has converged to the final nonlinear optimum. Such an approach is referred to as *Sequential Linear Programming* and experience has shown it to be a reasonably efficient method for engineering applications. However, in structural optimization, we have additional insights that allow us to make a better approximation to the original problem. Consider for example a simple truss element where we wish to calculate the stress. Then

$$\sigma = \frac{P}{A} \tag{5}$$

where σ is the stress, P is the force in the element, and A is the cross-sectional area, which is to be minimized. This leads to the simple optimization problem;

Minimize $\quad A$ (6)

Subject to; $\quad \dfrac{P}{A} \leq \bar{\sigma}$ (7)

While the objective function is linear, the constraint is clearly nonlinear in the design variable, A, so a direct linearization of this would not be accurate. However, we are free to choose another variable for design, so long as we link it to the cross-sectional area. Therefore, let the design variable be

$$X = \frac{1}{A} \tag{8}$$

from which the stress is now linear in X;

$$\sigma = PX \tag{9}$$

and the optimization problem becomes;

Minimize $\quad \dfrac{1}{X}$ (10)

Subject to; $\quad PX \leq \bar{\sigma}$ (11)

Here, we have made a trade where the objective is now nonlinear, but explicit, and the constraint is linear. Creating a linear approximation in X to this constraint is precise for determinate structures and is a significantly improved approximation for indeterminate structures. Therefore, relatively large changes may be made to the design during the approximate optimization stage before a new finite element analysis is needed. Also, if the objective function is the weight of the structure, it can be easily evaluated in its nonlinear form during the approximate optimization.

This basic concept is referred to as an approximation technique and was first introduced in the mid 1970's [3]. The methods are applicable to basic member sizing where the element stiffness matrix is the product of the original design variable (area or thickness) to some power, and a constant matrix. For structures where this is not true, such as beam elements or when geometric variables are considered, these methods are not directly applicable. One approach to the more general problem is to create a conservative, convex, approximation to the original problem [4]. The basic concept is that, by observing the algebraic sign on each component of the function gradient, an automatic switching may be performed between direct variables (Eq. 6) and reciprocal variables (Eq. 10). This creates an approximation that is conservative relative to a direct linearization and usually provides improved efficiencies.

Whatever approximation is used, the overall design process is summarized as follows:

1. Input user defined data.

2. Analyze the initial design.

3. Evaluate the objective function and constraints.

4. Sort the constraints and retain those that are critical or near critical.

5. Calculate the sensitivity of the responses with respect to some intermediate variables (e.g. Section Properties).

6. Create and solve the approximate optimization problem based on these sensitivities.

7. Analyze the proposed design.

8. Evaluate the objective function and constraints.

9. Check for convergence to the optimum. If satisfied, exit. Otherwise go to step 4.

The effect of this simple transformation to a high quality approximation is approximately two orders of magnitude improvement in design efficiency over previous methods.

3   Stress Approximations

Research in efficient approximation methods has continued, and recent results relative to stress constraints (the most common constraints considered in structural design) have demonstrated significant improvements over those obtainable by the approach of Eqs. 8 and 9 [5]. Consider again a simple bar element, but now instead of creating an approximation to the stress, approximate the force in the member;

$$P(X) \approx P(X^0) + \underline{\nabla} P(X^0) \cdot (X - X^0) \tag{12}$$

where here the design variable, X, is the original member size, A. Now proceed with the approximate optimization as before, except when stresses are needed they are calculated from

$$\sigma = \frac{P(X)}{A} \tag{13}$$

Using this approach, the approximate stress is nonlinear in the design variable, A, but is still explicit and easily evaluated. However, noting that displacements are proportional to load and stresses are related to the rate of change of displacements, it follows that Eq. 13 is a higher order approximation than Eq. 9. Also, it is no more difficult to calculate the sensitivity of element forces than to calculate stress sensitivities.

   For truss structures, this modification is almost trivial, but dramatically improves the design efficiency. For frame structures, this allows us to calculate member forces in terms of section properties, but treat the individual dimensions of the beam as design variables during the approximate optimization [5]. Thus for beam structures, the approximate optimization uses the following steps;

1. When the optimizer requires stresses, the member dimensions are provided to the approximate analysis program.

2. The section properties are calculated exactly from the member sizes.

3. The member end forces are approximated in a form like Eq. 12, except that the design variables are replaced by section properties since these provide a better approximation.

4. Finally the stresses are recovered in the usual FEM manner and returned to the optimizer.

   Note that this process requires the solution of numerous identical subproblems that are ideally suited for parallel processing.

   For more complicated elements such as three dimensional continuum elements, an additional step is required, which is also well suited for parallel computation. Having approximated the element nodal forces, it is necessary to determine the internal element displacement state, from which stresses are recovered. This entails first detecting and removing the rigid body degrees of freedom from the element and then solving a reduced set of equations at the element level. In the case of a 20 node brick element, this would require the solution of a 54 DOF set of equations. Following this the needed stresses are recovered in the usual manner [7].

To date, this method has been tested for truss, beam, and three dimensional continuum problems. It is interesting that this approach works especially well for shape optimization of trusses and frames and three dimensional continuum shape optimization. However, in the case of continuum structures, considerable computational effort is required during the approximate optimization phase which diminishes it's attractiveness here. Finally, for frame structures, calculation of sensitivities with respect to member properties, even though physical dimensions are used as design variables, has been found to lead to efficient and reliable results.

4 Frequency Approximations for Frame Structures

Frequency (eigenvalue) constraints pose a relatively difficult problem for approximation techniques. The source of this is seen by considering a simple one degree of freedom system where

$$\omega = \sqrt{\frac{K}{M}} \tag{14}$$

If the stiffness and mass are related to the design variables in the same manner, it is seen from Eq. 14 that the frequency will remain constant. Alternatively, if there are large non-structural masses, the denominator of Eq. 14 is more nearly constant and the frequency is proportional to the square root of the stiffness. Thus, as a first step, we can avoid some nonlinearity by constraining the eigenvalue rather than the frequency, since

$$\lambda = \frac{K}{M} \tag{15}$$

This simple observation, together with the idea that the approximation should be with respect to member properties, rather than the physical design variables, leads to acceptable efficiency for frame structures. The overall optimization algorithm is now much the same as for stress constraints so the approximate optimization proceeds in the following steps [6].

1. When the optimizer requires frequencies, the member dimensions are provided to the approximate analysis program.

2. The section properties are calculated exactly from the member sizes.

3. The eigenvalues are approximated in a form like Eq. 12, except that the design variables are replaced by section properties since these provide a better approximation.

4. Finally the frequencies are recovered and returned to the optimizer.

This simple algorithm has been found to be acceptably efficient for frame structures. Also, recently Canfield [8] has presented a new method based on approximation of the Rayleigh Quotient. In the case of frame structures, this new method will further enhance the approach given above. Furthermore, Canfield's method should significantly improve the quality of the approximation for a wide range of structures, and is considered a further step in the quest for high quality approximation techniques.

5   Design Examples

Examples are offered here to demonstrate the methods discussed above. The first is a relatively simple planar truss, demonstrating the state of the art in discrete element shape optimization. The second example is of a portal frame, demonstrating efficient sizing optimization of this type of structure. The third example is of a planar arch made of beam elements and designed for both sizing and shape and the fourth example is the three dimensional continuum shape optimization of a connecting rod. The detailed loading, material, and design/analysis information is provided in the references.

5.1   Case 1: Sizing and Shape Optimization of a Planar Tower

Figures 1a and 1b show the design of a planar tower designed to support three separate loading conditions [9]. Symmetry was imposed and there are a total of 27 independent area (sizing) design variables and 17 coordinate design variables. Figure 1a gives the initial configuration and Fig. 1b shows the final configuration. Constraints included stress limits and Euler buckling limits in the members. This design required a total of 11 analysis/gradient combinations with the associated solution of the approximate optimization problem; an order of magnitude improvement over previous methods for truss shape optimization. This example indicates the general state of the art in truss configuration optimization.

5.2   Case 2: Grillage Optimization

The grillage shown in Fig. 2 has been used as a common example in frame optimization. The structure is symmetric about the two vertical mid-planes and the elements are made of box beams as shown in Fig. 2. Stress, local buckling and displacement constraints are imposed and the beam dimensions are sought to minimize the weight of the structure. The iteration history is also shown in Fig. 2, where it is seen that convergence is achieved with only six detailed finite element analyses [5].

5.3   Case 3: Design of a Planar Arch for Optimum Shape

Figure 3 shows the results of the design of a planar arch for minimum weight [10]. The cross-sectional area and the nodal coordinates are found for minimum weight, subject to combined stress constraints. The bending moment of inertia was taken as a function of the area (this is not essential, but was a convenience for this preliminary study). The arch was fixed at each end and was required to support a uniform load. The optimum design was found with five detailed finite element analyses and the optimum shape was shown to be a parabola, which was intuitively expected, but now numerically proved.

5.4   Case 4: Design of a Continuum Structure for Optimum Shape

Figure 4 shows the design of an engine connecting rod [7]. This structure was modeled with 20-node solid elements and a 1/4th model used just over 2000 displacement degrees of freedom. Eight design variables were used to define the geometry and over 900 stress constraints were considered. As with the previous examples, this is a result of recent research wherein the member forces are approximated. The optimum was achieved after six design iterations. The initial and final shape are shown in Fig. 4. The iteration history is compared with that of reference 12 which used a direct stress approximation method. This design required just under one hour on a supercomputer. Some modifications in the optimization algorithm were made which reduced this time to about 20 minutes [11]. The research code used to solve this problem is based on the SAP IV program [13] and uses finite difference gradient calculations. For the real design of a connecting rod, several loading conditions must be considered and more geometric detail is needed. Also, the proper analysis of the interface between the connecting rod and the crankshaft requires nonlinearities be considered. Thus, even for this 'component' design problem, the computational needs will tax the power of our best supercomputers.

5.5   Case 5: Design of a Frequency Constrained Frame Structure

Figure 5 shows the idealized model of a helicopter tail section. The structure is made of tubular elements and their dimensions are found to minimize the structural weight subject to strength and frequency constraints. As seen from the iteration history, the results of the present method compare favorably with the best results of a recent method based on what is referred to as a 'convoluted Taylor series expansion' [14].

6 Summary

The state of the art in structural optimization is becoming relatively well developed and optimization is now included or is being added to most major commercial finite element programs. Much of the progress in this field has come from improved optimization algorithms and careful formulation of the problem based on approximation techniques. The purpose here has been to provide a brief overview of some recent developments in structural optimization. Relative to the size and complexity of the problems we expect to solve in the foreseeable future, the examples presented here are considered small and simple. Reference 15 offers a state of the art review of structural optimization up to the 1980 timeframe. There it was noted that, while is is difficult to predict the future, it was easy to state that at least structural optimization has a future. Today a part of that future is past and it can be said with some assurance that on the one hand structural optimization is a matured or at least maturing technology and on the other that past successes have only served to identify the need for continued research as well as the need for immense computational power. The key to success in such a computationally intensive area lies in the hands of continued research in efficient approximation methods and advances in computer technology.

References

[1] Haftka, R. T. and Kamat, M. P., (1985), *Elements of Structural Optimization*, Martinus Nijhoff, The Hague.

[2] Haug, E. J., Choi, K. K. and Komkov, V., (1985), *Design Sensitivity Analysis of Structural Systems*, Academic Press.

[3] Schmit, L. A. and Miura, H., (1976), 'Approximation Concepts for Efficient Structural Synthesis,' NASA CR-2552, March.

[4] Fleury, C. and Braibant, V., (1984), 'Structural Optimization—A New Dual Method Using Mixed Variables,' LTAS Report SA-115, University of Liege, Liege, Belgium, March.

[5] Vanderplaats, G. N. and Salajegheh, E., (1989) 'A New Approximation Method for Stress Constraints in Structural Synthesis,' *AIAA Journal*, **27**, 3, 352-358.

[6] Vanderplaats, G. N. and Salajegheh, E., (1988), 'An Efficient Approximation Technique for Frequency Constraints in Frame Optimization,' *Int. Journal for Numerical Methods* **26**, 1057–1069.

[7] Kodiyalam, S. and Vanderplaats, G. N., (1989), 'Shape Optimization of 3D Continuum Structures Via Force Approximation Technique,' *AIAA Journal*, **27**, 9, 1256-1263.

[8] Canfield, R. A., 'An Approximation Function for Frequency Constrained Structural Optimization,' *Proc. Symposium on Recent Advances in Multidisciplinary Analysis and Optimization*, NASA CP 3031, Part 2, pp. 937–953.

[9] Hansen, S. R. and Vanderplaats, G. N., (1990), 'An Approximation Method for Configuration Optimization of Trusses,' *AIAA Journal*, **28**, 1, 161-172.

[10] Vanderplaats, G. N. and Han, S. H., (to appear), 'Arch Shape Optimization Using Force Approximation Methods,' (submitted to *J. Structural Optimization*).

[11] Vanderplaats, G. N. and Kodiyalam, S., (1989), 'A Two Level Approximation Method for Stress Constraints in Structural Optimization,' *Proc. 1989 AIAA/ASME/ASCE/AHS Structures, Structural Dynamics and Materials Conference*, Mobile Alabama, April.

[12] Yang, R. J. and Botkin, M. E., (1987), 'A Modular Approach for Three Dimensional Shape Optimization of Structures,' *AIAA Journal* **25**, 3, March, 492–497.

[13] Bath, K. J., Wilson, E. L. and Peterson, F. E., 'SAP-IV: A Structural Analysis Program for Static and Dynamic Response of Linear Systems,' EERC, University of California, Berkeley, CA.

[14] Woo, C. H., (1986), 'Space Frame Optimization Subject to Frequency Constraints,' *Proc. 27th AIAA/ASME/ASCE/AHS Structures, Structural Dynamics and Materials Conference*, San Antonio, TX, May 19–21, AIAA Paper 86-0877.

[15] Vanderplaats, G. N., (1982), 'Structural Optimization—Past, Present and Future,' *AIAA Journal* **20**, 7, (July).

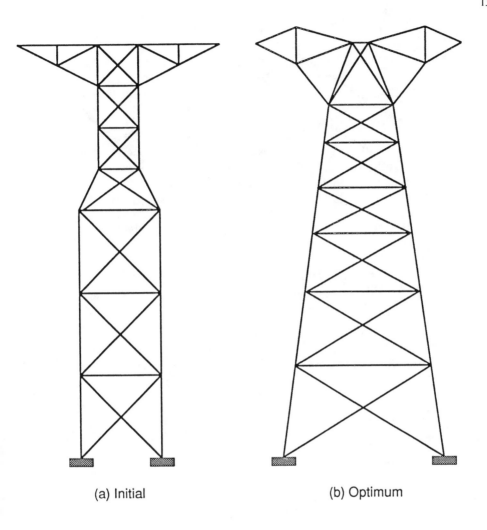

(a) Initial   (b) Optimum

Figure 1. *47-bar tower.*

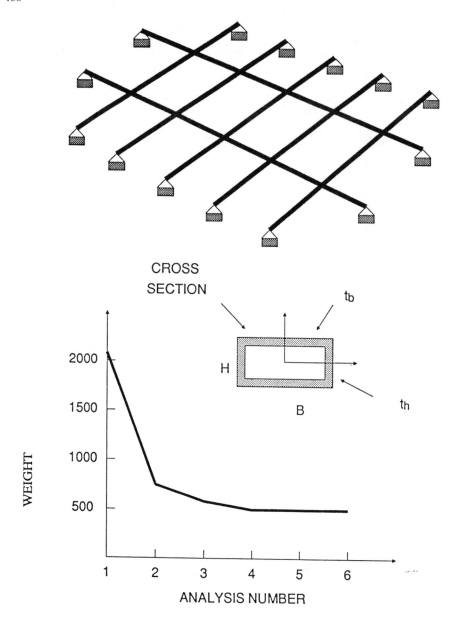

CROSS
SECTION

Figure 2. *Grillage.*

139

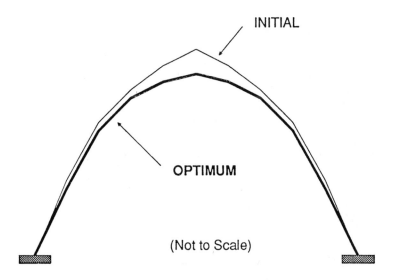

Figure 3. *Fixed planar arch.*

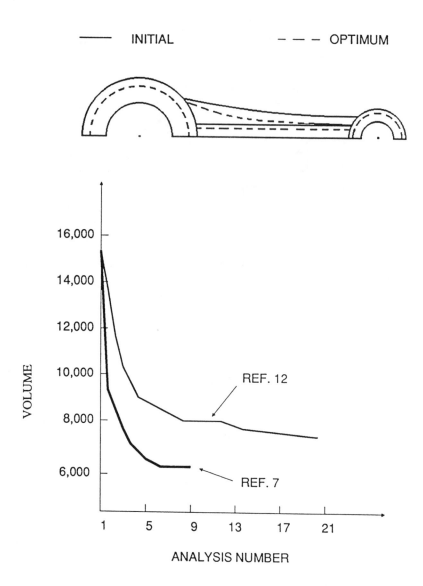

Figure 4. *Engine connecting rod.*

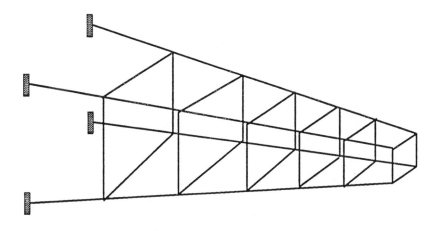

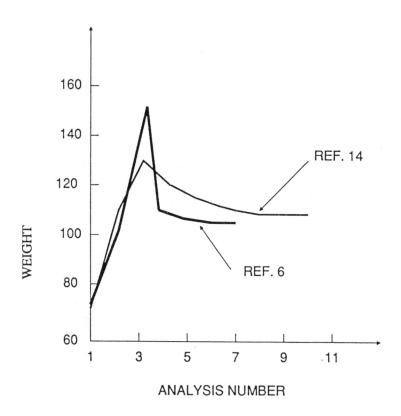

Figure 5. *Helicopter tail section.*

Classification and Comparison of LP Formulations for the Plastic Design of Frames [1]

T. K. H. Tam and A. Jennings
Department of Civil Engineering
The Queen's University of Belfast
Belfast
United Kingdom

Abstract The investigation of the design of structural frames to meet a collapse criterion according to rigid-plastic theory leads, under conventional assumptions, to a mathematical model requiring solution by linear programming (LP). However, there are many such different formulations in the literature. The main purpose of this paper is to classify all such methods into four types, each with an associated dual formulation. Relationships between the types of method are also presented and there is a brief discussion of their computational efficiencies and ease of automation.

Introduction

Alongside the development of plastic theory as a means of predicting the collapse load of steel frames, considerable effort was devoted to hand computational methods. One procedure was to identify all the possible collapse mechanisms and then, from virtual work considerations, establish the lowest load at which any of these will be activated. This has been described as the 'upper bound' or 'unsafe' method because, if all the possible collapse mechanisms have not been identified, the predicted collapse load may be above the theoretical value and hence unsafe to use in an analysis or design context. Alternative 'lower bound' or 'safe' procedures were also developed which involve examining force systems in equilibrium with the applied loads and which do not violate the yield conditions. The correct theoretical collapse load is the minimum of all possible unsafe solutions or the maximum of all possible safe solutions [1–3].

---

[1]This paper was originally published in *Engineering Structures* **11**, 163–178, 1989. and is reproduced here courtesy of Butterworths.

B. H. V. Topping (ed.),
Optimization and Artificial Intelligence in Civil and Structural Engineering, Volume I, 143–188.
© 1992 *All Rights Reserved. Printed in the Netherlands.*

Foulkes [4] was the first to recognise that, from the mathematical standpoint, plastic analysis reduces to a linear programming problem. This knowledge enables computational techniques developed for problems in other areas (in particular operational research; see, for instance, Taha [5]) to be utilized for the solution of problems in plastic analysis. One of the important aspects of plastic theory is that it is easily adopted to assignments in design as well as analysis. Where it can be assumed that the weight of each member of a frame is proportional to its plastic moment, the process of determining the minimum weight design for a frame of specified geometry can be shown to reduce to a linear programming exercise [4,6–11].

Although the possibility of obtaining such minimum weight designs automatically using plastic theory and linear programming has been appreciated for over 30 years, such techniques have not been used much in practice until recently. Many of the difficulties and prejudices relating to the use of plastic methods are disappearing, so it is likely that they may be much more heavily utilized in future. One reason for this is the vast improvements in computing power and availability of computers to design engineers that have taken place. Another reason is the greater knowledge of structural behaviour and the acceptance of plastic theory. The development of interactive user-friendly software and expert systems has also been a help, as has the move from safety factor to limit state design techniques [12].

Whereas the methods of formulating the mathematical equations relevant to elastic analysis of structural frames are very well presented in the literature, the formulations for plastic analysis and design are not well documented. This paper is aimed at clarifying this situation.

Problem Types in Analysis and Design

Figure 1a shows a fixed-base portal frame subject to proportional loading in which λ is the load factor and for which all members have the same cross-section with plastic moment $\overline{m}$. For this example it is usual, when employing the unsafe method, to analyse the first three of the four mechanisms shown in Fig. 2. A mechanism will be stable at any particular load only if the internal energy absorbed by the hinges is greater than or equal to the external work done in an incremental movement of the mechanism. For the four mechanisms (a)–(d) these conditions give:

$$
\left.
\begin{aligned}
&\text{(a) } \overline{m} \geq 150 \\
&\text{(b) } \overline{m} \geq 60 \\
&\text{(c) } \overline{m} \geq 140 \\
&\text{(d) } \overline{m} \geq 60
\end{aligned}
\right\} \tag{1}
$$

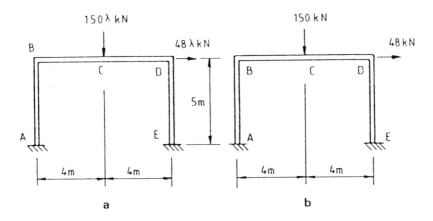

Figure 1. *A portal frame example with uniform member sectional properties: (a) the analysis problem; (b) the design problem.*

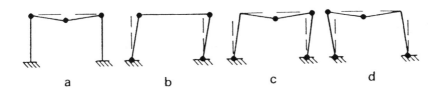

Figure 2. *Different possible collapse mechanisms.*

The critical mechanism is therefore (a), giving the first constraint as being the operative one.

If the value of plastic moment is given, the problem is one of analysis in which there is an upper bound of $\overline{m}/150$ for the sustainable load factor. On the other hand, if $\lambda = 1$ (as in Fig. 1b), the problem is one of design with this equation giving a lower bound of 150 kNm for the necessary plastic moment $\overline{m}$. The terminology 'upper bound' and 'lower bound' is thus confusing if design as well as analysis examples are being considered. Furthermore the terminology 'safe' and 'unsafe' is not appropriate in situations where the methods are pursued to their limit, thus obtaining the true theoretical solution. Instead, the terminology 'kinematic' will be used for methods in which virtual movements are examined and 'static' for methods in which equilibrium is utilized.

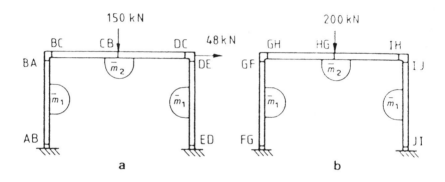

Figure 3. *A portal frame example using two types of member; (a) one loading case; (b) an additional loading case.*

Although the conversion from an analysis to a design problem is a trivial switch when considering the frame example shown in Fig. 1, other types of problem in analysis and design do not necessarily convert in this way. Consider the frame shown in Fig. 3a, originally solved by Heyman [13], in which the columns have a cross-section with plastic moment $\overline{m}_1$ but the beam cross-section has a different plastic moment $\overline{m}_2$. The usual problem in analysis is, given the values $\overline{m}_1$ and $\overline{m}_2$, to determine the load factor λ for collapse where λ is applied to both loads. Although there is a corresponding design problem in which the designer wishes to identify the minimum plastic moments for the members but keeping a constant ratio between $\overline{m}_1$ and $\overline{m}_2$, this is unlikely to be required very often. The more usual design case is where $\overline{m}_1$ and $\overline{m}_2$ are to be kept as unknowns with, in the absence of any other optimization criteria, their values to be chosen in such a way as to minimize the self-weight of the structure. Assuming that the weight of a cross-section is proportional to its plastic moment, the self-weight of the frame of Fig. 3a can be optimized by minimizing the function

$$w = l_1 \overline{m}_1 + l_2 \overline{m}_2 \tag{2}$$

where $\{l_1, l_2\} = \{10, 8\}$ gives the total lengths of column and beam members. (Note that w is not a true weight because it has the units of force $\times$ distance^2). This design problem has no obvious parallel in analysis.

Another difference between this minimum weight design and the problems in analysis and design discussed earlier is that a large number of cross-sections need to be taken into account. For the frame of Fig. 1 it is only necessary to allow for plastic hinges forming at any of the five joints A, B, C, D or E, whereas in the design of the frame, Fig. 3a, a plastic hinge forming at joint B or D will form at the end of the weakest member impinging on that joint. Because in minimum weight design it is not known a priori which of the two members is weaker, the possibility of plastic hinges forming in either member needs to be considered. The number of critical sections is therefore increased from five to seven, which may be designated AB, BA, BC, CB, DC, DE and ED.

A further complication arises when a frame needs to be designed to withstand two or more loading cases. In the minimum weight design (as opposed to the other analyses and designs discussed), it is not in general sufficient to optimize the structure for the loading cases separately. The way to obtain the theoretical optimum design in this case is to incorporate the various loading cases into a single optimization process. For instance, if the frame shown in Fig. 3a may also carry a vertical load of 200 kN at C, the required optimization is for the disjoint six-member frame shown in Figs. 3a and 3b. The links between the two parts of the compound frame is only through the common plastic moments $\overline{m}_1$ and $\overline{m}_2$.

The various mathematical models will be discussed in the first instance in relation to the design examples shown in Fig. 1b, which is very closely related to a standard type of analysis. By considering the design exercise it is possible to formulate the mathematics in a generalized way which permits immediate extension to the type of minimum weight design shown in Fig. 3. The other main case, being the load factor analysis of frames such as in Fig. 3a which have members of different sizes, can be obtained by a slight modification to the presented techniques.

Linear Programming and Duality

Consider the following mathematical form of a linear programming problem:

Minimize the function

$$h = c_1^T x_1 + c_2^T x_2 \tag{3a}$$

subject to

$$
\left.\begin{array}{l}
A_{11}x_1 + A_{12}x_2 \geq b_1 \\[2mm]
A_{21}x_1 + A_{22}x_2 = b_2
\end{array}\right\} \tag{3b}
$$

with

$$
x_1 \geq 0 \tag{3c}
$$

Here the variables x_1 are classified as 'constrained' as opposed to x_2 which are 'unconstrained' and the coefficient matrices A_{11}, A_{12}, A_{21} and A_{22} are in general rectangular. It is possible to convert all the equalities to constraints and all the unconstrained variables to constrained variables by expanding constraints (3b) in the form

$$
\left.\begin{array}{l}
A_{11}x_1 + A_{12}x_2^+ - A_{12}x_2^- \geq b_1 \\[2mm]
A_{21}x_1 + A_{22}x_2^+ - A_{22}x_2^- \geq b_2 \\[2mm]
-A_{21}x_1 - A_{22}x_2^+ + A_{22}x_2^- \geq -b_2
\end{array}\right\} \tag{4a}
$$

where

$$
x_2 = x_2^+ - x_2^- \tag{4b}
$$

such that

$$
x_2^+ \geq 0 \text{ and } x_2^- \geq 0 \tag{4c}
$$

Thus the problem described by system (3) can always be specified in the form:
Minimize

$$
h = c^T x \tag{5a}
$$

subject to

$$
Ax \geq b \tag{5b}
$$

and

$$
x \geq 0 \tag{5c}
$$

by setting

$$
c^T = \begin{bmatrix} c_1^T & c_2^T & -c_2^T \end{bmatrix} \tag{6a}
$$

$$
A = \begin{bmatrix}
A_{11} & A_{12} & -A_{12} \\
A_{21} & A_{22} & -A_{22} \\
-A_{21} & -A_{22} & A_{22}
\end{bmatrix} \tag{6b}
$$

$$b^T = \begin{bmatrix} b_1^T & b_2^T & -b_2^T \end{bmatrix} \tag{6c}$$

with

$$x^T = \begin{bmatrix} x_1^T & (x_2^+)^T & (x_2^-)^T \end{bmatrix} \tag{6d}$$

Below it will be shown that mathematical models arising from static formulations can always be specified in the form of system (3) and hence also in the form of system (5). In such cases the minimizing function h is the product of plastic moments contained in x and coefficients in c, which may be either non-dimensional or comprising member lengths.

It is a well known feature of linear programming that every solvable formulation has a dual [14,15]. The dual formulation for system (5) is:

Maximize

$$g = b^T y \tag{7a}$$

subject to

$$A^T y \le c \tag{7a}$$

and

$$y \ge 0 \tag{7a}$$

where, if $\bar{x}$ and $\bar{y}$ are the optimal set of variables for systems (5) and (7) respectively and $\bar{h}$ and $\bar{g}$ are the optimum values for the optimizing functions,

$$\bar{h} = c^T \bar{x} = \bar{y}^T A \bar{x} = \bar{y}^T b = \bar{g} \tag{8}$$

By solving either one of the two problems, systems (5) or (7), both the primal and dual solutions can be obtained. This is because the solution $\bar{y}$ of the maximization problem is given by the negative of the 'simplex multipliers' in the solution to the minimization problem. Conversely, the solution $\bar{x}$ of the minimization problem is given by the simplex multipliers in the solution to the maximization problem. (Simplex multipliers are also known as shadow prices, they appear in the coefficients of the slack and artificial variables of the objective function at the optimum).

The dual of each static formulation is a kinematic formulation in which the maximizing function, g, is the product of bending moments contained in b and rotations contained in y. For system (7) to have consistent units these rotations will be non-dimensional when c is non-dimensional but have units of length when c contains member lengths. In the latter case this apparent anomaly is acceptable because it is only the relative magnitudes of the rotations that are required in assessing collapse mechanisms.

Substituting equations (6) into the dual formulation system (7), it can be shown that the generalized form of the dual equivalent to system (3) is:

Maximize the function

$$g = b_1^T y_1 + b_2^T y_2 \tag{9a}$$

subject to

$$\begin{aligned} A_{11}^T y_1 + A_{21}^T y_2 &\le c_1 \\ A_{12}^T y_1 + A_{22}^T y_2 &= c_2 \end{aligned} \tag{9b}$$

with

$$y_1 \ge 0 \tag{9c}$$

It may be noted that the presence of equalities in the primal formulation gives rise to unconstrained variables in the dual and vice versa. This rule holds even if the kinematic formulation is considered as primal, in which case the static formulation is its dual.

Assumptions and Theorems

The fundamental assumption of simple plastic theory is that the material behaves in a rigid-plastic manner with no deformation taking place prior to yield. Where a member cross-section comprising such a material is loaded in pure bending, it will not distort until the plastic moment is reached, at which moment any value of curvature will be possible with the cross-section acting as a plastic hinge. The effects of shear force and axial load on the plastic moment have been ignored.

Although not fundamental to the theory, the assumption has also been made that members have equal plastic moments when loaded with positive or negative bending moments. This will normally be satisfied, the exception being a member which is unsymmetric about the bending axis and whose material has different yield stresses in tension and compression.

As a result of these assumptions it is found that the frame does not move until such time as a mechanism develops causing collapse. The investigation is uncomplicated by elastic deformations, which are ignored. Also the deformation mechanisms can be considered to be incremental and conforming to small deflection theory. It is assumed that buckling of any form does not take place. It is also assumed that the frame is loaded proportionally so that incremental collapse is not possible. Only straight members and concentrated loads are considered, in which case plastic hinges can be supposed to occur only at member ends or load points. In the case of design it is assumed that there is a continuous spectrum of available cross-sections for which the plastic moment is proportional to the self-weight.

The three basic theorems of plastic theory may be stated as follows:

1. there is a unique collapse load factor, λ_c;

2. if a mechanism requires a load factor λ to activate it then, if it is a correct collapse mechanism, $\lambda = \lambda_c$, otherwise $\lambda > \lambda_c$ (this is used in kinematic formulations);

3. if an equilibrium state at load factor λ has bending moments throughout the frame all less than or equal to the relevant plastic moments, then $\lambda \leq \lambda_c$ (this is used in static formulations).

The Method of Joint Equilibrium

The method of joint equilibrium due to Livesley [16,17] can be easily automated from standard types of input data for specifying frame geometries on a computer. It does, however, produce a large tableau (The tableau is the matrix requiring to be operated on during solution of the linear programming problem).

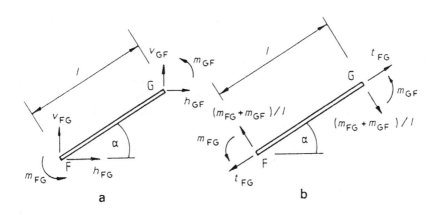

Figure 4. *Equivalent force system for a typical member: (a) end forces; (b) independent internal forces.*

In developing the equilibrium equations for a frame it is convenient to consider all loading points to be joints so that members are only loaded at their ends. Consider a typical member FG of inclination α and length l joining joints F and G and subject to end loads $q_{FG} = \{h_{FG}, v_{FG}, m_{FG}\}$ and $q_{GF} = \{h_{GF}, v_{GF}, m_{GF}\}$ at F and G respectively as shown in Fig. 4a. Because these forces need to satisfy three equilibrium equations, only three are independent. Considering, therefore,

$p_{FG} = \{t_{FG}, m_{FG}, m_{GF}\}$ as independent internal forces as shown in Fig. 4b, the end loads may be related to these according to

$$
\begin{bmatrix}
-\cos\alpha & -\sin\alpha/l & -\sin\alpha/l \\
-\sin\alpha & \cos\alpha/l & \cos\alpha/l \\
0 & 1 & 0 \\
\cos\alpha & \sin\alpha/l & \sin\alpha/l \\
\sin\alpha & -\cos\alpha/l & -\cos\alpha/l \\
0 & 0 & 1
\end{bmatrix}
\begin{bmatrix}
t_{FG} \\
m_{FG} \\
m_{GF}
\end{bmatrix}
=
\begin{bmatrix}
h_{FG} \\
v_{FG} \\
m_{FG} \\
h_{GF} \\
v_{GF} \\
m_{GF}
\end{bmatrix}
\tag{10}
$$

which may be expressed in submatrix form as

$$
\begin{bmatrix} H_{FG} \\ H_{GF} \end{bmatrix} [p_{FG}] = \begin{bmatrix} q_{FG} \\ q_{GF} \end{bmatrix}
\tag{11}
$$

Corresponding equations for the complete frame may be developed by using joint equilibrium equations of the form

$$
q_F = \sum_G q_{FG}
\tag{12}
$$

where $q_F = \{h_F, v_F, m_F\}$ are the external loads applied at a typical joint F and summation is for all members FG impinging on this joint. This gives

$$
Hp = q
\tag{13}
$$

where q is the vector of external joint loads whose typical set is q_F, p is a vector comprising all the internal forces whose typical set is p_{FG} and H is a force transformation matrix. This matrix may be constructed from a null matrix by adding the submatrices H_{FG} and H_{GF} corresponding to each member into the appropriate locations determined by the member number (for columns) and the impinging joint numbers when not built-in (for rows). For the portal frame shown in Fig. 1b, equation (13) takes the form shown below

$$
\left[
\begin{array}{ccc|ccc|ccc|ccc}
\multicolumn{3}{c|}{AB} & \multicolumn{3}{c|}{BC} & \multicolumn{3}{c|}{CD} & \multicolumn{3}{c}{DE} \\
0 & \frac{1}{5} & \frac{1}{5} & -1 & 0 & 0 & & & & & & \\
1 & 0 & 0 & 0 & \frac{1}{4} & \frac{1}{4} & & & & & & \\
0 & 0 & 1 & 0 & 1 & 0 & & & & & & \\
\hline
 & & & 1 & 0 & 0 & -1 & 0 & 0 & & & \\
 & & & 0 & -\frac{1}{4} & -\frac{1}{4} & 0 & \frac{1}{4} & \frac{1}{4} & & & \\
 & & & 0 & 0 & 1 & 0 & 1 & 0 & & & \\
\hline
 & & & & & & 1 & 0 & 0 & 0 & \frac{1}{5} & \frac{1}{5} \\
 & & & & & & 0 & -\frac{1}{4} & -\frac{1}{4} & 1 & 0 & 0 \\
 & & & & & & 0 & 0 & 1 & 0 & 1 & 0
\end{array}
\right]
\begin{bmatrix}
t_{AB} \\ m_{AB} \\ m_{BA} \\ t_{BC} \\ m_{BC} \\ m_{CB} \\ t_{CD} \\ m_{CD} \\ m_{DC} \\ t_{DE} \\ m_{DE} \\ m_{ED}
\end{bmatrix}
=
\begin{bmatrix}
0 \\ 0 \\ 0 \\ 0 \\ -150 \\ 0 \\ 48 \\ 0 \\ 0
\end{bmatrix}
\tag{14}
$$

Rearranging the vector p into the set of end moments m followed by the axial forces t gives

$$[H_1 \ H_2] \begin{bmatrix} m \\ t \end{bmatrix} = [q] \tag{15}$$

where H_1 and H_2 are the corresponding submatrices of H.

A static formulation for the frame of Fig. 1b also requires that $w = \overline{m}$ be minimized with

$$\begin{bmatrix} \overline{m} \\ \overline{m} \\ \overline{m} \\ \overline{m} \\ \overline{m} \\ \overline{m} \\ \overline{m} \\ \overline{m} \end{bmatrix} \geq \begin{bmatrix} m_{AB} \\ m_{BA} \\ m_{BC} \\ m_{CB} \\ m_{CD} \\ m_{DC} \\ m_{DE} \\ m_{ED} \end{bmatrix} \geq - \begin{bmatrix} \overline{m} \\ \overline{m} \\ \overline{m} \\ \overline{m} \\ \overline{m} \\ \overline{m} \\ \overline{m} \\ \overline{m} \end{bmatrix} \tag{16}$$

which may be specified as

$$J\overline{m} \geq m \geq -J\overline{m} \tag{17}$$

With J as a column vector of unit elements. Hence the optimization can be expressed in the generalized form:

Minimize

$$w = \begin{bmatrix} l^T & 0 & 0 \end{bmatrix} \begin{bmatrix} \overline{m} \\ m \\ t \end{bmatrix} \tag{18a}$$

subject to

$$\begin{bmatrix} J & -I & 0 \\ J & I & 0 \\ 0 & H_1 & H_2 \end{bmatrix} \begin{bmatrix} \overline{m} \\ m \\ t \end{bmatrix} \begin{matrix} \geq \\ \geq \\ = \end{matrix} \begin{bmatrix} 0 \\ 0 \\ q \end{bmatrix} \tag{18b}$$

with

$$(\overline{m} \geq 0) \tag{18c}$$

where in this case $l = [1]$ and the matrix for constraints (18b) is of order 25×13. The condition $\overline{m} \geq 0$ has been put in brackets because it can be ignored in the solution. Provided that J has non-negative coefficients, any solution for constraints (17) must also imply that $m \geq 0$. The staircase structure of the matrix of constraints (18b) makes it particularly suitable for decomposition algorithms [18].

The Dual of the Joint Equilibrium Method

The dual of the above method is:

Maximize

$$z = \begin{bmatrix} 0 & 0 & q^T \end{bmatrix} \begin{bmatrix} \theta^+ \\ \theta^- \\ u \end{bmatrix} \tag{19a}$$

subject to

$$\begin{bmatrix} J^T & J^T & 0 \\ -I & I & H_1^T \\ 0 & 0 & H_2^T \end{bmatrix} \begin{bmatrix} \theta^+ \\ \theta^- \\ u \end{bmatrix} \begin{matrix} \leq \\ = \\ = \end{matrix} \begin{bmatrix} l \\ 0 \\ 0 \end{bmatrix} \tag{19b}$$

with

$$\theta^+ \geq 0 \quad \text{and} \quad \theta^- \geq 0 \tag{19c}$$

Here the variables θ^+ and θ^- are the rotations of all possible plastic hinges at member ends, and u are those joint displacements corresponding to the forces implied by the right-hand side of constraints (18b). Also, if the condition $\overline{m} \geq 0$ is ignored in the primal formulation, this corresponds to specifying the inequality in constraints (19b) as an equality. An alternative specification for system (19) is:

Maximize

$$z = q^T u \tag{20a}$$

subject to

$$J^T \mid \theta \mid = l \tag{20b}$$

with

$$\theta = H_1^T u \tag{20c}$$

and

$$H_2^T u = 0 \tag{20d}$$

For the example of Fig. 1b the dual variables u describe a virtual displacement of the form $u = \{u_B, v_B, \theta_B, u_C, v_C, \theta_C, u_D, v_D, \theta_D\}$ where u, v and θ are horizontal displacement, vertical displacement and rotation respectively. Thus Eq. 20 gives:

Maximize

$$z = -150 \, v_C + 48 \, u_D \tag{21a}$$

subject to

$$| \theta_{AB} | + | \theta_{BA} | + | \theta_{BC} | + | \theta_{CB} | + | \theta_{CD} | + | \theta_{DC} |$$

$$+ | \theta_{DE} | + | \theta_{ED} | = 1 \qquad (21b)$$

with

$$
\begin{bmatrix}
\theta_{AB} \\
\theta_{BA} \\
\hline
\theta_{BC} \\
\theta_{CB} \\
\hline
\theta_{CD} \\
\theta_{DC} \\
\hline
\theta_{DE} \\
\theta_{ED}
\end{bmatrix}
=
\begin{bmatrix}
\frac{1}{5} & 0 & 0 & & & & & \\
\frac{1}{5} & 0 & 1 & & & & & \\
\hline
0 & \frac{1}{4} & 1 & 0 & -\frac{1}{4} & 0 & & \\
0 & \frac{1}{4} & 0 & 0 & -\frac{1}{4} & 1 & & \\
\hline
 & & 0 & \frac{1}{4} & 1 & 0 & -\frac{1}{4} & 0 \\
 & & 0 & \frac{1}{4} & 0 & 0 & -\frac{1}{4} & 1 \\
\hline
 & & & & & \frac{1}{5} & 0 & 1 \\
 & & & & & \frac{1}{5} & 0 & 0
\end{bmatrix}
\begin{bmatrix}
u_B \\
v_B \\
\theta_B \\
u_C \\
v_C \\
\theta_C \\
u_D \\
v_D \\
\theta_D
\end{bmatrix}
\qquad (21c)
$$

and

$$
\begin{bmatrix}
0 & 1 & 0 & & & & & & \\
-1 & 0 & 0 & 1 & 0 & 0 & & & \\
 & & & -1 & 0 & 0 & 1 & 0 & 0 \\
 & & & & & & 0 & 1 & 0
\end{bmatrix}
\begin{bmatrix}
u_B \\
v_B \\
\theta_B \\
u_C \\
v_C \\
\theta_C \\
u_D \\
v_D \\
\theta_D
\end{bmatrix}
=
\begin{bmatrix}
0 \\
0 \\
0 \\
0
\end{bmatrix}
\qquad (21d)
$$

The function being maximized in Eq. 21a is the external work done. Equation 21b sets the internal work/$\overline{m}$ to unity since θ_{AB}, θ_{BA}, etc., describe the rotations of all possible plastic hinges. Equations 21c define the hinge rotations as functions of the joint displacements and Eqs. 21d give constraints that are necessary to ensure compatibility at joints.

For a mechansim to become active, the internal and external work terms must be equal. Hence the maximum must correspond to the maximum value of $\overline{m}$ required to support any possible mechanism against collapse.

This method could have been derived as a kinematic formulation. The authors are not aware of any previous publication of the method.

The Method of Independent Mechanisms and its Dual

The method of independent mechanisms may be derived from the joint equilibrium method by eliminating the member axial forces, t, from the equilibrium equations (13) to give

$$Em = f \tag{22}$$

For instance, for the frame of Fig. 1b, if the second, first, seventh and eighth are used to eliminate t_{AB}, t_{BC}, t_{CD} and t_{DE} respectively from Eqs. 14, the following equilibrium equations result:

$$
\begin{bmatrix}
0 & 1 & 1 & 0 & 0 & 0 & 0 & 0 \\
\frac{1}{5} & \frac{1}{5} & 0 & 0 & 0 & 0 & \frac{1}{5} & \frac{1}{5} \\
0 & 0 & -\frac{1}{4} & -\frac{1}{4} & \frac{1}{4} & \frac{1}{4} & 0 & 0 \\
0 & 0 & 0 & 1 & 1 & 0 & 0 & 0 \\
0 & 0 & 0 & 0 & 0 & 1 & 1 & 0
\end{bmatrix}
\begin{bmatrix}
m_{AB} \\
m_{BA} \\
m_{BC} \\
m_{CB} \\
m_{CD} \\
m_{DC} \\
m_{DE} \\
m_{ED}
\end{bmatrix}
=
\begin{bmatrix}
0 \\
48 \\
-150 \\
0 \\
0
\end{bmatrix}
\tag{23}
$$

For this particular problem it is unnecessary to distinguish between m_{BA} and $-m_{BC}$, m_{CB} and $-m_{CD}$ and also between m_{DC} and $-m_{DE}$. Therefore using the first, fourth and fifth equations to eliminate m_{BC}, m_{CD} and m_{DE} gives

$$
\begin{bmatrix}
\frac{1}{5} & \frac{1}{5} & 0 & -\frac{1}{5} & \frac{1}{5} \\
0 & \frac{1}{4} & -\frac{1}{2} & \frac{1}{4} & 0
\end{bmatrix}
\begin{bmatrix}
m_{AB} \\
m_{BA} \\
m_{CB} \\
m_{DC} \\
m_{ED}
\end{bmatrix}
=
\begin{bmatrix}
48 \\
-150
\end{bmatrix}
\tag{24}
$$

Incorporating the yield constraints, (17), the generalised static formulation is:

Minimize

$$w = \begin{bmatrix} l^T & 0 \end{bmatrix} \begin{bmatrix} \overline{m} \\ m \end{bmatrix} \tag{25a}$$

subject to

$$
\begin{bmatrix}
J & -I \\
J & I \\
0 & E
\end{bmatrix}
\begin{bmatrix}
\overline{m} \\
m
\end{bmatrix}
\begin{matrix}
\geq \\
\geq \\
=
\end{matrix}
\begin{bmatrix}
0 \\
0 \\
f
\end{bmatrix}
\tag{25b}
$$

For the frame example of Fig. 1b, taking advantage of the small E matrix shown in equations (24) leads to a coefficient matrix in constraints (25b) of order 12×6. Equations (23) and (24) could, however, have been derived independently of the joint equilibrium method by applying the principle of virtual work to the sets of independent mechanisms shown in Figs. 5a and 5b respectively. The number of independent mechanisms is given by the difference between the number of critical sections and the degree of structural redundancy.

This formulation was first given by Chan [19], and the same approach has been adopted by Cohn et al. [20], Maier et al. [21] and others. For rectangular frames the set of independent mechanisms can be conveniently formed from beam, sway and joint mechanisms. (Examples of these three types of mechanism are the third, second and first mechanisms, respectively in Fig. 5a). These are known as elementary mechanisms [22] and have been generalized to frames of arbitrary geometry by Dorn and Greenberg [23].

An alternative set of independent mechanisms, called basic mechanisms, may be adopted [24]. Here, a structure is first rendered statically determinate by the insertion of d real hinges (where d is the degree of structural redundancy). This is done in such a way that no local mechanism is thereby created. Each basic mechanism is then obtained by the introduction of a further hinge at any one of the remaining sections. The relationship of the method of independent mechanisms with the joint equilibrium method has also been discussed by Gonzalez-Caro [25] and Watwood [26].

The dual of this method takes the form:

Maximize

$$z = \begin{bmatrix} 0 & 0 & f^T \end{bmatrix} \begin{bmatrix} \theta^+ \\ \theta^- \\ k \end{bmatrix} \tag{26a}$$

subject to

$$\begin{bmatrix} J^T & J^T & 0 \\ -I & I & E^T \end{bmatrix} \begin{bmatrix} \theta^+ \\ \theta^- \\ k \end{bmatrix} = \begin{bmatrix} l \\ 0 \end{bmatrix} \tag{26b}$$

with

$$\theta^+ \geq 0 \quad \text{and} \quad \theta^- \geq 0$$

This can be condensed to:

Maximize

$$z = f^T k \tag{27a}$$

158

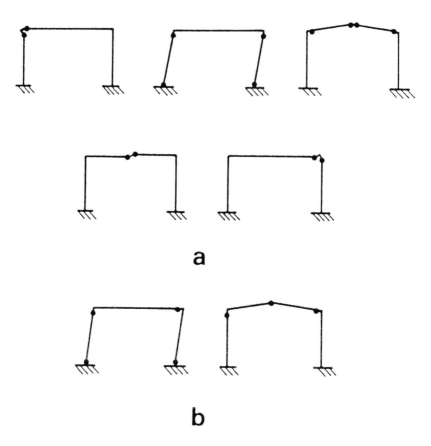

a

b

Figure 5. *Independent mechanisms for the frame of Fig. 1: (a) with eight critical cross-sections to yield Eqs. (23); (b) with five critical cross-sections to yield Eqs. (24).*

subject to

$$J^T \mid \theta \mid = l \tag{27b}$$

and

$$\theta = E^T k \tag{27c}$$

For the frame of Fig. 1b, using the condensed form of E from equations (24), the required optimization is:

Maximize

$$z = \begin{bmatrix} 48 & -150 \end{bmatrix} \begin{bmatrix} k_1 \\ k_2 \end{bmatrix} \tag{28a}$$

subject to

$$\mid \theta_{AB} \mid + \mid \theta_{BA} \mid + \mid \theta_{CB} \mid + \mid \theta_{DC} \mid + \mid \theta_{ED} \mid = 1 \tag{28b}$$

and

$$\begin{bmatrix} \theta_{AB} \\ \theta_{BA} \\ \theta_{CB} \\ \theta_{DC} \\ \theta_{ED} \end{bmatrix} \begin{bmatrix} \frac{1}{5} & 0 \\ \frac{1}{5} & \frac{1}{4} \\ 0 & -\frac{1}{2} \\ -\frac{1}{5} & \frac{1}{4} \\ \frac{1}{5} & 0 \end{bmatrix} \begin{bmatrix} k_1 \\ k_2 \end{bmatrix} \tag{28c}$$

Here k_1 and k_2 are coefficients governing the magnitudes of the two independent mechanisms shown in Fig. 5b.

Equations 28c defines the rotations of the plastic hinges in terms of these coefficients and equations (28a and 28b) relate to the external and internal virtual work functions respectively. In interpreting equations (28c) it should be noted that a positive θ means that the impinging member rotates clockwise relative to the joint.

This formulation was also presented by Chan [19], Cohn et al. [20], Maier et al. [21] and others.

The Method of Redundant Forces and its Dual

The equilibrium equations (13) can only be directly solved for the internal forces if the frame is statically determinate (which is a trivial case as far as plastic theory is concerned). In other cases this can be accomplished after introducing a suitable set of redundant forces as additional variables. If the internal forces relate to these redundant forces, r, according to

$$\overline{H}p = r \tag{29}$$

the compound equations

$$\hat{H}p = \begin{bmatrix} q \\ r \end{bmatrix} \tag{30a}$$

in which

$$\hat{H} = \begin{bmatrix} H \\ \overline{H} \end{bmatrix} \tag{30b}$$

will have a square coefficient matrix. If the redundant forces have been well chosen, this matrix will be non-singular and hence

$$p = \hat{H}^{-1} \begin{bmatrix} q \\ r \end{bmatrix} \tag{31}$$

If the vector p contains any axial forces these can be ignored because they play no part in the optimization. There may also be some bending moments which are not needed. The required relationship can thus be specified in the form

$$m = b + Br \tag{32}$$

where b can be formed by premultiplying q by the appropriate submatrix of $\hat{H}^{-1}$ and B comprises another submatrix of $\hat{H}^{-1}$ with the same rows but different columns. The variables m can now be eliminated by substituting for m in the constraints (17). Thus the complete optimization takes the form:

Minimize

$$w = l^T \overline{m} \tag{33a}$$

subject to

$$\begin{bmatrix} J & -B \\ J & B \end{bmatrix} \begin{bmatrix} \overline{m} \\ r \end{bmatrix} \begin{array}{c} \geq \\ \geq \end{array} \begin{bmatrix} b \\ -b \end{bmatrix} \tag{33b}$$

For the frame of Fig. 1b the choice of redundant forces shown in Fig. 6 gives

$$\overline{H} = \begin{bmatrix} 0 & 0 & 0 & 0 & -\frac{1}{5} & -\frac{1}{5} \\ & & & -1 & 0 & 0 \\ & & & 0 & 0 & 1 \end{bmatrix} \tag{34}$$

The equations in the form of equation (32) are therefore

$$\begin{bmatrix} m_{AB} \\ m_{BA} \\ m_{CB} \\ m_{DC} \\ m_{ED} \end{bmatrix} = \begin{bmatrix} 840 \\ -600 \\ 0 \\ 0 \\ 0 \end{bmatrix} + \begin{bmatrix} 0 & -8 & -1 \\ 5 & 8 & 1 \\ 5 & 4 & 1 \\ 5 & 0 & 1 \\ 0 & 0 & 1 \end{bmatrix} \begin{bmatrix} r_1 \\ r_2 \\ r_3 \end{bmatrix} \tag{35}$$

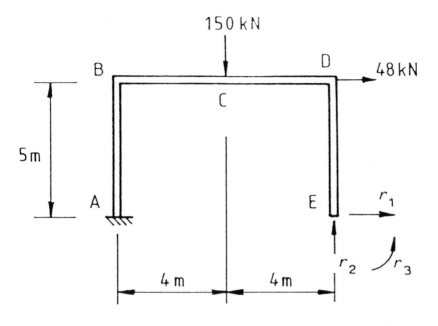

Figure 6. *A feasible choice of redundant forces for the frame of Fig. 1b.*

Thus the coefficient matrix for constraints (33b) has dimension 10×4.

For the given example, the derivation of Eqs. 35 by manipulation of the joint equilibrium equations is cumbersome compared with direct derivation from equilibrium of the statically determinate structure in the form of a bent cantilever. However, in order to adopt such a direct method for frames of arbitrary geometry, it is necessary to have access to data defining the loop topology [27,28]. In the case of the frame of Fig. 1 there is just one loop, ABCDE (counting EA as the closing link) requiring the introduction of three forces at one cut. However, for the frame in Fig. 7, there are four loops requiring the introduction of three redundant forces at each of four cuts, thus giving twelve redundant force variables in all.

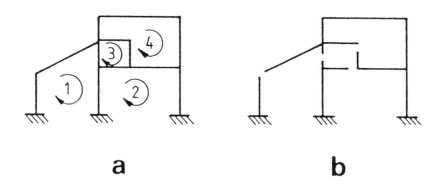

Figure 7. *A plane frame having four loops: (a) loop topology; (b) a cut structure which is statically determinate.*

An alternative to the above choice of redundant forces is to choose them all to be critical bending moments. In such cases, some rows of the B matrix comprise a unit matrix. If, for the example of Fig. 1b, M_{AB}, M_{CB} and M_{ED} are used as redundant forces $\bar{r}_1$, $\bar{r}_2$ and $\bar{r}_3$ respectively then

$$
\begin{bmatrix} m_{AB} \\ m_{BA} \\ m_{CB} \\ m_{DC} \\ m_{ED} \end{bmatrix} = \begin{bmatrix} 0 \\ -180 \\ 0 \\ -420 \\ 0 \end{bmatrix} + \begin{bmatrix} 1 & 0 & 0 \\ -\frac{1}{2} & 1 & -\frac{1}{2} \\ 0 & 1 & 0 \\ \frac{1}{2} & 1 & \frac{1}{2} \\ 0 & 0 & 1 \end{bmatrix} \begin{bmatrix} \bar{r}_1 \\ \bar{r}_2 \\ \bar{r}_3 \end{bmatrix} \tag{36}
$$

which could easily have been derived from the equilibrium equations (24) associated with independent mechanisms.

This method was first developed by Jennings [29] and Pearson [30]. The same approach was adopted by Toakley [31–33]. It was discussed in relation to the joint equilibrium method by Livesley [16] and in relation to the method of independent mechanisms by Sacchi et al. [34].

The dual of this method is:

Maximize

$$z = \begin{bmatrix} b^T & -b^T \end{bmatrix} \begin{bmatrix} \theta^+ \\ \theta^- \end{bmatrix} \qquad (37a)$$

subject to

$$\begin{bmatrix} J^T & J^T \\ -B^T & B^T \end{bmatrix} \begin{bmatrix} \theta^+ \\ \theta^- \end{bmatrix} = \begin{bmatrix} l \\ 0 \end{bmatrix} \qquad (37b)$$

with

$$\theta^+ \geq 0 \quad \text{and} \quad \theta^- \geq 0 \qquad (37c)$$

This may be simplified to:

Maximize

$$z = b^T \theta \qquad (38a)$$

subject to

$$J^T \mid \theta \mid = l \qquad (38b)$$

and

$$B^T \theta = 0 \qquad (38c)$$

For the example in Fig. 1b with redundant forces as shown in Fig. 6, Eq. 38c gives

$$\begin{bmatrix} 0 & 5 & 5 & 5 & 0 \\ -8 & 8 & 4 & 0 & 0 \\ -1 & 1 & 1 & 1 & 1 \end{bmatrix} \begin{bmatrix} \theta_{AB} \\ \theta_{BA} \\ \theta_{CB} \\ \theta_{DC} \\ \theta_{ED} \end{bmatrix} = \begin{bmatrix} 0 \\ 0 \\ 0 \end{bmatrix} \qquad (39)$$

These equations give the compatibility conditions that horizontal displacement, vertical displacement and rotation at the right-hand support are all zero (see Fig. 8).

This formulation was first given by Hoskin [35] for beams and pin-jointed frameworks, and later by Fenves and Gonzalez-Caro [36], Munro and Smith [37] and Davies [38].

164

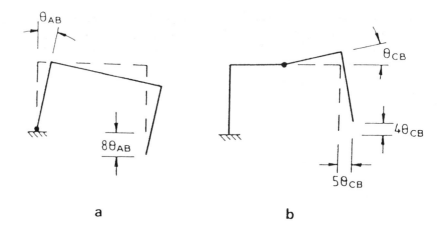

Figure 8. *Deflections of the cut structure shown in Fig. 6: (a) due to rotation θ_{AB}; (b) due to rotation θ_{CB}.*

The Method of all Mechanisms and its Dual

The method of all mechanisms may be obtained from the method of redundant forces by eliminating the redundancies. Consider the portal frame example with redundant forces as shown in Fig. 6. The method of redundant forces applied to this frame requires the following optimization:

Minimize

$$w = \overline{m} \tag{40a}$$

subject to

$$
\left.\begin{array}{llll}
\text{(i)} & \overline{m} \geq & 840 & -8r_2 - r_3(= m_{AB}) \geq -\overline{m} \\
\text{(ii)} & \overline{m} \geq -600 + 5r_1 + 8r_2 + r_3(= m_{BA}) \geq -\overline{m} \\
\text{(iii)} & \overline{m} \geq & 5r_1 + 4r_2 + r_3(= m_{CB}) \geq -\overline{m} \\
\text{(iv)} & \overline{m} \geq & 5r_1 & + r_3(= m_{DC}) \geq -\overline{m} \\
\text{(v)} & \overline{m} \geq & r_3(= m_{ED}) \geq -\overline{m}
\end{array}\right\} \tag{40b}
$$

Because there are four unknowns, it takes only four of these constraints to derive an absolute constraint on $\overline{m}$. Although there are five ways in which four constraints can be chosen from these five, two such choices give the same reduction. The four different results are as follows. With (i) or (v) missing, from $-m_{BA} + 2m_{CB} - m_{DC}$:

$$4\overline{m} \geq 600(\geq -4\overline{m}) \tag{41a}$$

With (iii) missing, from $m_{AB} + m_{BA} - m_{DC} + m_{ED}$:

$$4\overline{m} \geq 240(\geq -4\overline{m}) \tag{41b}$$

With (ii) missing, from $m_{AB} + 2m_{CB} - 2m_{DC} + m_{ED}$:

$$6\overline{m} \geq 840(\geq -6\overline{m}) \tag{41c}$$

With (iv) missing, from $-m_{AB} - 2m_{BA} + 2m_{CB} - m_{ED}$:

$$6\overline{m} \geq 360(\geq -6\overline{m}) \tag{41d}$$

These constitute all the tightest bounds and correspond to equation (1) with $\lambda = 1$. The ways in which the constraints need to be combined relate to the different possible collapse mechanisms shown in Fig. 2. For instance, the sway mechanism in Fig. 2b has $\theta_{AB} = \theta_{BA} = -\theta_{DC} = \theta_{ED}$ which matches with the formula $m_{AB} + m_{BA} - m_{DC} + m_{ED}$ for obtaining constraint (41b). Another feature is that where less constraints are required than the number of unknowns, as in the derivation of constraints (41a), the bending moment diagram at failure is unlikely to be unique (see Fig. 9), and there is likely to be only a partial collapse of the structure.

In general the optimization using all mechanisms can be specified as:

Minimize

$$w = l^T \overline{m} \tag{42a}$$

subject to

$$A\overline{m} \geq e \tag{42b}$$

where, for the above problem, both A and e are of order 4×1.

This static formulation was first derived by Foulkes [9] and later by Prager [39], Chan [19], Cohn et al. [20] and others.

The dual of this method is:

Maximize

$$z = e^T s \tag{43a}$$

subject to

$$A^T s = l \tag{43b}$$

with

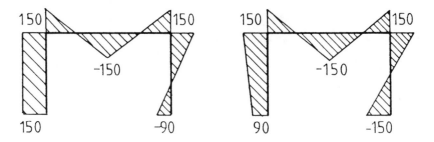

Figure 9. *Non-unique bending moment diagrams at collapse for frame of Fig. 1b.*

$$s \geq 0 \tag{43c}$$

which, for the frame of Fig. 1b, gives:

Maximize

$$z = 600\,s_1 + 240\,s_2 + 840\,s_3 + 360\,s_4 \tag{44a}$$

subject to

$$4s_1 + 4s_2 + 6s_3 + 6s_4 = 1 \tag{44b}$$

Here $s_1, \ldots, s_4$ are constrained variables which define the magnitudes of the various mechanisms in Fig. 2. The required plastic moment is the optimum value of z which occurs when $s = [1, 0, 0, 0]$. This kinematic formulation was first given by Maier et al. [21].

The inherent difficulty in using either the static or kinematic forms of the method of all mechanisms for all but the simplest problems is that the number of mechanisms to be examined may be very large indeed. The number of such mechanisms is equal to the number of different reductions possible of the constraint equations involving redundant forces. If partial collapse modes are not present, the number of such reductions is

$$^{c}C_{(c-d-1)} = {}^{c}C_{(d+1)} = \frac{c(c-1)\cdots(c-d)}{(d+1)!} \tag{45}$$

where c is the number of critical cross-sections and d is the number of redundant forces. With $c = 20$ and $d = 6$, for instance, this formula gives 77520 possible mechanisms. Alternatively, since any mechanism can be expressed as a linear combination of the $(c-d)$ independent mechanisms [23], by considering the presence (or absence) of each independent mechanism in the combination, the number of possible mechanisms may be estimated to be $2^{(c-d)} - 1$ from [34,40]. This gives 16383 mechanisms for the above example. Both of these estimates include a large number of inadmissible mechanisms as well as the admissible ones. The relationship of the method of all mechanisms with other techniques has also been discussed by Heyman [6], Foulkes [8], Gonzalez-Caro [25] and Chan [19]. Kawai [41] and also Bigelow and Gaylord [11] have examined hand methods of reducing the number of mechanisms to be considered. Rubinstein and Karagozian [42] have reduced the number of mechanisms constraints by using a weak beam, strong column model in rectangular frame problems such that plastic hinges are constrained to form only in the beams.

Static Formulations for Minimum Weight Design

The static formulations for the minimum weight design of the frame of Fig. 3a requires:

Minimize

$$w = 10\overline{m}_1 + 8\overline{m}_2 \tag{46a}$$

subject to

$$
\begin{bmatrix} \overline{m}_1 \\ \overline{m}_1 \\ \overline{m}_2 \\ \overline{m}_2 \\ \overline{m}_2 \\ \overline{m}_2 \\ \overline{m}_1 \\ \overline{m}_1 \end{bmatrix}
\geq
\begin{bmatrix} m_{AB} \\ m_{BA} \\ m_{BC} \\ m_{CB} \\ m_{CD} \\ m_{DC} \\ m_{DE} \\ m_{ED} \end{bmatrix}
\geq -
\begin{bmatrix} \overline{m}_1 \\ \overline{m}_1 \\ \overline{m}_2 \\ \overline{m}_2 \\ \overline{m}_2 \\ \overline{m}_2 \\ \overline{m}_1 \\ \overline{m}_1 \end{bmatrix}
\tag{46b}
$$

The bending moments are considered to be split into two groups, each associated with the different plastic moments $\overline{m}_1$ and $\overline{m}_2$. The joint equilibrium formulation (18) is still appropriate, however, provided that

$$\overline{m} = \{\overline{m}_1, \overline{m}_2\} \tag{47a}$$

$$l = \{10, 8\} \tag{47b}$$

and

$$J = \begin{bmatrix} 1 & 0 \\ 1 & 0 \\ 0 & 1 \\ 0 & 1 \\ 0 & 1 \\ 0 & 1 \\ 1 & 0 \\ 1 & 0 \end{bmatrix} \tag{47c}$$

One important difference for the portal frame example is that it is no longer possible to ignore m_{BC} and m_{DE} because, although they still are equal to $-m_{BA}$ and $-m_{DC}$ respectively, they are being compared to different plastic moments. In the method of independent mechanisms only m_{CD} can be eliminated from Eqs. 23 giving

$$\begin{bmatrix} 0 & 1 & 1 & 0 & 0 & 0 & 0 \\ \frac{1}{5} & \frac{1}{5} & 0 & 0 & 0 & \frac{1}{5} & \frac{1}{5} \\ 0 & 0 & -\frac{1}{4} & -\frac{1}{2} & \frac{1}{4} & 0 & 0 \\ 0 & 0 & 0 & 0 & 1 & 1 & 0 \end{bmatrix} \begin{bmatrix} m_{AB} \\ m_{BA} \\ m_{BC} \\ m_{CB} \\ m_{DC} \\ m_{DE} \\ m_{ED} \end{bmatrix} = \begin{bmatrix} 0 \\ 48 \\ -150 \\ 0 \end{bmatrix} \tag{48}$$

with the coefficient matrix for constraints (25b) of order 18×9.

If the redundant force method, constraint (40b) is replaced by

$$\left. \begin{array}{l} \overline{m}_1 \geq \quad 840 \quad\quad\; - 8r_2 - r_3 (= m_{AB}) \geq -\overline{m}_1 \\ \overline{m}_1 \geq -600 + 5r_1 + 8r_2 + r_3 (= m_{BA}) \geq -\overline{m}_1 \\ \overline{m}_2 \geq \quad 600 - 5r_1 - 8r_2 - r_3 (= m_{BC}) \geq -\overline{m}_2 \\ \overline{m}_2 \geq \quad\quad\; 5r_1 + 4r_2 + r_3 (= m_{CB}) \geq -\overline{m}_2 \\ \overline{m}_2 \geq \quad\quad\; 5r_1 \quad\quad + r_3 (= m_{DC}) \geq -\overline{m}_2 \\ \overline{m}_1 \geq \quad\quad -5r_1 \quad\quad - r_3 (= m_{DE}) \geq -\overline{m}_1 \\ \overline{m}_1 \geq \quad\quad\quad\quad\quad\quad\; r_3 (= m_{ED}) \geq -\overline{m}_1 \end{array} \right\} \tag{49}$$

with the coefficient matrix for constraints (33b) of order 14×5.

Consider now the method of all mechanisms. Forming from constraints (49), the constraints associated with $-m_{BA} + 2m_{CB} + m_{DE}$ and $m_{AB} + 2m_{CB} + 2m_{DE} + m_{ED}$ give

$$2\overline{m}_1 + 2\overline{m}_2 \geq 600(\geq -2\overline{m}_1 - 2\overline{m}_2) \tag{50a}$$

and

$$4\overline{m}_1 + 2\overline{m}_2 \geq 840(\geq -4\overline{m}_1 - 2\overline{m}_2) \tag{50b}$$

Hence $\overline{m}_1 = 120$ and $\overline{m}_2 = 180$ is a solution which satisfies the five constraints associated with m_{AB}, m_{BA}, m_{CB}, m_{DE} and m_{ED}. This is the solution which gives the minimum weight although in general, with this number of critical cross-sections, there may be 21 sets of absolute constraints similar to (50) to examine before the minimum weight can be confirmed.

The minimum weight design corresponds to $r = \{-48, 75, 120\}$ with the bending moment diagram as shown in Fig. 10. This gives a 2.22% saving in weight over the optimum structure having a uniform cross-section throughout with $\overline{m} = 150$.

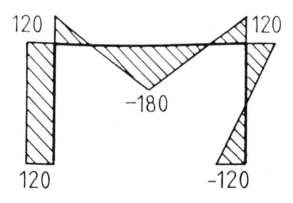

Figure 10. *Unique bending moment diagram at collapse for frame of Fig. 3a.*

In general, if there are c critical cross-sections, d redundant forces and g member groups, the maximum number of absolute constraint sets which may need to be examined are

$$^{c}C_{(c-d-g)} = \frac{c(c-1)\cdots(c-d-g+1)}{(d+g)!} \tag{51}$$

Kinematic Formulations for Minimum Weight Design

When a minimum weight design is expressed in kinematic form, Eq. 20b, $J^T \mid \theta \mid$ $= l$ relates the rotations of the plastic hinges to the lengths associated with each of the member groups. The need to satisfy these equations was first recognised by Foulkes [8] and is embodied in Foulkes' theorem. For the design example shown in Fig. 3a this gives

$$\mid \theta_{AB} \mid + \mid \theta_{BA} \mid + \mid \theta_{DE} \mid + \mid \theta_{ED} \mid \; = 10 \tag{52a}$$

$$\mid \theta_{BC} \mid + \mid \theta_{CB} \mid + \mid \theta_{DC} \mid \; = 8 \tag{52b}$$

(Here CD has been omitted from the list of critical cross-sections).

The kinematic optimization problem is therefore to find, amongst all mechanisms satisfying the compatibility and Foulkes equations, the one which has the maximum value for the external virtual work. Foulkes equations cannot normally be satisfied by any one of the conventional mechanisms used for plastic analysis, such as those in Fig. 2, but require a combination of g such mechanisms where g is the number of groups. For the frame shown in Fig. 3a the rotations satisfying these equations and giving the optimum solution are

$$\theta = \left\{ 1 \quad -3 \quad 0 \quad 8 \quad 0 \quad 5 \quad 1 \right\} \tag{53}$$

This is a combination of two simple mechanisms as shown in Fig. 11, both of which will be activated when the maximum design loading is reached. Once an optimal solution of the form shown in equation (53) has been found, an automatic procedure can be used for evaluating all the simple collapse mechanisms having a single degree of freedom [28].

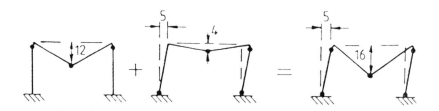

Figure 11. *Derivation of a Foulkes mechanism from two simple mechanisms.*

Design for Multiple Loading Cases

Where a frame is subject to more than one loading case, the minimum weight design is not necessarily the same as the optimum for any of the loading cases considered separately. The minimum weight design for the frame shown in Fig. 3, for instance, has $\overline{m} = \{60, 340\}$ with bending moment diagrams for the two loading cases as shown in Fig. 12, as opposed to $\overline{m} = \{120, 180\}$ and $\overline{m} = \{0, 400\}$ if the loading cases were considered separately. In this case, because both loading cases affect the optimum design, the Foulkes mechanism is a combination of one simple mechanism from each of the parts of the disjoint frame representing the two loading cases. The plastic hinge positions for these mechanisms are shown in Fig. 12. However, if the second loading case in Fig. 3 is modified by reducing the magnitude of the applied load to 120 kN say, this loading case would no longer be critical and both mechanisms contributing to the Foulkes mechanism would relate to the first loading case.

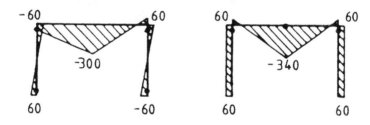

Figure 12. *Bending moment diagrams and plastic hinge locations for minimum weight frame to carry loads shown in Fig. 3; (a) loading case 1; (b) loading case 2.*

The main effect of the inclusion of multiple loading cases within the optimization is to make the relevant matrices very much larger [28,43]. For the frame shown in Fig. 3 the number of members, joints, critical cross-sections, redundant forces and independent mechanisms all double because of the extra loading case. Only the number of groups remains unaltered. However, because the structure being analysed is disjoint, the matrices H, E and B can be arranged in block diagonal form.

The first LP formulation for optimization over multiple loading cases was given

172

by Foulkes [9], who employed the method of all mechanisms. Similar formulations were presented by Bigelow and Gaylord [11] and Cohn et al. [20]. LP formulations for multiple loading cases using the other three methods were given by Pearson [30], Anderheggen and Thurlimann [44], Chan [19,45], Cohn et al. [20] and Smith and Munro [46]. Some of these make use of the generalized Foulkes mechanism conditions for multiple loading cases developed by Prager [47,48], Chan [19] and Polizzotto [49].

Computational Aspects using Standard LP Algorithms

Since at the optimum of a linear programming solution the optimum values of both the primal variables (such as bending moments) and the dual variables (such as plastic hinge rotations) are available, a complete optimal plastic design solution can be obtained by solving any of the eight formulations presented above.

Standard computer algorithms for solving linear programming problems almost universally use the simplex method [14,15]. Problems normally need to be formulated with all variables constrained to be non-negative and the solution techniques involves moving between vertices in the space defined by these variables; first of all, in phase 1, seeking a feasible solution and then, in phase 2, trying to improve the optimizing function.

The minimum weight problem shown in Fig. 3a has been analysed in this way using both static and kinematic formulations for each of the four methods described above. In each case the trial values for the variables were initially all zero. Table 1 gives information about all the solutions. The quoted computer times were for a BBC microcomputer with a 6502 8-bit microprocessor. In considering the tableau sizes logical variables such as slack, surplus and artifical variables have not been included and all the variables have been constrained (e.g. by replacing θ with $\theta^+ - \theta^-$).

The simplest of these solutions to interpret is the static formulation for the method of all mechanisms because a graph can be drawn with the only two variables, $\overline{m}_1$ and $\overline{m}_2$, as coordinates. Fig. 13 shows lines defining the six significant constraints and also the region of permissible design, which is also the feasible region for the linear programming algorithm. The path chosen by the simplex algorithm is o–a–b–c–d–e–f–z. In this case several phase 1 steps are required to reach the permissible region but no further optimization in phase 2 is required because the first vertex reached in the permissible region happens to be the one of minimum weight. (Note that the contours of equal weight are all parallel to the one shown, labelled $w = 2640$, with weight increasing linearly with distance from the origin).

For static formulations of the other methods, the design space involves other

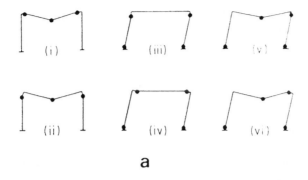

a

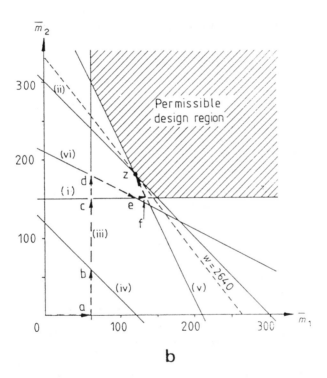

b

Figure 13. *Collapse mechanisms and permissible design region for frame of Fig. 3a: (a) collapse mechanisms; (b) design space showing mechanism constraint lines and optimisation path for method of all mechanisms.*

Table 1. *Optimization for frame of Fig. 3a using different LP formulations.*

Formulation		Design path (Fig. 14)		Tableau size	No. of iterations	Time (s)
		Impermissible design	Permissible design			
Joint equilibrium	Static	o–n–g	p–k–z	25 × 26	15	1298
	Kinematic	*Not shown*	z	14 × 34	18	356
Independent mechanisms	Static	o–g	h–i–j–k–z	18 × 16	13	407
	Kinematic	*Not shown*	z	9 × 22	15	111
Redundant forces	Static	o–l	m–l–z	14 × 18	9	145
	Kinematic	*Not shown*	z	5 × 14	9	21
All mechanisms	Static	o–a–b–c–d–e–f	z	6 × 2	8	20
	Kinematic	o–q–r	z	2 × 6	4	3

variables as well as $\overline{m}_1$ and $\overline{m}_2$. Hence the iteration points for the simplex algorithm do not necessarily appear as vertices on this graph. The paths adopted for all the static solutions are shown on Fig. 14.

In the kinematic formulations, the Foulkes equations have been modified to constraints of the form

$$| \theta_{AB} | + | \theta_{BA} | + | \theta_{DE} | + | \theta_{ED} | \leq 10 \qquad (54a)$$

and

$$| \theta_{BC} | + | \theta_{CB} | + | \theta_{DC} | \leq 8 \qquad (54b)$$

This is admissible because it is equivalent, in the static formulation, to making $\overline{m}_1$ and $\overline{m}_2$ constrained rather than unconstrained (see Eq. 18c). The effect of this change for the kinematic form of the method of all mechanisms is to make the initial trial solution with all variables equal to zero feasible as far as the linear programming algorithm is concerned and so only phase 2 of the optimization is needed. The feasible region is no longer coincident with the permissible design region and, indeed can only touch it along the optimum weight contour. Although $\overline{m}_1$ and $\overline{m}_2$ are not variables, their values for any intermediate point on the design path can be obtained and are shown in Fig. 14. The optimization progresses with weight increasing monotonically until the permissible design region is reached at the optimum. For other kinematic formulations the optimization paths are not readily interpretable in the design space of Fig. 14 and so have not been shown.

In Table 2 the numbers of constraints and variables (excluding logical variables) have been specified for all the methods in terms of parameters relating to the

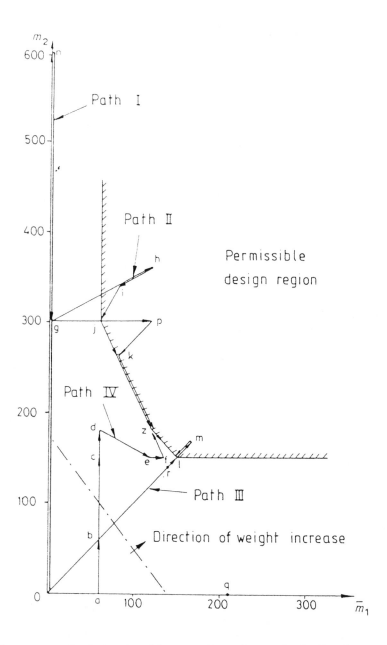

Figure 14. *Optimization history of static methods for frame of Fig. 3a: Path I—method of joint equilibrium; Path II—method of independent mechanisms; Path III—method of redundant forces; Path IV—method of all mechanisms.*

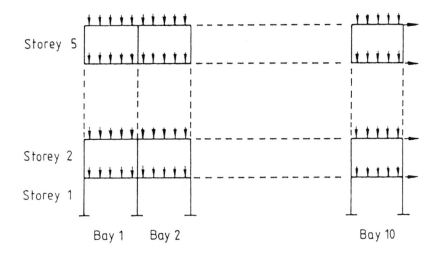

Figure 15. *A large frame problem with* $c = 460$, $d = 150$, $h = 310$, $n = 316$ *and* $s = 355$.

configuration of the structure. Each kinematic formulation has fewer constraints than the corresponding static formulation. Because the amount of computation required in linear programming is mainly associated with the number of constraints rather than the number of variables, the kinematic formulations appear to be more effective than static formulations, a fact which is confirmed by the computing times shown in Table 1. Computational experience has shown that, if a feasible starting solution is not readily available, the total number of iterations for the explicit form of the revised simplex method is roughly 1.7 times the number of constraints and the total number of floating point operations is approximately equal to the cube of the number of constraints [50]. The last two columns of Table 2 relate to the five storey, 10 bay frame shown in Fig. 15 for which it has been assumed that the number of groups, g, is 10. The tableau sizes are given and also an estimate, based on the above criterion, of the relative amounts of computation required if no advantage is taken of sparsity in the tableaux.

The fact that the method which is most easily automated for frames of arbitrary geometry, namely the joint equilibrium method, is computationally the most cumbersome means that the choice of method is not clear cut. Some considerations are as follows:

1. Where it is only required to design frames of a specific topology (e.g. if all are single bay, single storey portal frames), the construction of the tableau for the method of redundant forces may be easily accomplished.

2. Where it is required to design frames of arbitrary topology, it may be more efficient to evaluate by hand and input loop data in addition to the normal geometric data so that the method of redundant forces can be easily implemented.

3. An alternative to (2) above is to construct matrices for the joint equilibrium method and then reduce these to yield a formulation for either the method of independent mechanisms or of redundant forces. If the reduction is to be carried out completely automatically the method of reduction needs to be developed in such a way that it is robust and reliable. This means that the independent mechanisms or the redundant forces need to be well chosen [16,25,26,51,52].

4. If a reduction as in (3) above is to be implemented, it is likely to be more efficient to introduce joints along each member corresponding to loading points after the reduction has been carried out rather than beforehand.

5. The efficiency of the kinematic method of redundant forces can be improved by establishing a full starting basis for the optimization (thus avoiding the need for artificial variables and hence phase 1 of the simplex method). This is possible by first expressing the Foulkes equations as inequalities (similar

to constraints (54)). All the variables θ^+, θ^- are initially set to zero. The set of starting basic variables will consist of the g slack variables associated with the Foulkes equations, plus d suitably chosen θ variables. The tableau needs to be reduced by applying Gauss-Jordan steps such that the columns associated with these basic θ variables together with those associated with the slack variables form a unit submatrix. In this way the number of iterations is reduced to approximately the number of constraints [50].

6. The kinematic form of the method of all mechanisms may be particularly efficient when the number of mechanisms requiring investigation is large. However, no effective way has been devised to generate quickly all the admissible mechanisms. The method of Krajcinovic [53] requires the solution of numerous sets of simultaneous equations and has the disadvantage that it also generates inadmissible mechanisms.

The Use of Different LP Algorithms

The above comparison has been based on the use of the standard simplex method for solving the LP problem. More efficient LP algorithms are also available such as the revised simplex method [54], which employs a matrix inverse computed in product form. However, this method is numerically less stable than the standard simplex method. Numerically more stable algorithms using the inverse concept but employing LU or LQ factorization techniques have also been employed [55,56]. Decomposition algorithms [57,58] determine the optimal solution by first decomposing the problem into small subproblems and then solving these subproblems almost independently. These methods, and also those described in Dantzig et al. [18] are particularly effective where the LP tableau is large and sparse. The projection technique by Karmarkar [59] may also prove to be efficient for certain large sparse LP problems.

It is also possible, however, to develop special purpose algorithms aimed at the particular forms of LP problems derived from plastic analysis and design of structures, and these will be discussed below.

In casting the static formulation of the method of redundant forces, Eq. 33, into standard LP format, it is necessary to duplicate the variables r. Toakley [31] avoids this by replacing each unconstrained variable r_i by

$$r_i = r_i^u - r_i'$$ (55)

where r_i^u is chosen such that the variables r_i' should never be negative and hence can be treated as constrained. Furthermore, only one of the two sets of yield constraints in system (33) is stored explicitly during implementation of the algorithm. The dual simplex method [60] is employed for solution. The efficiency of Toakley's method is similar to that of optimization by the standard simplex method of solving a problem having $(d + g)$ constraints and hence may be similar in efficiency to the kinematic form of the method of redundant forces (system (37)) using a standard LP algorithm.

A compact simplex algorithm has been developed by the authors which also avoids duplication of variables and constraints in the static form of the method of redundant forces [27,61,62]. In this technique the structural equations are operated on directly in a way which can be interpreted physically. Experiments with this method indicate that it is approximately 50 times more efficient than the corresponding solution using a standard LP algorithm and requires one-tenth of the storage space. It is better than the kinematic form of the method of redundant forces, but by a much smaller margin. When multiple loading cases are involved, even greater efficiencies relative to both the static and kinematic LP formulations have been recorded [28]. A corresponding compact kinematic procedure is under investigation.

Other Design Considerations

If a nonlinear relationship is used between the weight and plastic moment of a member or if a cost function is employed which is related in a nonlinear manner to the variables, the resulting optimization would no longer be of the LP type. Marcal and Prager [63] and Prager and Shield [64] included lower bounds on plastic moments by the use of convex cost functions. The use of convex programming for optimal plastic design was further investigated by Save [65]. In the mathematical programming formulation of Ridha and Wright [66] and Chan [45], nonlinear cost functions were utilized.

In the design formulation it has been assumed that a continuous range of sections is available. This is true in steel structures where members are fabricated from plate. However, if rolled sections are to be used, only a limited number of discrete cross-sections will be available. Bigelow and Gaylord [11] and Horne and Morris [67] considered semi-automatic methods of making this choice. Full automation can be achieved by means of integer programming [32] or dynamic programming [68], but these methods require a large amount of computation.

Techniques of revising rigid-plastic solutions to take account of the reduction in the plastic moment of resistance of a structural member due to the presence of axial force were developed by Toakley [33,69]. Alternatively, the interaction of the axial force and bending moment in determining the condition for formation of a plastic hinge may be included directly in the design formulation. This was investigated by Fenves and Gonzalez-Caro [36], Yokoo et al. [70] and Grierson and Aly [71].

The inclusion of deflection constraints in the rigid-plastic formulations, which causes the equations to be of nonlinear programming form, has been investigated by Toakley *et al.* [72]. Approximate methods of including linearized sway deflection constraints in the LP formulations were developed by Rubinstein and Karagozian [42] and Horne and Morris [67].

In order to include change of geometry and instability effects in an analysis, it is possible to use elastic-plastic techniques (e.g. Jennings and Majid [73]). Near optimal iterative elastic-plastic design procedures have been devised by Murray and Ostapenko [74] and Majid and Anderson [75]. Toakley *et al.* [72] produced optimal designs by iterative cycles of rigid-plastic design and elastic-plastic analysis. Checks required for secondary design constraints such as the shear capacity of members, local and web buckling and lateral torsional buckling have been considered by Bigelow and Gaylord [11] and Toakley *et al.* [72].

The use of tapered members in plastic design was first investigated by Vickery [76]. Tam and Jennings [43] have incorporated facilities to take account of tapered members and pinned connections into both the standard LP and their compact formulations. Absolute minimum weight design methods using continuously varying sections have been developed by Horne [77], Heyman [78,79], Prager and Shield [64] and Mayeda and Prager [80]. The inclusion of structural self-weight in the design formulation has been considered by Maier *et al.* [21]. The effect of connections on the overall cost has been discussed by Ridha and Wright [66] and Witteveen and Tolman [81]. Ridha and Wright [66] and also Morris [82] obtained near optimal designs for large multistorey frames by using partitioning techniques to reduce the amount of computation.

Where it is required to avoid incremental collapse in the presence of variable repeated loading, it is necessary to ensure that the structure 'shakes down.' Iterative LP techniques for shakedown design have been developed by Anderheggen and Thurlimann [44], Cohn *et al.* [20] and Davies [38]. Konig [83] presented nonlinear programming formulations for the direct solution of shakedown design problems. The plastic design of a structure subjected to moving loads corresponds to one having an infinite number of loading cases. This has been investigated by Gross and Prager [84], Save and Prager [85] and Sacchi *et al.* [34].

Multi-criteria optimal plastic design procedures have been given by Cohn [86], embracing ultimate limit state and service limit state constraints. Probability-based methods have been presented by Thevendran and Thambiratnam [87].

The inclusion of most of the above facilities introduces nonlinearities into the formulations which increase substantially the complexity and computational requirements. In view of this it would seem advantageous to use one of the basic rigid-plastic formulations presented in this paper (or an enhancement) to produce an initial design which is revised as necessary to satisfy other requirements.

Table 2. *Tableau dimensions and computational factors for different LP formulations.*

Formulation		No. of constraints	No. of variables	Frame of Fig. 15	
				Tableau size	Estimated computational factor
Joint equilibrium	Static	$3n + 4s$	$6s + g$	2368×2140	3242
	Kinematic	$3s + g$	$6n + 4s$	1075×3316	303
Independent mechanisms	Static	$h + 2c$	$2c + g$	1230×930	454
	Kinematic	$c + g$	$2h + 2c$	470×1540	25
Redundant forces	Static	$2c$	$2d + g$	920×310	190
	Kinematic	$d + g$	$2c$	160×920	1
All mechanisms	Static	a	g	$? \times 10$	?
	Kinematic	g	a	$10 \times ?$	?

$a = $ no. of mechanisms, $c = $ no. of critical cross sections, $d = $ no. of redudant forces, $g = $ no. of groups, $h = $ no. of independent mechanisms, $n = $ no. of joints, $s = $ no. of beam segments.

Conclusion

It has been shown that all existing methods of plastic analysis and design can be categorized into four basic techniques, each of which has both a static and a kinematic form. Whereas the method of joint equilibrium is the easiest to automate for frames of arbitrary geometry, it is the most cumbersome to implement. However, other methods can be derived from it. Within each basic technique the kinematic formulation is the most efficient when using standard linear programming algorithms for optimization. However, efficient compact algorithms have also been developed for the static form of the method of redundant forces.

References

[1] Horne, M. R., (1950), 'Fundamental propositions in the plastic theory of structures,' *J. ICE* **34**, 174–177.

[2] Greenberg, H. J. and Prager, W., (1951), 'Limit design of beams and frames,' *Proc. ASCE* **77**, 1–12. (Also in *Trans ASCE*, **117**, 447, 1952, with discussions).

[3] Baker, J. F. and Heyman, J., (1969), *Plastic Design of Frames I-Fundamentals*, Cambridge University Press, Cambridge, UK.

[4] Foulkes, J., (1953), 'Minimum weight design and theory of plastic collapse,' *Quart Appl Math*, **10**, 347–358.

[5] Taha, H. A., (1971), *Operations Research—An Introduction*, MacMillan, New York.

[6] Heyman, J., (1951), 'Plastic design of beams and plane frames for minimum material consumption,' *Quart. Appl. Math* **8**, 373–381.

[7] Heyman, J., (1953), 'Plastic design of plane frames for minimum weight,' *Struct. Engr.* **31**, 125–129.

[8] Foulkes, J., (1954), 'The minimum weight design of structural frames,' *Proc. Roy. Soc. A* **223**, 482–494.

[9] Foulkes, J., (1955), 'Linear programming and structural design,' *Proc. 2nd Symp. Linear Programming*, Washington DC, H. A. Antosiewicz (Ed.), National Bureau of Standards and USAF, 2, 177–184.

[10] Vargo, L. G., (1956), 'Non-linear minimum weight design of planar structures,' *J. Aero. Sci.* **23**, 956–960.

[11] Bigelow, R. H. and Gaylord E. H., (1967), 'Design of steel frames for minimum weight,' *Proc ASCE* **93**, ST6, 109–131.

[12] Jennings, A. and Tam T. K. H., (1985), 'The minimum weight concept in interactive plastic design,' *Proc. 2nd Int. Conf. on Civil and Struct. Eng. Computing*, London, B. H. V. Topping (Ed.), Engineering Technics Press, Edinburgh, UK, 363–368.

[13] Heyman, J., (1971), *Plastic Design of Frames II—Applications*, Cambridge University Press, Cambridge, UK.

[14] Hadley, G., (1962), *Linear Programming*, Addison-Wesley, Reading, MA, USA.

[15] Dantzig, G. B., (1963), *Linear Programming and Extensions*, Princeton University Press, Princeton, New Jersey, USA.

[16] Livesley, R. K., (1967), 'The selection of redundant forces in structures, with an application to collapse analysis of frameworks,' *Proc. Royal. Soc. A* **301**, 493–505.

[17] Livesley, R. K., (1975), *Matrix Methods of Structural Analysis*, Pergamon, Oxford, UK.

[18] Dantzig, G. B, Dempster M. A. H. and Kallio M. J. (Eds), (1981), *Large-Scale Linear Programming Vols I and II*, IIASA Collaborative Proceedings Series, CP-81-51, Inter. Inst. Appl. Sys. Anal., Laxenburg, Austria.

[19] Chan, H. S. Y., (1969), 'On Foulkes mechanism in portal frame design for alternative loads,' *J. Appl. Mech.* **36**, 73–75.

[20] Cohn, M. Z., Ghosh, S. K. and Parimi, S. R., (1972), 'Unified approach to theory of plastic structures,' *Proc. ASCE* **98**, EM15, 1133–1158.

[21] Maier, G., Srinivasan R. and Save, M. A., (1976), 'On limit design of frames using linear programming,' *J. Struct. Mech.* **4**, 349–378.

[22] Neal, B. G. and Symonds, P. S., (1952), 'The rapid calculations of the plastic collapse load for a framed structure,' *Proc. ICE* **1**, Pt 3, 58–100.

[23] Dorn, W. S. and Greenberg, H. J., (1956), 'The mechanism technique and some questions in the plastic collapse of frames,' *Symposium on Plasticity in Structural Engineering*, Varenna, N. Zanichelli Publ., Bologna, Italy, 93–108.

[24] Munro, J., (1965), 'The elastic limit analysis of planar skeletal structures,' *Civ. Eng. and Public Works Review* **60**, 671–677.

[25] Gonzalez-Caro, A., (1968), 'A network-topological formulation of the analysis and design of rigid-plastic framed structures,' Ph.D. dissertation, Univ of Illinois, Urbana, USA.

[26] Watwood, V. B., (1979), 'Mechanism generation for limit analysis,' *Proc. ASCE* **105**, ST1, 1–15.

[27] Jennings, A. and Tam, T. K. H., (1986), 'Automatic plastic design of frames,' *Eng. Struct.* **8**, 138–147.

[28] Tam, T. K. H., (1987), 'Computer methods for optimal plastic design of frames,' Ph.D. dissertation, The Queen's University of Belfast, UK.

[29] Jennings, A., (1954), 'The minimum weight of steel structures with reference to the problems of automatic design,' MSc dissertation, Manchester Univ., UK.

[30] Pearson, C. E., (1958), 'Structural design by high–speed computing machines,' *Proc. ASCE 1st Conf. Electronic Computation*, Kansas City, USA, 417–437.

[31] Toakley, A. R., (1968), 'Some computational aspects of optimum rigid plastic design,' *Int. J. Mech. Sci.* **10**, 531–537.

[32] Toakley, A. R., (1968) 'Optimum design using available sections,' *Proc. ASCE* **94**, ST5, 1219–1241.

[33] Toakley, A. R., (1969), 'The optimum plastic design of unbraced frameworks,' *Civ. Eng. Trans. Inst. Engrs. Australia*, CE11, 111–116.

[34] Sacchi, G, Maier, G. and Save, M. A., (1975), 'Limit design of frames for movable loads by linear programming,' *Proc IUTAM Symp 'Optimisation in Struct Design,'* Warsaw, Poland, A. Sawczuk and Z. Mroz (Eds.), Springer-Verlag, Berlin, 415–432.

[35] Hoskin, B. C., (1960), 'Limit analysis, limit design and linear programming,' Struct. and Mater Rep 274, Aero. Res. Lab. Aust. Defence Scientific Service, Melbourne.

[36] Fenves, S. J. and Gonzalez-Caro, A., (1971), 'Network-topological formulation of analysis and design of rigid-plastic framed structures,' *Int. J. Num. Meth. Eng.* **3**, 425–441.

[37] Munro, J. and Smith D. L., (1972), 'Linear programming duality in plastic analysis and synthesis,' *Proc. Int. Symp. Computer-Aided Struct Design*, Univ. of Warwick, UK, A1, 22–54.

[38] Davies, J. M., (1972), 'A new formulation of the plastic design problem for plane frames,' *Int. J. Num. Meth. Eng.* **5**, 185–192.

[39] Prager, W., (1957), 'Linear programming and structural design,' RAND Res Memo RM-2021; ASTIA Docu AD 150661.

[40] Cohn, M. Z. and Grierson D. E., (1971), 'An automatic approach to the analysis of plastic frames under fixed and variable loading,' *Struct. Engr.* **49**, 291–297.

[41] Kawai, T., (1965), 'Plastic analysis and minimum weight design of multi-storey frames,' *Proc. Conf. Plastic Design of Multistorey Frames*, Lehigh Univ., Bethlehem, PA, USA.

[42] Rubinstein, M. F. and Karagozian, J., (1966), 'Building design using linear programming,' *Proc. ASCE* **92**, ST6, 223–245.

[43] Tam, T. K. H. and Jennings, A., (1988), 'Optimal plastic design of frames with tapered members,' *Computers and Structures* **30**, 537–544.

[44] Anderheggen, E. and Thurlimann, B., (1966), 'Optimal design using linear programming,' *Pub. Int. Assoc. Bridge. and. Struct. Engrs* **26**, 555–571.

[45] Chan, H. S. Y., (1968), 'Mathematical programming in optimal plastic design,' *Int. J. Solids. Struct.* **4**, 885–895.

[46] Smith, D. L. and Munro, J., (1976), 'Plastic analysis and synthesis of frames subjected to multiple loadings,' *Eng. Optimisation* **2**, 145–157.

[47] Prager, W., (1967), 'Optimum plastic design of a portal frame for alternative loads,' *J. Appl. Mech.* **34**, 772–773.

[48] Prager, W., (1971), 'Foulkes mechanism in optimal plastic design for alternative loads,' *Int. J. Mech. Sci.* **13**, 971–973.

[49] Polizzotto, C., (1974), 'Optimal design for multiple sets of loads,' *Meccanica* **9**, 206–213.

[50] Wolfe, P. and Cutler, L., (1963), 'Experiments in linear programming' in *Recent Advances in Mathematical Programming*, R. L. Graves and P. Wolfe (Ed.), McGraw-Hill, New York, USA, 177–200.

[51] Robinson, J. and Regl R. R., (1963), 'An automated matrix analysis for general frames,' *J. Amer. Helicopter Soc.* **11**, 16–35.

[52] Robinson, J., (1965), 'Automatic selection of redundancies in the matrix force method,' *Canadian Aero. and Space J.* **11**, 9–12.

[53] Krajcinovic, D., (1969), 'Limit analysis of structures,' *Proc. ASCE* **95**, ST9, 1901–1909, (see also Anton, J. M., Discussion of 'Limit analysis of structures' by Krajcinovic, D., (1969), *Proc. ASCE*, 1970, **96**, ST4, 871–872.)

[54] Orchard-Hays, W., (1968), *Advanced Linear Programming Computing Techniques*, McGraw-Hill, New York.

[55] Murty, K. G., (1976), *Linear and Combinational Programming*, Wiley, New York.

[56] Gill, P. E., Murray W. and Wright M. H., (1981), *Practical Optimization*, Academic Press, New York.

[57] Dantzig, G. B. and Wolfe, P., (1960), 'A decomposition principle for linear programming,' *Op. Res.* **8**, 101–111.

[58] Himmelblau, D. M. (Ed.), (1973), 'Decomposition of large scale problems,' *Proc. NATO Inst. on Decomposition*, North Holland.

[59] Karmarker, N., (1984), 'A new polynomial-time algorithm for linear programming,' AT & T Bell Laboratories, Murray Hill, New Jersey, 1–68.

[60] Lemke, C. E., (1954), 'The dual method of solving the linear programming problem,' *Naval Res. Logistics Quart.* **1**, 36–47.

[61] Jennings, A., (1983), 'Adapting the simplex method to plastic design,' *Proc. Instability and Plastic Collapse of Steel Structures*, Manchester Univ., Morris (Ed.), Granada, London, 164–173.

[62] Tam, T. K. H., (1987), 'Computer programs for plastic design of frames,' Civil Eng. Dept. Rep., Queen's University of Belfast, UK.

[63] Marcal, P. V. and Prager, W., (1964), 'A method of optimal plastic design,' *J. de Mecanique* **3**, 509–530.

[64] Prager, W. and Shield R. T., (1967), 'A general theory of optimal plastic design,' *Trans. ASME, J. Appl. Mech.* **34**, 184–186.

[65] Save, M. A., (1972), 'A unified formulation of the theory of optimal plastic design with convex cost function,' *J. Struct. Mech.* **1**, 267–276.

[66] Ridha, R. A. and Wright R. N., (1967), 'Minimum cost design of frames,' *Proc. ASCE* **93**, ST4, 165–183.

[67] Horne, M. R. and Morris L. J., (1973), 'Optimum design of multistorey rigid frames,' in *Optimum Structural Design—Theory and Applications*, R. H. Gallagher and O. C. Zienkiewicz (Eds.), Wiley, New York.

[68] Palmer, A. C., (1968), 'Optimal structural design by dynamic programming,' *Proc. ASCE* **94**, ST8, 1887.

[69] Toakley, A. R., (1970), 'Axial load effects in optimum rigid plastic design,' *Build. Sci.* **5**, 111–115.

[70] Yokoo, Y., Nakamuru T. and Keii M., (1975), 'The minimum weight design of multistorey building frames,' *Proc. IUTAM Symp Optimisation in Struct Design*, Warsaw, A. Sawczuk and Z. Mroz (Eds.), Springer-Verlag, Berlin, 294–312.

[71] Grierson, D. E. and Aly A. A., (1980), 'Plastic design under combined stresses,' *Proc. ASCE* **106**, EM4, 585–607.

[72] Toakley, A. R., Batten D. F. and Wilson B. G., (1975), 'Optimum plastic design of unbraced frameworks,' *Proc. IUTAM Symp. Optimisation in Structural Design*, Warsaw, A. Sawczuk and Z. Mroz (Eds.), Springer-Verlag, Berlin, 294–312.

[73] Jennings, A. and Majid K. I., (1965), 'Elastic-plastic analysis by computer of framed structures loaded up to collapse,' *Struct. Engr.* **43**, 407–412.

[74] Murray, T. M. and Ostapenko A., (1966), 'Optimum design of multi-storey frames by plastic theory,' Fritz Eng. Lab. Rep. No. 273-42, Lehigh Univ, Bethlelem, PA, USA.

[75] Majid, K. I. and Anderson, D., (1968), 'Elastic-plastic design of sway frames by computer,' *Proc. ICE* **41**, 705–729.

[76] Vickery, B. J., (1962), 'The behaviour at collapse of simple steel frames with tapered members,' *Struct. Eng.* **40**, 365–376.

[77] Horne, M. R., (1953), 'Determination of the shape of fixed-ended beams for maximum economy according to the plastic theory,' Final Report, *Int. Assoc. Bridge. and Struct. Eng.*, 4th Cong., Cambridge and London.

[78] Heyman, J., (1959), 'On the absolute minimum weight design of framed structures,' *Quart. J. Mech. and Appl. Maths.* **12**, 314–324.

[79] Heyman, J., (1960), 'On the minimum weight design of a simple portal frame,' *Int. J. Mech. Sci.*, **1**, 121–134.

[80] Mayeda, R. and Prager, W., (1967), 'Minimum weight design of beams for multiple loading,' *Int. J. Solids and Struct.* **3**, 1001–1011.

[81] Witteveen, J. and Tolman F. P., (1975), Discussions of Toakley, A. R. *et al.*, (Ref 72).

[82] Morris, L. J., (1972), 'Automatic design of sway frames,' *Proc Int. Symp Computer-Aided Struct Design*, Univ of Warwick, UK, B1.1–B1.15.

[83] Konig, J. A., (1975), 'On optimum shakedown design,' *Proc IUTAM Symp. 'Optimisation in Struct Design,'* Warsaw, A. Sawczuk and Z. Mroz (Ed.), Springer-Verlag, Berlin, 405–414.

[84] Gross, O. and Prager, W., (1962), 'Minimum weight design for moving loads,' *Proc. 4th US Cong. Appl. Mech.*, ASCE, New York, 2, 1047–1051.

[85] Save, M. A. and Prager, W., (1963), 'Minimum weight design of beams subjected to fixed and moving loads,' *J. Mech. Phys. Solids*, **11**, 255–267.

[86] Cohn, M. Z., (1979), 'Multi-criteria optimal design of frames,' *Plasticity by Mathematical Programming*, M. Z. Cohn and G. Maier (Eds.), Pergamon Press, (Proc. NATO Adv. Study Inst. Eng.), Univ. of Waterloo, Ontario, Canada.

[87] Thevendran, V. and Thambiratnam D. P., (1985), 'Optimum design of beams in multi-storey steel frames using the LRFD criteria,' *Eng. Optimisation* **9**, 21–36.

On Some Properties of
Optimum Structural Topologies

U. Kirsch
Department of Engineering Science and Mechanics
Virginia Polytechnic Institute and State University
Blacksburg, Virginia
United States of America
on leave from:
The Department of Civil Engineering
Technion, Haifa, Israel

Abstract Toplogical design, where the member connectivity is sought in addition to member sizing, is perhaps the most challenging of the structural optimization tasks. Due to the basic difficulties involved in the solution process, various simplifications and approximations are often considered. The present article introduces some typical characteristics and properties of the problem.

The topology of discrete structures is optimized by assuming zero lower bounds on cross-sections. It is shown that the optimal topology might correspond to a singular solution even for simple structures.

Assuming the force method analysis formulation, the problem can be stated in a linear programming form under certain assumptions. It is then possible to derive analytically some conditions related to optimal topologies. In addition, the difficulty of singular optimal solutions is eliminated.

The effect of compatibility conditions on optimal topologies is studied. It is shown that for particular geometries or loading conditions, where some element forces change from tension to compression or vice versa, the optimal topology might represent an unstable structure. Analytical conditions are derived to obtain geometries of multiple optimal topologies. It is shown how new optimal topologies are introduced from existing basic optimal topologies by combination rather than the common approach of elimination.

1 Introduction

A general optimal design probem often involves non-convex and highly nonlinear functions. Considerable work has been done in the last two decades on numerical optimum design of structures. Most of this work is on optimization of cross-sections

B. H. V. Topping (ed.),
Optimization and Artificial Intelligence in Civil and Structural Engineering, Volume I, 189–210.
© 1992 *Kluwer Academic Publishers. Printed in the Netherlands.*

and relatively little work is related to optimization of the topology. It is recognised, however, that optimization of the structural topology can greatly improve the design. That is, potential savings affected by layout optimization are generally more significant than those resulting from fixed-layout optimization. The fact that topological optimization did not enjoy the same degree of progress as fixed-layout optimization may be attributed to some basic difficulties involved in the solution process. These make the topological design problems, where the member connectivity is sought in addition to member sizings, perhaps the most challenging of the structural optimization tasks (Topping [1], Kirsch [2]).

Most of the investigations in structural optimization have been directed in two aspects:

(a) Fundamental properties of structures studied by means of analytical methods and simple idealized mathematical representations. Work on analytical methods, although sometimes lacking the practicality of being applied to realistic structures, is nonetheless of fundamental importance. It provides insight into the design problem and it often provides the theoretical lower bound optimum against which more practical designs may be judged. Analytical methods are most suited for fundamental studies of simple structural systems. The structural design is usually represented by a number of unknown functions and the goal is to find the form of these functions.

(b) Application of numerical methods to discretized models of practical design problems. Numerical methods usually employ mathematical programming techniques. A near optimal design is automatically generated in an iterative manner. An initial guess is used as starting point for a systematic search for better designs. The search is terminated when certain criteria are satisfied, indicating that the current design is sufficiently close to the true optimum.

The form of a discrete structure can usually be described by three types of design variables:

(a) The topological variables, defining the pattern of connection of members or the number and spatial sequence of elements, joints and supports in a structural system.

(b) The geometrical or configurational variables, describing the structural geometry (nodal coordinates).

(c) The sizing variables, representing the cross-sectional dimensions or member sizes.

Both the topological and the geometrical variables define the layout of the structure. While sizing variables may be optimized under either fixed or variable layout, the optimization of layout must always be accompanied or followed by sizing optimization. This necessity is due to changes in the analysis equations resulting from the change in structural layout. Because of the wide differences in the numerical behaviour of the three types of variables, their simultaneous inclusion in a single design problem is often avoided.

Layout optimization problems for discrete structures can broadly be divided into two classes:

Geometric optimization where coordinates of the joints and cross-sectional areas are treated as design variables. In general, the design variables are assumed to be continuous and numerical search algorithms are used to find the optimum. The topology is usually assumed to be fixed, unless some of the joints coalesce during the solution process. Some methods allow topological consideration at certain points during the design process.

Topological optimization where members are removed from a highly connected structure to derive an optimum topology with the corresponding cross-sections. That is, in a topological optimization problem both the topological and the sizing variables are optimized simultaneously. While the topological variables might be integer or discrete, the geometrical and sizing variables are usually assumed to be continuous.

The present article deals only with topological optimization. The geometric optimization problem is not considered here. One basic problem in topological optimization is that the structural model is itself allowed to vary during the design process. Discrete structures are generally characterized by the fact that the finite element model of the structure is not modified during the optimization process. In topological design, however, since members are added or deleted during the design process both the finite element model and the set of design variables change. These phenomena greatly complicate the design and analysis interactions.

Another difficulty is that the number of possible element-joint connectivities grows dramatically as the number of possible joint locations is increased. This number might be very large particularly in structures of practical size. One solution to this problem is to specify a reduced set that does not include all possible element-joint connectivities. However, a fundamental disadvantage of this approach is that the optimal solution may not be included in the specified set.

An additional difficulty that might be encountered during elimination of members is that the problem can have singular global optima that cannot be reached by assuming a continuous set of variables. This suggests that it may be necessary to represent some design variables as integer variables and to declare the existence or absence of a structural element. An example of an integer topological variable is a truss member joining two nodes which is limited to the values 1 (the member exists), or 0 (the member is absent). Other examples of integer topological variables include the number of spans in a bridge, the number of columns supporting a roof system, or the number of elements in a grillage system. While methods for mixed integer programming have been developed, these methods are still computationally costly for practical engineering applications.

2 Problem Statement

The above mentioned considerations have lead to various approximations and simplifications in the problem formulation. These include:

- approximate analysis models (e.g. rigid plastic);

- consideration of only certain constraints (e.g. stress constraints);

- simplified objective function (e.g. weight);

- consideration of simple structural systems (e.g. trusses);

- simplified sizing variables (e.g. cross-sectional areas);

- consideration of a limited number of loadings.

The most commonly used approximate topological optimization formulation is the linear programming (LP). While the displacement method formulation is the prevalent structural analysis tool in current computational practice, the force method formulation is adopted in the present work due to several reasons:

(a) The effect of compatibility conditions can be studied directly.

(b) The analysis model is most convenient to investigate properties of optimal topologies.

(c) A linear programming formulation can be obtained under certain assumptions.

Assuming the force method analysis formulation, the redundant forces N are given in terms of the design variables by the compatibility equations:

$$FN = \delta \tag{1}$$

and the elements forces A are given in terms of N by the equilibrium equations:

$$A = A_L + A_N N \tag{2}$$

In this formulation F is the flexibility matrix; δ are displacements corresponding to redundants due to the applied loads; A_L and A_N are forces due to the applied loads and unit values of redundants, respectively.

Most topological optimization studies are based on the assumption of an initial ground structure that contains many joints and members connecting them. Member areas are allowed to reach zero and hence can be deleted automatically from the structure. This permits elimination of uneconomical members during the optimization process (Dorn et al. [3], Hemp [4], Reinschmidt and Russell [5]).

In general, the problem of optimizing the topological and sizing variables subject to general constraints and objective function can be stated in a nonlinear programming form. However, due to the difficulties discussed earlier the following simplifications are often assumed:

(a) The objective function represents the weight and can be expressed in linear terms of the sizing variables.

(b) Only linear stress constraints and side constraints on the sizing variables are considered.

(c) Elastic compatibility conditions are temporarily neglected.

Using these approximations an LP formulation can be obtained. The LP solution is usually not the final optimum and some modifications to account for elastic compatibility and other constraints might be required. However, this solution may be viewed as a lower bound on the optimum. Some procedures that can be used to modify the lower bound solution are discussed elsewhere (Kirsch [2]). In certain cases, the LP solution may provide the final optimal topology.

Some of the advantages of the LP formulation are:

(a) Problems of singular optimal topologies are eliminated.

(b) Large structures with large numbers of members and joints can efficiently be solved.

(c) The global optimum is reached in a finite number of steps.

(d) Some properties of optimal topologies can be investigated conveniently.

Assume a grid of points that may be connected by many potential truss members to form a ground structure with a finite number of members. Considering the above mentioned simplifications, the primal LP problem is: Find the cross-sectional areas X and the member forces A such that

$$Z = l^T X \to \min \qquad \text{(objective function)} \qquad (3)$$

$$-\sigma^L X \le A \le \sigma^U X \qquad \text{(stress constraints)} \qquad (4)$$

$$KA = P \qquad \text{(eqilibrium conditions)} \qquad (5)$$

in which l is the vector of members lengths; σ^L and σ^U are allowable stresses; K is a matrix of member direction cosines; and P is the applied load vector.

To further simplify the solution process, the member forces can be expressed in terms of a set of redundant forces by the equilibrium equation (2). Substituting Eq. (2) into Eq. (4), the LP problem (3)–(5) becomes: Find X and N such that

$$Z = l^T X \to \min \qquad (6)$$

$$-\sigma^L X \le A_L + A_N N \le \sigma^U X \qquad (7)$$

This formulation is superior to the former one in terms of problem size. Specifically, the equalities (5) have been eliminated and the number of variables is reduced (only the redundant forces are considered as independent variables instead of all member forces). The stress constraints (7) are formulated for all given loading conditions. Side constraints on both cross-sectional areas and members forces can be included in the LP formulation.

Axis transfer techniques can be applied to ensure nonnegativity of the force variables. The LP has the ability to eliminate unnecessary members, thereby allowing optimization of the structural topology. Some properties of the LP solutions are useful in studying the corresponding optimal topologies.

3 Types of Structure

Assuming an arbitrary initial topology, the resulting optimal design might represent one of the following classes of structure:

- statically indeterminate structure (SIS)

- statically determinate structure (SDS)

- unstable structure (USS)

Elimination of elements from the structure will change the numbers of variables and potentially active constraints.

It should be noted that if buckling constraints are considered, the areas of compression elements might not converge to zero. In a complete nonlinear programming formulation, as the area of a compression element decreases, the buckling stress also decreases, until it becomes critical; then the area of the element will increase.

3.1 Statically Determinate Structures

Assuming an initial SIS and solving the LP for zero lower bounds on cross-sections, it is possible that the resulting optimal design will represent an SDS. That is, for certain selection of redundant forces one can obtain at the optimum $N = 0$. In this case the element forces A are constant, compatibility conditions can always be satisfied and the LP solution is the final optimum. It has been shown (Dorn *et al.* [3]) that for structures subjected to a single loading condition with $X^L = 0$, the optimal solution indeed will represent an SDS. This results while not guaranteed, might be obtained also for structures subjected to multiple loading conditions.

Assume that I is the number of elements and K is the degree of statical indeterminacy. In the case of a single loading condition, the number of variables for the initial SIS is $I + K$. If each element is assumed to be fully stressed, the number of critical constraints at the optimum is I. For a nondegenerate optimal solution the number of basic (nonzero) variables at the optimum is I, and the number of nonbasic (zero) variables is K. For a degenerate optimal solution, the number of nonzero variables at the optimum is smaller than I. Both cases might represent a SDS.

3.2 Unstable Structures

The number of members eliminated during the LP solution might be larger than the degree of statical indeterminacy and the optimal topology might represent an USS. The optimal structure in this case is satisfying equilibrium and stress constraints but there are unstable members. This situation might occur for certain geometries or loading conditions, where the forces in some elements change from tension to compression or vice versa. These elements are not required to maintain equilibrium for that particular geometry or loading condition, and will be eliminated during the LP solution. A possible solution to this problem is to add members to the optimal configuration. They are not needed to satisfy equilibrium, but they are required to satisfy the necessary relationship which exists between the joints and the members in a stable structure.

The truss shown in Fig. 1 is subjected to three loads, P_1, P_2, and $P_1 - P_2$, acting simultaneously. Optimal topologies for various P_1/P_2 ratios are shown in Fig. 2. The main observations that can be made are as follows (Kirsch [6]):

- For $P_2 < P_1/2$ and $P_1/2 < P_2 < P_1$, the optimal topologies are SDS (Figs 2(a) and 2(c)).

- For $P_2 = P_1/2$ (Fig. 2(b)) the optimal topology is an USS that can be viewed as a particular case of the former two topologies. The diagonal element (Fig. 2(a)) or the vertical element (Fig 2(c)) become zero for this particular loads ratio.

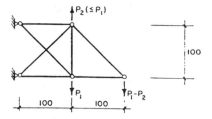

Figure 1. *Truss example.*

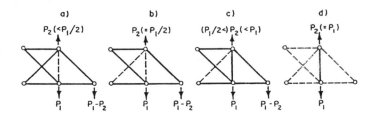

Figure 2. *Optimal topologies.*

- For $P_1 = P_2$ (Fig. 2(d)) the optimal topology is reduced to a single unstable vertical element.

These results illustrate how unstable topologies might be obtained for particular loading cases.

3.3   Statically Indeterminate Structures

In cases where the optimal LP solution $\tilde{X}_{LP}$, $\tilde{N}_{LP}$, represents an SIS, compatibility conditions might not be satisfied. The vector Δ_{LP}, defined by:

$$\Delta_{LP} = \tilde{F}_{LP}\tilde{N}_{LP} - \tilde{\delta}_{LP} \tag{8}$$

indicates the discrepancy in satisfying the compatibility conditions by the optimal LP solution. The tilde and subscripts LP in (8) denote optimal values of the LP problem. For certain geometries or loading conditions it is possible that

$$\Delta_{LP} = 0 \tag{9}$$

In this case compatibility conditions are satisfied for the optimal SIS and the LP solution is the final optimum. A case of particular interest is one where the LP problem possesses an infinite number of optimal solutions representing multiple optimal topologies. Such geometries will be discussed in the next section.

If the condition (9) is not satisfied for the optimal SIS, it is still possible to maintain compatibility by applying a set of prestressing forces (Kirsch [7]). This can always be done for structures subjected to a single loading condition. In the case of multiple loading conditions, a single set of prestressing forces does not ensure that compatibility conditions will be satisfied at the optimum for all loading conditions. Therefore, the optimal LP topology might be different from that obtained by solving the complete problem. Yet, the optimal LP topology is a lower bound on the optimum and may be used as a reference for comparison with other solutions.

In order to compare between the LP solution and the true optimum it is necessary to find the elastic force distribution corresponding to the LP optimum $\tilde{N}_{EL}$. This can be done by analysis of the optimal structure (Eq. (1)). It should be emphasised that a certain deviation from elastic force distribution is often allowed on account of the inelastic behaviour. The allowable deviation can be defined by the set of linear constraints

$$\alpha^L \tilde{N}_{EL} \leq \alpha \tilde{N}_{LP} \leq \alpha^U \tilde{N}_{EL} \qquad (10)$$

where α, α^L, and α are matrices of given parameters. In cases where the constraints (1) are not satisfied for any loading condition, modifications of the LP optimum are required. Several procedures have been proposed for this purpose (Kirsch [2], Reinschmidt and Russell [5], Schmit et al. [8,9]).

4   Singular and Local Optima

One reason for excluding compatibility conditions from optimization of the ground structure layout is that the computational effort in the solution process is considerably reduced. An additional difficulty is that the optimal topology might correspond to a singular point in the design space. This occurs since a change in the structural topology (elimination of one or more elements) will result in a corresponding modification of the compatibility equations and elimination of some constraints previously included in the problem statement. That is, a change in the structural tolology might change the design space. If the optimal solution is a singular point in the design space, it might be difficult or even impossible to arrive at the true optimum by numerical search algorithms. The singularity of the optimal topology in truss structures was first shown by Sved and Ginos [10]. Singular optima of grillages have been studied recently (Kirsch [11,12]).

Another difficulty is that several local optima representing different optimal topologies might exist. These problems of computational effort, singularity and local optima have motivated applications of the LP approach where compatibility

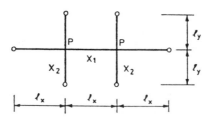

Figure 3. *Simple grillage.*

conditions are not considered in optimization of the ground structure layout. If the optimal layout represents a statically indeterminate structure, satisfaction of compatibility equations can be checked.

The problems of local optima and singular solutions are demonstrated by the simple grillage shown in Fig. 3 (Kirsch [12]). Assume $P = 10.0$ acting perpendicular to the grillage plane, $\ell_x = 1.0$, $\ell_y = 1.4$, stress constraints $|\mathbf{M}| \leq \mathbf{M}^A$ ($\mathbf{M}^A =$ allowable moments) and linear objective function. Results of the complete problem (considering compatibility conditions) are demonstrated for two cases:

(a) $X_i =$ depth of cross-section, width of cross-section $= 1.0$ (Fig. 4a).

(b) $X_i =$ width of cross-section, depth of cross-section $= \sqrt{6}$ (Fig. 4b).

Three local optima representing three different topologies have been obtained in case (a), with objective function values 80.6, 144.7 and 125.7. In case (b) the constraints $X_2 \geq \epsilon_x$ is assumed (ϵ_x being some small value). The resulting optimum is $\tilde{X}_1$ with the constraint $M_2 \leq M_2^A$ being active. Assuming $\epsilon_x = 0$ the modified solution is $\tilde{X}_2$, which is a singular global optimum, with the latter constraint being deleted. This global optimum can be reached by the LP formulation without compatibility conditions.

5 Multiple Optimal Topologies

Solving the LP problem, it is possible that for certain geometries identical optimal objective function values are obtained for an infinite number of force distributions in the structure. The various optimal force distributions usually correspond to several different optimal topologies. Some properties of such particular geometries where the optimal objective function value becomes independent of the force distribution are discussed subsequently.

For purpose of investigation, assume a structure subjected to a single loading condition such that each element is fully stressed. In addition, no lower bounds

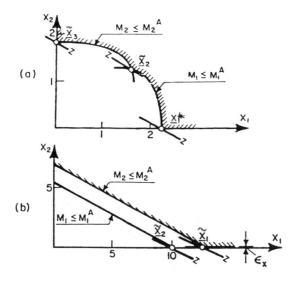

Figure 4. *Local and singular optima.*

are imposed on cross-sections. From (7) one can obtain for the i-th element

$$A_i = A_{Li} + A_{Ni}N = \sigma_i^A X_i \tag{11}$$

in which σ_i^A is the allowable stresses ($\sigma_i^A = \sigma_i^U$ or $\sigma_i^A = -\sigma_i^L$).

Substituting X_i from (11) into (6) yields

$$Z = \sum_{i=1}^{I} \frac{\ell_i}{\sigma_i^A}\left(A_{Li} + A_{Ni}N\right) \tag{12}$$

Define the constant Z_0 by

$$Z_0 = \sum_{i=1}^{I} \frac{\ell_i}{\sigma_i^A} A_{Li} \tag{13}$$

Substituting (13) into (12) yields

$$Z = Z_0 + \sum_{i=1}^{I} \frac{\ell_i}{\sigma_i^A} A_{Ni}N \tag{14}$$

The objective function will become independent of the force distribution in the structure if

$$\sum_{i=1}^{I} \frac{\ell_i}{\sigma_i^A} A_{Ni} = 0 \tag{15}$$

The conditions (15) form a system of K equations expressed in terms of the geometric parameters. If the number of equations is equal to the number of geometric parameters then it might be possible to find a geometry satisfying the conditons (15). In this case the objective function is independent of N, and the various force distributions corresponding to different combinations of eliminated elements are optimal. That is, multiple optimal topologies (MOT) are obtained for this particular geometry.

Solving the LP for a specific geometry, it is assumed that the optimal solution will correspond to a certain vertex A, with an optimal force distribution $\tilde{N}_A$. In cases where an infinitesimal change in the geometric parameters will result in a new optimal vertex B, one can find the geometry where the transition in the set of active constraints at the optimum occurs. The objective function contours for this geometry are parallel to the boundary of the feasible region, and the LP problem possesses an infinite number of optimal solutions and several corresponding optimal topologies.

Consider an initial ground structure with I members and K redundant forces, subjected to L loading conditions. The constraints (7) for the i-th member and any specific loading condition become

$$X_i + \sum_{k=1}^{K} a_{ik}^U N_k \geq b_i^U \tag{16}$$

$$X_i + \sum_{k=1}^{K} a_{ik}^L N_k \geq b_i^L \tag{17}$$

in which the constants a_{ik}^U, b_i^U, a_{ik}^L, b_i^L are given by

$$\left.\begin{array}{ll} a_{ik}^U = -\dfrac{A_{Nik}}{\sigma_i^U} & b^U = \dfrac{A_{Li}}{\sigma_i^U} \\[2mm] a_{ik}^L = \dfrac{A_{Nik}}{\sigma_i^L} & b^U = -\dfrac{A_{Li}}{\sigma_i^L} \end{array}\right\} \tag{18}$$

Only one of the two constraints (16), (17) might be active, that is, the number of active constraints at the optimum cannot exceed $L \cdot I$.

The constraints (16), (17) can be expressed as

$$X + a_l N_l - S_l = b_l \qquad l = 1, \ldots, L \tag{19}$$

in which l is a subscript denoting the loading condition; the elements of a_l and b_l are constant; and S_l are vectors of surplus variables. Assume a basic feasible solution of the LP (14), (19), given in a canonical form

$$\begin{Bmatrix} X_B \\ N_B \\ S_B \end{Bmatrix} + a \begin{Bmatrix} X_N \\ N_N \\ S_N \end{Bmatrix} = b \tag{20}$$

$$Z = Z_0 + C_1^T X_N + C_2^T N_N + C_3^T S_N \tag{21}$$

where the elements of a, b, C and Z_0 are constant, and subscripts B and N denote basic and nonbasic variables, respectively. The LP solution is optimal if $C_j \geq 0$ $(j = 1, 2, 3)$. If the latter conditions hold and at least one C_j equals zero, then the problem might have multiple basic optimal solutions representing MOT. The basic optimal topologies may be used to introduce new optimal topologies, not corresponding to the basic solutions. If $C_j < 0$ at least for one nonbasic variable, an improved basic feasible solution with a corresponding topology and a modified set of active constraints can be introduced.

Assume an MOT geometry where all basic optimal solutions with a corresponding set of T topologies have been determined. It is possible now to introduce linear combinations of the basic feasible solutions

$$\left. \begin{aligned} X &= \sum_{t=1}^{T} \alpha_t X_t \\ N &= \sum_{t=1}^{T} \alpha_t N_t \end{aligned} \right\} \tag{22}$$

in which α_t are coefficients satisfying

$$\left. \begin{aligned} 0 &\leq \alpha_t \leq 1 \qquad t = 1, \ldots, T \\ \sum_{t=1}^{T} \alpha_t &= 1 \end{aligned} \right\} \tag{23}$$

Equations (22), (23) provide multiple optimal cross sectional areas and force distribution corresponding to all optimal topologies. New combined optimal topologies are obtained for α_t values other than 0 and 1. Based on these definitions, any specific basic optimal topology is given by a certain α_t equals unity and all the remaining α_t equal to zero. The combined optimal topologies are defined by the linear combination

$$E = \sum_{t=1}^{T} \alpha_t E_t \tag{24}$$

where E_t are vectors of 0/1 elements, describing the existence or absence of members. For any assumed values of α_t satisfying Eq. (23) one can check for each member i its existence $(E_i > 0)$ or absence $(E_i = 0)$. The cross sectional areas and the force distribution for the combined optimal topologies may vary in the range determined by the α_t values. The combined topologies usually represent SIS, but compatibility condition might be satisfied for some α_t values.

One important property of the MOT geometry is that new optimal topologies are introduced from existing basic optimal topologies by combination rather than the common ground structure approach of elimination. Part of the optimal topologies are usually statically determinate structures, therefore the LP solution is the final elastic optimum.

The grillage shown in Fig. 5 is subjected to 21 equal loads P, acting at all joints. Assuming $\ell_x \cdot \ell_y = 1.0$ and a different section for each element, the optimal topologies for the LP formulation and various ℓ_y/ℓ_x ratios are shown in Fig. 6 (Kirsch [12]). Evidently, for small ℓ_y/ℓ_x ratios the optimal structure consists of only ℓ_y beams (topology I). Increasing this ratio gradually, the central ℓ_x beam is first included (topology II). Then the two side beams in this direction are retained (topology III) while the ℓ_y beams are eliminated (topologies III-VI), with only the ℓ_x beams retained for large ℓ_y/ℓ_x ratios (topology VI). None of the optimal topologies is similar to the original one (Fig. 5). Since the number of geometric parameters (ℓ_x, ℓ_y) is two and the number of redundants is eight, Eq. (15) cannot be solved and there is no MOT geometry. Modifying the structure to one with a single uniform beam in the ℓ_x direction and seven beams with different lengths in the ℓ_y direction (Fig. 7), the following MOT geometry is obtained by (15)

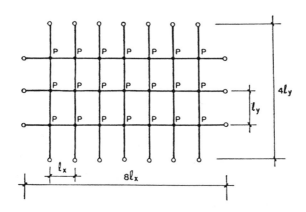

Figure 5. *Grillage example.*

$$\ell_{y1} = 2\,\ell_x \qquad \ell_{y2} = 2.2828\,\ell_x \qquad \ell_{y3} = 3.464\,\ell_x \qquad \ell_{y4} = 4\,\ell_x$$

The objective function becomes independent of the redundant forces

$$Z = 64\,P\ell_x^2$$

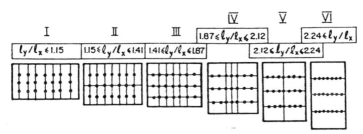

Figure 6. *Optimal topologies.*

The various optimal topologies corresponding to the basic feasible solutions of the LP problem are shown in Fig. 8. All the sixteen topologies give an identical optimal Z value. While the original topology (Fig. 7) does not correspond to any basic feasible solution, it possesses an identical optimal Z value. Since some of the optimal topologies represent statically determinate structures (topologies 1 and 16), no modifications are required to account for elastic compatibility.

The 15-bar truss shown in Fig. 9 is subjected to three loads P acting simultaneously. Assume non-uniform depth with a single geometric variable Y, $\sigma^U = -\sigma^L = 20.0$ and $P = 10.0$ (all dimensions are in kips and inches). Two SDS basic optimal topologies are obtained for $Y = 100$ (Figs 10a, 10b), and a SIS combined optimal topology (Fig. 10c) is introduced by (equations (22)–(24), Kirsch [13])

$$X = \alpha_a X_a + \alpha_b X_b \qquad\qquad 0 < \left\{ \begin{matrix} \alpha_a \\ \alpha_b \end{matrix} \right\} < 1$$

$$N = \alpha_a N_a + \alpha_b N_b$$

$$E = \alpha_a E_a + \alpha_b E_b \qquad\qquad \alpha_a + \alpha_b = 1$$

where $E_a^T = \{1,0,1,1,1,1,0,1\}$ and $E_b^T = \{1,1,0,1,1,1,1,0\}$. Variation of $\tilde{Z}$ with Y and the corresponding optimal topologies are shown in Fig. 11. An unstable topology (Fig. 10d) is obtained for $Y = 75$, where some active constraints of topology a (in members 3 and 8) change from tension to compression. It can be noted that Z is not sensitive to changes in Y. Assuming uniform depth Y (Fig. 12), MOT have been obtained for any positive Y (Kirsch [13]).

For two fixed supports, nineteen optimal topologies can be introduced (Fig. 13). The five SDS basic optimal topoloogies are shown in Fig. 13a, four of them being unstable and two are non-symmetric structures. However these two types of structure are needed to introduce the complete set of fourteen SIS combined optimal

204

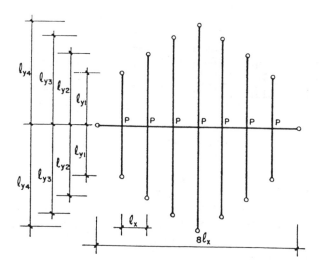

Figure 7. *Modified grillage example.*

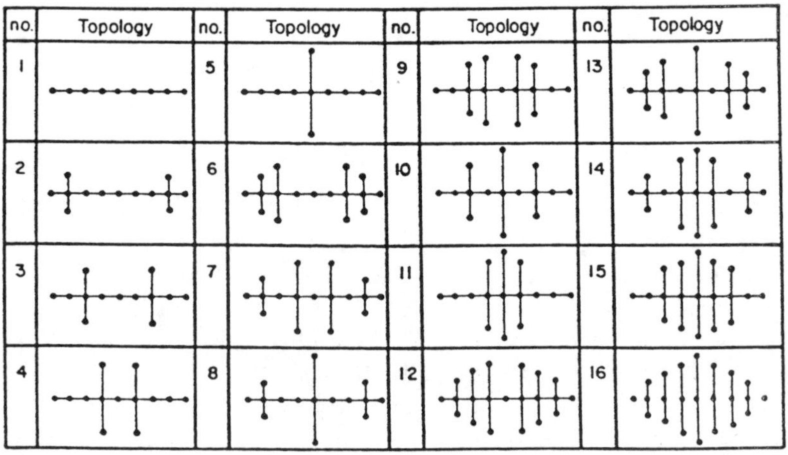

no.	Topology	no.	Topology	no.	Topology	no.	Topology
1		5		9		13	
2		6		10		14	
3		7		11		15	
4		8		12		16	

Figure 8. *Optimal topologies.*

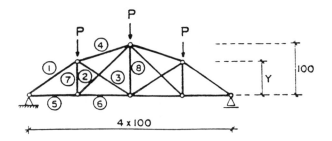

Figure 9. *Fifteen-bar truss.*

topologies (Fig. 13b), five of which also being unstable structures. Variations of Z with Y is shown in Fig. 14. The assumption of two fixed supports significantly improved the optimum ($\tilde{Z} = 632.4$ versus $\tilde{Z} = 836.7$ for a single fixed support). In both cases Z is not sensitive to changes in Y near the optimum. It has been found that compatibility conditions are satisfied for the SIS combined optimal topologies (Kirsch [13]).

6   Concluding Remarks

It has been noted that some basic difficulties greatly complicate the design and analysis interactions and make the topological design problem perhaps the most challenging of the structural optimization tasks. To overcome the difficulties involved in the solution process, various simplifications and approximations have been proposed. These include approximate analysis models, consideration of only simple structural systems and a limited number of loadings, consideration of only certain constraints, simplified variables and objective function. Despite all the difficulties, it is recognised that substantial savings can be achieved in topological optimization compared with sizing optimization.

A major difficulty in solving problems of optimal topology is that the solution might represent a singular point in the design space. It has been shown that this may occur even in simple trusses subjected to a single loading condition. In such cases it might be difficult or even impossible to arrive at the true optimum by numerical search algorithms used to solve the general nonlinear programming problem.

Neglecting compatibility conditions and formulating the problem as a linear program, it is possible to overcome this difficulty and to find a lower bound optimal topology. In cases where this topology represents an SDS, compatibility conditions must not be considered and the LP solution is the final optimum. If the optimal LP topology is an SIS, compatibility conditions might not be satisfied and the solution

206

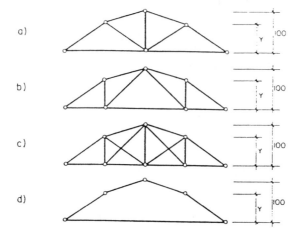

Figure 10. *Optimal topologies.*

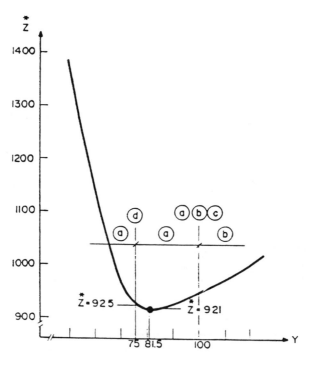

Figure 11. *Variation of $\tilde{Z}$ with Y.*

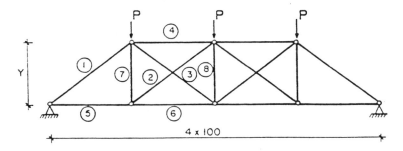

Figure 12. *Uniform depth truss.*

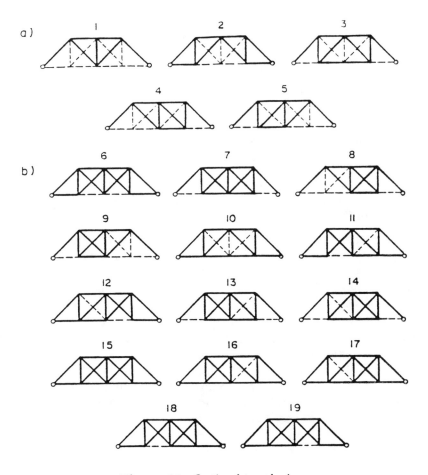

Figure 13. *Optimal topologies.*

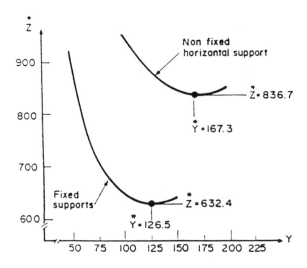

Figure 14. *Variation of $\tilde{Z}$ with Y.*

should be modified accordingly.

It has been shown that for particular geometries or loading conditions the optimal topology might represent an USS. In such cases forces in some elements change from tension to compression or vice versa. A possible solution to this problem is to add members to the optimal configuration.

For certain geometries the optimal objective function value might become independent of the force distribution in the structure. For these particular MOT geometries, there is a transition in the set of active constraints at the optimum. The various basic optimal solutions represent several corresponding basic topologies that can be identified by changing the basic variables. A procedure to introduce additional optimal topologies by linear combinations of the basic topologies is demonstrated. The forces and the cross-sectional areas for the combined topologies may vary in a range determined by the corresponding basic topologies. Some of the latter topologies usually respresent SDS, therefore compatibility conditions can always be satisfied. In cases where compatibility conditions are not satisfied for an SIS topology, the LP solution can be viewed as a lower bound on the optimum. One significant property of MOT geometries is that new optimal topologies are introduced from the known basic topologies by combination rather than the common appraoch of elimination.

Potential applications of topological optimization techniques are numerous. However, these methods are still in the stage of early development and progress is much needed in this area.

References

[1] Topping, B. H. V., (1983), 'Shape optimization of skeletal structures: A review,' *Jnl. of Structural Engineering, ASCE* **109**, 1933–1951.

[2] Kirsch, U., (1989), 'Optimal topologies of structures,' *Applied Mechanics Review*, (In press).

[3] Dorn, W. S, Gomory R. E. and Greenberg, H. J., (1964), 'Automatic design of optimal structures,' *Jnl. de Mecanique* **3**, 25–52.

[4] Hemp, W. S., (1973), *Optimum Structures*, Clarendon Proess, Oxford, UK.

[5] Reinschmidt, K. F. and Russell, A. D., (1974), 'Applications of linear programming in structural layout and optimization,' *Computers and Structures* **4**, 855–869.

[6] Kirsch, U., (1988), 'Optimal topologies of truss structures,' *Computer Methods in Applied Mechanics and Engineering*, **72**, 15-28.

[7] Kirsch, U., (1989), 'The effect of compatibility and prestressing on optimized trusses,' *Jnl. of Structural Engineering, ASCE* **115**, 724–737.

[8] Farshi, B. and Schmit, L. A., (1974), 'Minimum weight design of stress limited trusses,' *Jnl. of the Structural Div., ASCE* **100**, 97–107.

[9] Sheu, C. Y. and Schmit, L. A., (1972), 'Minimum weight design of elastic redundant trusses under multiple static loading conditions,' *AIAA Jnl.* **10**, 155–162.

[10] Sved, G. and Ginos, Z., (1968), 'Structural optimization under multiple loading,' *Int. Jnl. Mech. Sci.* **10**, 803–805.

[11] Kirsch, U. and Taye, S., (1986), 'On optimal topology of grillage structures,' *Engineering with Computers* **1**, 229–243.

[12] Kirsch, U., (1987), 'Optimal topologies of flexural systems,' *Engineering Optimization* **11**, 141–149.

[13] Kirsch, U., (1989), 'On the relationship between optimum structural geometries and topologies,' (submitted for publication).

The Theorems of
Structural and Geometric Variation
for Engineering Structures

B.H.V. Topping
Department of Civil Engineering
Heriot-Watt University, Riccarton,
Edinburgh, EH14 4AS, United Kingdom

Abstract   This paper reviews the application of the theorems of structural and geometric variation to optimization, design and non-linear analysis.

1 Introduction

The repeated use of matrix displacement analysis for determining design sensitivities during an optimization procedure is both time consuming and costly. This is the case when the problem is particularly large and insensitive or large numbers of simultaneous equations have to be solved repeatedly. To avoid such repeated analysis a number of re-analysis algorithms [3, 5] have been developed. The objective [5] of these algorithms is:

> *"to find the response of the modified structure using the original response of the structure such that the computational effort is less than that required for a fresh analysis."*

One technique which is particularly attractive, as it may be interpreted in a physical sense, is that which uses the theorems of structural variation. These were developed originally for use with trusses [10, 12] and rigidly jointed frames [15]. The theorems have since been further developed [17, 19] for use with finite element problems.

More recently, the theorems have been employed to account for variations in the coordinates of structural joints [1, 22, 23]. These geometric theorems of variation have been shown to be a generalisation of the structural theorems of variation. They may be applied both to geometric and material non-linear analysis and optimization problems.

B. H. V. Topping (ed.),
Optimization and Artificial Intelligence in Civil and Structural Engineering, Volume I, 211–222.
© 1992 *Kluwer Academic Publishers. Printed in the Netherlands.*

212

2 The Theorems of Structural Variation

These theorems may be used to predict the internal member forces and the joint displacements throughout the structure when the material or cross-sectional properties of the member are altered or when one or more members are removed altogether. It may be shown [10, 12] that the internal force in member j of a pin-jointed structure, in which the cross-sectional area of member i has been modified, is given by:

$$P_j^* = P_j + r_i f_{ji} \tag{1}$$

where: $j \neq i; P_j^*, P_j$ are the internal forces in member j under the applied loads in the modified and original structure repectively; and f_{ji} is the internal force in member j of the original structure caused by unit loads applied at the ends of the member i (acting parallel to and in a direction to put member i in tension). The internal force in the modified member i is given by:

$$P_i^* = P_i + r_i(1 + \alpha f_{ii}) \tag{2}$$

where the change in member i is defined by $\alpha = dA_i/A_i$ and the new area is $A' = A_i + dA_i = (1 + \alpha)A_i$. For total removal of member i, $A_i' = 0$ and $dA_i = -A_i$ and hence $\alpha = -1$.

The scale factor r_i, called the variation factor, is determined by equilibrium conditions. These may be established by the combination of the analyses of the original structure under the applied and unit loads. Such equilibrium conditions at the ends of the modified member give the following equation:

$$r_i = -\alpha P_i/(1 + \alpha f_{ii}) \tag{3}$$

The displacements in the modified structure, δ^*, may be determined by superposition thus:

$$\delta^* = \delta + r_i \delta_i \tag{4}$$

where δ and δ_i are the displacements of the original structure subject to the applied and unit loads respectively.

It is important to note that this procedure only allows for the modification of one member at a time. Subsequent modifications may be performed provided all the f and δ vectors are modified for changes to the $i - th$ member.

Bakry [7] has shown that simultaneous modification of two or more members may be undertaken using the theorems. Bakry's approach may be rewritten using the above notation as follows:

If n members are to be modified by $\alpha_i, \alpha_j, \alpha_k, ...\alpha_n$, then n equilibrium conditions, at the ends of each of the members, must be considered using the analysis for

n pairs of unit loads scaled with the n factors $r_i, r_j, r_k, \ldots\ldots r_n$. These n equilibrium conditions yield a series of simultaneous linear equations that may be expressed in the following general form and solved to determine the variation factors:

$$\alpha_i P_i = -r_i - \alpha_i \sum_{j=1}^{n} r_j f_{ij} \qquad\qquad i = 1, \ldots n \qquad\qquad (5)$$

The internal forces in the modified members may be calculated using the expression:

$$P_i^* = (1 + \alpha_i)P_i + (1 + \alpha_i) \sum_{j=1}^{n} r_j f_{ij} \qquad\qquad (6)$$

For unaltered members the internal forces are given by:

$$P_m^* = P_m + \sum_{j=1}^{n} r_j f_{mj} \qquad\qquad (7)$$

and the joint displacements may be calculated using:

$$\delta^* = \delta + \sum_{j=1}^{n} r_j \delta_j \qquad\qquad (8)$$

Again, if members are to be further modified, the analysis under the pairs of unit loads must be adjusted using the above procedure for the alterations to the n members.

Atrek [6] derived a simplified form of equation (6) using matrix notation. In the simplified form the relationship between the forces in the modified members and the variation factors is uncoupled. The resulting equation involves less computations and may lead to considerable economy when the number of modified members is large. A mathematical study of the theorems was undertaken by Filali [9].

The application of the these theorems to rigidly jointed frames is formulated in a similar manner [15] and for finite element problems may be determined by analogy with pin-jointed trusses [17, 19]. A generalised formulation is discussed in reference [8].

The theorems of structural variation [10, 12] may be summarised thus: *The internal forces and nodal displacements of a modified structure may be determined from the factored analysis of the original structure subject to the applied loading and a number of unit loading cases.* The method is therefore of the influence coefficient category. It is important to note that the modifications may include removal of members.

2.1 Applications of the Theorems of Structural Variation

These theorems have been used in the optimization of trusses [10, 13] and rigidly jointed frames [14, 15]. An important application is in the design of structures of variable topology. One approach developed by Majid and Elliott [10, 12] for trusses and by Majid and Saka [14] for frames is to combine several candidate topologies to form a 'ground structure'. Optimization procedures may then be used to determine the optimum member cross-sectional properties and to remove structurally inefficient members by using the theorems to provide re-analysis information and design sensitivities. Another approach to selecting optimum topology is to move the coordinates of structural joints to their most efficient positions during the design procedure. Both these approaches to the automated design of the topology of skeletal structures are well established [20]. In the second approach, in which geometric variables are included among the design variables, the re-analysis cannot be accomplished using the theorems of structural variation. The re-analysis may be accomplished, however, by using the theorems of geometric variation [22].

The theorems have also been used for the nonlinear analysis of space trusses [16] and the elasto-plastic analysis of frames [11]. The application of the theorems of structural variation to linear and nonlinear finite element problems was discussed in references [1, 2, 21]. It was concluded, in the second of these papers, that there are a number of difficulties in applying the structural theorems to nonlinear finite element problems. Firstly, there is the difficulty of defining for a particular element. If the elasticity matrix, [D], is not changing linearly then it is not possible to define a single coefficient for each element to specify the change in structural properties. For example, in elasto-plastic analysis, each coefficient of [D] will generally reduce by different amounts of the plastic part of the matrix [D]. Secondly, the structural theorems may become inefficient, requiring the calculation of internal forces that are not generally determined during a finite element analysis. For nonlinear finite element analysis problems the structural theorems appear to be limited. For linear finite analysis problems such as iterative design or the analysis of modified structures, the structural theorems may be used efficienctly [1, 21] for a range of types of idealisations. Care must, however, be exercised to ensure the most suitable element types are used for the idealisation.

Numerous different design alternatives may be considered, at little extra computational expense, by using the theorems of structural variation to provide an 'exact' analysis.

3 The Theorems of Geometric Variation

The formulation of the theorems of geometric variation will be given here to enable comparison with the structural theorems discussed above.

If the coordinates of a skeletal structure are to be varied, then the member lengths and angles between the members connected to the joints will also be varied. The latter variation is termed rotation since the clockwise angle of the member to the vertical axis changes. The former variation is termed elongation since the member length changes.

If end j of the $i - th$ member in a truss is varied the member undergoes rotation and elongation. The internal force carried by the the ith member may be determined, if first the original structure is analysed for unit loads applied at each degree of freedom of the $j - th$ and $k_t h$ ends of member i. There are a total of four unit loads for each modified member as a result of joint variation. The internal force carried by member i under the action of a horizontal unit load at joint j is given by:

$$f'_{i,jh} = E_i A_i \{(L_i^* - L_i')/L_i'\} \tag{9}$$

and

$$L_i' = \sqrt{\{(X_k - X_j)^2 + (Y_k - Y_j)^2\}} \tag{10}$$

$$L_i^* = \sqrt{\{[(X_k + \delta_{kh,jh}) - (X_j + \delta_{jh,jh})]^2 + [(Y_k + \delta_{kv,jh}) - (Y_j + \delta_{jv,jh})]^2\}} \tag{11}$$

where:

$f'_{i,jh}$ = internal force of the $i - th$ modified member resulting from a unit horizontal load;

L_i' = length of the $i - th$ member after joint variation;

L_i^* = is termed the 'stretch' length, assuming that the displacements are the same as before variation;

$X_j,\ Y_j,\ X_k,\ Y_k,$ = coordinates of joint j and k respectively after joint variation;

$\delta_{jh,jh},\ \delta_{jv,jh},$ = displacements at joint j arising from a unit horizontal load at joint j of the original structure; and

$\delta_{kh,jh},\ \delta_{kv,jh},$ = displacements at joint k arising from a unit horizontal load at node j of the original structure.

A similar relationship also holds for the unit vertical load at end j. In addition, the member suffers a change in rotation from Θ_{ik} to Θ_{ik}' owing to the joint variation. The subscript ik denotes that the angle is measured at end k of the $i - th$ member clockwise from the positive vertical axis. The elongation and rotation will result in unbalanced forces and these are termed compensating forces. The compensating forces at joint k corresponding to the unit horizontal load at joint j are:

216

$$C_{kh,jh} = f_{i,jh} \sin\Theta_{ik} - f'_{i,jh} \sin\Theta'_{ik} \tag{12}$$

$$C_{kv,jh} = f_{i,jh} \cos\Theta_{ik} - f'_{i,jh} \cos\Theta'_{ik} \tag{13}$$

where:

$C_{kh,jh}$ = horizontal compensating force at joint k, arising from the modification of member i under a unit horizonatal load applied at joint j;

$C_{kv,jh}$ = vertical compensating force at joint k, arising from the modification of member i under unit horizontal load applied at joint j;

Θ_{ik} = original clockwise angle to the positive vertical axis at node k of the $i-th$ member; and

Θ'_{ik} = modified clockwise angle to the vertical at node k of the $i-th$ member.

Compensating forces at end j arising from the rotation of member i under the action of a unit horizontal load at joint j may be calculated similarly. The compensating forces for other unit loads jv, kh and kv are evaluated using similar equations to those above. The various compensating forces for a member are tabulated below. For solution, a total of 16 compensating forces must be evaluated using equations (12) and (13).

Table 1 Compensating Forces

Compensating Forces		Horiz	Vert	Horiz	Vert
At joint j	Horiz	$C_{jh,jh}$	$C_{jh,jv}$	$C_{jh,kh}$	$C_{jh,kv}$
	Vert	$C_{jv,jh}$	$C_{jv,jv}$	$C_{jv,kh}$	$C_{jv,kv}$
At joint k	Horiz	$C_{kh,jv}$	$C_{kh,jv}$	$C_{kh,kh}$	$C_{kh,kv}$
	Vert	$C_{kv,jh}$	$C_{kv,jv}$	$C_{kv,kh}$	$C_{kv,kv}$

Thus far the discussion has been based on a single modified member. If there are N members connected to joint k, all the N members will sustain elongations and rotations. Unit load analyses at the other ends of the N members are therefore also required. The net compensating forces at a typical joint k is the sum of the member contributions. This is given by:

$$C_{kh,jh} = \sum_{i=1}^{N}(f_{i,jh} \sin\Theta_{ik} - f'_{i,jh} \sin\Theta'_{ik}) \tag{14}$$

$$C_{kv,jh} = \sum_{i=1}^{N}(f_{i,jh} \cos\Theta_{ik} - f'_{i,jh} \cos\Theta'_{ik}) \tag{15}$$

If the applied loads are at the joints of the modified members, the equilibrium equations may be formed. This may be undertaken by using the principle of superposition of unit load analyses and applied loads. The compensating forces are evaluated from the unit load analyses and if these are scaled the equilibrium equations are:

$$
\begin{aligned}
(1 + C_{jh,jh})r_{jh} &+ C_{jh,jv}r_{jv} &+ \cdots + C_{jh,jh}r_{kv} &= F_{jh} \\
C_{jv,jh}r_{jh} &+ (1 + C_{jh,jh})r_{jv} &+ \cdots + C_{jv,kv}r_{kv} &= F_{jv} \\
\vdots & \qquad \vdots & \qquad \vdots & \quad \vdots \\
C_{kv,jh}r_{jh} &+ C_{kv,jv}r_{jv} &+ \cdots + (1 + C_{kv,kv})r_{kv} &= F_{kv}
\end{aligned}
\tag{16}
$$

where:

r_{jh}, is the horizontal scale factor for joint j; and

F_{jh}, is the applied horizontal load for joint j.

The subscripts j and k denote the joints of the modified members. These subscripts vary from $j = 1$ to $j = k$, where k is now the total number of joints of the N modified members. The total number of equilibrium equations is therefore equal to the total number of degrees of freedom of the joints of the N members. These joints are termed the affected joints.

The foregoing concepts may be generalised by considering the case where there are a affected joints. The total number of affected joints is the sum of the varied joints together with all joints directly connected to the joints that are varied. Similarly, the number of members affected is equal to the total number of members connected to the varied joints. At the a affected joints there are applied loadings as for example in equation (16). There may, however, be other joints not included in a which carry applied loads. These are termed the u unaffected joints. The equilibrium equations for each degree of freedom at the $(a + u)$ joints are formed by superposition. For a plane truss there are $2(a + u)$ equations. The equilibrium equations in matrix form are:

$$
\left[
\begin{bmatrix}
1 & \vdots & 0 \\
\cdots & \cdots & \cdots \\
0 & \vdots & 1
\end{bmatrix}
+
\begin{bmatrix}
C_{aa} & \vdots & C_{au} \\
\cdots & \cdots & \cdots \\
C_{ua} & \vdots & C_{uu}
\end{bmatrix}
\right]
\left\{
\begin{matrix}
r_a \\
\cdots \\
r_u
\end{matrix}
\right\}
=
\left\{
\begin{matrix}
F_a \\
\cdots \\
F_u
\end{matrix}
\right\}
\tag{17}
$$

where:

$[C_{aa}]$ = matrix of compensating forces at the a joints arising from unit loadings at the a joints;

$[C_{au}]$ = matrix of compensating forces at the a joints arising from unit loadings at the u joints;

218

$[C_{ua}]$ = matrix of compensating forces at the u joints arising from unit loadings at a joints;

$[C_{uu}]$ = matrix of compensating forces at the u joints due to unit loadings at the u joints;

r_a = vector of scale factors for the a joints;

r_u = vector of scale factors for the u joints;

F_a = vector of applied loads at the a joints; and

F_u = vector of applied loads at the u joints.

If there is no applied loading at the u joints, equation (17) reduces to (16). The compensating forces at the u joints are zero, as they are not involved in any changes. Therefore:

$$[C_{ua}] = [0] \tag{18}$$

$$[C_{uu}] = [0] \tag{19}$$

Substitution of equations (18) and (19) into equation (17) and considering only the lower half gives:

$$\{r_u\} = \{F_u\} \tag{20}$$

Substitution of equation (20) into the upper half of equation (17) gives:

$$\{r_a\} = ([I] + [C_{aa}])^{-1}(\{F_a\} - [C_{au}]\{r_u\}) \tag{21}$$

When the scale factors have been solved, the member forces and displacements of the modified structure are obtained as follows:
Forces in the modified members may be calculated using:

$$P_i^* = \sum_{j=1}^{a+u}(r_{jh}f'_{i,jh} + r_{jv}f'_{i,jv}) \tag{22}$$

and for the unaltered members:

$$P_i^* = \sum_{j=1}^{a+u}(r_{jh}f_{i,jh} + r_{jv}f_{i,jv}) \tag{23}$$

Equations (22) and (23) represent the first theorem of geometric variation, and the second theorem is given by the modified displacements (for each component) which are:

$$\delta^* = \sum_{j=1}^{a+u} (r_{jh}\delta_{jh} + r_{jv}\delta_{jv}) \tag{24}$$

where:

δ_{jh} = displacement arising from a unit horizontal load at joint j; and

δ_{jv} = displacement arising from a unit vertical load at node j.

This formulation of the theorems of geometric variation is a generalisation of the the theorems of structural variation which enables not only the structural properties of the members to be varied but also the geometry of the structure. The geometric theorems have the added versatility that members may be added as well as deleted from the structure. The application of the geometric theorems to rigidly-jointed frames may be readily developed [25].

3.1 The Application of the Theorems of Geometric Variation

The application of the geometric theorems to the re-analysis of trusses and rigidly jointed frames was considered in references [23] and [25] respectively. The application of the geometric theorems to the design of trusses was demonstrated in in reference [23] and to the non-linear analysis of rigidly jointed frames in reference [24]

The application of the geometric theorems to finite element problems has been considerd in references [1, 4, 22]. In references [1, 21], it was shown that the equilibrium equations for nodes of modified elements arising from the variation of the nodal coordinates may be established using the theorems. The efficiency of the formulation was investigated and it was shown that this full and assymmetric set of equilibrium equations may be assembled and solved efficiently provided the higher order elements are used and the number of modified elements is not large. A number of examples were used to illustrate the application of the geometric theorems to nonlinear analysis and design problems [1, 4, 22]. The application of the geometric theorems to nonlinear finite element analysis was considered in references [1, 4]. It was noted that the geometric theorems may be easily incorporated into existing computer codes with little modification.

4   Conclusion and Discussion

The theorems of geometric Variation may be used to predict joint displacements and internal member forces throughout a pin-jointed structure when:

- One or more of the joint coordinates are varied.

220

- One or more of the member strucural properties are varied.

- One or more members are added or deleted from the structure.

- Any combination of of the above takes place simultaneously.

The geometric theorems enable geometric as well as structural variations to be made to the structure. The theorems may also be used with rigidly-jointed frames and structures idealised using finite elements.

The object of the theorems of variation is to reduce the number of equations solved in the re-analysis. Unfortunately the system of equilibrium equations (5) or (17) is full and asymmetric. This will not generally lead to inefficiency unless the number of modifications is large. For structural modifications alone the theorems of structural variation will be more efficient as fewer equilibrium equations need to be assembled and solved.

The design studies and efficiency tests presented in references [1, 20, 21, 22, 23] indicate that the theorems of structural and geometric variation may be used efficiently for the re-analysis of structures. Potential applications include: nonlinear analysis; interactive design studies; optimization and other structural sensitivity calculations.

References

[1] Abu Kassim, A.M., "The Theorems of Structural and Geometric Variation for Linear and Nonlinear Finite Element Analysis", A thesis submitted for the degree of Doctor of Philosophy, Department of Civil Engineering and Building Science, University of Edinburgh, 1985.

[2] Abu Kassim, A.M., Topping, B.H.V., "The Theorems of Structural Variation for Linear and Nonlinear Finite Element Analysis", Proceedings of the Second International Conference on Civil and Structural Engineering Computing, London, vol.2, 159-171, Civil-Comp Press, Edinburgh, 1985.

[3] Abu Kassim, A.M., Topping, B.H.V., "Static Reanalysis: A Review", Journal of Structural Engineering, American Society of Civil Engineers, v.113, n.5, 1029-1045, 1987.

[4] Abu Kassim, A.M., Topping, B.H.V., "The Theorems of Geometric Variation for Nonlinear Finite Element Analysis", Computers and Structures, v.25, n.6, 877-893, 1987.

[5] Arora, J.S., "Survey of Structural Reanalysis Techniques", Journal of the American Society of Civil Engineers, Structural Division, vol.102, n.ST4, 783-802, 1976.

[6] Atrek, E., " Theorems of Structural Variation: A simplification", Int. J. for Numerical Methods in Engineering, vol.21, 481-485, 1985.

[7] Bakry, M.A.E., "Optimal Design of Transmission Line Towers", A thesis submitted for the degree of Doctor of Philosophy, University of Surrey, 1977.

[8] Celik, T., Saka, M.P., "The Theorems of Structural Variation in the Generalised Form", Proceedings of the Southeastern Conference on Theoretical and Applied Mechanics, University of South Carolina, 726-731, 1986.

[9] Filali, A.A., "Theorems of Structural Variation - concerning the inverse of the stiffness matrix", Proceedings of the Second International Conference on Civil & Structural Engineering Computing, vol.2, 149-152, Civil-Comp Press, Edinburgh, 1985.

[10] Majid, K.I., "Optimum Design of Structures", Newnes Butterworths, London, 1974.

[11] Majid, K.I., Celik, T., "The Elastic-Plastic Analysis of Frames by the Theorems of Structural Variation", Int. J. for Numerical Methods in Engineering, vol.21, 671-681, 1985.

[12] Majid,K.I., Elliott, D.W.C., "Forces and Deflections in Changing Structures", The Structural Engineer, vol.51, n.3, 93-101, 1973.

[13] Majid,K.I., Elliott, D.W.C., "Topological Design of Pin Jointed Structures by Non-Linear Programming", Proceedings of the Institution of Civil Engineers, vol.55, part 2, 129-149, 1973.

[14] Majid, K.I., Saka,M.P., "Optimum Shape Design of Rigidly Jointed Frames", Proceedings of the Symposium on Computer Apllications in Engineering, University of Southern California, vol.1, 521-531, 1977.

[15] Majid, K.I., Saka, M.P., Celik, T., "The Theorems of Structural Variation Generalised for Rigidly Jointed Frames", Proceedings of the Institution of Civil Engineers, Part 2, vol.65, 839-856, 1978.

[16] Saka, M.P., Celik, T., "Nonlinear Analysis of Space Trusses by the Theorems of Structural Variation", Proceedings of the Second International Conference on Civil and Structural Engineering Computing, V.2, 153-158, Civil-Comp Press, Edinburgh,1985.

[17] Topping, B.H.V.,"The Application of Dynamic Relaxation to the Design of Modular Space Structures", A thesis submitted for the degree of Doctor of Philosophy, The City University, London, 1978.

[18] Topping, B.H.V., "The Theorems of Structural and Geometric Variation: A Review", Engineering Optimization, v.11, 239-250, 1987.

[19] Topping, B.H.V., "The Application of the Theorems of Structural Variation to Finite Element Problems", International Journal for Numerical Methods in Engineering, vol.19, 141-144, 1983.

[20] Topping, B.H.V., "Shape Optimisation of Skeletal Structures: A Review", Journal of Structural Engineering, American Society of Engineers, vol.119, n.8, 1933-1951, 1983.

[21] Topping, B.H.V., Abu Kassim, A.M., "The Use and Efficiency of the Theorems of Structural Variation for Finite Element Analysis", International Journal for Numerical Methods in Engineering, v.24, 1901-1920, 1987.

[22] Topping, B.H.V., Abu Kassim, A.M., "The Theorems of Geometric Variation for Finite Element Analysis", International Journal for Numerical Methods in Engineering, v.26, 2577-2606, 1988.

[23] Topping, B.H.V., Chan, H.F.C, "Theorems of geometric variation for engineering structures", Proceedings of Instn of Civ Engrs, part2, v.87, 469-486, 1989.

[24] Topping, B.H.V., Khan, A.I., "Non-linear analysis of rigidly jointed frames using the geometric theorems of variation", to be published.

[25] Topping, B.H.V., Yuen, E.C.Y., "The theorems of geometric variation for rigidly jointed frames", to be published.

Optimal Design of Large Discretized Systems by Iterative Optimality Criteria Methods

G. I. N. Rozvany, W. Gollub and M. Zhou
Essen University
Essen
Federal Republic of Germany

Abstract   After explaining the discrepancy between analysis and optimization capability of currently available soft- and hardware, a short history of continuum-based optimality criteria (COC) is given and certain fundamental concepts are introduced. The analytical COC approach is illustrated by three simple examples in which the differences between optimal plastic and optimal elastic strength design are also explained and it is shown that the solution is in general not fully stressed for elastic systems with a stress constraint. The third analytical example demonstrates the proposed technique for problems with vanishing members. The rest of the text deals in detail with the COC formulation for elastic systems with stress and displacement constraints and its iterative solution for large discretized systems. It is shown in worked examples that the proposed approach can handle up to one million elements and one million variables and a number of local and global constraints of the same order of magnitude. One particularly ill-conditioned example is chosen to demonstrate that in such problems an accurate solution does require a very large number of elements in the discretization. Other examples include various geometrical constraints, two-dimensional systems, allowance for selfweight and cost of reactions as well as problems with permissible normal and shear stresses.

1 Introduction

Structural optimization deals with the optimal design of all systems that (a) consist, at least partially, of solids and (b) are subject to stresses and deformations. This integrated discipline plays an increasingly important role in all branches of technology. It has two main branches (Fig. 1): one dealing with *discrete parameter problems*, in which the unknown quantities are finite dimensional vectors $\mathbf{x}_i \in I\!\!R^n$; and the other one with *distributed parameter problems* in which the unknowns are infinite dimensional vectors. With the exception of layout problems, the latter represent unknown generalized functions or mappings of the type $f_i \in I\!\!\Gamma := \{f_i | f_i : I\!\!R^n \to I\!\!R^1\}$ $(n = 1, 2)$.

B. H. V. Topping (ed.),
Optimization and Artificial Intelligence in Civil and Structural Engineering, Volume I, 223–288.
© 1992 *Kluwer Academic Publishers. Printed in the Netherlands.*

224

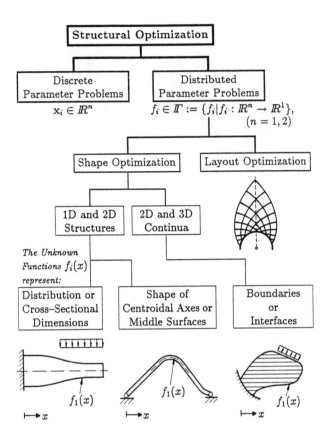

Figure 1. *A classification of structural optimization problems.*

Distributed parameter optimization can be divided further into classes of problems involving shape and layout optimization. In *shape optimization* we may consider *one-dimensional structures* (e.g. bars, beams, arches, rings, cables etc.) or *two-dimensional structures* (plates, shells, grillages, shellgrids, cable nets etc.), for which the unknown functions $f_i(x)$ describe either the *distribution of cross-sectional dimensions* (Fig. 1) or the *shape of centroidal axes or middle surfaces*. In the case of *two- or three-dimensional continua*, the unknown functions may represent *boundaries* of solids or *interfaces* between different materials.

The other main branch of distributed parameter optimization is concerned with *layout problems*, in which the unknown quantities are: (a) the topology or configuration, describing the spatial sequence or connectivity of members and joints; (b) the location of the joints and the shape of the centroidal axes or middle surfaces in between them; and (c) the cross-sectional dimensions of the members. In the literature dealing with discretized formulations of layout optimization, these three problem solving tasks are referred to as (a) 'topological' and (b) 'geometrical' optimization, and (c) 'sizing.'

The current lecture will be dealing with optimization of the distribution of cross-sectional dimensions (Fig. 1), in the context of very large discretized systems. In the second lecture by this speaker, layout optimization will be discussed.

The *aim of structural optimization* is to select the geometry (and possibly the materials) of a structure in such a way that (a) the latter satisfies *certain requirements* with regard to *structural response* or *behaviour* (stresses, deformations, load capacity, vibrations etc.) as well as *restrictions on the geometry* and (b) a quantity $\Phi \in \mathbb{R}^1$ termed *cost* or objective *function(al)* takes on a *minimum* value.

Considering now *methods of structural optimization*, we distinguish between two basic approaches: *direct cost minimization methods*, e.g. mathematical programming (MP) methods, in which in each successive step we attempt to reduce the cost value until a local minimum is reached; and *indirect methods* such as optimality criteria (OC) methods.

Optimality criteria are necessary (and in some cases sufficient) conditions of cost minimality. *Applications of OC-methods* may deal with *relatively simple, idealised systems*, for which an *analytical solution* is sought, with a view to determining

(a) fundamental features of optimal solutions,

(b) the range of validity and applicability of various numerical methods, and

(c) the relative economy of realistic designs (in which case the exact theoretical solution is used as a basis for comparison).

Alternatively, OC methods may be applied to *large, real systems*, for

(a) checking the optimality of solutions already determined by some other method, or

226

(b) to develop efficient interactive re-sizing strategies.

The current lecture is dealing with the last application (b) above.

Since closed form analytical solutions are only available for relatively simple problems, most real design problems require *discretization* [e.g. finite element (FE), finite difference (FD) or boundary element (BE) methods]. Because the accuracy of the above methods usually increases with an increase in the degree of freedom (DF) of the discretized model, it is desirable to use a relatively fine FE (or FD) mesh. In the case of mathematical programming (MP) methods, however, there exists a considerable *discrepancy* between the *analysis capability* and *optimization capability*. Berke and Khot (1987) point out that with currently available hard- and software it is possible to analyse systems with many thousand degrees of free- dom but optimization is restricted to a couple of hundred variables if the problem is highly nonlinear and nonseparable. The same authors observe correctly that the rate of convergence of optimality criteria (OC) methods exhibits weak or no dependence on the number of design variables and hence their optimization capa- bility is limited only by the analysis capability of the FE software. In fact, it will be shown in this lecture that properly formulated OC methods enable the designer to optimize simultaneously even one million variables (or systems with one million elements) subject to hundreds of thousands of local and global constraints.

Optimality criteria methods fall into one of the following two categories: *dis- cretized optimality criteria* (DOC) methods which have been developed since the late sixties mostly by aero-space engineers (e.g. Berke, Khot, Venkayya etc.) and use the Kuhn-Tucker condition in a *finite dimensional design space*, expressed in terms of *nodal forces*; and *continuum-based optimality criteria* (COC) *methods* which use cost minimality conditions in *infinite dimensional design spaces*, ex- pressed in terms of *generalized stresses* and re-interpreted by introducing the con- cept of an *adjoint system*. The latter type optimality criteria are derived by using the calculus of variations, control-theory, functional analysis or energy-principles of mechanics.

In *iterative COC-methods*, each iteration consists of two basic steps:

(a) the *analysis* of the real and adjoint systems and

(b) *updating* of the cross-sectional dimensions.

Since the updating operation usually involves uncoupled and explicit equations, the storage and CPU time requirements for the latter are only a fraction of that required for the analysis phase.

The first application of the iterative COC method was concerned with the shape optimization of arch-grids. In this problem (Rozvany and Prager, 1979) the *analysis* phase involved a set of simultaneous equations representing the equilibrium and compatibility (i.e. equal elevation) conditions for the system and the *updating* phase consisted of the fulfilment of nonlinear but explicit and uncoupled equations consisting of optimality criteria. The latter required that the mean square slope of each arch be unity

$$\frac{\int_0^L \left(\frac{dy}{dx}\right)^2 dx}{L} = 1 , \tag{1}$$

where y is the arch elevation, x the horizontal coordinate and L is the arch span. Satisfaction of (1) was achieved by a suitable scaling of the arch shape (i.e. changing the horizontal reactions of a funicular for given vertical loads). In the updating phase, the equal elevation condition for intersecting arches is temporarily violated but it is restored in the analysis phase. After the analysis phase, the optimality condition in (1) is temporarily violated but it is restored in the updating phase. Repeating the above two sub-steps, an excellent convergence was achieved even for relatively large systems, as can be seen from Fig. 2.

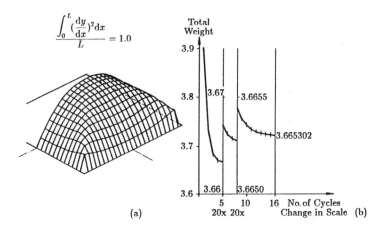

Figure 2. *The first application of the iterative continuum-based optimality criteria (COC) method by Rozvany and Prager in 1979: shape optimization of gridshells.*

2 A Short History of Continuum-Based Optimality Criteria

Using Prager's terminology (e.g. Prager, 1974), the *basic variables of structural mechanics* are generalized stresses $\mathbf{Q}(\mathbf{x}) = [Q_1(\mathbf{x}), \ldots, Q_n(\mathbf{x})]$, generalized strains $\mathbf{q}(\mathbf{x}) = [q_1(\mathbf{x}), \ldots, q_n(\mathbf{x})]$, loads $\mathbf{p}(\mathbf{x}) = [p_1(\mathbf{x}), \ldots, p_m(\mathbf{x})]$ and displacements $\mathbf{u}(\mathbf{x}) = [u_1(\mathbf{x}), \ldots, u_m(\mathbf{x})]$ where $\mathbf{x}$ are the spatial coordinates. A generalized stress may represent a local stress or stress resultant (e.g. a bending moment or shear force) and a generalized strain may refer to a local strain or to an entire cross-section (e.g. curvature or twist of a bar).

The first optimality criterion for a continuum was proposed by Michell (1904), who considered minimum weight trusses. Since his solutions usually consist of an infinite number of members, Prager (1974) termed Michell's frameworks 'truss-like continua.' Michell's optimality condition takes the form

$$\epsilon = k \operatorname{sgn} N \qquad (\text{for } |N| > 0) , \tag{2}$$

$$|\varepsilon| \leq k \qquad (\text{for } N = 0) , \tag{3}$$

where N is the member force, ε is the longitudinal strain in the members and k is a constant. The 'sgn' function has the usual meaning (sgn $N = 1$ for $N > 0$ and sgn $N = -1$ for $N < 0$). Expressed in words, Michell's criterion requires the axial strain to be of constant absolute value (k) in directions of non-zero axial forces, and in the directions of zero axial forces the absolute value of the strain must not exceed the above value (k).

The next optimality condition was introduced half a century later by Foulkes (1954) for the optimal plastic design of least-weight prismatic beams. His criterion can be stated as

$$\frac{\sum_i |\theta|}{L_i} = k , \tag{4}$$

where i denotes a segment of constant cross-section, θ are the plastic hinge rotations, L_i the segment length and k a constant. This means that for segment-wise prismatic frames the *average absolute curvature* for each segment must be the same ($\overline{\kappa}_{av\,i} = k$).

Foulkes' optimality criterion was extended to plastic beams with freely varying rectangular cross-section of given depth but varying width by Heyman (1959), who obtained an optimality condition somewhat similar to that of Michell:

$$\overline{\kappa} = k \operatorname{sgn} M \qquad (\text{for } M > 0) , \tag{5}$$

$$|\overline{\kappa}| \leq k \qquad (\text{for } M = 0) , \tag{6}$$

where $\bar{\kappa} = -\bar{u}''$ is the 'curvature' of the beam deflection $\bar{u}$, M is the beam bending moment and k is a constant. The inequality condition in (6) was not stated explicitly by Heyman and was used only later by the first author (e.g. Rozvany, 1976) and others. Whereas $\bar{u}$ is a fictitious or 'adjoint' deflection, it also represents one possible velocity field at plastic collapse. In the case of Michell's optimality condition in (2) and (3), the overbar for the symbol ε was omitted, because for that problem the real and adjoint strains are identical.

A general optimality criterion for the optimal plastic design of structures with freely varying cross-sectional dimensions was proposed by Prager and Shield (1967) who generalized an earlier optimality condition by Marcal and Prager (1964). In order to explain the Prager-Shield condition in greater detail, we review here briefly the *lower bound theorem of plastic design*. According to this theorem, the actual collapse load is always greater than or at most equal to the calculated load if the corresponding stress field $\mathbf{Q}$ (a) is statically admissible (satisfying the equilibrium equations and static boundary conditions), and (b) fulfils the yield inequality at all cross-sections.

On the basis of the above theorem, we can adopt the following *plastic design procedure* for redundant structures (Fig. 3a):

(i) Make the structure statically determinate by releasing all redundant connections (Fig. 3b, in which the clamping action at the left hand end of the beam has been removed). The resulting structure is termed 'primary structure.'

(ii) Apply arbitrary actions (moments or forces) corresponding to the redundancies released in step (i), see Fig. 3c.

(iii) Determine the generalized stresses (e.g. moments) caused by the external load in the primary structure (Fig. 3d, top).

(iv) Calculate the generalized stresses caused by the arbitrarily chosen redundant actions (Fig. 3d, middle).

(v) Superimpose the generalized stresses obtained in steps (iii) and (iv) (Fig. 3d, bottom). These generalized stresses are now statically admissible and are denoted by $\mathbf{Q}^S$ (in the case of a beam, by M^S).

(vi) Adopt cross-sectional dimensions such that the yield inequality is satisfied everywhere (Fig. 3e). For a plastic beam designed for flexural stresses only, the yield inequality is

$$|M^S| \leq M_p ,\tag{7}$$

where M_p is the plastic moment capacity of the beam. At the latter moment value, the entire cross-section is in yield, if a rigid-perfectly plastic behaviour is assumed

(Fig. 3g). For the considered beam, an M^S diagram with only positive moments could also give an admissible (safe) design (Fig. 3f), but the latter would be clearly uneconomical.

In an optimal plastic design for freely varying cross-sectional dimensions, the above yield inequality is replaced by the equality

$$|M^S| = M_p \, . \tag{8}$$

Clearly, a design with $|M^S| < M_p$ at some cross-sections cannot be optimal if the cross-sectional dimensions are not subjected to geometrical constraints.

In the general formulation of Prager and Shield (1967), the total cost Φ to be minimized can be expressed as the integral of the specific cost ψ over the structural domain D. The specific cost (which may represent the cross-sectional area), in turn, can be expressed as a function of the statically admissible generalized stress vector

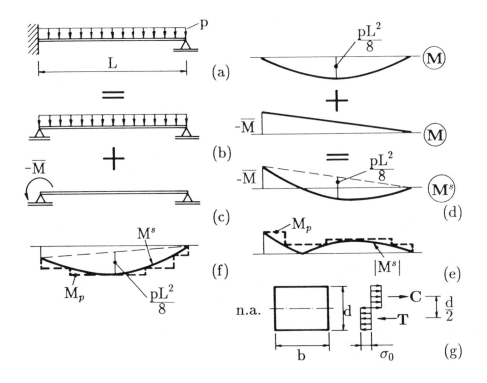

Figure 3. *An example illustrating the plastic design procedure based on the lower bound theorem of limit design.*

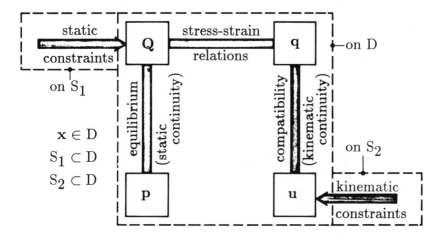

$\mathbf{Q}(\mathbf{x})$: generalized stresses
$\mathbf{q}(\mathbf{x})$: generalized strains
$\mathbf{p}(\mathbf{x})$: generalized loads
$\mathbf{u}(\mathbf{x})$: generalized displacements

Figure 4. *Fundamental relations of structural mechanics.*

$\mathbf{Q}^S$ at the considered cross-section $\psi = \psi(\mathbf{Q}^S)$. It follows that the optimal plastic design problem can be formulated as

$$\min \Phi = \int_D \psi(\mathbf{Q}^S)\, d\mathbf{x} , \qquad (9)$$

where $\mathbf{x}$ are the spatial coordinates.

Before stating the Prager-Shield condition, we review briefly the *fundamental relations of structural mechanics*. Considering an *elastic* system, for example, these relations are shown in Fig. 4. On the structural domain D, we must satisfy the equilibrium (or static continuity) conditions $(\mathbf{p}, \mathbf{Q})$, the compatibility (or kinematic continuity) conditions $(\mathbf{u}, \mathbf{q})$ and the generalized stress-strain relations $(\mathbf{Q}, \mathbf{q})$. In addition, on some subset S_1 of the domain D static constraints (or static boundary conditions) and on some other subset $S_2 \subset D$ kinematic constraints (or kinematic boundary conditions) must be fulfilled.

The above relations are illustrated in the context of a *Bernoulli-type cantilever beam* in Fig. 5, for which the generalized stress is the bending moment $(\mathbf{Q} \rightarrow$

M) and the generalized strain the curvature ($\mathbf{q} \rightarrow \kappa$). Then the equilibrium, compatibility and generalized strain-stress relations are

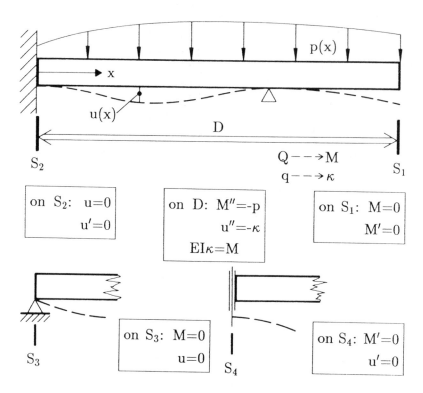

Figure 5. *An example illustrating the fundamental relations of structural mechanics.*

$$M'' = -p \,, \tag{10}$$

$$u'' = -\kappa \,, \tag{11}$$

$$EI\kappa = M \,, \tag{12}$$

where p is the load, E is Young's modulus, I is the moment of inertia and other symbols have been defined earlier. At the free end of the beam (S_1), only static boundary conditions need to be observed

$$M = M' = 0 \,, \tag{13}$$

and at the clamped end (S_2) the kinematic boundary conditions are

$$u = u' = 0 \,. \tag{14}$$

At the bottom of Fig. 5, supports representing some mixed boundary conditions are indicated.

The requirements for (lower bound) *plastic design* are shown in Fig. 6, in which the stress field must satisfy static boundary conditions (constraints), equilibrium (static continuity) conditions and the yield inequality $Y(\mathbf{Q}) \leq Y_0$.

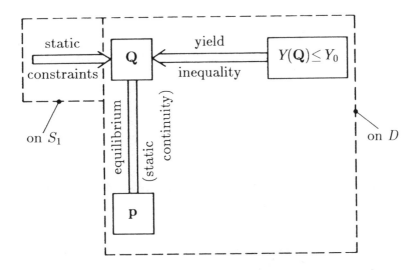

Figure 6. *Fundamental relations of plastic design.*

In *optimal plastic design* (Fig. 7), we are still required to satisfy static boundary conditions and equilibrium, but the yield inequality is replaced with the cost minimality condition in which the 'specific cost function' $\psi(\mathbf{Q})$ is based on the assumption that the yield equality is satisfied at all cross-sections.

Optimal plastic design via Prager-Shield *optimality criteria* is represented schematically in Fig. 8. The above optimality conditions were derived by Prager and Shield (1967) from energy theorems, by the author (e.g. Rozvany, 1976) from variational principles and in a discretized form by Thierauf (1978). In Fig. 8, we have the actual loads $\mathbf{p}$ and generalized stresses $\mathbf{Q}$ of the real structure and the displacements $\bar{\mathbf{u}}$ and generalized strains $\bar{\mathbf{q}}$ of the adjoint structure. The latter two must satisfy the usual compatibility conditions. For structures with rigid and costless supports, the kinematic boundary conditions for $\bar{\mathbf{u}}$ are the same as for a usual

234

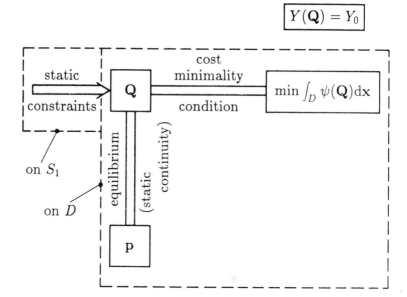

Figure 7. *Fundamental relations of optimal plastic design by direct cost minimization.*

structure, but for non-zero support cost the adjoint kinematic boundary conditions are given in Fig. 8, in which $\Omega(\mathbf{R})$ is the cost of the reaction $\mathbf{R} = (R_1, \ldots, R_m)$. In the optimal strain-stress relation (Fig. 8), the adjoint strains $\bar{\mathbf{q}}$ must be kinematically admissible (denoted by $\bar{\mathbf{q}}^K$) and are given by the gradient of the specific cost function $\psi(\mathbf{Q})$, in which the stress vector must be statically admissible $(\mathbf{Q}^S)$:

$$\bar{\mathbf{q}}^K = \mathcal{G}[\psi(\mathbf{Q}^S)] . \tag{15}$$

The optimality criterion in (15) represents a considerable extension of the original Prager-Shield (1967) condition. Both in the adjoint boundary conditions for $\bar{\mathbf{u}}$ and in the strain-stress relation $(\mathbf{Q}, \bar{\mathbf{q}})$ in Fig. 8, we make use of the generalized gradient or 'G-gradient operator' $\mathcal{G}$ (e.g. Rozvany, 1976, 1981). For differentiable functions $\psi(\mathbf{Q})$ or $\Omega(\mathbf{R})$, the G-gradient has the meaning

$$\mathcal{G}[\psi(\mathbf{Q})] = \mathbf{grad}\ \psi(\mathbf{Q}) = \left(\frac{\partial \psi}{\partial Q_1}, \ldots, \frac{\partial \psi}{\partial Q_n}\right) , \tag{16}$$

$$\mathcal{G}[\Omega(\mathbf{R})] = \left(\frac{\partial \Omega}{\partial R_1}, \ldots, \frac{\partial \Omega}{\partial R_m}\right) , \tag{17}$$

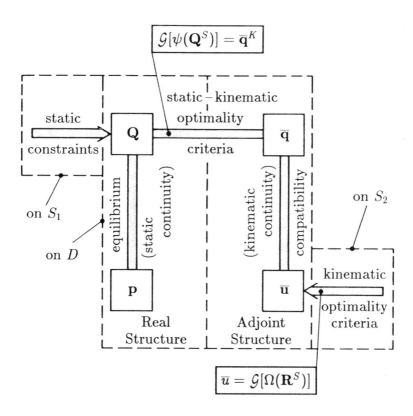

Figure 8. *Fundamental relations of optimal plastic design via opti-mality criteria.*

but for non-differentiable and discontinuous functions the G-gradient has an extended meaning. This is shown, in the context of *functions with a single variable* (Q), in Fig. 9. It can be seen that for *differentiable* (smooth) *functions* the adjoint strain $\bar{q}$ (scalar quantity in this case) is given by

$$\bar{q} = \frac{d\psi}{dQ} , \tag{18}$$

implying (Fig. 9a)

$$\psi = \int_0^{Q^*} \bar{q}\,dQ , \tag{19}$$

where Q^* is some arbitrary stress value. In Fig. 9a, we also define the 'complementary cost'

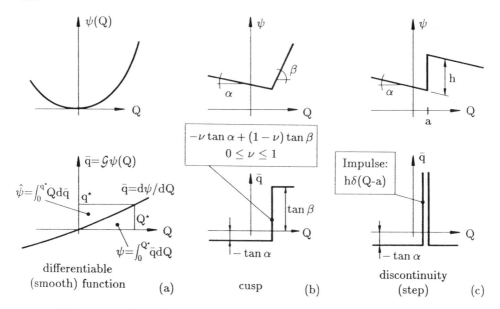

Figure 9. *G-gradient operator for functions and generalized functions with a single variable.*

$$\hat{\psi} = \int_0^{q^*} Q d\bar{q} \, , \tag{20}$$

where the strain q^* corresponds to the stress value Q^*. The concept of complementary cost will be used in duality principles [see the relation (21) later].

In the case of a *cusp* of a piece-wise differentiable function, the G-gradient, and the corresponding adjoint strain value, are non-unique and equal the convex combination of the adjacent slopes (gradients).

Finally, in the case of a *discontinuity* (step) in the specific cost function, the G-gradient, and the adjoint strain, contain an impulse (Dirac distribution) δ whose (Lebesque) integral equals the magnitude of the step.

The optimality criteria approach in Fig. 8 converts an optimization problem (Fig. 7) into a problem of analysis which has both conceptual and computational advantages. Whilst in plastic optimal design only the static aspects of the structure are 'real' and only the kinematic aspects of the structure are 'adjoint,' we shall see later that in optimal elastic design we must consider a complete real and a complete adjoint structure (see Fig. 14).

For convex specific cost functions $\psi(\mathbf{Q})$, the minimum total cost can be calculated either from the primal formula in (9) or in the following *dual formula*

$$\Phi_{\min} = \int_D [\mathbf{p}\mathbf{u}^K - \hat{\psi}(\mathbf{q}^K)]\mathrm{d}\mathbf{x} \,, \tag{21}$$

in which $\mathbf{u}$ and $\mathbf{q}$ satisfy the Prager-Shield condition (15). Moreover, (9) and (21) furnish *upper and lower bounds* on Φ for any statically admissible stress field $\mathbf{Q}^S$ and any kinematically admissible strain/displacement field $(\mathbf{q}^K, \mathbf{u}^K)$.

The Prager-Shield optimality condition has been extended to a number of other problems in *optimal plastic design*, including multi-constraint optimization (Rozvany and Cohn, 1970), alternate loading conditions and multi-component systems (Charett and Rozvany, 1972), non-convex and discontinuous specific cost functions (Rozvany, 1973a, 1974a), allowance for cost of supports (Rozvany, 1974b), generalized specific cost functions (Rozvany, 1976), optimization of support location (Rozvany, 1975; Mroz and Rozvany, 1975; Prager and Rozvany, 1975), allowance for cost of connections (Rozvany and Mroz, 1975), allowance for self-weight (Rozvany, 1977b), bounded spatial gradients (Rozvany, 1984) as well as continuous and segment-wise linear cost-distribution (Rozvany, Menkenhagen and Spengemann, 1989; Rozvany, Spengemann et al., 1989).

Moreover, general continuum-based optimality criteria have been derived for *optimal elastic design*, including optimal segmentation (Masur, 1975b; Rozvany, 1975), optimal location of supports (Masur, 1975a), compliance and deflection constraints (Rozvany, 1976), given buckling load (Prager and Taylor, 1968; Rozvany and Mroz, 1977), stress constraints (Rozvany, 1977a, 1978), segmentation (Rozvany, Ong and Karihaloo, 1986), bounded spatial gradients (Rozvany, Yep, Ong and Karihaloo, 1988), combined deflection and stress constraints (Rozvany, Booz and Ong, 1987). A comprehensive review of continuum-based optimality criteria (COC) and their applications is given in a recent book by the author (Rozvany, 1989), which also includes a very large number of worked examples, a brief introduction to optimal layout theory, a short review of useful variational principles together with exercises and a bibliography listing over one thousand publications.

3 Illustrative Examples and a Comparison of Optimal Plastic and Optimal Elastic Strength Designs

The Prager-Shield (1967) optimality condition in (15) and Fig. 8 will be illustrated by some elementary examples in this section.

3.1 Clamped Beam with a Point Load—Identical Plastic and Elastic Optimal Designs

We consider the beam in Fig. 10a with a rectangular cross-section of given depth d and variable width b (Fig. 10g). Since the plastic moment capacity of this beam is $M_p = \sigma_0 bd^2/4$, where σ_0 is the yield stress, the cross-sectional area is $\psi = bd$ and the yield condition is $|M| \leq M_p$, the specific cost function for a cross-section at yield is

$$\psi = k|M| , \tag{22}$$

with $k = 4/(\sigma_0 d)$. Then (15) implies the optimality criteria in (5) and (6) which are represented graphically in Figs. 10e and f. The moment (M) and adjoint displacement $(\bar{u})$ diagrams in Figs. 10b and c satisfy the above optimality criteria as well as equilibrium and compatibility. This means that, out of all statically admissible moment diagrams, the one with points of contraflexure at the quarter-span will be the optimal one, if *plastic design* is used. The variation of the moment capacity M_p is shown in Fig. 10d, in which it is assumed that some negligibly small minimum moment capacity (δ) is specified.

The optimal plastic design solution in Fig. 10d is the best available *statically admissible* solution for the specific cost function $\psi = k|M|$. It also represents an optimal elastic design, provided that the corresponding elastic generalized strains (here: curvatures) are *kinematically admissible*. Considering a permissible stress condition ($\sigma \leq \sigma_p$ where σ is the flexural stress in the beam and σ_p is the permissible maximum value of the latter), we have

$$\sigma_{\max} = \frac{|M|}{\frac{bd^2}{6}} \leq \sigma_p . \tag{23}$$

Moreover, the flexural stiffness of the beam is

$$s = \frac{E\,bd^3}{12} . \tag{24}$$

Combining (23) and (24), we have the strength condition for the elastic beam:

$$k|M| \leq s , \tag{25}$$

with $k = \frac{Ed}{2\sigma_p}$. For fully stressed cross-sections, (25) is fulfilled as an equality:

$$s = k|M| . \tag{26}$$

Since s is proportional to the cross-sectional area, we may minimize the quantity $\Phi = \int_D s\,dx$ in a scaled problem. Then it follows from (22) and (26) that we have the same specific cost function in both the plastic and elastic design problems under consideration.

As mentioned above, the solution in Fig. 10d becomes also the best elastic design if it fulfils the compatibility conditions for the moments in Fig. 10b. These two diagrams furnish elastic curvatures of the magnitude

$$\kappa = \frac{M}{s} = \frac{|M|\operatorname{sgn} M}{k|M|} = \frac{\operatorname{sgn} M}{k} , \tag{27}$$

239

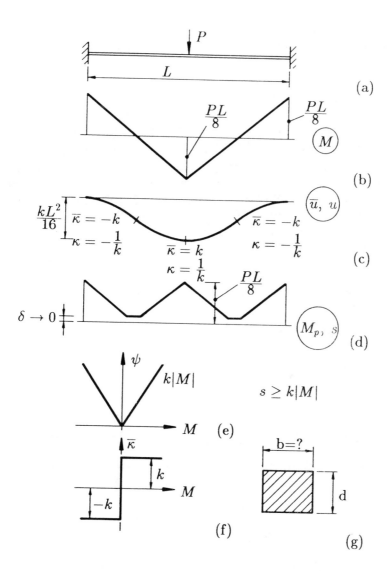

Figure 10. *First elementary example illustrating the analytical COC approach: identical optimal plastic and optimal elastic strength design.*

which are also represented by Fig. 10c, if we replace $\overline{\kappa} = k$ and $\overline{\kappa} = -k$, respectively, with $\kappa = \frac{1}{k}$ and $\kappa = -\frac{1}{k}$. Since both the real and adjoint displacement fields are kinematically admissible and the optimality condition in (5) and (6) or in Figs. 10e and f are also satisfied by the adjoint displacements, we have established that in this case the solution in Fig. 10d is also optimal for elastic design. A more direct proof of the optimality of this solution will be given in Section 4.3.1.

The considered example illustrates a rather special class of problems for which the optimal elastic design is *fully stressed*. It will be seen from Section 3.2.2, however, that this is not the case in general.

The optimal total cost (= beam weight) can be calculated from either the primal [see (9) with (22)] or the dual [see (21)] formulation. It can be seen from Figs. 9a and 10f that the complementary cost is zero ($\hat{\psi} = 0$) for the considered specific cost function if $\kappa \leq k$. Then (9) and (22) with Fig. 10b and (21) with Fig. 10c furnish:

$$\Phi_{\min} = \int_D k|M|\,\mathrm{d}x = \frac{kPL}{8}L\frac{1}{2} = \frac{kPL^2}{16} , \qquad (28)$$

$$\Phi_{\min} = \int_D p\overline{u}\,\mathrm{d}x = P\overline{u}_{\max} = \frac{kPL^2}{16} , \qquad (29)$$

which show a complete agreement.

3.2 Clamped Beam with a Point Load and Selfweight—Differing Plastic and Elastic Designs

3.2.1 Plastic Design In the case of *optimal plastic design with allowance for selfweight* (ψ), the modified Prager-Shield condition becomes (Rozvany, 1989, p. 46)

$$\overline{\mathsf{q}}^K = (1 + \overline{u})\mathcal{G}[\psi(\mathbf{Q}^S)] . \qquad (30)$$

Considering the beam of given depth but variable width in Fig. 11a with selfweight, (30) with $\psi = k|M|$ reduces to

$$\tilde{\overline{\kappa}} = \mathrm{d}^2\tilde{\overline{u}}/\mathrm{d}\tilde{x}^2 = k(1 + \tilde{\overline{u}})\,\mathrm{sgn}\,M , \qquad (31)$$

where $(\tilde{\kappa}, \tilde{u})$ are kinematically admissible and M is statically admissible. Denoting the span by $2\tilde{a}$, the distance of the zero moment point from the midspan by $\tilde{b}$ and the horizontal coordinates by $\tilde{z}$ and $\tilde{x}$, we introduce the nondimensional notation:

$$a = k^{1/2}\tilde{a}, \quad b = k^{1/2}\tilde{b}, \quad z = k^{1/2}\tilde{z}, \quad x = k^{1/2}\tilde{x} . \qquad (32)$$

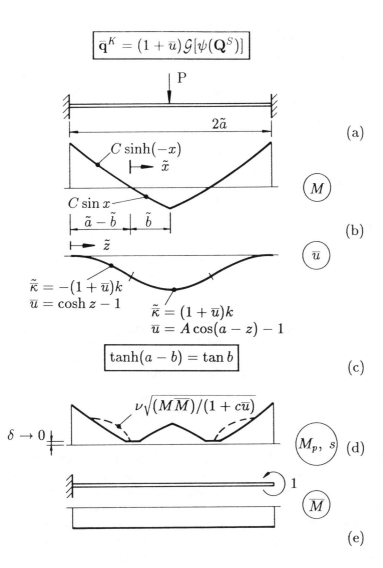

Figure 11. *Second elementary example: beam with selfweight and external load for which the optimal plastic and optimal elastic strength designs are different.*

It should be noted that ψ in this problem is the beam weight per unit length and hence the cost factor k in the specific cost function $\psi = k|M|$ converts force $\times$ length into force/length. It follows that k has a dimension length^{-2} and $k^{1/2}$ a dimension length^{-1}. The adjoint deflection $\tilde{\bar{u}}$ is non-dimensional ($\tilde{\bar{u}} = \bar{u}$) and the dimension of $-\bar{u}'' = \tilde{\kappa}$ is length^{-2}.

The nondimensional version of (31) then becomes

$$\bar{\kappa} = \mathrm{d}^2\bar{u}/\mathrm{d}x^2 = (1 + \bar{u})\,\mathrm{sgn}\,M\,. \tag{31a}$$

Equation (31a) implies for the outer segments (with $0 \le \tilde{z} \le \tilde{a} - \tilde{b}$ in Fig. 11) of the beam with sgn $M = -1$ and the end conditions $\bar{u}(0) = \bar{u}'(0) = 0$

$$\bar{u} = \cosh z - 1\,. \tag{33}$$

For the inner segment (with $\tilde{a} - \tilde{b} \le \tilde{z} \le \tilde{a}$ in Fig. 11), we have sgn $M = 1$ and thus the end condition $\bar{u}'(\tilde{a}) = 0$ implies

$$\bar{u} = A\cos(a - z) - 1\,, \tag{34}$$

where A is an unknown constant. Then the continuity and slope continuity requirements at $\tilde{z} = \tilde{a} - \tilde{b}$ imply

$$\cosh(a - b) = A\cos b, \quad \sinh(a - b) = A\sin b, \tag{35}$$

or

$$\tanh(a - b) = \tan b\,. \tag{36}$$

The relation (36) gives the optimal location of the zero moment point. For very small nondimensional spans (with $a \to 0$) (36) furnishes $b/a = 0.5$ as in Example 3.1, because for those spans the selfweight is negligibly small. For $a = 2$, for example, the optimal value of b/a is 0.3549. This is to be expected because for a significant selfweight effect more weight is shifted towards the ends of the beam. The adjoint deflection field $\bar{u}$ is shown in Fig. 11c and the moments M in Fig. 11b.

The latter is furnished by the equilibrium equations:

$$\text{(for } M < 0) \quad \psi'' = k\psi, \qquad \text{(for } M > 0) \quad \psi'' = -k\psi\,, \tag{37}$$

implying for the end conditions $\psi(0) = 0$, $\psi'(b) = kP/2$ (for details, see Rozvany, 1989, p. 210):

$$\text{(for } x < 0) \quad \psi = C\sinh(-x)\,, \tag{38}$$

$$\text{(for } x > 0) \quad \psi = C\sin x\,, \tag{39}$$

with $C = k^{1/2}P/(2\cos b)$. Then the total cost becomes

$$\Phi = 2\int_0^{\tilde{a}} k|M|\,dz = \frac{P}{\cos b}\left[\int_0^b \sin x\,dx + \int_0^{-(a-b)} \sinh(-x)\,dx\right]$$

$$= P\left[\frac{\cosh(a-b)}{\cos b} - 1\right]. \tag{40}$$

Check by Differentiation. The results obtained by the optimality criteria method in (36) can now be verified by using the stationarity condition $d\Phi/db = 0$ and (40), which yield:

$$\frac{-\sinh(a-b)\cos b + \sin b\cosh(a-b)}{\cos^2 b} = 0. \tag{41}$$

The latter implies (36).

Check by Dual Formulation. Since in this problem the complementary cost $\hat{\psi}$ is again zero, the total cost in (40) is also given by (21) as

$$\Phi = \int_D p\bar{u}\,dx = P\bar{u}_{max}, \tag{42}$$

in which $\bar{u}_{max} = \bar{u}|_{z=a}$ is given by (34):

$$\bar{u}_{max} = A - 1. \tag{43}$$

The value of A can be evaluated from the first relation under (35):

$$A = \frac{\cosh(a-b)}{\cos b}. \tag{44}$$

Substitution of (43) and (44) into (42) confirms the total cost in (40). A detailed treatment of the above problem can be seen in the following references: Rozvany (1977b), Rozvany (1989, pp. 208–211).

3.2.2 Elastic Design   It can be shown readily that in the current problem the optimal elastic design differs from the optimal plastic design. Considering, as in Example 3.1, the elastic strength condition in (25), the curvature for fully stressed regions of the real structure will be [cf. (27)]

$$\kappa = (\text{sgn } M)/k. \tag{45}$$

This means that for a fully stressed solution based on the moments in Fig. 11b, we would have a curvature of $(-1/k)$ over a length $(a-b)$ and a curvature of $(1/k)$ over a length b. Since for finite spans (36) gives $(a-b) > b$, the above curvature field would clearly violate the compatibility condition $\int_0^a \kappa\,dx = 0$, since

$$\int_0^a \kappa\,dx = (a-b)(-1/k) + b(1/k) \neq 0. \tag{46}$$

In order to restore compatibility, the flexural stiffness $s(x)$ will have to be increased in the outer segments (having a negative bending moment). It will be shown in Section 4.3.2 that the optimal elastic solution in this case is the one shown in Fig. 11d in which $\overline{M}$ has the following meaning. The zero rotation requirement at the right hand end can be expressed by means of a virtual work equation of the form

$$\int_D \frac{M\overline{M}}{s} \mathrm{d}x = 0 , \qquad (47)$$

where $\overline{M}$ is caused by a unit couple at the end where the rotation is to be zero (after removing the redundant support at that end, see Fig. 11e).

In this problem, the optimal plastic and optimal elastic designs clearly differ if slope continuity is preserved at points of contraflexure by prescribing a small mimimum stiffness value (δ). The understressed regions (broken lines in Fig. 11d) are governed by the zero displacement constraint provided by the redundancy of the beam. However, the optimal elastic design becomes identical with the optimal plastic one if $hinges$ are permitted at points of zero mement and hence the structure becomes statically determinate.

3.3 An Elementary Layout Problem with Vanishing Beam Regions

It was already stated in (6) that for a vanishing beam region with $M = 0$ we still have a curvature requirement for the adjoint system, namely $|\overline{\kappa}| \leq k$. In general, the Prager-Shield optimality condition in (15) does give such non-unique values of the generalized strains for $\mathbf{Q} = 0$ and this property of the adjoint system forms the basis of the optimal layout theory which will be explained in the second lecture of this speaker.

In order not to obscure the principles involved by computational complexities, we consider a quite elementary problem (Fig. 12). Fig. 12a shows a simple system of two interconnected beams in plan view. The beams are simply supported and a point load P is applied at their intersection. We use again the simple specific cost function $\psi = k|M|$ (Fig. 12h) and the corresponding optimal moment-curvature requirement furnished by the Prager-Shield condition (15) is shown in Fig. 12i. Note that for $M = 0$ the value of $\overline{\kappa}$ is non-unique.

The obvious optimal solution in the problem consists of a short beam AA carrying the entire load P and a long beam BB of zero cross-sectional area, carrying a zero load. The corresponding load and moment diagrams are shown in Figs. 12b, c, e and f. If we want to $prove$, however, that the above solution is really optimal, we can do so by using the Prager-Shield optimality condition (here Fig. 12i). For the short beam AA, the moments are everywhere positive and hence by Fig. 12i or (5) we have $\overline{\kappa} = k$. The coresponding adjoint deflection field $\overline{u}$ is shown in Fig. 12d.

245

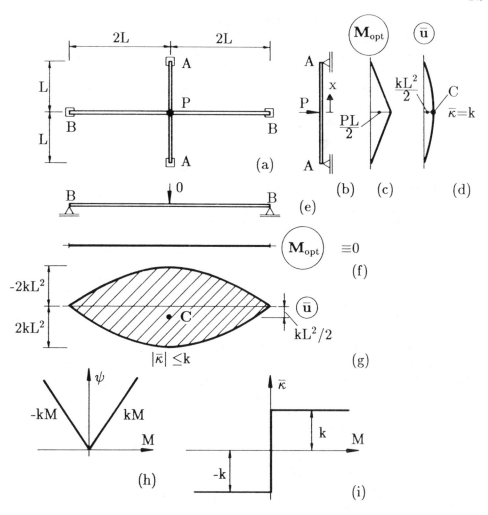

Figure 12. *Third elementary example: a simple layout problem with vanishing members.*

For the long beam BB, the adoint curvature, by Fig. 12i or (6), has the non-unique value $|\overline{\kappa}| < k$. Since at the supports the deflection must be zero due to the kinematic admissibility requirement, the non-unique adjoint displacement field $\overline{u}$ for that beam is given by the shaded area in Fig. 12g. As the latter contains the point C which corresponds to the maximum deflection of the short beam, compatibility of the two beam deflection fields is satisfied and hence all requirements of the Prager-Shield criterion are fulfilled. Because the latter is also a *sufficient* condition for convex specific cost functions (see Fig. 12h), we have now established that the considered solution is optimal.

The above example demonstrates, how *vanishing members* can be handled by continuum-based optimality criteria. Moreover, elimination of non-optimal members from a system constitutes a simple case of *layout optimization*.

4 The COC Method for Linearly Elastic Structures with Stress and Deflection Constraints

4.1 General Formulation

We consider structures for which the cross-sectional geometry is partially prescribed in such a way that the cross-section is fully defined by a finite number of variables

$$\mathbf{z}(\mathbf{x}) = [z_1(\mathbf{x}), \dots, z_r(\mathbf{x})] , \qquad (48)$$

termed *cross-sectional parameters*. Examples of such parameters are given in Fig. 13. In earlier publications, it was usual to consider rectangular cross-sections in which either the height, or the width or the height/width ratio was kept constant (Fig. 13a). The current formulation can handle much more general cross-sectional geometries, such as the one shown in Fig. 13b.

The specific cost ψ (e.g. the weight, volume or material/construction costs per unit length or unit area) is a function of the cross-sectional parameters:

$$\psi(\mathbf{x}) = \psi[z_1(\mathbf{x}), \dots, z_r(\mathbf{x})] . \qquad (49)$$

The structure is subject to w alternate loads

$$\mathbf{p}_1(\mathbf{x}), \dots, \mathbf{p}_k(\mathbf{x}), \dots, \mathbf{p}_w(\mathbf{x}) , \qquad (50)$$

which equilibrate the (statically admissible) generalized stresses

$$\mathbf{Q}_1^S(\mathbf{x}), \dots, \mathbf{Q}_k^S(\mathbf{x}), \dots, \mathbf{Q}_w^S(\mathbf{x}) . \qquad (51)$$

Each cross-section referred to the coordinates $\mathbf{x}$ is subject to t strength (or stress) conditions:

$$S_\ell[\mathbf{Q}_k^S(\mathbf{x}), \mathbf{z}(\mathbf{x})] \geq 0 \qquad (\ell = 1, \dots, t; \quad k = 1, \dots, w) . \qquad (52)$$

Moreover, the structure is subject to v displacement constraints, which can be expressed through the principle of virtual work in the form:

247

$$\int_D \mathbf{Q}_k^{S,K} \bullet [\mathbf{F}]\overline{\mathbf{Q}}_j^S \, dx \begin{cases} \leq \Delta_j \\ = \Delta_j \end{cases} \quad (j=1,\ldots,v; \quad k=1,\ldots,w),$$ (53)

where $[\mathbf{F}]$ is the local flexibility matrix converting the stress vector into a strain vector

$$\mathbf{q}_k = [\mathbf{F}]\mathbf{Q}_k \,,$$ (54)

and $\overline{\mathbf{Q}}_j^S$ is the virtual stress vector equilibrating some virtual load $\overline{\mathbf{p}}_j$. In the case of a displacement constraint at a single point, the virtual load is a unit ('dummy') load at that point in the direction of the prescribed displacement. The displacement constraint may refer, however, to a weighted combination of several displacements, in which case the virtual load consists of several forces or couples. Finally, if in one displacement condition j the prescribed quantity is the sum of the products of the external forces and the corresponding displacements (i.e. the 'compliance' C):

$$\Delta = \int_D \mathbf{p}_k \bullet \mathbf{u}_k \, dx = C \,,$$ (55)

then the real and virtual loads are the same:

$$\mathbf{p}_k = \overline{\mathbf{p}}_j \,, \quad \mathbf{Q}_k = \overline{\mathbf{Q}}_j \,.$$ (56)

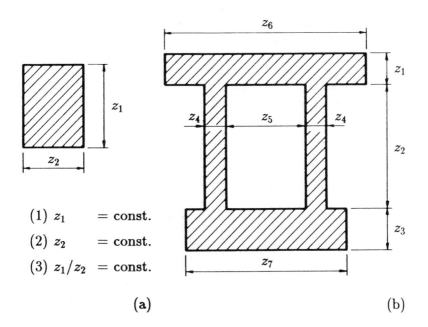

(1) z_1 = const.
(2) z_2 = const.
(3) z_1/z_2 = const.

(a) (b)

Figure 13. *Examples of cross-sectional parameters z.*

Displacement constraints fall into one of the following categories:

(i) *Physical* displacement constraints or redundancies which are always equalities.

(ii) *Operational* displacement constraints prescribing certain displacements or their maximum permissible value at the design loads. These can be equalities but they are usually inequalities.

4.2 General Optimality Criteria

The COC approach for a *single deflection constraint* (without strength constraints) is shown schematically in Fig. 14 in which the equilibrium, compatibility and generalized strain-stress relations as well as the static constraints (static boundary conditions) are the same for the real and adjoint structures. For rigid and costless supports, the kinematic constraints (kinematic boundary conditions) are also identical; however, the latter two differ, if the cost of the supports is non-zero (see Fig. 8). For given values of the real and adjoint generalized stresses, the optimality criteria at the top of Fig. 14 furnish the optimal values of the cross-sectional parameters.

For a *displacement and a strength (stress) constraint* and a single load condition, the COC approach is summarized in Fig. 15. The main difference in comparison to Fig. 14 is that (i) the optimality criteria are somewhat more complicated and (ii) the adjoint strains $\overline{\mathbf{q}}$ are dependent on both the real and adjoint stresses $(\mathbf{Q}, \overline{\mathbf{Q}})$.

In the case of several load conditions $(k = 1, \ldots, w)$, strength constraints $(\ell = 1, \ldots, t)$ and displacement constraints $(i = 1, \ldots, v)$, the following relations replace the ones in Fig. 15:

$$\overline{\mathbf{q}}_k^K = \sum_\ell \lambda_{\ell k}(\mathbf{x}) \mathcal{G}_{,\mathbf{Q}} \left[S_\ell(\mathbf{z}, \mathbf{Q}_k^{S,K}) \right] + \sum_j \nu_{jk}[\mathbf{F}]\overline{\mathbf{Q}}_j^{S,K} , \qquad (57)$$

$$\mathcal{G}_{,\mathbf{z}}\left[\psi(\mathbf{z}) \right] + \sum_\ell \sum_k \lambda_{\ell k}(\mathbf{x}) \mathcal{G}_{,\mathbf{z}} \left[S_\ell(\mathbf{z}, \mathbf{Q}_k^{S,K}) \right]$$

$$+ \sum_j \sum_k \nu_{jk}\overline{\mathbf{Q}}_j^{S,K} \bullet \left\{ \mathcal{G}_{,\mathbf{z}}[\mathbf{F}] \right\} \mathbf{Q}_k^{S,K} = 0 . \qquad (58)$$

The superscripts (S, K) indicate that both the real and the adjoint stresses must be statically admissible and must produce kinematically admissible strains.

If $c\psi(\mathbf{z})$ denotes the weight per unit length or area and we make allowance for *selfweight*, then the first term in (58) must be premultiplied by $(1 + c\overline{u})$:

$$(1 + c\overline{u})\mathcal{G}_{,\mathbf{z}}\left[\psi(\mathbf{z}) \right] \ldots , \qquad (59)$$

(see Rozvany, 1989, p. 72 in which $c = 1$ was adopted). If a stress or displacement constraint is inactive for a particular loading condition, then the corresponding Lagrangian ($\lambda_{\ell k}$ or ν_{jk}) in (57), (58) and (59) becomes zero:

$$\nu_{jk} \geq 0, \quad \text{and } \nu_{jk} > 0 \text{ only if } \int_D \overline{Q}_j^{S,K} \bullet [F] Q_k^{S,K} \, dx = \Delta_j \,, \tag{60}$$

$$\lambda_{\ell k}(x) \geq 0, \text{ and} \lambda_{\ell k}(x) > 0 \text{only if } S_\ell \left[z(x), \, Q_k^{S,K}(x) \right] = 0 \,. \tag{61}$$

Moreover, if limits on the value of a cross-sectional parameter z_i are prescribed

$$z_{i\,\min} \leq z_i \leq z_{i\,\max} \,, \tag{62}$$

then (58) is replaced by

$$F_i = 0 \quad (\text{for } z_{i\,\min} < z_i < z_{i\,\max}) \,, \tag{63}$$

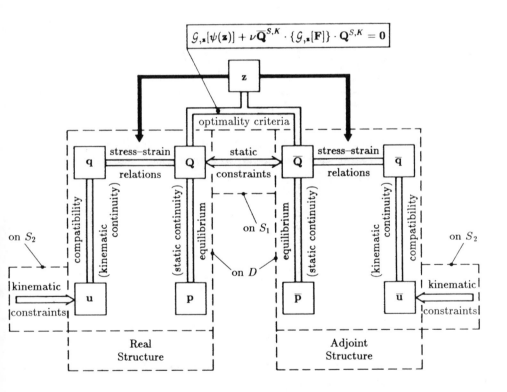

Figure 14. *Fundamental relations of optimal elastic design with a deflection constraint using the COC approach.*

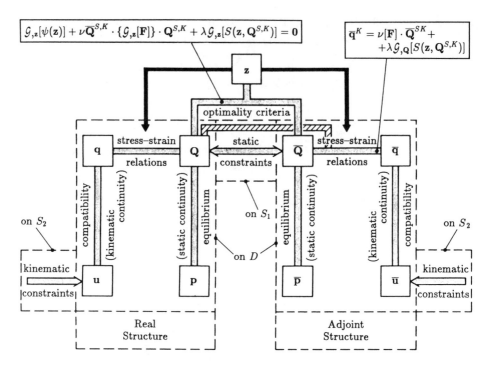

Figure 15. *Fundamental relations of optimal elastic design with a strength and a deflection constraint using the COC approach.*

$$F_i \geq 0 \qquad \text{(for } z_i = z_{i\,\text{min}}\text{)} \,, \tag{64}$$

$$F_i \leq 0 \qquad \text{(for } z_i = z_{i\,\text{max}}\text{)} \,, \tag{65}$$

where F_i denotes one scalar component of the LHS of (58) corresponding to the cross-sectional parameter z_i.

For a derivation of the above optimality criteria, see the book by Rozvany (1989, pp. 70–71).

4.3 Application to Bernoulli Beams of Given Depth and an Independent Proof by Variational Methods

4.3.1 Derivation of the Optimality Criteria from the General Formulae in Figs. 14 and 15.
In the case of a beam of given depth $d = $ const., but variable width $b = b(x)$, the flexural stiffness s is proportional to the cross-sectional area A

$$s(x) = \frac{Ed^3}{12}b(x) \propto A = d\,b(x),$$

(66)

and hence we can adopt the stiffness as the specific cost:

$$\psi = s, \qquad \Phi = \int_D s\,dx.$$

(67)

Moreover, the generalized stresses become

$$Q = M, \quad \overline{Q} = \overline{M},$$

(68)

if we consider a single load condition and a single displacement constraint. The flexibility matrix and its gradient in this problem reduce to

$$[\mathbf{F}] = \frac{1}{s}, \qquad \frac{d[\mathbf{F}]}{d\,s} = -\frac{1}{s^2},$$

(69)

and the single strength condition is (considering only axial stresses)

$$S = -s + k|M| \leq 0,$$

(70)

where $k = Ed/2\sigma_p$ (see Section 3.1). Considering first a displacement constraint only (without strength constraint), the optimality condition on the top of Fig. 14 furnishes

$$1 = \frac{\nu M\overline{M}}{s^2} \Rightarrow s = \sqrt{\nu M\overline{M}},$$

(71)

which was originally derived by Barnett (1961) and was also obtained in a discretized form by Berke (1970). If a minimum value is prescribed for s (i.e. $s \geq s_{\min}$), then by (63) and (64) we have

$$(\text{for } \nu M\overline{M} \leq s_{\min}^2) \qquad s = s_{\min},$$

(72)

$$(\text{for } \nu M\overline{M} > s_{\min}^2) \qquad s = \sqrt{\nu M\overline{M}}.$$

(73)

If we impose both the strength constraint in (70) and a displacement constraint, then the optimality condition at the top of Fig. 15 (with $z_1 \rightarrow s$) yields

$$1 - \frac{\nu M\overline{M}}{s^2} - \lambda = 0,$$

(74)

and the adjoint strain-stress relation in Fig. 15 implies (with $\overline{q} \to \overline{\kappa}$)

$$\overline{\kappa} = \frac{\nu \overline{M}}{s} + \lambda k \, \text{sgn} \, M \quad \text{(for } M \neq 0) \, . \tag{75}$$

For *stress-controlled regions* with $s = k|M| = kM \text{sgn} \, M$, we have by (74) and (75)

$$\lambda = -\frac{\nu M \overline{M}}{s^2} + 1 = -\frac{\nu \overline{M}}{ks} \, \text{sgn} \, M + 1 \, , \tag{76}$$

$$\overline{\kappa} = \frac{\nu \overline{M}}{s} - \frac{\nu \overline{M}}{s} + k \, \text{sgn} \, M = k \, \text{sgn} \, M \, . \tag{77}$$

Returning to the simple example in Fig. 10 (Section 3.1), the adjoint curvatures $\overline{\kappa}$ in Fig. 10c clearly satisfy the optimality criterion (77).

For *deflection controlled regions* with $s > k|M|$, we have

$$(61) \Rightarrow \lambda = 0, \quad (74) \Rightarrow s = \sqrt{\nu M \overline{M}}, \quad (75) \Rightarrow \overline{\kappa} = \frac{\nu \overline{M}}{s} \, . \tag{78}$$

If *selfweight (cs)* is also taken into consideration, then by (59) we have

$$s = \sqrt{\frac{\nu M \overline{M}}{1 + c \overline{u}}} \, . \tag{79}$$

4.3.2 Derivation of the Optimality Criteria Using the Calculus of Variations

The considered beam problem can be stated as

$$\min \Phi = \int_D \left[s + \frac{\nu M \overline{M}}{s} + \lambda(-s + k|M| + \beta) + \overline{u}(M'' + p) \right.$$

$$\left. + u(\overline{M}'' + \overline{p}) \right] dx - \nu \Delta + \nu \alpha \, , \tag{80}$$

where ν is a Lagrangian multiplier (constant), $\lambda(x)$, $\overline{u}(x)$ and $u(x)$ are Lagrangian functions, α is a slack variable and β is a slack function. The expressions in brackets after $\overline{u}$ and u, respectively, represent the equilibrium conditions for the real and virtual loads. Then for variation of s, M and $\overline{M}$ we obtain the following Euler-Lagrange equations:

$$\delta s: \quad 1 - \frac{\nu M \overline{M}}{s^2} - \lambda = 0 \, , \tag{81}$$

$$\delta M: \quad \frac{\nu \overline{M}}{s} - \lambda \kappa \, \text{sgn} \, M = -\overline{u}'' \, , \tag{82}$$

$$\delta \overline{M} : \quad \frac{\nu M}{s} = -u'' \ . \tag{83}$$

With the notation $\overline{\kappa} = -\overline{u}''$, (81) and (82) are identical to the optimality criteria (74) and (75) derived from the general formulae given earlier. Moreover, the equation in (83) (within the factor ν) is the usual elastic moment-curvature relation for the real system.

Considering the functional $\Phi = \int_D f(s, M, \overline{M}) \, dx$ in (80), transversality conditions for variation of M and $\overline{M}$ give the same end conditions for u and $\overline{u}$ as in the case of an elastic beam, provided that the supports are rigid and costless. At a clamped end (B), for example, both the moment M and the shear force $V = -M'$ are variable and hence the following transversality conditions must be satisfied (e.g. Rozvany, 1989, p. 378):

$$\delta M' : \quad (f,_{M''})_B = 0 \ , \tag{84}$$

$$\delta M : \quad [f,_{M'} - (f,_{M''})']_B = 0 \ , \tag{85}$$

where commas indicate partial derivatives with respect to the symbol in the subscript. Combining (84) and (85) with (80), we have

$$\delta M' : \quad \overline{u}_B = 0 \ , \tag{86}$$

$$\delta M : \quad \overline{u}'_B = 0 \ , \tag{87}$$

which are the usual end conditions for a clamped end. At a simple support, only the shear force $V = -M'$ is variable and hence only end condition (86) applies, as in the case of an elastic beam. If we also consider the cost of an end reaction $R = V = -M'$ at a point B, then we have a mixed variational problem with a functional

$$\Phi = \int_D f(s, M, \overline{M}) \, dx + \Omega[M'(x_B)] \ . \tag{88}$$

Then the transversality condition for the above mixed problem (e.g. Rozvany, 1989, p. 390) requires

$$\delta M' : \quad \left[f,_{M''} + \Omega,_{(-M')} \right]_B = 0 \ , \tag{89}$$

or by (80) and (89)

$$\overline{u}_B = \Omega,_R \ , \tag{90}$$

which follows also from the more general formula for the kinematic boundary condition of the adjoint system in Fig. 8. This confirms that the boundary conditions for the adjoint displacements can be different from those for the real displacements.

The beam weight per unit length is $w = bd\gamma$ where γ is the specific weight. This can be represented as $w = cs$ where $c = 12\gamma/(d^2 E)$ is a constant and $s = bd^3 E/12$ is the flexural stiffness. If we include the selfweight in the loading, then the equilibrium term $\bar{u}(M'' + p)$ in (80) is replaced by $\bar{u}(M'' + p + cs)$. Then (81) becomes

$$1 + c\bar{u} - \frac{\nu M \overline{M}}{s^2} - \lambda = 0 , \tag{91}$$

and for deflection controlled regions (with $\lambda = 0$) (91) implies

$$s = \sqrt{\frac{\nu M \overline{M}}{1 + c\bar{u}}} \tag{92}$$

as in Fig. 11d. The same result can be derived from (59).

5 An Iterative COC Algorithm for Large Discretized Systems

As mentioned in the Introduction, the system of equations given in Figs. 14 and 15 can be solved iteratively by using the following two steps in each iterative cycle:

(a) We *analyse* the *real* and *adjoint* systems using some available FE software. For only deflection constraints, both the real and the adjoint systems involve only linear equations.

(b) We *update* the cross-sectional parameters z using the optimality criteria at the top of Figs. 14 and 15 or their extensions in (58) and (59). A comprehensive list of optimality criteria for other design conditions is given in the author's recent book (Rozvany, 1989).

Apart from the above main steps, some other computations are necessary. This can be seen from Fig. 16 which shows the flowchart of the iterative COC algorithm for a deflection constraint, in the context of Bernoulli-beams of variable width. The same procedure can be used for *any elastic structure* with a deflection constraint. In Fig. 16 the symbol i denotes the i-th element and z_i denotes its flexural stiffness. We stipulate a minimum stiffness constraint

$$z_i \geq z_a \qquad \text{(for all } i\text{)} . \tag{93}$$

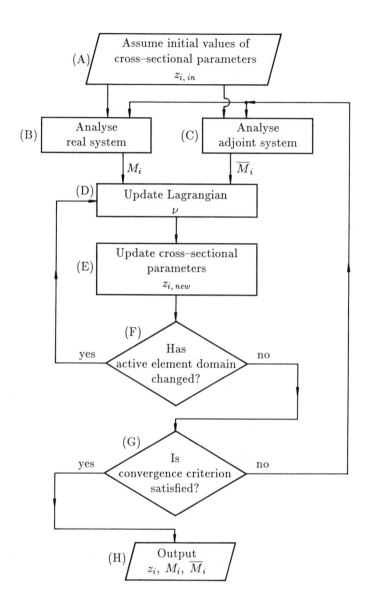

Figure 16. *Iterative COC method for a deflection and prescribed minimum stiffness constraint in the context of beam optimization.*

Elements with $z_i > z_a$ shall be termed *active elements* and those with $z_i = z_a$ *passive elements*, as in earlier work by Berke and Khot (1988). The computational steps in Fig. 16 involve the following operations:

(A) The same constant initial stiffness value z_{in} can be adopted for all elements if a better estimate is not available.

(B,C) Any standard FE software can be used for analysing the real and adjoint systems. The adjoint beam is subject to the virtual load associated with the deflection constraint.

(D) The updated value of the Lagrangian ν can be calculated from the discretized work equation

$$\Delta = \sum_i \frac{M_i \overline{M}_i L_i}{z_i} , \tag{94}$$

where M_i and $\overline{M}_i$ are the real and adjoint moment values at the middle of the beam element i and L_i is the length of the element i.

The RHS of (94) can be split into active (A) and passive (P) element domains in the form

$$\Delta = \sum_A \frac{M_i \overline{M}_i L_i}{\sqrt{\nu M_i \overline{M}_i}} + \sum_P \frac{M_i \overline{M}_i L_i}{z_a} , \tag{95}$$

which implies the following relation for the updating of the Lagrangian ν:

$$\sqrt{\nu} = \frac{\sum\limits_A L_i \sqrt{M_i \overline{M}_i}}{\Delta - \sum\limits_P \frac{M_i \overline{M}_i L_i}{z_a}} . \tag{96}$$

The relation (95) represents an approximation because the moments in the numerator of the first fraction are from the *current* cycle, while the stiffness $\sqrt{\nu M_i \overline{M}_i}$ is the value in the *forthcoming* cycle. This error becomes negligibly small in later iterations because the change in the moment values becomes insignificant. The fact that in (94) the expression $\sum_i \int_{D_i} [M(x)\overline{M}(x)/z(x)]dx$ is replaced by $\sum_i M_i \overline{M}_i L_i/z_i$ is also an approximation, but the error involved is relatively small when a large number of elements are used. It has been shown (Rozvany et al., 1989), for example that for a clamped beam with various numbers (N) of elements the following relative errors (E) occur if (94) is used:

N	E
50	0.001517
200	0.000092
1000	0.000004

(E) The new beam stiffness is calculated from the relations based on (72) and (73):

$$\left(\text{for } \nu M_i \overline{M}_i \leq z_a^2\right) \qquad z_i = z_a \,, \tag{97}$$

$$\left(\text{for } \nu M_i \overline{M}_i > z_a^2\right) \qquad z_i = \sqrt{\nu M_i \overline{M}_i} \,. \tag{98}$$

(F) The reason for this step is as follows. If some elements have changed from the active to the passive set or vice versa, then (96) would give an incorrect estimate of ν and hence steps (D) and (E) must be repeated.

(G) The tolerance test can be based on the change in the total cost

$$\frac{\Phi_{\text{new}} - \Phi_{\text{old}}}{\Phi_{\text{new}}} \leq \overline{E} \,, \tag{99}$$

where $\overline{E}$ is a given tolerance value. A more reliable additional criterion is a check on the cross-sectional parameters

$$\left(\text{for all } i\right) \qquad \frac{z_{i,\text{new}} - z_{i,\text{old}}}{z_{i,\text{new}}} \leq E_1 \,. \tag{100}$$

6 Example of an Ill-Conditioned Problem: Beam with a Deflection Constraint, No Limits on the Cross-Sectional Parameters

The method described in Figs. 14–16 is suitable for optimizing any linearly elastic system with a deflection constraint, if the parameters z fully define the cross-sectional geometry and they vary freely along the centroidal axes (subject to minimum values z_a).

In the case of elastic structures, it is useful to know the absolute lower limit of the structural weight when the prescribed minimum cross-sectional area is zero ($z_a = 0$). Since we still want to avoid concentrated rotations at locally vanishing cross-sections, we consider the limiting case of a sequence of solutions in which the prescribed minimum cross-section has first a finite value but then it approaches zero

$(z_a \to 0)$. This can be evaluated analytically, but in iterative COC calculations it is necessary to specify some small value (for example 10^{-7} times the expected maximum value of z_i) for z_a. It will be seen, however, that for such low values of z_a,

(a) we require a very large number of elements for a reasonable accuracy, and

(b) the problem becomes relatively ill-conditioned and hence it requires a much larger number of iterations.

It will be demonstrated subsequently that the considered problems can be handled much more effectively by the COC approach than by MP methods, since the latter could handle only a very small number of elements due to a computational instability caused by the near-singularity in the solution.

6.1   Problem Description and Analytical Solution

We consider a clamped, uniformly loaded beam (Fig. 17a) of given depth d but variable width z for which the deflection at midspan (A) has an upper limit

$$u_A \leq \Delta .$$
(101)

The virtual (adjoint) load $\bar{p}$, and the real and adjoint moment diagrams $(M, \overline{M})$ are shown, respectively, in Figs. 17b, c and d. We introduce the nondimensional notation $\tilde{x} = x/a$, $\tilde{z} = z/a$, $\tilde{M} = M/pa^2$, $\tilde{\overline{M}} = \overline{M}/a$, $\tilde{u}_A = u_A/\Delta$, $\tilde{s} = 12s/(d^3aE) = \tilde{z}$, $\tilde{\Phi} = \Phi/(da^2\gamma)$, where x is the longitudinal coordinate, E is Young's modulus, s is the flexural stiffness, Φ is the total beam weight, γ is the specific weight of the beam material and other symbols have been defined earlier. The nondimensional width $\tilde{z}$ and stiffness $\tilde{s}$ have been made identical since they are linearly interdependent in this problem. The real and adjoint bending moment diagrams (Figs. 17c and d) are represented by

$$\tilde{M} = \tilde{M}_0 - \tilde{x}^2/2, \quad \tilde{\overline{M}} = \tilde{M}_1 - \tilde{x}/2 .$$
(102)

The considered problem is to be solved for the case when the prescribed minimum cross-sectional area approaches zero: $z_a \to 0$. This means that apart from regions of infinitesimal length with $z = z_a$, our relevant optimality condition by (73) is:

$$\tilde{z} = \sqrt{\nu \tilde{M} \tilde{\overline{M}}} .$$
(103)

In addition, the compatibility conditions for the real and adjoint systems and the nondimensional displacement condition imply:

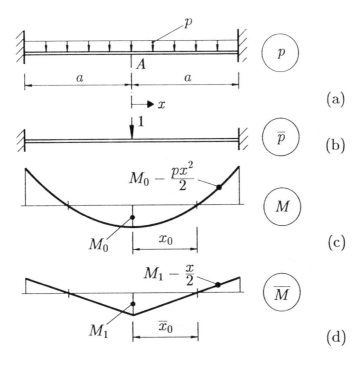

Figure 17. *Beam example—given deflection at midspan and no lower limit on the cross-sectional area.*

$$\int_0^1 \frac{\tilde{M}}{\tilde{z}}\mathrm{d}\tilde{x} = 0, \quad \int_0^1 \frac{\tilde{\overline{M}}}{\tilde{z}}\mathrm{d}\tilde{x} = 0, \quad 2\int_0^1 \frac{\tilde{M}\tilde{\overline{M}}}{\tilde{z}}\mathrm{d}\tilde{x} = 1 , \tag{104}$$

where $\tilde{M}$, $\tilde{\overline{M}}$ and $\tilde{z}$ are given by (102) and (103). Moreover, (103) gives a non-imaginary (real) value for $\tilde{z}$ only if in Fig. 17c and d

$$x_0 = \overline{x}_0 . \tag{105}$$

The relations under (104) and (105), after substitution of (102) and (103), represent four simultaneous equations and the unknown quantities involved are $\tilde{M}_0$, $\tilde{M}_1$ and ν. It can be shown easily that the above equations cannot be satisfied simultaneously. The paradoxical nature of this state of affairs was pointed out by Masur in discussion of his paper (Masur, 1975) and is considered in detail elsewhere (Rozvany, Rotthaus et al., 1990). As is shown in the above paper, the explanation of this paradox is that in the exact analytical solution for $z_a \to 0$, the beam stiffness

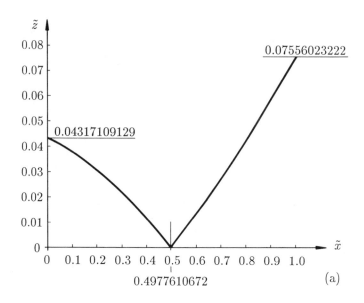

Figure 18a. *Optimal stiffness distribution in the beam example: analytical and COC solutions.*

(nondimensionally: $\tilde{s} = \tilde{z}$) takes on a second order infinitesimal stiffness value over a first order infinitesimal beam length with $z = z_a$. The rotations (i.e. integrated curvatures) over such a length tend to finite values $(\theta, \overline{\theta})$ which have been shown to satisfy the relation

$$\theta \overline{V} = \overline{\theta} V \ , \tag{106}$$

where V and $\overline{V}$ are the shear forces in the real and adjoint beams at the considered location. It will be seen in Fig. 18b that discretized COC solutions with 100000 elements fully confirm the type of singularity predicted by the analytical solution.

The complete analytical solution (Rozvany, Zhou et al., 1989, Rozvany, Rotthaus et al., 1990) yields the following optimal stiffness variation for the above problem

$$\tilde{z} = \left[(\tilde{x}_0)^{5/2}(2^{9/2} - 7) - (\tilde{x}_0 + 1)^{3/2}(7\tilde{x}_0 - 3)\right](\tilde{x}_0 - \tilde{x})\sqrt{\tilde{x}_0 + \tilde{x}}/15 \ , \tag{107}$$

and the optimal $\tilde{x}_0$ value is furnished by the cubic equation

$$\tilde{x}_0^3\left[2^9 - 7(2)^{11/2}\right] - 63\tilde{x}_0^2 - 15\tilde{x}_0 - 1 = 0 \ . \tag{108}$$

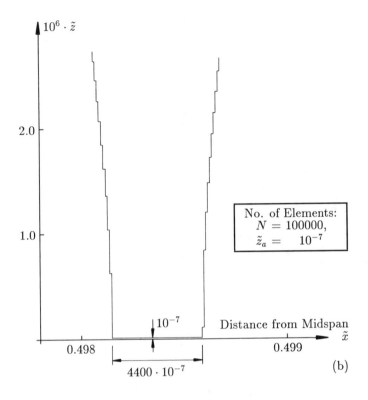

Figure 18b. *Enlarged detail of the stiffness distribution in the vicinity of the singularity (COC solution).*

Moreover, the optimal total weight becomes

$$\Phi = 4\left[\tilde{x}_0^{5/2}(2^{9/2} - 7) - (\tilde{x}_0 + 1)^{3/2}(7\tilde{x}_0 - 3)\right]^2/225 . \tag{109}$$

The relation (108) furnishes the optimal $\tilde{x}_0$−value of

$$\tilde{x}_0 = 0.49776107 , \tag{110}$$

and then (109) and (110) imply

$$\tilde{\Phi}_{\text{opt}} = 0.060448186 . \tag{111}$$

The optimal variation of the stiffness $\tilde{z}(\tilde{x})$ is shown graphically in Fig. 18a.

262

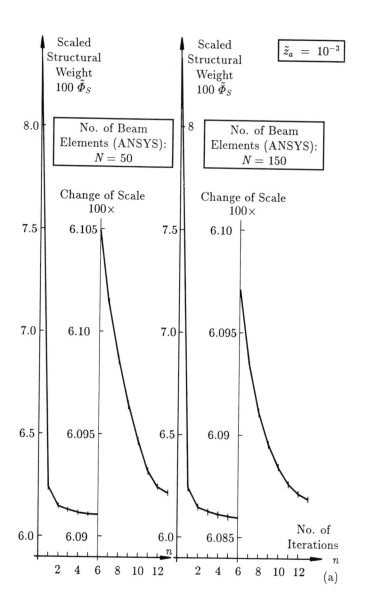

Figure 19a. *Scaled total structural weight $\tilde{\Phi}_s$ as a function of the number of iterations (n) in the beam example for various numbers (N) of elements.*

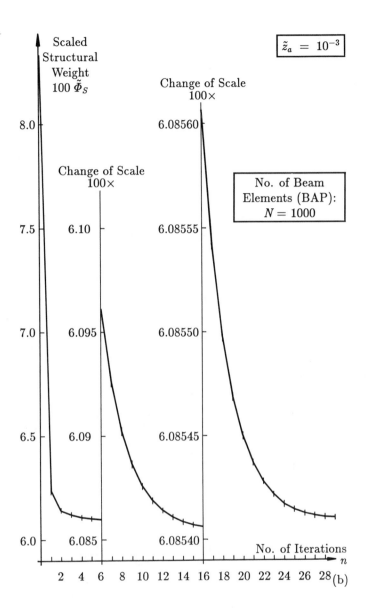

Figure 19b. *See Fig. 19a.*

264

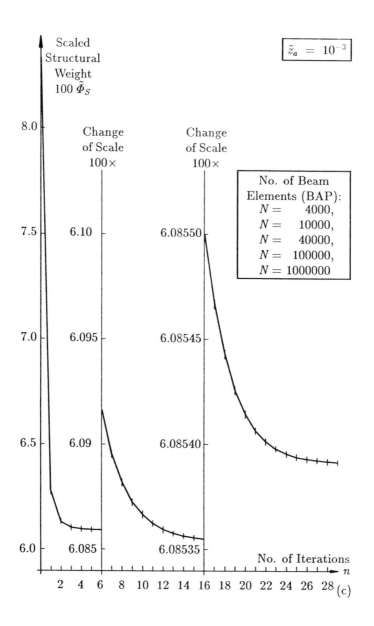

Figure 19c. *See Fig. 19a.*

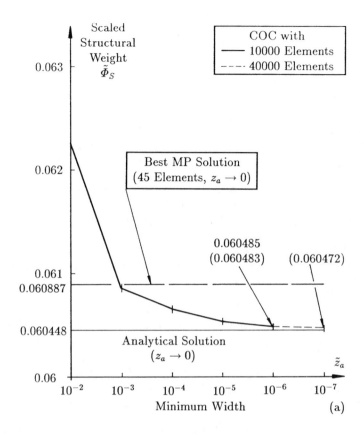

Figure 20a. *A comparison of structural weight values $\tilde{\Phi}_s$ obtained analytically, by an MP method and by the COC method using various prescribed minimum width values $(\tilde{z}_a)$ and either 10000 elements (continuous line) or 40000 elements (broken line).*

6.2 Iterative COC Solutions

Using the COC procedure described in Section 5 (Figs. 14 and 16), discretized numerical solutions were obtained by using various numbers of elements ranging from fifty to one million. The structural analysis of the real and adjoint systems was carried out using the following programs:

The special purpose program BAP for beam analysis was found, as expected, to be much more efficient computationally than a general purpose FE program.

ANSYS   (number of elements $N \leq 150$),

BAP   (number of elements $N \geq 50$).

Both programs yielded identical results for $N = 50, 100$ and 150. Using the COC approach, the following conclusions were reached:

(i) The convergence in the considered example is found fully monotonic and almost uniform, and its rate largely independent of the number of elements (N) used (see Figs. 19a–c).

(ii) In spite of an unusual singularity in the analytical optimal solution, the total weight in the best COC solution (100000 elements, $\tilde{z}_a = 10^{-8}$, $\tilde{\Phi} = 0.060452$) differs only by 0.006 per cent from the analytical solution ($\tilde{\Phi} = 0.060448$).

(iii) For problems with near-singularities in the discretized formulation, a very large number of elements are required for a reasonably high accuracy (see Figs. 20a and b).

(iv) The type of singularity predicted by the analytical solution (stiffness of higher order infinitesimal over a length of first order infinitesimal) is indicated by the discretized COC solution in the vicinity of the singularity (see Fig. 18b).

(v) It can be seen from Fig. 21 that with a decreasing value of $\tilde{z}_a$, the problem becomes more and more ill-conditioned due to the near-singularity at $\tilde{z}_i = \tilde{z}_a$ and hence the convergence becomes correspondingly slower.

(vi) The considered problem was also investigated by using mathematical programming (MP) methods (Fig. 23). Both Lawo (Karlsruhe, FRG) and Wang (Singapore) have found that with $\tilde{z}_a \to 0$, signs of instability appear when the number of elements (N) exceeds 50. The best MP solution (with $N = 45$) yielded a total weight of $\tilde{\Phi} = 6.088672$ which represents an error of about 0.7 per cent in comparison with the exact analytical solution.

(vii) In solving similar problems, Berke and Khot (1987, 1988) used the recurrence relation

$$(z_i)_{n+1} = (z_i)_n \{\nu M_i \overline{M}_i / z_i^2\}_n^{1/r}, \tag{112}$$

where r is a 'step size parameter.' For $r = 2$, (112) reduces to the optimality criterion (98) used in this lecture. The number (n) of iterations required to fulfil the convergence criterion for a relatively stable problem (with $\tilde{z}_a = 10^{-3}$) is shown in Fig. 22 in dependence on the step size parameter in (112). In the latter, either a convergence criterion of the type in (99) is used, or a fixed limit of $\tilde{\Phi} < 0.060856$. The latter type seems to eleminate the 'noise' in the convergence polygons. It can

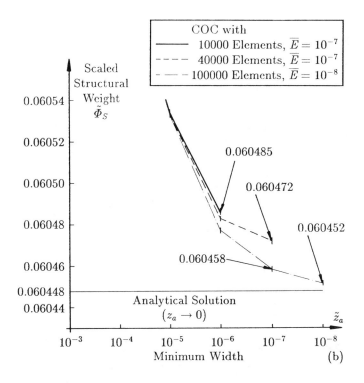

Figure 20b. *A comparison of COC results using 10000, 40000 and 100000 elements.*

be seen from Fig. 22 that for the most stable convergence polygon $(\overline{E} = 10^{-6})$ there is a relatively small difference between the number of iterations for the optimal overrelaxation $(r = 1.3,\ n = 15)$ and the 'natural' relaxation $(r = 2,\ n = 21)$ used herein. Moreover, the r-value for the former is difficult to predict and is very close to an unstable region $(r \leq 1.2)$.

7 COC Procedures for Various Additional Constraints

7.1 Aims

The main aim of this section is to demonstrate the power and versatility of the COC algorithm through applications to various classes of design problems including

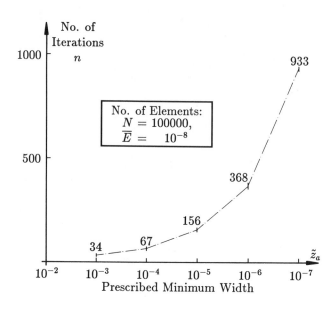

Figure 21. *Number of iterations necessary for the convergence crite-rion as a function of the prescribed minimum width $(\tilde{z}_a)$.*

several types of geometrical constraints. An additional aim is to show some real-life design situations where this technique is particularly useful. The problems to be solved herein are summarized in Fig. 24, in which square brackets refer to the subsection numbers to be used.

7.2 Deflection Constraint and Lower Limit on the Cross-Sectional Parameters

The following notation will be used for various regions in the optimal solution:

For regions governed by the normal stress constraint, a superscript may indicate the sign of the bending moment for beam elements (R_σ^+, R_σ^-). In Section 6, we considered a problem without a lower limit on the cross-sectional area, but in

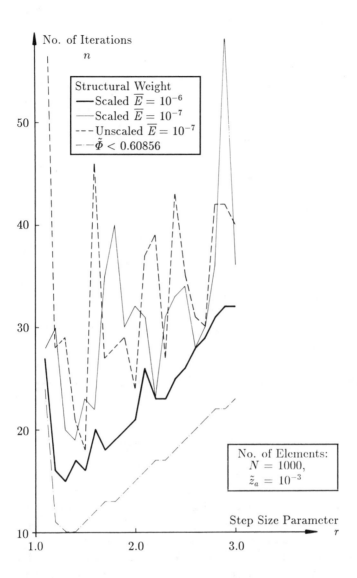

Figure 22. *Beam example: number (n) of iterations required to fulfil the convergence criterion as a function of the Berke-Knot step size parameter.*

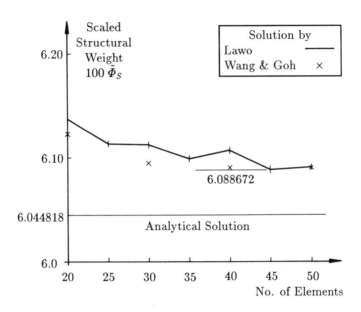

Figure 23. *A comparison of structural weights obtained analytically and by various MP methods.*

numerical calculations we had to impose such a lower limit. The corresponding optimality criteria were given in (58), (63) and (64), and the equivalent discretized optimality conditions in (72) and (73). Considering again the beam in Fig. 17a but with a relatively large lower limit on the non-dimensional stiffness ($z_a \geq 0.01$), the COC algorithm has yielded the optimal stiffness distribution in Fig. 25. As in this stable problem the element number does not effect significantly the results, only 10000 elements were used. The total beam weight for various geometrical constraints is compared with the absolute minimum weight below in Table 1.

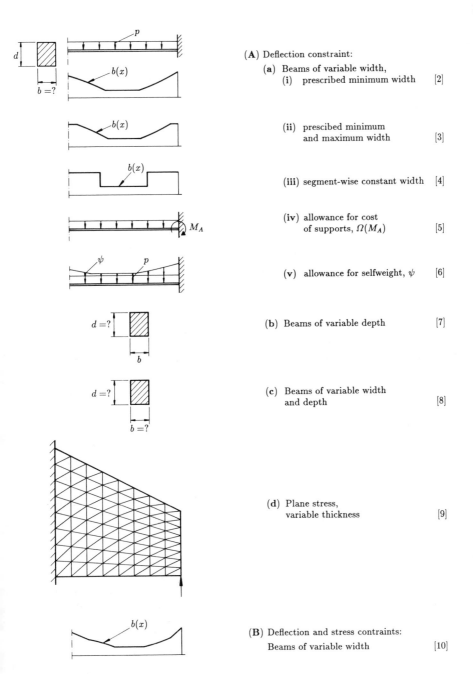

(**A**) Deflection constraint:

    (**a**) Beams of variable width,
        (**i**)   prescribed minimum width   [2]

        (**ii**) prescibed minimum
             and maximum width   [3]

        (**iii**) segment-wise constant width [4]

        (**iv**) allowance for cost
             of supports, $\Omega(M_A)$ [5]

        (**v**) allowance for selfweight, ψ [6]

    (**b**) Beams of variable depth   [7]

    (**c**) Beams of variable width
        and depth   [8]

    (**d**) Plane stress,
        variable thickness   [9]

(**B**) Deflection and stress contraints:
    Beams of variable width   [10]

Figure 24. *Further classes of test problems solved by the COC procedure.*

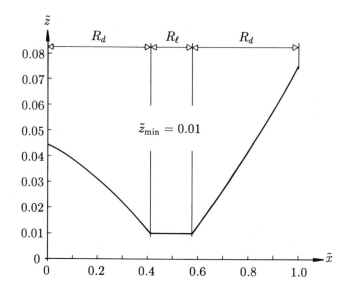

Figure 25. *Optimal stiffness distribution of a clamped beam with a minimum stiffness and a deflection constraint ($\tilde{z}_{min} = 0.01$).*

7.3 Deflection Constraint, Lower and Upper Limits on the Cross-Sectional Parameters

If both upper and lower limits are imposed on the cross-sectional parameters, then (58) with (63)–(65) applies. This means that for beams of variable width, for example, (72) and (73) are replaced by

$$(\text{for } \nu M\overline{M} \leq s_{\min}^2) \qquad s = s_{\min} , \qquad\qquad (113)$$

$$(\text{for } \nu M\overline{M} \geq s_{\max}^2) \qquad s = s_{\max} , \qquad\qquad (114)$$

$$(\text{for } s_{\max}^2 > \nu M\overline{M} > s_{\min}^2) \qquad s = \sqrt{\nu M\overline{M}} . \qquad\qquad (115)$$

In the actual COC algorithm (Section 5), the following changes are necessary. In (95) and (96), the passive set (P) includes all elements goverend by (113) and (114). Moreover, (97) and (98) must be changed in accordance with (113)–(115). COC solutions for $\tilde{z}_{\min} = 0.01$ with $\tilde{z}_{\max} = 0.06$ and $\tilde{z}_{\max} = 0.05$ are given in Figs. 26 and 27 and the corresponding beam weights in Table 1. It can be seen that $\tilde{z}_{\max} = 0.05$ has a more severe effect on both the stiffness distribution (Fig. 27) and beam weight (Table 1).

7.4 Deflection Constraint for Segmented Structures

Let the structural domain D be divided into segments D_α $(\alpha = 1, 2, \ldots, \omega)$ and let on each segment D_α the distribution of the cross-sectional parameters z_i $(i = 1, 2, \ldots, r)$ be prescribed in the form $z_i = \Lambda_{\alpha i}\Gamma_{\alpha i}(\mathbf{x})$ where $\Lambda_{\alpha i}$ are unknown constants and $\Gamma_{\alpha i}$ give shape functions. In a vectorial form, the above relations can be stated as

$$(\text{on } D_\alpha) \qquad \mathbf{z} = \mathbf{\Lambda}_\alpha \mathbf{\Gamma}_\alpha . \qquad\qquad (116)$$

Then the optimality criterion (58) for a single load condition is replaced by

$$\int_{D_\alpha} \left\{ \mathcal{G}_{,\Lambda_\alpha\Gamma_\alpha} [\Psi(\mathbf{\Lambda}_\alpha\mathbf{\Gamma}_\alpha)]\mathbf{\Gamma}_\alpha + \sum_\ell \lambda_\ell \ \mathcal{G}_{,\Lambda_\alpha\Gamma_\alpha} [S_\ell(\mathbf{\Lambda}_\alpha\mathbf{\Gamma}_\alpha, \mathbf{Q})]\mathbf{\Gamma}_\alpha \right.$$

$$\left. + \sum_j \nu_j \overline{\mathbf{Q}} \bullet (\mathcal{G}, {}_{\Lambda_\alpha\Gamma_\alpha}[\mathbf{F}]\mathbf{\Gamma}_\alpha)\mathbf{Q} \right\} dx = 0 \quad (\alpha = 1, 2, \ldots, \omega) . \qquad (117)$$

For beams of variable width and segment-wise constant cross-sections, we have

$$\left. \begin{array}{l} \psi \to s, \quad \mathbf{\Gamma}_\alpha \to 1 \quad (\alpha = 1, \ldots, \omega), \quad S \to (s - k|M|), \\[2mm] \overline{Q} \to \overline{M}, \quad [\mathbf{F}] \to 1/s, \mathbf{Q} \to M, \quad \mathbf{\Gamma}_\alpha \to s_\alpha , \end{array} \right\} \qquad (118)$$

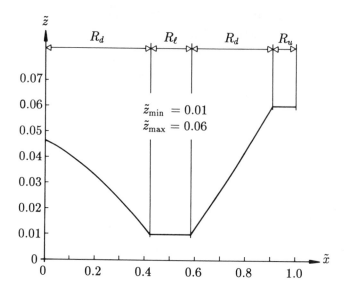

Figure 26. *Optimal stiffness distribution of a beam with* $\tilde{z}_{\min} = 0.01$
and $\tilde{z}_{\max} = 0.06$.

where s_α is the flexure stiffness of the segment α. Then (117) reduces to:

$$L_\alpha + L_\alpha \lambda_\ell - \int_{D_\alpha} \sum_j \nu_j (M\overline{M} / s_\alpha^2) \, \mathrm{d}x = 0 , \qquad (119)$$

where L_α is the length of the segment α. Finally, if no stress constraint is active and we have only one deflection constraint, then (119) reduces to

$$L_\alpha = \nu \int_{D_\alpha} (M\overline{M} / s_\alpha^2) \, \mathrm{d}x, \qquad s_\alpha = \left[\frac{\nu \int_{D_\alpha} M\overline{M} \, \mathrm{d}x}{L_\alpha} \right]^{1/2} , \qquad (120)$$

which replaces the relation in (72). The corresponding discretized optimality criterion is

$$z_\alpha = \sqrt{\frac{\sum\limits_{D_\alpha} \nu M_i M_i L_i}{L_\alpha}} , \qquad (121)$$

which replaces the one in (98). Using the relation (121) and the usual COC algorithm, the solution in Fig. 28 was obtained for given segment lengths of $L_1 =$

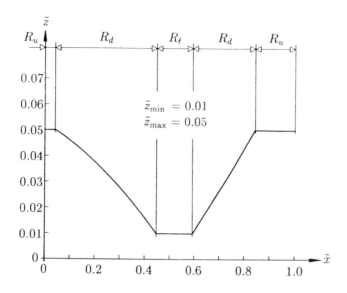

Figure 27. *Optimal stiffness distribution of a beam with $\tilde{z}_{min} = 0.01$ and $\tilde{z}_{max} = 0.05$.*

0.3, $L_2 = 0.4$ and $L_3 = 0.3$ for each half-beam. The corresponding beam weight is shown in Table 1.

Note. The iterative COC results in Sections 7.2–7.4 were confirmed completely by independent calculations. In the case of prescribed non-zero minimum stiffness $(s_{min} > 0)$, the real and adjoint displacement fields $(u, \bar{u})$ contain no singularities and hence the values of M_0, M_1 (Fig. 10) and the Lagrangian ν can be calculated from the three equations under (104) together with (102) and (103). The additional equation (105) does not apply in this case. Since the integrals involved are elliptic, the equations under (104) were solved by numerical integration involving 50000 intervals and a Newton-Raphson procedure. The results in Sections 7.2 and 7.3 were thus confirmed to at least six significant digits (Rozvany et al., 1990). The result in Section 7.4 was confirmed analytically by using differentiation with respect to the three segment stiffnesses.

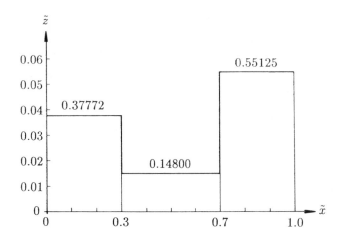

Figure 28. *Optimal stiffness distribution of a beam with three prismatic segments over each halfspan.*

7.5 Deflection Constraint with Allowance for the Cost of Supports

It was already explained in Fig. 8 that the kinematic boundary conditions of the adjoint system must be modified if the cost of the supports (reactions) is taken into consideration. In both plastic and elastic design, the displacements of the adjoint system along such supports $(S_2 \subset D)$ are given by (Rozvany, 1989, p. 73)

$$(\text{on } S_2) \qquad \bar{u} = \mathcal{G}_{,\mathbf{R}} \left[\Omega \{\mathbf{R}^S\} \right], \tag{122}$$

where $\mathbf{R}$ is the reaction vector and Ω is its cost. Considering, for example, the beam in Fig. 17 with the reaction cost $\Omega = \beta |M_e|$ where β is a given constant and M_e is the end moment, (122) furnishes a rotation (slope) of

$$\bar{u}' = \pm \beta \tag{123}$$

at the ends of the beam. Optimal solutions for beams with such reaction costs have been determined (Rozvany, Zhou et al., 1990).

7.6 Deflection Constraint, Allowance for Selfweight

The general optimality criteria for an external load with selfweight were given in (58) with (59) and for beams of variable width with a deflection constraint in (79) or (92). A discretized version of the latter was used (Rozvany, Zhou et al., 1990) in iterative COC procedures, replacing (98). In calculating the updated stiffness from (79), for example, the adjoint deflection values ($\overline{u}$) were taken from the prior cycle.

7.7 Defelction Constraint with Nonlinear Specific Cost/Stiffness/Strength Functions

7.7.1 Beams of Variable Depth For rectangular cross-sections of variable depth z but constant width b, we have

$$\psi = bz , \quad S = kz^2 - |M| , \quad [F] = 1/(cz^3) , \tag{124}$$

where $k = b\sigma_p/6$, $c = bE/12$, σ_p is the permissible stress and E is Young's modulus. Substituting (124) into the optimality criteria and adjoint stress-strain relations in Figs. 14 and 15, we obtain the appropriate equations for the COC procedure. In the case of a deflection constraint only, for example, Fig. 14 and (124) yield the optimality criterion (with $b \to 1$ and $3\nu/c \to \nu$ for a normalized formulation):

$$1 - \nu M\overline{M}/z^4 = 0, \quad z = \sqrt[4]{\nu M\overline{M}} , \tag{125}$$

which replaces (73) in iterative COC applications (Rozvany, Zhou et al., 1990).

7.7.2 Beams of Constant Width/Depth Ratio For rectangular cross-sections of variable width B and depth z with $B = bz$, (124) is replaced by

$$\psi = bz^2 , \quad S = kz^3 - |M| , \quad [F] = 1/cz^4 , \tag{126}$$

where again $k = b\sigma_p/6$, $c = bE/12$. Then for a deflection condition Fig. 14 and (126) yield the optimality criterion (after normalizations)

$$2bz - \nu M\overline{M}/z^5 = 0 , \quad z = \sqrt[6]{\nu M\overline{M}/2b} , \tag{127}$$

which replaces (73) in iterative COC applications (Rozvany, Zhou et al., 1990).

7.8 Deflection Constraint with Non-Separable Specific Cost Functions

In general, the specific cost function is non-separable in terms of the cross-sectional parameters **z**. A simple example of this class of problems is a beam with independently variable width z_1 and depth z_2 and bending in both horizontal and vertical directions (M_H, M_V). In that case we have

$$\mathbf{Q} = (M_H,\ M_V)\ , \overline{\mathbf{Q}} = (\overline{M}_H,\ \overline{M}_V)\ , \left.\begin{array}{c} \\ \\ \\ \\ \end{array}\right\}$$

$$[\mathbf{F}] = \left[\begin{array}{ccc} \frac{1}{cz_1^3 z_2} & 0 & 0\frac{1}{cz_1 z_2^3} \end{array}\right]\ ,\ \psi = z_1 z_2\ ,\ c = E/12\ . \qquad (128)$$

We consider a single deflection constraint which is some linear combination of horizontal and vertical displacements:

$$\Delta \geq \int_D \left(\frac{M_H \overline{M}_H}{c\ z_1^3\ z_2} + \frac{M_V \overline{M}_V}{c\ z_1\ z_2^3}\right) \mathrm{d}x\ . \qquad (129)$$

Then Fig. 14 and (128) furnish the optimality conditions

$$z_2 - \nu \left(\frac{3M_H \overline{M}_H}{c\ z_1^4\ z_2} + \frac{M_V \overline{M}_V}{c\ z_1^2\ z_2^3}\right) = 0\ , \left.\begin{array}{c} \\ \\ \\ \\ \end{array}\right\}$$

$$z_1 - \nu \left(\frac{M_H \overline{M}_H}{c\ z_1^3\ z_2^2} + \frac{3M_V \overline{M}_V}{c\ z_1\ z_2^4}\right) = 0\ , \qquad (130)$$

which imply after normalization $(\nu/c \to \nu)$:

$$z_1^4 z_2^4 = \nu \left(3M_H \overline{M}_H z_2^2 + M_V \overline{M}_V z_1^2\right) = \nu \left(M_H \overline{M}_H z_2^2 + 3M_V \overline{M}_V z_1^2\right) \quad (131)$$

or

$$z_1^2 z_2^4 = 4\nu M_V \overline{M}_V\ , \qquad z_1^4 z_2^2 = 4\nu M_H \overline{M}_H \qquad (132)$$

implying

$$z_1^3 = 2\nu^{1/2} \frac{M_H \overline{M}_H}{\sqrt{M_V \overline{M}_V}}\ , \qquad z_2^3 = 2\nu^{1/2} \frac{M_V \overline{M}_V}{\sqrt{M_H \overline{M}_H}}\ . \qquad (133)$$

The latter replaces (73) in COC applications (Rozvany, Zhou et al., 1990).

7.9 Two-Dimensional Systems

The proposed COC algorithm can be readily extended to two-dimensional systems (e.g. plane stress, plates, shells). In collaboration with MBB GmbH in the FRG, the authors investigated various problems which can have practical implications in the aero-space technology. The simplest of these, a test-example, is shown in Fig. 24d in which a plate of variable thickness (z) in *plane stress* is subject to a point load at one free corner and the vertical deflection is constrained at the other free corner. In the considered case we have

$$\left.\begin{array}{l} \mathbf{Q} = (N_x,\ N_y,\ N_{xy})\ , \qquad \overline{\mathbf{Q}} = \left(\overline{N}_x,\ \overline{N}_y,\ \overline{N}_{xy}\right)\ , \\[2em] [\mathbf{F}] = \dfrac{1}{z}\begin{bmatrix} c_1 & -c_2 & 0 \\ -c_2 & c_1 & 0 \\ 0 & 0 & c_3 \end{bmatrix}\ , \\[3em] \psi = z\ , \quad c_1 = \dfrac{1}{E}\ , \quad c_2 = \dfrac{\nu}{E}\ , \quad c_3 = \dfrac{1}{G}\ . \end{array}\right\} \tag{134}$$

Then Fig. 14 with (134) implies the optimality criterion

$$\left.\begin{array}{l} 1 = (\nu/z^2)\left[c_1(N_x\overline{N}_x + N_y\overline{N}_y) - c_2(N_x\overline{N}_y + \overline{N}_x N_y) + c_3 N_{xy}\overline{N}_{xy}\right]\ , \\[2em] z = \sqrt{\nu\left[(c_1(N_x\overline{N}_x + N_y\overline{N}_y) - c_2(N_x\overline{N}_y + \overline{N}_x N_y) + c_3 N_{xy}\overline{N}_{xy}\right]}\ , \end{array}\right\} \tag{135}$$

which replaces (73) in the COC algorithm. For the case of a system with 3200 elements, the layout of various optimal regions is shown in Fig. 29 and the actual thickness variation for the R_d-region is represented graphically in Fig. 30. It can be seen that rib-like concentrations of material occur in the right half of the system. For further details, the reader is referred to the paper by Rozvany, Zhou et al. (1990).

7.10 Combined Deflection and Stress Constraints

For *beams of variable width*, this problem was already discussed in Section 4.3.1 [see (77) and (78)] and some further problems were treated in general in Sections 7.7 and 7.8.

In applying these optimality criteria in an iterative COC algorithm, it is necessary to test in the updating operation (Step E in Fig. 16) whether a particular element is governed by a prescribed minimum value constraint or a stress constraint and then the appropriate optimality condition must be selected. Moreover, in updating the Lagrangian in Step D of Fig. 16, we must regard all stress-controlled elements as being passive.

Figures 31–33 show the optimal stiffness distribution for the beam in Fig. 17 with the stress constraint $\tilde{z} \geq k|\overline{M}|$ for various values of the stress factor k. For values outside the range $0.2 \leq k \leq 0.24$, the solution is less interesting because either the deflection constraint or the stress constraint governs most of the element stiffnesses.

The same procedure was used for developing a program for the optimization of *long-span laminated timber beams of variable depth*, using the German design

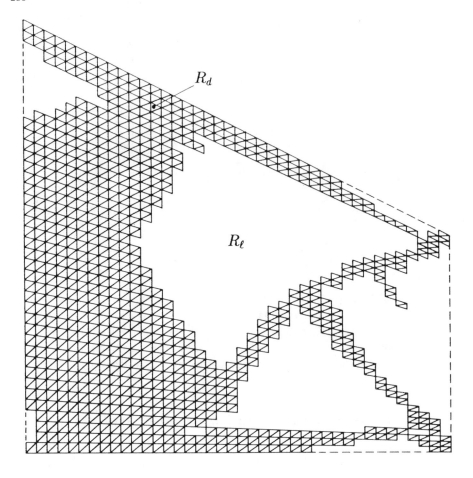

R_d

R_ℓ

Figure 29. *Optimal regions of a plate in plane stress subject to a displacement constraint.*

code (DIN 1052) for constraints on deflections, axial stresses and shear stresses. A typical optimal design and the corresponding moment (M) and shear force (Q) diagrams are shown in Fig. 34. In a real design situation, the top of the beam cross-sections would follow a plane and the underside of the beam would form a curved surface.

In the above problem, the location of the maximal displacement is not known and hence in the iterative procedure the adjoint load always corresponds to the location of the maximal displacement in the prior cycle. Using an MP method, one would have to specify that the displacement at *each element* (100000 points in

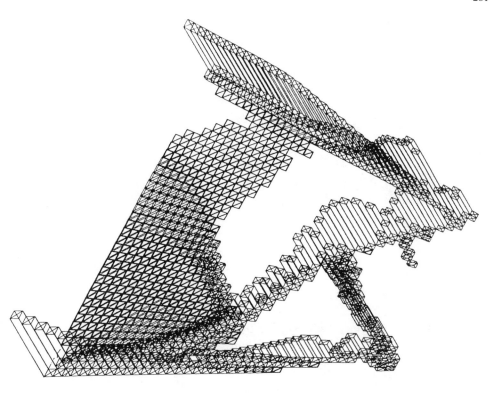

Figure 30. *Optimal stiffness distribution of a plate in plane system subject to a displacement constraint.*

this case) is smaller than a prescribed maximum value. Using however the strategy mentioned above, a very large number of alternative displacement constraints can be handled efficiently, even when several displacement constraints are simultaneously active.

For the beam in Fig. 34, the admissible prismatic beam is shown in broken line for a comparison. The latter would have a volume of 4.05 m^3, whereas that of the optimal beam is 2.56 m^3. This means that a prismatic design has *58% higher volume* than the optimal one.

The considered program is being extended to allowance for selfweight [see the optimality criterion in (59)] and built-up cross-sections. For longer spans, considerably higher savings are expected than the one indicated above.

282

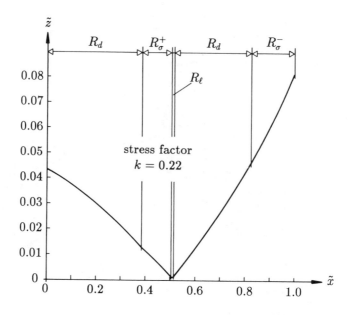

Figure 31. *Optimal stiffness distribution of a beam subject to a dis-*
placement and a stress constraint (k = 0.20).

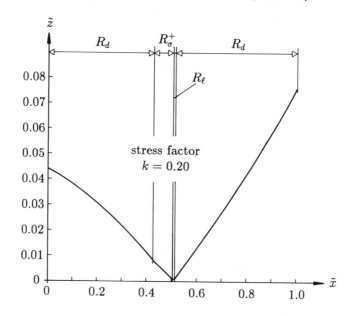

Figure 32. *Optimal stiffness distribution of a beam subject to a dis-*
placement and a stress constraint (k = 0.22).

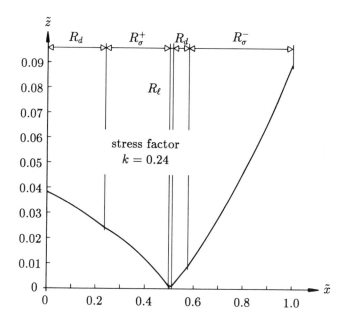

Figure 33. *Optimal stiffness distribution of a beam subject to a displacement and a stress constraint (k = 0.24).*

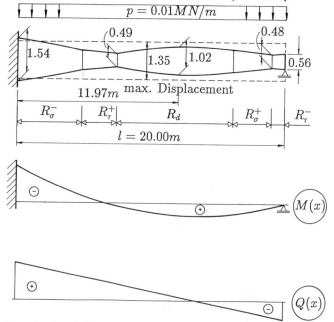

Figure 34. *Optimal design of a laminated timber beam using the provisions of the German design code DIN 1052.*

8 Concluding Remarks

(a) A comprehensive set of analytical continuum-based optimality criteria (COC) for various design problems are presented in a recent book by the first author (Rozvany, 1989).

(b) It has been established herein that the iterative use of the COC algorithm enables us to optimize discretized systems with a very large number of variables and local as well as global constraints. The optimization capability is limited only by the analysis capability of the FE program used. With the help of special beam analysis programs, it was demonstrated that even one million variables with a similar number of constraints can easily be optimized by the COC technique.

(c) The rate of convergence was found largely independent of the number of elements used.

(d) In contrast to the COC algorithm, usual mathematical programmings (MP) methods can handle only a few hundred variables, causing a significant discrepancy between analysis and optimization capability.

(e) The COC approach is being implemented for the optimization of large real systems (e.g. aero-space structures, long-span laminated beams, etc.) with a number of design constraints.

Acknowledgment The authors are indebted to the Deutsche Forschungsgemeinschaft (DFG) for financial support (Proj. No. Ro 744/1-1); to Dr. L. Berke (NASA), Dr. N. S. Khot (NASA), Prof. M. Bendsøe (Techn. Univ. Denmark) and Dr. G. Kneppe (MBB GmbH) for useful suggestions; to Mrs. S. Rozvany for editing the text; to Mrs. A. Fischer for processing the manuscript; and to Mrs. E. Becker for plotting and processing the diagrams.

References

Barnett, R.L., (1961), 'Minimum-weight designof beams for deflection', *J. Eng. Mech. Div. ASCE*, **87**, EM1, 75-100.

Berke, L., (1970), 'An Efficient Approach to the minimum weight design of deflection limited structures', *AFFDL-TM-70-4-FDTR*.

Berke, L. and Khot, N. S., (1987), 'Structural optimization using optimality criteria,' in C. A. Mota Soares (Ed.), *Computer aided optimal design: structural and mechanical systems*, pp. 271-312. Springer–Verlag, Berlin, Heidelberg, New York.

Berke, L. and Khot, N. S., (1988), 'Performance characteristics of optimality criteria methods,' in G. I. N. Rozvany, B. L. Karihaloo (Eds.), *Structural optimization*, (Proc. IUTAM Symposium, Melbourne, 1988), pp. 39–46, Kluwer Acad. Publ., Dordrecht.

Charrett, D. E. and Rozvany, G. I. N., (1972), 'Extensions of the Prager-Shield theory of optimal plastic design,' *Int. J. Non-Linear Mech.* **7**, 1, 51–64, (February).

Foulkes, J., (1954), 'The minimum-weight design of structural frames,' *Proc. Royal Soc.* **223**, No. 1155, 482–494, (May).

Heyman, J., (1959), 'On the absolute minimum weight design of framed structures,' *Quart. J. Mech. Appl. Math.* **12**, 3, 314–324, (August).

Marcal, P. V. and Prager, W., (1964), 'A method of optimal plastic design,' *J. de Mécan.* **3**, 4, 509–530.

Masur, E. F., (1975a), 'Optimality in the presence of discreteness and discontinuity,' in A. Sawczuk, Z. Mróz (Eds.), *Optimization in structural design*, (Proc. IUTAM Symp. held in Warsaw, Aug. 1973), pp. 441–453, Springer-Verlag, Berlin.

Masur, E. F., (1975b), 'Optimal placement of available sections in structural eigenvalue problems,' *J. Optimiz. Theory Appl.* **15**, 1, 69–84, (January).

Michell, A. G. M., (1904), 'The limits of economy of material in frame-structures,' *Phil. Mag.* **8**, 47, 589–597, (November).

Mróz, Z. and Rozvany, G. I. N., (1975), 'Optimal design of structures with variable support conditions,' *J. Optimiz. Theory Appl.* **15**, 1, 85–101, (January).

Prager, W., (1974), *Introduction to structural optimization.*, (Course held Int. Centre for Mech. Sci., Udine, CISM, 212), Springer-Verlag, Vienna.

Prager, W. and Rozvany, G. I. N., (1975), 'Plastic design of beams: optimal locations of supports and steps in yield moment,' *Int. J. Mech. Sci.* **17**, 12, 627–631, (October).

Prager, W. and Shield, R. T., (1967), 'A general theory of optimal plastic design,' *J. Appl. Mech.* **34**, 1, 184–186, (March).

Prager, W. and Taylor, J. E., (1968), 'Problems of optimal structural design,' *J. Appl. Mech.* **35**, 102–106, (March).

Rozvany, G. I. N., (1973a), 'Non-convex structural optimization problems,' *J. Eng. Mech. Div. ASCE* **99**, EM1, 243–248, (February).

Rozvany, G. I. N., (1973b), 'Optimal plastic design for partially preassigned strength distribution,' *J. Optimiz. Theory Appl.* **11**, 4, 421–436, (April).

Rozvany, G. I. N., (1974a), 'Optimal plastic design with discontinuous cost functions,' *J. Appl. Mech.* **41**, 1, 309–310, (March).

Rozvany, G. I. N., (1974b), 'Optimization of unspecified generalized forces in structural design,' *J. Appl. Mech.* **41**, 4, 1143–1145, (December).

Rozvany, G. I. N., (1975), 'Analytical treatment of some extended problems in structural optimization, Part I,' *J. Struct. Mech.* **3**, 4, 359–385.

Rozvany, G. I. N., (1976), *Optimal design of flexural systems*, Pergamon Press, Oxford, (Russian translation: Stroiizdat, Moscow, 1980).

Rozvany, G. I. N., (1977a), 'Elastic versus plastic optimal strength design,' *J. Engrg. Mech. Div. ASCE*, **103**, EM1, 210–215, (February).

Rozvany, G. I. N., (1977b), 'Optimal plastic design: allowance for self-weight,' *J. Engrg. Mech. Div. ASCE* **103**, EM6, 1165–1170, (December).

Rozvany, G. I. N., (1978), 'Optimal elastic design for stress constraints,' *Comp. Struct.* **8**, 3, 455–463, (May).

Rozvany, G. I. N., (1981), 'Variational methods and optimality criteria,' in E. J. Haug and J. Cea (Eds.), *Optimization of distributed parameter structures*, (Proc. NATO ASI held in Iowa City), pp. 82–111, Sijthoff and Noordhoff, Alphen aan der Rijn, The Netherlands.

Rozvany, G. I. N., (1984), 'Prager-Shield optimality criteria with bounded spatial gradients,' *J. Engrg. Mech. Div. ASCE* **110**, EM1, 129–136, (January).

Rozvany, G. I. N., (1989), *Structural design via optimality criteria*, Kluwer Acad. Publ., Dordrecht.

Rozvany, G. I. N., Booz, W. and Ong, T. G., (1987), 'Optimal layout theory: multiconstraint elastic design,' in K. L. Teo, H. Paul, K. L. Chew, C. M. Wang, *et al.* (Eds.), *Proc. Int. Conf. on Optimization: Techniques and Applications*, (held in Singapore, April 1987), pp. 138–151, Nat. Univ. Singapore.

Rozvany, G. I.N. and Cohn, M. Z., (1970), 'Lower-bound optimal design of concrete structures,' *J. Eng. Mech. Div. ASCE* **96**, EM6, 1013–1030, (December).

Rozvany, G. I. N., Menkenhagen, J. and Spengemann, F., (1989), 'Prager-Shield optimality criteria for structures with linear segmentation,' *J. Eng. Mech. ASCE* **114**, 1, 203-209, (January).

Rozvany, G. I. N. and Mróz, Z., (1975), 'Optimal design taking cost of joints into account,' *J. Eng. Mech. Div. ASCE* **101**, 6, 917–921, (December).

Rozvany, G. I. N. and Mróz, Z., (1977), 'Column design: optimization of support conditions and segmentation,' *J. Struct. Mech.* **5**, 3, 279–290.

Rozvany, G. I. N. and Ong, T. G., (1986), 'A general theory of optimal layouts for elastic structures,' *J. Engrg. Mech. Div. ASCE* **112**, 8, 851–857, (August).

Rozvany, G. I. N. , Ong, T. G., Karihaloo, B.L., (1986), 'A general theory of optimal elastic design for structures with segmentation,' *J. Appl. Mech. ASME* **53**, 2, 242–248, (June).

Rozvany, G. I. N. and Prager, W., (1979), 'A new class of structural optimization problems: optimal archgrids,' *Comp. Meth. Appl. Mech. Engrg.* **19**, 1, 127–150, (June).

Rozvany, G. I. N., Rotthaus, M., Spengemann, F., Gollub, W., Zhou, M., Lawo, M., Goh, C. J. and Wang, C. M., (1990), 'On the Masur paradox,' *Mech. Struct. Machines* **18**, 3 (scheduled for publication).

Rozvany, G. I. N., Spengemann, F., Menkenhagen, J. and Wang, C. M., (1989), 'Extension of Heyman's and Foulkes' theorems to structures with linear segmentation,' *Int. J. Mech. Sci.* **31**, 2, 87–106.

Rozvany, G. I. N., Yep, K. M., Ong T. G. and Karihaloo, B. L., (1988), 'Optimal design of elastic beams under multiple design constraints,' *Int. J. Solids Struct.* **24**, 4, 331–349, (April).

Rozvany, G. I. N., Zhou, M., Rotthaus, M., Gollub, W. and Spengemann, F., (1989), 'Continuum-type optimality criteria methods for large finite element systems with a displacement constraint, Part I,' *Struct. Optim.* **1**, 1, 47–72.

Rozvany, G. I. N., Zhou, M., Gollub, W. and Rotthaus, M., (1990), 'Continuum-type optimality criteria methods for large finite element systems with a displacement constraint, Part II', *Struct. Optim.* **2**, 1, 77-104.

Thierauf, G., (1978), 'A method for optimal limit design of structures with alternative loads,' *Comp. Meth. Appl. Mech. Engrg.* **16**, 135–149.

Symbol:	Active Constraint:
R_ℓ	lower limit on cross-section
R_u	upper limit on cross-section
R_d	displacement constraint
R_σ	normal stress constraint
R_τ	shear stress constraint

Table 1. *Comparison of beam weights for various constraints with the absolute minimum weight for a deflection constraint only. The symbol k denotes the factor in the stress constraint.*

Beam weight	Difference %	Constraint
0.060448	0	deflection only
0.062264	3.00	$\tilde{z}_{\min} = 0.01$
0.062529	3.44	$\tilde{z}_{\min} = 0.01$, $\tilde{z}_{\max} = 0.06$
0.063832	6.00	$\tilde{z}_{\min} = 0.01$, $\tilde{z}_{\max} = 0.05$
0.067578	11.80	3 prismatic segments
0.060934	0.80	$\tilde{z}_{\min} = 0.001$, $k = 0.20$
0.061015	0.94	$\tilde{z}_{\min} = 0.001$, $k = 0.22$
0.061666	2.01	$\tilde{z}_{\min} = 0.001$, $k = 0.24$

Approximate Models for Structural Optimization

U. Kirsch

Department of Engineering Science and Mechanics
Virginia Polytechnic Institute and State University
Blacksburg, Virginia
United States of America
on leave from:
The Department of Civil Engineering
Technion, Haifa, Israel

Abstract   Practical methods for optimal design of structures are usually based on approximation concepts. Some approximate models for structural optimization are presented in this article. Two classes of such models are discussed: (a) Approximate equilibrium models, where approximations of the displacements in terms of the design variables will lead to an optimal solution which does not necessarily satisfy equilibrium. (b) Approximate compatibility models, where approximations of the forces will lead to an optimal solution which does not necessarily satisfy compatibility. High quality approximations for the two classes are presented. The relationships between the various approximations are derived and some models are shown to be equivalent under certain assumptions. It is shown that the improved behaviour models proposed in this study often provide sufficient results with a moderate computational effort. The possibility of improving the optimal solution by prestressing and passive control devices is demonstrated.

1   Introduction

Applications of approximation concepts in structural optimization have been motivated by the following characteristics of the design problem:

- The problem size (number of variables and constraints) is usually large. Each member involves at least one variable, and various failure modes under each of several load conditions must be considered.

- The constraints are implicit functions of the design variables. That is, evaluation of the constraints for any given design involves solution of a set of simultaneous equations.

B. H. V. Topping (ed.),
Optimization and Artificial Intelligence in Civil and Structural Engineering, Volume I, 289–325.
© 1992 *Kluwer Academic Publishers. Printed in the Netherlands.*

In general, the solution of the optimal design problem is iterative and consists of repeated analyses followed by redesign steps. The number of redesigns (or repeated analyses), which is a function of the problem dimensionality, is usually large and the total computational effort involved in practical design problems may become prohibitive. This is the case particularly if each redesign involves extensive calculations such as solution of a large system of equations, evaluation of the constraint functions, calculation of constraint derivatives, etc. Means which may be used to alleviate this difficulty include:

- Reduction of the number of independent variables by linking or basis reduction (Pickett et al. [1], Schmit et al. [2–4]).

- Reduction of the number of constraints considered at a given iteration using regionalization or truncation techniques (Schmit et al. [2–4]).

- Introduction of approximate models of the structural behaviour in terms of the independent variables (Schmit et al. [2–4], Kirsch et al. [5–14]).

It is recognized that only methods which do not involve many implicit analyses are practical for optimal design applications. In this article some approximate models of the design problem are presented. Elastic analysis is assumed throughout and both displacement method and force method formulations are considered. It is shown that approximations of the nodal displacements, often used in optimal design, essentially lead to solutions which do not necessarily satisfy equilibrium. Similarly, approximations of the redundant forces may lead to solutions which do not satisfy compatibility. Some approximations of both the displacements and the forces in terms of the independent variables are presented and assumptions leading to simplified formulations such as linear programming, constant force models, stress ratio, and scaling are discussed.

1.1   Problem Statement

The general design problem considered in this article can be stated as the following nonlinear programming problem:

Find the design variables X such that

$$Z = F(X) \rightarrow \min \qquad \text{(objective function)} \qquad (1)$$

$$X^L \leq X \leq X^U \qquad \text{(side constraints)} \qquad (2)$$

$$\sigma^L \leq \sigma \leq \sigma^U \qquad \text{(stress constraints)} \qquad (3)$$

$$r^L \leq r \leq r^U \qquad \text{(displacement constraints)} \qquad (4)$$

in which L and U are subscripts denoting lower and upper bounds, respectively; and σ and r are vectors of stresses and displacements, respectively. Both the stresses and displacements are usually implicit functions of the design variables, given by the analysis equations.

The number of independent design variables is often reduced by assuming several elements to have prescribed ratios between their sizes. In many optimal design problems, the number of elements needed in the analysis is much larger than the number of design variables required properly to describe the design problem. Frequently, it is neither necessary nor desirable for each element to have its own independent design variable. Design variable linking or basis reduction fixes the relative size of some preselected group of elements so that some independent variables control the size of all elements. This can be accomplished by relating the original design variables X to the independent variables X_I according to the expression

$$X = LX_I \tag{5}$$

where L is the matrix of linking constants giving the predetermined ratios between variables X and X_I. In Eq. 5 the variables X are taken as a linear combination of X_I. In many cases in which only simple design variable linking is used, the matrix L takes on a special form, in which each row contains only one nonzero element. The reduced-basis concept further reduces the number of independent design variables by expressing the vector X_I as a linear combination of s basis vectors B_i, giving

$$X_I = \sum_{i=1}^{s} y_i B_i = By \tag{6}$$

Substituting Eq. 6 into 5 gives

$$X = LBy = Ty \tag{7}$$

where T is the matrix of prelinked basis vectors and y is the vector of a reduced set of design variables.

The number of inequality constraints in optimal design problems may be very large, particularly in structures consisting of many elements and subjected to multiple loading conditions. Constraint deletion techniques can be used to reduce the number of constraints. It is recognised that, during each stage of an iterative design process, it is only necessary to consider critical or potentially critical constraints. On the basis of analysis of the structure, all the inequality constraints may be evaluated. Constraint deletion techniques are then used to temporarily eliminate redundant and noncritical constraints that are not likely to influence significantly the design process during the subsequent stage. For each constraint type the most critical constraint value is identified using regionalization and truncation techniques (Schmit [4]). An example of the former technique is that only the most critical stress constraint in each region in each load condition is retained. The regionalization idea works well provided the design changes made during a stage are small enough that they do not result in a shift of the critical constraint location within a region. A truncation technique, on the other hand, involves temporary deletion of constraints for which the ratio of the stress to its allowable value is so low that the constraint will clearly be inactive during the stage. Evidently, none of the constraints included in the original problem statement are permanently deleted unless they are strictly redundant.

The selection of an objective function can be one of the most important decisions in the whole optimal design process. In some situations, an obvious objective function exists. For example, if the cost of the structure is assumed to be proportional to its weight, then the objective function will represent the weight. In general, the objective function represents the most important single property of a design, but it may represent also a weighted sum of a number of properties. Weight is the most commonly used objective function due to the fact that it is readily quantified, although most optimization methods are not limited to weight minimization. The weight of the structure is often of critical importance, but the minimum weight is not always the cheapest. Cost is of wider practical importance than weight, but it is often difficult to obtain sufficient data for the construction of a real cost function.

1.2 Approximate Behaviour Models

In most optimal design procedures the behaviour of the structure must be evaluated many times for successive modifications in the design variables. This operation, which involves much computational effort, is one of the main obstacles in applying optimization methods to large structural systems. Approximate behaviour models used in structural optimization can usually be classified as follows:

(a) Reanalysis methods, intended to analyse efficiently new designs using information obtained from previous ones (Arora [5], Kirsch [5], Abu Kassim and Topping [16]). These methods are usually based on series expansion and require less computational effort than exact reanalysis. One problem often encountered is that the accuracy of the solution may not be sufficient.

(b) Approximations based on temporarily neglecting the implicit analysis equations. This can be done, e.g. by assuming fixed internal forces (Kirsch [5], Berke and Venkayya [17], Khot, Berke and Venkayya [18]) or neglecting the compatibility conditions (Reinschmidt et al. [19,20], Frashi and Schmit [21], Kirsch [23]).

Both classes of approximations are discussed in this article.

In general, two conflicting factors should be considered in choosing an approximate behaviour model for a specific optimanl design problem:

(a) the accuracy of the calculations, or the quality of the approximation; and

(b) the computational effort involved, or the efficiency of the method.

1.2.1 Approximate Equilibrium Considering the displacement method of analysis, the displacements r are first computed by solving the set of simultaneous equilibrium equations

$$Kr = R \qquad (8)$$

in which K is the system stiffness matrix and R is the load vector. The stresses σ can then be determined directly by

$$\sigma = Sr \qquad (9)$$

Here S is the system stress transformation matrix.

Considering the analysis formulation Eqs. 8 and 9, the equilibrium equations are the only implicit conditions. Thus any approximation of the displacements will lead to an optiman solution X_{opt}, r_{opt}, which does not necessarily satisfy equilibrium. However, this solution can be viewed as the exact optimum of a structure subjected to a set of imaginary loads R_0 given by

$$R_0 = K_{opt}\, r_{opt} \qquad (10)$$

The difference between the real loadings and the imaginary loadings,

$$\Delta R = R - R_0 \qquad (11)$$

indicates the discrepancy in satisfying the original equilibrium conditions by the optimal solution of the approximate problem.

While different approximations of the displacements are possible, the object is usually to obtain explicit expressions $r(X)$ in terms of the design variables

$$r \simeq r(X) \qquad (12)$$

instead of the implicit Eqs. 8. The result is that the general problem can be formulated as the following explicit approximate equilibrium problem (EAEP): Find X such that

$$Z = F(X) \to \min \qquad (13)$$

$$X^L \le X \le X^U \qquad (14)$$

$$\sigma^L \le Sr(X) \le \sigma^U \qquad (15)$$

$$r^L \le r(X) \le r^U \qquad (16)$$

Various approximations can often be improved by using intervening variables. These are defined by

$$Y_i = Y_i(X) \tag{17}$$

One of the more popular intervening variables is the reciprocal of X_i

$$Y_i = \frac{1}{X_i} \tag{18}$$

The reason for this is that displacement and stress constraints for determinate structures are often linear functions of the receiprocal variables. For statically indeterminate structures, the use of these variables still proves to be a useful device to obtain better approximations. It is often assumed that the elements of the stiffness matrix K are linear functions of the design variables and the elements of the load vector R are independent of X. From Eqs. 8 and 18 it can be observed that for the j-th displacement r_j,

$$\left. \begin{array}{l} r_j(\Lambda X) = \dfrac{1}{\Lambda} r_j(X) \\[2mm] r_j(\Lambda Y) = \Lambda r_j(Y) \end{array} \right\} \tag{19}$$

in which Λ is a scaling multiplier $(\Lambda > 0)$. The significance of these properties is that a given design can easily be scaled by modifying Λ so that any desired displacement is equal to its limiting value. This operation, called scaling of the design, is used in many optimal design methods. In cases where the stiffness matrix elements are nonlinear functions of the design variables (for example, in frame elements where both moments of inertia and cross-sectional areas are considered), still linear approximations may be used for the nonlinear terms.

If the elements of matrix S are constant (in truss structures, for example), then from Eq. 9 we find for the j-th stress, σ_j,

$$\left. \begin{array}{l} \sigma_j(\Lambda X) = \dfrac{1}{\Lambda} \sigma_j(X) \\[2mm] \sigma_j(\Lambda Y) = \Lambda \sigma_j(Y) \end{array} \right\} \tag{20}$$

In any case the stresses are explicit functions of X and r. Some possibilities of introducing the explicit approximations $r(X)$ and the relationship between the various methods will be discussed later in this article.

The significance of the EAEP formulation is that the constraint functions can readily be evaluated for any assumed design. If first-order (linear) approximations are considered, a linear programming problem is obtained. The main advantage is that large problems can be efficiently solved by standard computer programs. Since the solution, X_{opt}, of the EAEP is only an approximation to the true optimum, it is necessary to check its adequacy. One possibility is to analyse the optimal design of the EAEP by an exact method. A constrained feasible design is then calculated by scaling and the objective function value is evaluated. Finally, the approximate constraint functions are modified and the EAEP is solved again, if necessary. These steps are repeated until convergence. In the case of linear approximations, the well-known sequence of linear programs formulation is obtained. Various means, such as move limits on X can be applied to improve the convergence.

1.2.2 Approximate Compatibility Using the force method of analysis the redundant forces N can be computed for any given design by solving the set of compatibility equations

$$FN = \delta \qquad (21)$$

where F is the flexibility matrix and δ is the vector of displacements corresponding to redundants due to loads on the primary structure. The internal forces A and the displacements r are then computed by

$$A = A_L + A_N N \qquad (22)$$

$$r = r_L + r_N N \qquad (23)$$

in which A_L, r_L are the vectors of forces and displacements, respectively, due to loads, and A_N, r_N are the matrices of forces and displacements, respectively, due to unit value or redundants. All these quantities are computed for the primary structure.

If only a small number of displacements must be analysed, it might be convenient to use the virtual-load method. It is often assumed that the displacements r can be expressed as

$$r = D\frac{1}{g(X)} \qquad (24)$$

where the elements of $1/g(X)$ represent the cross-sectional properties to be considered (for example, $g(X)$ are the cross-sectional areas in truss elements, or the moments of inertia in beam elements); and the elements of matrix D are given by

$$D_{ij} = \frac{1}{E} \int_j A_j A_{ij}^Q \, \mathrm{d}l_j \qquad (25)$$

A_j is the force in the j-th member due to the actual loads; A_{ij}^Q is the force in the j-th member due to a virtual load $Q_i = 1.0$ applied in the i-th direction; l_j is the member length; and E is the modulus of elasticity.

It should be noted that Eq. 24 is based on the assumption that a single force (such as axial force or bending moment) is sufficient to describe the response behaviour of each member. However, the present formulation can be extended to the more general case of multiple force members. The design variables X usually represent the cross-sectional areas (in truss structures) or the moments of inertia (in flexural systems). In such cases

$$\frac{1}{g(X)} = \frac{l}{X} \tag{26}$$

and Eq. 24 becomes

$$r = D\frac{1}{X} = DY \tag{27}$$

Considering the analysis formulation of Eqs. 21–23, the compatibility equations (Eqs. 21) are the only implicit conditions. Explicit approximations of the redundant forces in terms of the design variables

$$N \simeq N(X) \tag{28}$$

can be used for elimination of these conditions and for simple evaluations of the constraint functions for any given design. Any approximation $N(X)$ will lead to an optimal solution X_{opt}, N_{opt} which does not necessarily satisfy compatibility. This solution can be viewed as the exact optimum of a structure with displacements in the direction of the redundants δ_0 given by

$$\delta_0 = F_{\text{opt}} N_{\text{opt}} \tag{29}$$

The difference

$$\Delta\delta = \delta - \delta_0 \tag{30}$$

indicates the discrepancy in satisfying the original compatibility conditions by the optimal solution of the approximate problem.

Based on Eq. 28, the general design problem can be stated as the following Explicit Approximate Compatibility Problem (EACP):

Find X such that

$$Z = F(X) \rightarrow \min \tag{31}$$

$$X^L \leq X \leq X^U \tag{32}$$

$$\sigma^L X \leq A_L + A_N N(X) \leq \sigma^U X \tag{33}$$

$$r^L \leq r_L + r_N N(X) \leq r^U \tag{34}$$

Equation 33 is based on the assumed stress force relation

$$A = \sigma X \tag{35}$$

In many design problems, the elements of the flexibility matrix F and the displacement vectors δ are linear functions of the reciprocal variables Y. From Eq. 21 it can be shown that in this case

$$N(\Lambda X) = N(X) \tag{36}$$

That is, the force distribution in the structure is not affected by scaling of the design. The stresses and displacements become explicit functions of X, and under certain assumptions can be shown to be equivalent to those of Eqs. 19, 20.

2 Approximate Equilibrium Models

2.1 Problem Statement

To preserve the efficiency, the presentation is concentrated on models which do not involve matrix inversion. Only the decomposed stiffness matrix, known from exact analysis of the initial design, is required to obtain the approximate expressions. The presented models are based on a single exact analysis.

The equilibrium equations for a given design variables vector $\tilde{X}$ are

$$\tilde{K}\tilde{r} = R \tag{37}$$

where $\tilde{K}$ is the stiffness matrix corresponding to the design $\tilde{X}$; R is the load vector whose elements are assumed, in the development, to be independent of the design variables; and $\tilde{r}$ is the nodal displacements vector computed at the design $\tilde{X}$. Assuming a change ΔX in the design variables so that the modified design is

$$X = \tilde{X} + \Delta X \tag{38}$$

the corresponding stiffness matrix will be

$$K = \tilde{K} + \Delta K \tag{39}$$

Here ΔK is the change in the stiffness matrix due to the change ΔX.

The object in this section is to introduce explicit models for efficient calculation of the displacements r corresponding to designs X, obtained by changing the value of the design variables. It is assumed that the displacements $\tilde{r}$ are known from analysis of the initial design. It is also assumed that $\tilde{K}$ can be decomposed as follows

$$\tilde{K} = \tilde{U}^T \tilde{U} \tag{40}$$

where $\tilde{U}$ is an upper triangular matrix.

Approximations along given lines in the design space are often required in optimal design procedures. This problem is common to many mathematical programming methods such as feasibile directions or penalty functions. In general, the lines (or direction vectors) are determined successively by the optimization method used. In each of the given directions, it is usually necessary to evaluate the constraint functions (or to repeat the anlysis) many times. A line in the design space can be defined in terms of a single independent variable by

$$X = \tilde{X} + \gamma \Delta \tilde{X} \tag{41}$$

in which $\tilde{X}$ is a given initial design, $\Delta \tilde{X}$ is a given vector in the design space, and the variable γ determines the step size. Approximations along a line require much less computations since only a single variable is inolved.

The elements of the stiffness matrix are usually some functions of γ. One common case is when the modified stiffness matrix can be expressed as

298

$$K = \tilde{K} + f(\gamma)\Delta\tilde{K} \tag{42}$$

In truss structures where X represents the cross-sectional areas, or in beam elements where the momens of inertia are chosen as design variables, the elements of the stiffness matrix are linear functions of X. Equation 39 then becomes

$$K = \tilde{K} + \sum_i K_i^0 \Delta X_i \tag{43}$$

where K_i^0 is a matrix of constant coefficients, representing the contribution of the i-th element. For the line of Eq 41 we obtain

$$K = \tilde{K} + \gamma\Delta\tilde{K} \tag{44}$$

If the elements of K are functions of aZ_i^b (Z_i being the naturally chosen design variables, and a and b being given constants) we may use the transformation

$$X_i = aZ_i^b \tag{45}$$

and obtain the linear relationship (Eq. 43). The expression (Eq. 45) is suitable, for example, for standard joists (Clarkson [23]). In cases where such transformations are not possible (for example, in frame elements where the stiffness matrix is a function of both moments of inertia and cross-sectional areas), linear approximations may be used for the non-linear terms of the stiffness matrix.

2.2 First Order Approximations

Writing the equilibrium equations for the modified design as

$$\left(\tilde{K} + \Delta K\right)\left(\tilde{r} + \Delta r\right) = R \tag{46}$$

and substituting Eq. 37 yields

$$\Delta K\tilde{r} + \tilde{K}\Delta r + \Delta K\Delta r = 0 \tag{47}$$

Ignoring the term $\Delta K\Delta r$ and rearranging, one can obtain

$$r(X) = \tilde{r} + \Delta r = \tilde{r} - \tilde{K}^{-1}\Delta K\tilde{r} = (I - B)\tilde{r} \tag{48}$$

where

$$B \equiv \tilde{K}^{-1}\Delta K \tag{49}$$

Equation 48 is a First Order Approximation (FOA) of the displacements.

Premultiplying Eq. 46 by $\tilde{K}^{-1}$ and substituting Eqs. 37, 49 and $r = \tilde{r} + \Delta r$ yields

$$(I + B)r = \tilde{r} \tag{50}$$

Premultiplying by $(I + B)^{-1}$ and expanding

$$(I + B)^{-1} = I - B + B^2 - \cdots \tag{51}$$

equation 50 becomes equivalent to the following Binomial Series Approximation (BSA)

$$r = \left(I - B + B^2 - \cdots \right) \tilde{r} \tag{52}$$

Defining

$$\left. \begin{aligned} r_1 &\equiv -B\tilde{r} \\ r_2 &\equiv -Br_1 \end{aligned} \right\} \tag{53}$$

etc. the BSA can be expressed as

$$r = \tilde{r} + r_1 + r_2 + \cdots \tag{54}$$

For the given triangularization (Eq. 40) the calculation of the coefficient vectors r_1, r_2, ... involves only forward and back substitution. The solution is based on the recurrence relation

$$\tilde{K} r_k = R_k \tag{55}$$

in which k denotes the term in the series and

$$\left. \begin{aligned} R_k &= -\Delta K r_{k-1} \\ r_0 &= \tilde{r} \end{aligned} \right\} \tag{56}$$

It can be noted that if only the first two terms of the BSA (Eq. 52) are considered, then this approximate model becomes equivalent to the FOA (Eq. 48).

Taylor series expansion is one of the most commonly used explicit behaviour models in optimal design. Expanding the displacements about $\tilde{X}$ yields

$$r = \tilde{r} + \sum_i \frac{\partial \tilde{r}}{\partial X_i} \Delta X_i + \frac{1}{2} \sum_i \sum_j \frac{\partial^2 \tilde{r}}{\partial X_i \partial X_j} \Delta X_i \Delta X_j + \cdots \tag{57}$$

Since the calculation of high-order derivaties is usually not practicable, linear approximations are often used. These require evaluation of the first derivatives which can readily be calculated (see Section 2.3).

To improve the quality of the first order Taylor series expansion, the displacements can be expressed in terms of the intervening variables Y_j

$$r(Y) \simeq \tilde{r} + \nabla \tilde{r}_y \left(Y - \tilde{Y} \right) = \tilde{r} + \sum_j \left(\frac{\partial \tilde{r}}{\partial X_j} \Big/ \frac{\partial Y_j}{\partial X_j} \right) \left(Y_j - \tilde{Y}_j \right) \tag{58}$$

in which $\nabla \tilde{r}_y$ is the matrix of first derivatives of r with respect to Y at $\tilde{Y}$. For the reciprocal approximation (Eq. 18), Eq. 58 becomes

$$r(X) \simeq \tilde{r} + \sum_j \left(\frac{\tilde{X}_j}{X_j}\right) \frac{\partial \tilde{r}}{\partial X_j} \left(X_j - \tilde{X}_j\right) \tag{59}$$

One disadvantage of the reciprocal approximation is that it becomes infinite if any X_j is zero. This difficulty can be overcome by the simple transformation (Haftka and Kamat [24])

$$Y_j = \frac{1}{X_j + \delta X_j} \tag{60}$$

where δX_j is taken as a small fraction of a typical value of X_j.

It can be shown (Kirsch [12]) that if the elements of the stiffness matrix are linear functions of X (Eq. 43), then the first order Taylor series expansion is equivalent to the FOA (Eq. 48).

Consider the case where the displacements and the stresses are homogeneous functions of degree -1, that is, the conditions (Eqs. 19, 20) hold. Fuchs [25] has shown that, based on Euler's theorem on homogeneous functions

$$\left.\begin{aligned}
\tilde{\nabla} r_X \tilde{X} = -\tilde{r} \\
\tilde{\nabla} \sigma_X \tilde{X} = -\tilde{\sigma}
\end{aligned}\right\} \tag{61}$$

and the first order Taylor series expansion becomes

$$\left.\begin{aligned}
r = 2\tilde{r} + \nabla \tilde{r}_X X \\
\sigma = 2\sigma + \nabla \tilde{\sigma}_X X
\end{aligned}\right\} \tag{62}$$

For the reciprocal variables (Eq. 18) the corresponding expressions are

$$\left.\begin{aligned}
r = \tilde{\nabla} r_y Y \\
\sigma = \tilde{\nabla} \sigma_y Y
\end{aligned}\right\} \tag{63}$$

It has been shown that the linearized functions (Eq. 63) are exact for any point along the scaling line through $\tilde{Y}$, $\Lambda \tilde{Y}$. This property illustrates the advantage in using linearized constraints in the inverse variables. The approximations (Eq. 63) can be shown to be equivalent to the assumption of constant internal forces (Kirsch [5]). The latter subject will be discussed later in Section 3.4.

2.3 Design Sensitivity Analysis

Calculation of derivatives of constraint functions with respect to design variables, called design sensitivity analysis, is required in using Taylor series approximations and most of the efficient optimization methods.

Assuming the constraint

$$g \equiv r_j - r^u \leq 0 \qquad (64)$$

and a single design variable X, then

$$\frac{dg}{dx} = \frac{\partial g}{\partial X} + Z^T \frac{dr_j}{dX} = \frac{dr_j}{dX} \qquad (65)$$

where $\frac{\partial g}{\partial X} = 0$ and Z is a vector having unit value at the j-th location and zeros elsewhere

$$Z = \frac{\partial g}{\partial r} = I_j \qquad (66)$$

Differentiation of Eq. 8 with respect to X gives

$$K \frac{dr}{dX} = \frac{dR}{dX} - \frac{dK}{dX} r \qquad (67)$$

Premultiplying Eq. 67 by $Z^T K^{-1}$ yields

$$Z^T \frac{dr}{dX} = Z^T K^{-1} \left(\frac{dR}{dX} - \frac{dK}{dX} r \right) \qquad (68)$$

Numerically, the calculation of $Z^T \frac{dr}{dX}$ may be performed in two different ways:

(a) The direct method, where Eq. 67 is solved for $\frac{dr}{dX}$ and then taking the corresponding component $\frac{dr_j}{dX}$.

(b) The adjoint method where the adjoint vector λ is first computed by solving

$$K\lambda = Z \qquad (69)$$

and then Eq. 65 is written as (Eqs. 67, 69)

$$\frac{dg}{dX} = \frac{\partial g}{\partial X} + \lambda^T \left(\frac{dR}{dX} - \frac{dK}{dX} r \right) \qquad (70)$$

where use has been made of the symmetry of K.

It has been noted (Arora and Haug [26]) that the diect method is more efficient than the adjoint method when the number of design variables is smaller than the number of displacement constraints; the adjoint method is more efficient when the number of design variables is larger than the number of these constraints. In both methods the previously calculated factors of $\tilde{U}$ and $\tilde{U}^T$ (Eq. 40) are used for efficient solution of Eqs. 67 or 69.

Derivatives of r with respect to Y can be obtained by similar methods.

2.4 Improved Approximations by Scaling

Series expansions are based on information of a single design. As a result, the quality of the approximations might be sufficient only for a limited region. Improved approximations by scaling may be introduced, to achieve the following advantages (Kirsch [10]):

(a) the series coefficients can readily be calculated by a simple procedure, thus preserving the efficiency of the method;

(b) the quality of the approximations can be improved by scaling of the initial design.

The series (Eq. 52) converges if, and only if

$$\lim_{k \to \infty} B^k = 0 \tag{71}$$

A sufficient criterion for the convergence of the series is that

$$\|B\| \leq 1 \tag{72}$$

where $\|B\|$ is the norm of B. A method for improving the series convergence is proposed in this section, based on scaling of the initial design.

It can be observed from Eq. 37 that analyzing a design with stiffness matrix

$$K_\Lambda = \Lambda \tilde{K} \tag{73}$$

where Λ is a positive scalar multiplier, the resulting displacements are

$$r_\Lambda = \frac{1}{\Lambda} \tilde{r} \tag{74}$$

The stiffness matrix of a modified design K can be expressed as (see Eqs. 39 and 73)

$$K = \tilde{K} + \Delta K = K_\Lambda + \Delta K_\Lambda = \Lambda \tilde{K} + \Delta K_\Lambda \tag{75}$$

Substituting this expression into the analysis equations of the modified design (Eq. 47), premultiplying by $\tilde{K}^{-1}$ and rearranging, yields

$$\left(I + \frac{1}{\Lambda} \tilde{K}^{-1} \Delta K_\Lambda \right) \Lambda r = \tilde{r} \tag{76}$$

Substituting ΔK_Λ from Eq. 75 into Eq. 76 gives

$$\left(I + \frac{1-\Lambda}{\Lambda} I + \frac{1}{\Lambda} \tilde{K}^{-1} \Delta K \right) \Lambda r = \tilde{r} \tag{77}$$

Defining (see Eq. 49)

$$B_\Lambda \equiv \frac{1-\Lambda}{\Lambda}I + \frac{1}{\Lambda}\tilde{K}^{-1}\Delta K = \frac{1-\Lambda}{\Lambda}I + \frac{1}{\Lambda}B \tag{78}$$

substituting into Eq. 77 and expanding $(I + B_\Lambda)^{-1}$, we obtain the following series for r in terms of Λ

$$r = \frac{1}{\Lambda}\left(I - B_\Lambda + B_\Lambda^2 - \cdots\right)\tilde{r} \tag{79}$$

One possible criterion for evaluating Λ is that the Euclidean norm of B_Λ is minimized, that is

$$\|B_\Lambda\| = \left(\sum_{i=1}^{N}\sum_{j=1}^{N}B_{\Lambda ij}^2\right)^{\frac{1}{2}} \to \min \tag{80}$$

in which N denotes the dimension of matrix B.

One drawback of using the criterion (Eq. 80) is that the elements of matrix B must be calculated. Since this operation involves much computational effort, we may use an alternative criterion which minimizes the Euclidean norm of the first term in the series Eq. 79, that is

$$\|B_\Lambda \tilde{r}\| \to \min \tag{81}$$

Substituting $B_{\Lambda ij}$ into Eq. 81, differentiating and setting the result equal to zero, yields

$$\Lambda = \frac{a}{b} \tag{82}$$

where

$$\left.\begin{aligned} a &= \sum_{i=1}^{N}\left(\tilde{r}_i - r_{1i}\right)^2 \\ b &= \sum_{i=1}^{N}\left(\tilde{r}_i^2 - \tilde{r}_i r_{1i}\right)^2 \end{aligned}\right\} \tag{83}$$

and the elements r_{1i} can readily be computed by Eq. 53. The effect of Λ on the quality of the approximations will be demonstrated subsequently.

2.4.1 Thirteen Bar Truss (Kirsch and Toledano [9]) The initial geometry and the loadings are shown in Fig. 1a. The single geometric variables represents the span. Choosing the line

$$Y = 60 + 93\alpha \qquad 0 \le \alpha \le 1$$

the geometry for $\alpha = 1$ is shown in Fig. 1b. The cross-sectional areas are:

304

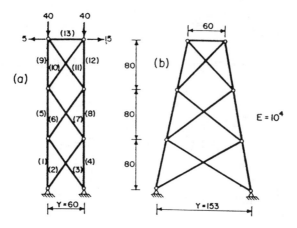

Figure 1. *Thirteen-bar truss.*

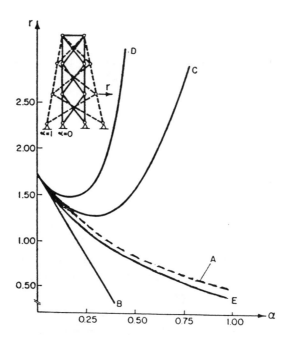

Figure 2. *Results, thirteen-bar truss.*

$$X^T = \{1.42, 0.10, 0.61, 2.91, 1.67, 0.29, 0.10,$$

$$2.87, 1.76, 0.10, 0.99, 2.24, 0.28\}$$

Displacements are calculated by the following methods: Exact (A), first (B) and second (C) Taylor series, Binomial series (D) and improved approximations (E). The results shown in Fig. 2 indicate that methods B, C, D provide poor approximations for the present example. Method E provides adequate approximations for large changes in the geometry. The scaling improves the convergence of the Binomial series; $\|B\tilde{r}\| = 28.3$ is reduced to $\|B_A\tilde{r}\| = 0.56$ with the scaling multiplier being $\Lambda = 1.72$.

3 Approximate Compatibility Models

3.1 First Order Approximations

The object of this section is to present high quality approximations of structural member forces by the force method of analysis. In statically indeterminate structures, the analysis unknowns when using this method are redundant member forces. Since the number of these variables is often low and generally member forces are less sensitive to design changes than displacements are, application of the force method contributes to the efficiency and to the quality of approximations in an optimization process.

The problem discussed herein can be stated as follows. Given a design X_A and the corresponding redundant forces N_A, the object is to determine efficiently the forces N corresponding to various changes ΔX in the design without repeating the exact solution procedure that involves solving the implicit compatibility Eqs. 21

$$FN = \delta \tag{84}$$

Here, F is the flexibility matrix and δ is the displacement compatibility vector. Once the redundant forces are determined, all other member forces can be obtained from statical equilibrium Eqs. 22. Similar to the displacement method formulation, solution of Eq. 84 is often carried out through decomposition of the matrix F into two triangular matrices

$$F = U_F^T U_F \tag{85}$$

where U_F is an upper triangular matrix. The forces N are then computed by forward and back substitution. One advantage of this procedure, to be illustrated later, is that U_F can be used for repeated approximate analyses of the structure.

Assume that for a given initial design X_A, the corresponding F_A and N_A (Eq. 84), are known and F_A is given in the decomposed form (Eq. 85). For a change ΔX in the design, the modified analysis equations (Eqs. 84) become

$$(F_A + \Delta F)(N_A + \Delta N) = \delta_A + \Delta \delta \tag{86}$$

where ΔF, ΔN and $\Delta \delta$ are the corresponding changes in the F, N and δ. Rearranging Eq. 86, neglecting the second order term $\Delta F \cdot \Delta N$ and defining, for brevity

$$\left. \begin{aligned} C &= F_A^{-1} \Delta F \\ N_\delta &= F_A^{-1} \Delta \delta \\ \tilde{N} &= N_A + N_\delta \end{aligned} \right\} \tag{87}$$

The First Order Approximations (FOA) model is obtained

$$N = N_A + \Delta N = \tilde{N} - C N_A \tag{88}$$

The quality of approximations by the FOA model has been demonstrated for some examples (Kirsch and Taye [14]).

Another approximate model can be obtained by the binomial series expansion. Rewriting Eq. 86 in the form

$$(F_A + \Delta F) N = \delta_A + \Delta \delta \tag{89}$$

premultiplying Eq. 89 by F_A^{-1} and substituting Eq. 87, one obtains

$$(I + C)N = \tilde{N} \tag{90}$$

Premultiplying Eq. 90 by $(I + C)^{-1}$ and expanding the latter expression under certain conditions as

$$(I + C)^{-1} = I - C + C^2 - \cdots \tag{91}$$

Equation 90 becomes equivalent to the following Binomial Series Approximations (BSA) model

$$N = (I - C + C^2 - \cdots) \tilde{N} \tag{92}$$

More accurate results can be obtained by the BSA when several terms of the series are retained.

To calculate the terms of the series (Eq. 92) define

$$N_1 = -C\tilde{N} \tag{93}$$

$$N_2 = -C N_1 \tag{94}$$

etc. The series then becomes

$$N = \tilde{N} + N_1 + N_2 + \cdots \tag{95}$$

For the given triangularization (Eq. 85), the determination of the vectors N_1, N_2, etc. involves only forward and back substitutions. The solution is based on the recurrence relation

$$F_A N_k = -\Delta F N_{k-1} \qquad (96)$$

in which k denotes the term in the series and

$$N_0 = \tilde{N} \qquad (97)$$

It can be noted that if only the first two terms of the BSA (Eq. 92) are considered, then this approximate model becomes equivalent to the FOA (Eq. 88). Besides this, these two models under this last condition can be shown to be equivalent to the first order Taylor series model (Kirsch [6]) when the elements of F and δ are linear functions of Y. It will be shown that the FOA may provide poor results for large changes in the design. The BSA might also diverge for such changes. Improved approximations by scaling are presented subsequently.

3.2   Improved Approximations by Scaling

A procedure that greatly improves the quality of the binomial series (Eq. 92) approximations is presented in this section. The method (Kirsch and Taye [14]) is based on scaling the initial flexibility matrix F_A as

$$F_f = f F_A \qquad (98)$$

and employing norm minimization concepts to determine the optimal scaling factor f in Eq. 98 such that the quality of approximations is improved over those of the previous two models.

The matrix F of the modified design can be expressed as

$$F = F_A + \Delta F = f F_A + \Delta F_f \qquad (99)$$

Substitute for ΔF from Eq. 99 into Eq. 89 and premultiply by F_A^{-1} to obtain

$$\left(I + \frac{1}{f} F_A^{-1} \Delta F_f \right) f N = \tilde{N} \qquad (100)$$

Substituting for ΔF_f from Eq. 99 into Eq. 100 and noting also Eq. 87 gives

$$\left(I + \frac{1-f}{f} I + \frac{1}{f} C \right) f N = \tilde{N} \qquad (101)$$

Define, for brevity, C_f as

$$C_f = \frac{1-f}{f} I + \frac{1}{f} C \qquad (102)$$

Using this last expression in Eq. 101 and expanding $(I + C_f)^{-1}$, the following Improved Binomial Series Approximations (IBSA) model of N in terms of f is obtained

$$N = \frac{1}{f}\left(I - C_f + C_f^2 - \cdots\right)\tilde{N} \tag{103}$$

For $f = 1$, C_f becomes equivalent to C and the IBSA is reduced to the BSA model of Eq. 92.

The object now is to select f such that the convergence properties and approximation qualities of the series (Eq. 103) are improved over those of the series Eq. 92. It has been shown (Kirsch [9,10]) that different selections of f for approximations of displacements may lead to improved results. Similar to Eq. 81, a criterion that minimizes the Euclidean norm of the second term in the series (Eq. 103) is used

$$\| C_f\tilde{N} \|\to \min. \tag{104}$$

This does not involve a prior determination of elements of matrix C. By this criterion, the scaling factor f is determined from

$$f = \frac{\sum\left(\tilde{N}_i + \hat{N}_i\right)^2}{\sum\left(\tilde{N}_i^2 + \tilde{N}_i \cdot \hat{N}_i\right)} \tag{105}$$

in which the vector $\hat{N}$ is defined by

$$\hat{N} = C\tilde{N} \tag{106}$$

The advantage of this criterion is that all terms in Eq. 105 can readily be computed.

3.2.1 Twenty-Five-Bar Truss (Kirsch and Taye [14]) The truss shown in Fig. 3 has been analysed for designs given by

$$X_A^T = \{10\}$$

$$\Delta X^T = \{95, 100, 82, 83, 80, 99, 88, 95, 96, 88, 97, 82,$$
$$81, 94, 88, 88, 96, 93, 91, 86, 95, 87, 92, 84, 85\}$$

The chosen redundants are the forces in members 19–25, and the changes in the variables are about 900% of the initial design. Approximate reanalysis results by each of the models and the set of corresponding exact values are given in Table 1. Convergence histories by the BSA (Eq. 92) and IBSA (Eq. 103) models are shown in Figs. 4a and b respectively.

It can be seen that poor results have been obtained by the FOA model. The series (Eq. 92) and (Eq. 103) converge to the exact solution. It is to be noted, however, that only two terms are sufficient to obtain good results by the IBSA while more than 50 terms are required by the BSA to reach a similar accuracy level.

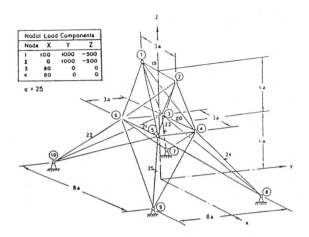

Figure 3. *Twenty-five-bar truss.*

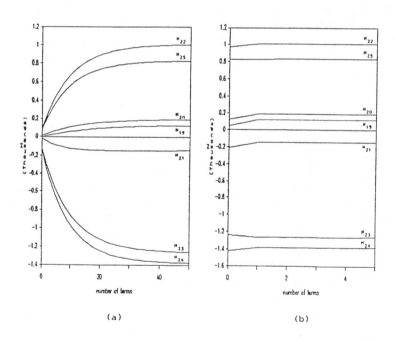

(a)   (b)

Figure 4. *Results, twenty-five-bar truss.*

Table 1. *Redundant forces, 25-bar truss.*

Mem.	Eq. 88	Eq. 92		Eq. 103 $f = 0.1$		Exact
		Two†	Five†	Two†	Five†	
19	77.5	10.5	27.1	125.9	126.7	126.7
20	168.8	25.7	59.1	193.1	193.3	193.2
21	−136.6	−39.2	−80.7	−147.2	−146.1	−146.0
22	999.5	186.4	403.5	1007.9	1008.4	1008.4
23	−1242.1	−236.5	−510.8	−1261.6	−1262.0	−1261.9
24	1400.1	−270.8	−580.8	−1380.5	−1381.8	−1382.0
25	875.8	157.8	340.0	829.3	830.6	830.7

† Terms

3.3 Linear Programming Formulation

Neglecting the compatibility conditions (Eq. 84), X and N can be considered as independent variables. Assuming that the objective function and the allowable member forces can be expressed in linear terms of X and considering only stress and side constraints, the EACP (Eqs. 31–34) can be stated as the following LP problem: Find X and N such that

$$Z = \ell^T X \to \text{min.} \tag{107}$$

$$X^L \leq X \leq X^U \tag{108}$$

$$\sigma^L X \leq A_L + A_N N \leq \sigma^U X \tag{109}$$

in which ℓ is a vector of constants. Under certain circumstances the LP formulation can be preserved even if displacement constraints are considered. This is the case, for example, in continuous beams if Eq. 26 holds. To illustrate the linear displacement constraints, we note first that the internal forces A are linear functions of N. Also, a statically equivalent internal force system corresponding to the virtual loads Q_i (Eq. 25) may be selected so that the forces A_{ij}^Q are constant. Thus, the elements D_{ij} can be expressed as

$$D_{ij} = D_{ij0} + \sum_k D_{ijk} N_k \tag{110}$$

where D_{ij0} and D_{ijk} are constant coefficients. From Eqs. 27, 110

$$r_i = \sum_j \frac{D_{ij0} + \sum_k D_{ijk} N_k}{X_j} \tag{111}$$

Choosing a statically equivalent system with hinges assumed over the intermediate supports, the displacement r_i in the h-th span can be expressed as

$$r_i = \frac{D_{ih0} + D_{ih1}N_1 + D_{ih2}N_2}{X_h} \tag{112}$$

in which N_1, N_2 are the bending moments over the supports of the member under consideration. Substituting this equation into the displacement constraint $r_i \leq r_i^U$ and rearranging yields

$$D_{ih0} + D_{ih1}N_1 + D_{ih2}N_2 - r_i^U X_h \leq 0 \tag{113}$$

These linear displacement constraints can be added to the LP problem.

The LP formulation is most suitable for optimal design of trusses. If no lower bounds on cross-sectional areas are considered, that is

$$X^L = 0 \tag{114}$$

the LP solution may lead to elimination of unnecessary members from the structure. If the optimal solution represents a statically determinate structure, the compatibility conditions need not be considered. Also, if the optimal structure represents an unstable configuration, members can be added to satisfy the necessary relationship between joints and members in a stable structure.

3.4 Constant Force Models

The assumption of constant internal forces during a certain stage of the solution process is common to many optimal design procedures. If the conditions (Eq. 26) holds, then the displacements can be expressed as Eq. 27

$$r = DY \tag{115}$$

Assuming constant internal forces then

$$D = \tilde{D} \tag{116}$$

Substituting Eq. 116 into Eq. 115 the constant force approximations become

$$r(Y) = \tilde{D}Y \tag{117}$$

It can be shown (Kirsch [5]) that under the present assumptions

$$\tilde{D} = \tilde{\nabla}r_y \tag{118}$$

That is, the assumption of constant internal forces is equivalent to first order Taylor series expansions of the displacements (Eq. 63) in terms of the inverse variables. Therefore, this assumption may lead to either approximate compatibility or approximate equilibrium models.

The well known stress ratio technique is also based on the assumption of constant internal forces during the redesign steps. Assuming that the relation (Eq. 35) holds, the design variables in the $(k + 1)$-th cycle are calculated from

$$X^{(k+1)} = A^{(k)} \frac{1}{\sigma^U} \tag{119}$$

Here $A^{(k)}$ is a diagonal matrix. The inverse variables, accordingly, are given by

$$Y^{(k+1)} = \sigma^U \frac{1}{A^{(k)}} \tag{120}$$

This method can be shown to be equivalent to neglecting the force redistribution for first order Taylor series approximations in the inverse variables (Kirsch [11]).

In constant force models the constraint surfaces are approximated by planes normal to the coordinate axes in the variables space.

3.5   Effect of Compatibility

Solutions of the EACP usually will not satisfy the compatibility conditions. If the resulting design is statically determinate, consideration of these conditions is not necessary and the solution of the EACP is the final optimum. As noted earlier, this would be the case in truss optimal design, for example, if members are eliminated by the LP solution to obtain a statically determinate truss. Several procedures have been proposed to obtain a solution satisfying the compatibility conditions. The object is to modify the optimal solution of the EACP so that (Eq. 30)

$$\Delta \delta = \delta - \delta_0 = \delta - F_{opt} N_{opt} = 0 \tag{121}$$

in which δ_0 is computed from the EACP solution (Eq. 29).

To study the effect of compatibility conditions on the optimum, solutions of two formulations have been compared (Kirsch [27]). The NLP, where compatibility conditions are considered and the LP where the latter conditions are neglected. The objective function is expressed in linear terms of the design variables and only linear stress constraints are considered. Since compatibility conditions are not considered in the LP formulation, the optimal solution $\tilde{Z}_{LP}$ will be at least as good as that of the NLP.

$$\tilde{Z}_{LP} \leq \tilde{Z}_{NLP} \tag{122}$$

Therefore, $\tilde{Z}_{LP}$ can be viewed as a lower bound on $\tilde{Z}_{NLP}$. It is possible now to show under what circumstances optimal values of objective function or force distribution in the structure are identical for the two problems, that is

$$\tilde{Z}_{LP} = \tilde{Z}_{NLP} \tag{123}$$

or

$$\tilde{N}_{LP} = \tilde{N}_{NLP} \tag{124}$$

In cases where conditions (Eq. 123) and (Eq. 124) are satisfied, the compatibility conditions do not affect the optimal solution. One such class of problems is the statically determinate structures, where

$$N = 0 \tag{125}$$

The forces A are constant, compatibility must not be considered and the two formulations become identical.

In the presentation that follows, statically indeterminate structures will be considered. A distinction will be made between the following geometries of the structure:

Y_A the usual case where the optimal objective function values are not identical:

$$\tilde{Z}_{LP} < \tilde{Z}_{NLP} \tag{126}$$

The optimal force distributions ($\tilde{N}_{LP}$ and $\tilde{N}_{NLP}$) may or may not be identical in this case.

Y_B where the two optimal objective functions are identical, but the ELP problem possesses multiple optimal force distributions, including the one corresponding to the optimal NLP force distribution:

$$\left. \begin{array}{c} \tilde{Z}_{LP} = \tilde{Z}_{NLP} \\ \tilde{N}_{LP} \epsilon \tilde{N}_{NLP} \end{array} \right\} \tag{127}$$

Y_C the two solutions are identical:

$$\left. \begin{array}{c} \tilde{Z}_{LP} = \tilde{Z}_{NLP} \\ \tilde{N}_{LP} = \tilde{N}_{NLP} \end{array} \right\} \tag{128}$$

For the usual case (Eq. 126) it is possible that the difference between the two optimal solutions is small. In order to study the nature of the various cases, define the following geometries with particular properties:

Y_{TL} a geometry for which there is a transition in the set of active constraints at the LP optimum ($\tilde{X}_{LP}$)

Y_{TN} a geometry for which there is a transition in the set of active constraints at the NLP optimum ($\tilde{X}_{NLP}$).

The objective contours for $Y = Y_{TL}$ are parallel to the boundary of the feasible region. That is, the LP problem possesses an infinite number of optimal solutions and corresponding force distributions, bounded by $\tilde{N}_{LP}^L$ and $\tilde{N}_{LP}^U$. If the optimal NLP force distribution for this geometry, $\tilde{N}_{NLP}$, is within this range

$$\tilde{N}_{LP}^L \leq \tilde{N}_{NLP} \leq \tilde{N}_{LP}^U \tag{129}$$

then the conditions (Eq. 127) are satisfied and

$$Y_{TL} = Y_B \tag{130}$$

In such cases. the optimal LP and NLP solutions are identical for the transition geometry Y_{TL}.

For the NLP formulation, the number of active constraints at the optimum is usually smaller than that of the LP formulation. This is due to the fact that the number of independent variables (X) in the NLP problem is smaller than the number of variables $(X$ and $N)$ in the LP problem. For the transition geometry $Y = Y_{TN}$, the number of active constraints at the NLP optimum is usually increased; therefore there is a better prospect that the two optimal force distributions will become identical and the conditions (Eq. 128) will be satisfied. In such cases

$$Y_{TN} = Y_C \tag{131}$$

that is the LP and NLP solutions are identical for Y_{TN}.

3.5.1 Two Beam Grillage (Kirsch [27]) The grillage shown in Fig. 5 is subjected to the following loading conditions

Loading $L1$: $P_1 = P_2 = 10$ (four loads)

Loading $L2$: $P_1 = 10$ (two loads), $P_2 = 5$ (two loads)

Owing to symmetry, only four potentially critical stress constraints (sections A, B C and D) are considered. The numbers of variables for a single loading condition are three (LP formulation) and two (NLP formulation). The geometrical parameters are limited by the constraint $Y \leq 2.0$ and N (the interaction force between the two beams) is limited in sign.

Active constraints at the optimum for various geometries are given in Table 2. The transition geometries for each of the loading conditions are as follows:

Loading $L1$ Loading $L2$

$Y = 0.985 = Y_{TN}$ $Y = 1.006 = Y_{TN}$

$Y = 1.657 = Y_{TL}$

$Y = 2.000 = Y_{TN}$

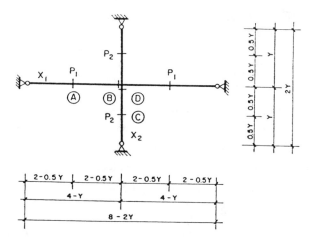

Figure 5. *Two-beam grillage.*

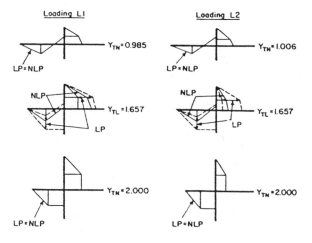

Figure 6. *Optimal moment distribution.*

Table 2. *Active constraints at the optimum for various geometries and loadings.*

	LP formulation		NLP formulation	
Loading	Geometry	Active constraints	Geometry	Active constraints
L1	$Y < 1.657$	A, B, D	$Y < 0.985$	B, D
	$Y_{TL} = 1.657$	A, D/A, B, D/	$Y_{TN} = 0.985$	A, B, D
		A, B, C, D	$0.985 < Y < 2.0$	A, D
	$1.657 < Y \leq 2.0$	A, B, C, D	$Y_{TN} = 2.0$	A, B, C, D
L2	$Y < 1.657$	A, B, D	$Y < 1.006$	B, D
	$Y_{TL} = 1.657$	A, D/A, B, D/	$Y_{TN} = 1.006$	A, B, D
		A, B, C, D	$1.006 < Y < 2.0$	A, D
	$1.657 < Y \leq 2.0$	A, B, C, D	$Y_{TN} = 2.0$	A, B, C, D

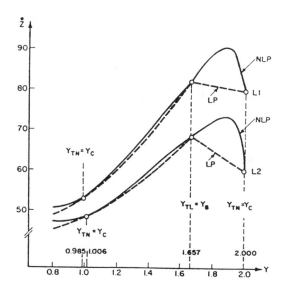

Figure 7. *Variation of $\tilde{Z}$ with Y.*

Optimal moment distributions for the transition geometries are shown in Fig. 5 and variations of $\tilde{Z}_{LP}$ and $\tilde{Z}_{NLP}$ with Y are illustrated in Fig. 7. The following geometries are identified from Figs. 6 and 7:

Loading $L1$	Loading $L2$
$Y = 0.985 = Y_C$	$Y = 1.006 = Y_C$
	$Y = 1.657 = Y_B$
	$Y = 2.000 = Y_C$

Based on these results it can be seen that the conditions Eq. 130 or 131 are satisfied for all listed geometries. Furthermore, for two of the geometries ($Y = 1.657$ and $Y = 2.000$) these conditions are satisfied for both loadings.

Although $\tilde{Z}_{LP} = \tilde{Z}_{NLP}$ for Y_B and Y_C, it is instructive to note that there is a considerable difference in $\tilde{Z}$ for some geometries. For $Y = 1.9$ one can obtain

$$\frac{\tilde{Z}_{NLP}}{\tilde{Z}_{LP}} = 1.13 \qquad \text{(loading } L1)$$

$$\frac{\tilde{Z}_{NLP}}{\tilde{Z}_{LP}} = 1.19 \qquad \text{(loading } L2)$$

3.6 Effect of Prestressing

Prestressing by 'lack of fit' may be applied to maintain compatibility at the LP optimum (Kirsch [28]). To study the effect of prestressing on the optimal solution assume the general case where the optimal LP solution $\tilde{X}_{LP}$, $\tilde{N}_{LP}$ represents a statically indeterminate structure (SIS). Elastic analysis of the optimal design will provide the corresponding elastic force distribution $\tilde{N}_E$. If the conditions

$$\Delta_{LP} \equiv \tilde{F}_{LP}\tilde{N}_{LP} - \tilde{\delta}_{LP} = 0 \tag{132}$$

are not satisfied, then $\tilde{N}_{LP} \neq \tilde{N}_E$, and a set of prestressing forces, $\tilde{N}_P$, given by

$$\tilde{N}_P = \tilde{N}_{LP} - \tilde{N}_E \tag{133}$$

may be applied to maintain compatibility at the LP optimum. In the case of a single loading condition, the forces $\tilde{N}_P$ can easily be determined by Eq. 133. For multiple loading conditions, different values of $\tilde{N}_P$ might be obtained for the various loadings, i.e., at least for some i,j.

$$N_{pi} \neq N_{pj} \qquad i,j = 1, ..., J \tag{134}$$

Here i, j = loading condition; and J = number of loadings. It should be noted, however, that the assumption of applying different sets of prestressing forces for various loading conditions is not practical.

It is possible now to identify situations where the LP and NLP optimal solutions are identical, i.e.,

$$\tilde{Z}_{LP} = \tilde{Z}_{NLP} \tag{135}$$

In cases where the LP optimal design represents a statically determinate structure (SDS), then for a certain selection of redundant forces, one can obtain

$$\tilde{N}_{LPj} = \tilde{N}_{NLPj} = 0 \qquad j = 1, \ldots, J \tag{136}$$

For statically, indeterminate structures, it is also possible that the optimal force distributions of both problems are identical, i.e.

$$\tilde{N}_{LPj} = \tilde{N}_{NLPj} \neq 0 \qquad j = 1, \ldots, J \tag{137}$$

In this case, prestressing forces are not required to maintain compatibility, i.e.,

$$N_{Pj} = 0 \qquad j = 1, \ldots, J \tag{138}$$

In cases where a single set of prestressing forces is sufficient to maintain compatibility for all loading conditions, we have

$$N_{Pi} = N_{Pj} \neq 0 \qquad i, j = 1, \ldots, J \tag{139}$$

The various possible cases of LP solution with the corresponding necessary conditions are characterized in the following:

1. For a single loading and $X^L = 0$, the resulting optimal structure will be an SDS. The values of $\tilde{Z}_{LP}$ and $\tilde{Z}_{NLP}$ are identical (Eq. 135), no redundant forces exist (Eq. 136) and no prestressing forces are required (Eq. 138).

2. For multiple loadings and $X^L = 0$, the optimal solution may represent an SDS. This might occur for a geometry of multiple optimal topologies where compatibility conditions must not be satisfied for those topologies representing SDS. In such cases, prestressing forces are not required.

3. For an SIS and a single loading, it is possible to apply a set of prestressing forces such that the LP solution will satisfy compatibility (Eqs. 135 and 139).

4. For certain situations of case 3, compatibility conditions are satisfied at the optimal LP solution without prestressing (Eqs. 135, 137, 138).

5. For the general case of SIS subjected to multiple loadings a single set of prestressing forces might be sufficient to maintain compatibility at the LP optimum (Eqs. 135 and 139).

6. For the general case, if compatibility conditions are satisfied at the LP solution for all loading conditions, prestressing forces are not required to maintain compatibility (Eqs. 135, 137 and 138).

7. For the general case, it might not be possible to maintain compatibility at the LP optimum for all loading conditions by appying a single set of prestressing forces. However, it is possible to satisfy compatibility by applying different sets of prestressing forces for the various loading conditions (Eqs. 134 and 135).

3.6.1 Eleven Bar Truss (Kirsch [28]) In the simple example demonstrated subsequently the allowable stresses for all elements are

$$\sigma^U = -\sigma^L = 20.0$$

and the assumed concentrated loads are

$$P = P_1 = P_2 = 10.0$$

Conditions of symmetry are considered to reduce the problems' dimensionality. The lower bound on cross-sections is $X^L = 0$. No upper bounds on X have been assumed.

The truss shown in Fig. 8 is subjected to two loading conditions (P_1 and P_2 respectively). Two selections of redundants have been assumed: forces in elements 1 and $\bar{1}$, and force in element 4. Results are summarized in Tables 3 and 4, and the optimal design is shown in Fig. 9. The optimal objective function value is $\tilde{Z} = 125$. Note that the optimal structure is statically indeterminate, and prestressing by lack of fit can be used to maintain compatibility at the optimum for the two loading conditions. The required prestressing force is -0.71 for elements 1 and $\bar{1}$ or 1.00 for element 4 (see Table 4).

3.7 Effect of Passive Control

Another possibility to maintain compatibility at the LP optimum is to use passive control devices (Kirsch [29]). Assuming control variables C corresponding to the redundant forces N, the LP is first solved with $C = 0$ to obtain a lower bound on the optimum. The control variables are then selected such that compatibility conditions will be maintained. If the latter conditions cannot be satisfied, a solution is selected such that some measure of the constraint violation is minimized. The LP solution is then modified and the control variables are revised until a solution satisfying compatibility is achieved. It has been observed that the LP solution can often be maintained by passive control devices.

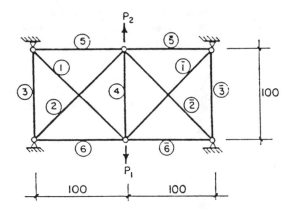

Figure 8. *Eleven-bar truss.*

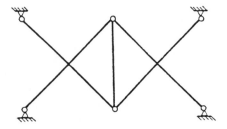

Figure 9. *Optimal topology.*

Table 3. *Optimal solution, eleven bar-truss.*

Element	$\tilde{A}_i$		$\check{X}_i$
	Load 1	Load 2	
1	3.54	-3.54	0.177
2	3.54	3.54	0.177
3	0	0	0
4	5.00	5.00	0.250
5	0	0	0
6	0	0	0

Table 4. *Prestressing forces, eleven-bar truss.*

Redundant	Load 1			Load 2		
elements	$\tilde{N}_{\text{LP}}$	$\tilde{N}_{\text{E}}$	$\tilde{N}_{\text{P}}$	$\tilde{N}_{\text{LP}}$	$\tilde{N}_{\text{E}}$	$\tilde{N}_{\text{P}}$
1 and $\bar{1}$	3.54	4.25	−0.71	−3.54	−2.83	−0.71
4	5.00	4.00	1.00	5.00	4.00	1.00

3.7.1 Four Beam Grillage (Kirsch [29]) The simple example illustrated subsequently is mainly intended to demonstrate the potential savings in the weight of the structure. Although various types of control devices may be considered, only two possibilities will be demonstrated:

1. A Linear Spring Device (LSD) whose flexibility (or stiffness) is designed to achieve the desired displacements under the applied loads.

2. A Limited Displacement Device (LDD) where the maximum displacement is limited by the control device.

Once basic difference between the two device types is that for the LSD a unique statical scheme is considered, whereas for the LDD this scheme is changed after the maximum displacement occurs. A possible problem with the LSD is that flexibility and strength are two conflicting factors that must be considered in the spring design.

The grillage shown in Fig. 10a is subjected to a single load, P, acting at point C. The structure is simply supported at points A, E, F and J. A uniform cross-section with

$$\alpha = \frac{EI}{GJ} = 2.5$$

has been assumed, where EI is the flexural rigidity and GJ is the torsional rigidity. The assumed design variable, X, represents the bending moment capacity and the objective function is $Z = 36X$. The assumed control devices are either an LSD (Fig. 10b) or an LDD (Fig. 10c) at point C. Results for the LP, OC and NLP problems are shown in Fig. 11 and in Table 5 (OC = optimal control, NLP = optimum without control). The optimal solution for the LSD is obtained for $(EI)C = 18.75$ (C is the spring flexibility), where stress constraints at sections B (left), C, and D (right), and at element DHI become active. The maximum displacement for an LDD occurs under a load of 0.385 P. It can be noted that the NLP optimal weight is 62% heavier than the OC optimal weight which is identical to the LP optimum.

322

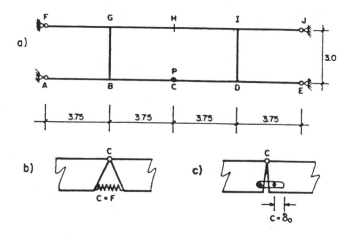

Figure 10. *Four-beam grillage.*

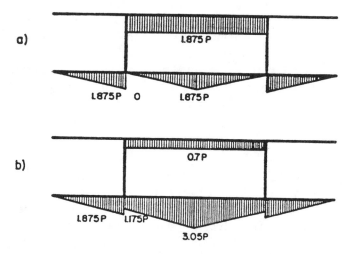

Figure 11. *Optimal moment distributions (a) LP and OC (b) NLP.*

Table 5. *Summary of results, grillage example.*

Problem	$\tilde{X}$	$\tilde{Y}$
LP	1.875 P	67.5 P
OC	1.875 P	67.5 P
NLP	3.050 P	109.8 P

4 Concluding Remarks

It has been noted that using the displacement analysis formulation, approximations of the displacement functions in terms of the design variables will lead to a solution which does not necessarily satisfy equilibrium, but can be viewed as an exact optimum for a different set of external loads. Similarly, using the force method formulation, approximations of the redundant force functions will lead to a solution which does not necessarily satisfy compatibility, but can be viewed as an exact optimum for a different set of displacements in the direction of the redundants. Based on these physical interpretations, solution procedures of the explicit approximate problem can be introduced.

Some improved approximate behaviour models have been presented and the relationship between the various approximations have been derived. It is shown that these models become equivalent under certain assumptions. High quality approximation can be achieved with a relatively small computational effort. Simplified models based on linear programming formulations and constant forces are discussed and the effect of compatibility conditions on the optimum as been illustrated. The possibility of applying prestressing by lack of fit and passive control forces to maintain compatibility at the optimum is demonstrated. Approximate models, such as those pesented in this article, are essential in optimal design of large structural systems and may lead to efficient solution procedures.

References

[1] Pickett, R. M., Rubinstein, M. F. and Nelson, R. B., (1973), 'Automated structural synthesis using a reduced number of design coordinates,' *AIAA J.* **11**, 489–494.

[2] Schmit, L. A. and Farshi, B., (1974), 'Some approximation concepts for structural synthesis,' AIAA J. **12**, 692–699.

[3] Schmit, L. A. and Fleury, C., (1979), 'An improved analysis/synthesis capability based on dual methods—ACCESS 3,' *AIAA/ASME/ASCE/AMS Twentieth Conference*, St Louis, MO.

[4] Schmit, L. A., (1984), 'Structural optimization, some key ideas and insights,' In Atrek, E., Gallagher, R. H., Ragsdell, K. M. and Zienkiewicz, O. C. (Eds.), *New Directions in Optimum Structural Design*, John Wiley and Sons, New York.

[5] Kirsch, U., (1981), *Optimal Structual Design—Concepts, Methods and Applications*, McGraw-Hill, New York

[6] Kirsch, U., (1981), 'Approximate structural analysis based on series expansion,' *Comp. Meth. Appl. Mech. Eng.* **26**, 205–223.

[7] Kirsch, U., (1982), 'Approximate structural reanalysis for optimization along a line,' *Int. J. Numerical Methods in Eng.* **18**, 635–651.

[8] Kirsch, U., (1982), 'Optimal design based on approximate scaling,' *J. Struct. Div., ASCE* **108**, 888–909.

[9] Kirsch, U. and Toledano, G., (1983), 'Approximate reanalysis for modifications of structural geometry,' *Computers and Structures* **16**, 269–279.

[10] Kirsch, U., (1984), 'Approximate behaviour models for optimum structural design,' In Atrek, E., Gallagher, R. H., Ragsdell, K. M. and Zienkiewicz, O. C. (Eds.) *New Directions in Optimum Structural Design*, John Wiley and Sons, New York.

[11] Kirsch, U., (1985), 'Approximation concepts for optimum structural design,' In Gero, J. S. (Ed.), *Design Optimization*, Academic Press, New York.

[12] Kirsch, U., (1985), 'On some simplified models for optimal design of structural systems,' *Com. Meth. Appl. Mech. Eng.* **48**, 155–169.

[13] Kirsch, U., (1988), 'Applications of mathematical programming to the design of civil engineering structures,' in Levari, R. (Ed.), *Engineering Design, Better Results through Operations Research Methods*, North-Holland.

[14] Kirsch, U. and Taye, S., (1988), 'High quality approximations of forces for optimum structural design,' *Computers and Structures* **30**, 519–527.

[15] Arora, J. S., (1976), 'Survey of structural reanalysis techniques,' *J. Struct. Div., ASCE* **102**, 783–802.

[16] Abu Kassim, A. M. and Topping, B. H. V., (1987), 'Static reanalysis of structures: A review,' *J. Struct. Eng., ASCE* **113**, 1029–1045.

[17] Berke, L and Venkayya, V. B., (1974), 'Review of optimality criteria approaches to structural optimization,' Presented at ASME Winter Annual Meeting, New York.

[18] Khot, N. S., Berke, L. and Venkayya, V. B., (1978), 'Comparison of optimality criteria algorithms for minimum weight design of structures,' *AIAA/ASME/SAE 19th Conf.*, Bethesda, Md

[19] Reinschmidt, K. F. and Russell A. D., (1974), 'Applications of linear programming in structural layout and optimization,' *Computers and Structures* **4**, 855–869.

[20] Reinschmidt, K. F. and Norabhoompipat, T., (1975), 'Structural optimization by equilibrium linear programming,' *J. Struct. Div., ASCE* **101**, 921–938.

[21] Farshi, B. and Schmit, L. A., (1974), 'Minimum weight desigh of stress limited trusses,' *J. Struct. Div., ASCE* **100**, 97–107.

[22] Kirsch, U., (1980), 'Optimal design of trusses by approximate compatibility,' *Computers and Structures* **12**, 93–98.

[23] Clarkson, T., (1965), *The Elastic Analysis of Flat Grillages*, Cambridge University Press, Cambridge, Mass.

[24] Hafka, R. T and Kamat, M. P., (1985), *Elements of Structural Optimization*, Martinus Nighoff, The Hague.

[25] Fuchs, M. B., (1980), 'Linearized homogeneous constraints in structural design,' *Int. J. Mech. Sciences* **22**, 33–40

[26] Arora, J. S and Haug, E. J., (1979), 'Methods of design sensitivity analysis in structural optimization,' *AIAA J.* **17**, 970–974.

[27] Kirsch, U., (1987), 'The effect of compatibility conditions on optimal design of flexural structures,' *Int. J. for Numerical Methods in Engrg.* **24**, 1173–1185.

[28] Kirsch, U., (1989), 'The effect of compatibility and prestressing on optimized trusses,' *J. of Structural Eng., ASCE* **115**, 724–737.

[29] Kirsch, U., (1988), 'Improved optimum structural design by passive control,' *Engineering with Computers* **4**.

Computer-Automated Optimal Design of Structural Steel Frameworks

Professor Donald E. Grierson
Department of Civil Engineering
Solid Mechanics Division
University of Waterloo, Ontario
Canada

Abstract This lecture concerns the computer-automated design of least-weight structural frameworks. First, steel structures under static loads are considered and members are automatically sized using commercial standard steel sections in full conformance with steel design standard provisions for strength/stability and stiffness. The necessary features and functions of such a design system are identified, and the underlying synthesis strategy is discussed. This aspect of the design system is illustrated using a corresponding professional-practise software code that is applied for the least-weight design of a steel mill crane framework comprised of a variety of member types and subject to a number of load effects.

The capabilities of the system are then extended to allow for the least-weight design of structural frameworks under both service and ultimate loading conditions. Here, acceptable elastic stresses and displacements are ensured at the service-load level while, simultaneously, adequate safety against plastic collapse is ensured at the ultimate-load level. This aspect of the design system is illustrated using a prototype software code that is applied for the least-weight design of an aluminum tunnel-ring frame subjected to gravity and internal pressure loading. Finally, the computer-based design methodology is extended to the least-weight design of structural frameworks subjected to dynamic loading. Constraints are placed on dynamic displacements, dynamic stresses, natural frequencies and member sizes. Extreme responses are identified so as to convert the time-parametric design problem in an explicit non-parametric form. The method is shown to be capable of designing structures under simultaneous static and dynamic loading. A prototype software code is applied for the least-weight design of a steel trussed arch subjected to non-structural masses and an impulse force.

B. H. V. Topping (ed.),
Optimization and Artificial Intelligence in Civil and Structural Engineering, Volume I, 327–353.
© 1992 *Kluwer Academic Publishers. Printed in the Netherlands.*

1 Introduction

This lecture is an overview of work carried out at the University of Waterloo concerning the development of a general computer code for the least-weight design of structural frameworks. Initial work was undertaken by Grierson and Schmit [1], who considered thin-walled structures comprised of bar, membrane and shear panel elements under uniaxial stress. The work was subsequently extended by Chiu [2] and Grierson and Chiu [3] to include frameworks comprised of beam and column members under combined axial and bending stresses. These studies accounted for service and ultimate performance criteria simultaneously, but considered the sizes of the cross-sections for the members of the structure as continuous design variables to the synthesis process. Lee [4] and Grierson and Lee [5] extended the design method such that the cross-section sizes could be taken as discrete variables, thereby allowing for design using commerical standard sections. Further studies by Cameron [6] and Grierson and Cameron [7] extended the work to automatically size trusses using standard sections subject to the provisions of steel design standards. Work by Hall *et al.* [8,9] extended the method to account for second-order $(P - \Delta)$ effects for frameworks under static loading. Kramer [10] and Kramer and Grierson [11] extended the method to allow for design under dynamic loading. Finally, recent work conducted by Cameron [12] introduced the use of expert system techniques into the design process. (This latter topic is dealt with by the writer in another lecture of the NATO-ASI.)

This lecture summarizes the work described in the foregoing in three areas concerning the least-weight design of: (1) steel building frameworks in professional practise; (2) structural frameworks under both service and ultimate static loading conditions; and (3) structural frameworks under dynamic loading.

2 Steel Building Frameworks

A routine design task in structural engineering offices is to size the members of a steel framework of given topology using commercially available standard steel sections that conform with the performance requirements of the governing steel design standard. Traditionally, the design involves a trial-and-error process wherein repeated analysis and design of the structure is conducted until the performance criteria have been met and some measure of economy has been achieved. However, even when the analysis is conducted using a computer, such a procedure is often quite cumbersome because of the many facets of the design problem that must be accommodated. Very often, in fact, a less-than-optimal final design is found after a tedious (re)analysis/(re)design process that owes its termination only to the skilled intuition and patience of the designer.

Computer-based optimization techniques can be effectively applied to overcome the noted difficulties for the synthesis process described in the foregoing. This section of the lecture addresses this issue with particular emphasis on dealing with the broad nature of the structural steel design problem encountered in professional practise. The features and functions required for such a comprehensive design system are identified and the underlying synthesis strategy is discussed. A corresponding professional-practice structural optimization software code is applied for the least-weight design of a steel mill crane framework.

2.1 Features and Functions of Software System

The necessary features and functions of a professional-practise software system applicable for the least-weight design of structural steel frameworks are briefly presented in the following from a variety of viewpoints. First, the overall capabilities required of such a system are identified. Second, a number of numeric-based capabilities necessary for the proper operation of the system are noted. Finally discussed are a number of additional features that enhance the usefulness of the system in the professional design office.

As the underlying basis for design, the system should be capable of conducting first-order and second-order analysis of planar [9] and space [13] frameworks. Account should be taken of the strength/stability of individual members and of the stiffness of the structure as a whole. Verification of the design should be in conformance with the strength/stability and stiffness provisions of the governing steel design standard, e.g., [14,15]. Members should be sized using commercially available standard steel sections. The system should have an optimization capability to determine a least-weight design of the structure.

With a view to computational effectiveness, the system should also have the capabilities described in the following. User-identified member and node indices should be automatically re-indexed internally for solution by a banded-matrix solver. User-specified pinned releases that introduce stiffness singularities should be automatically modified internally to restore structure stability. The local buckling classification of each member cross-section should be internally calculated. The in-plane and out-of-plane effective length factors for each member should be internally calculated, as should be the unbraced compression flange length for each flexural member. The system should have a discrete optimization capability that automatically sizes members by selecting from a complete database of standard steel sections. All strength and stiffness constraints for the design should be simultaneously accounted for by the weight optimization process.

The practical usefulness of a software system for optimal structural steel design is enhanced by the additional features listed in the following. The user-interface format should be a full-screen spreadsheet that is addressable by both keyboard and mouse. Data input should allow for both Imperial and Metric units, at least in North America, and corresponding material-properties default values should be given for Young's modulus, yield and ultimate stresses, thermal expansion coefficient, etc. The user should be allowed to use descriptive alphanumeric names to identify members and nodes. A broad range of section profiles should be accounted for in the standard section database (e.g., W, T, hollow-box, single and double angle, etc.) Fixed, pinned and roller supports should be accounted for, as well as pinned releases at connections, bolted connections for truss members and various gusset-plate thicknesses for back-to-back double angle sections. The user should have the option to consider members having symmetrical sections in either strong-axis or weak-axis bending. So as to satisfy conventional fabrication practise, the user should be allowed to specify groups of members to have common section properties, to impose limitations on section depths, and to fix the section properties of selected members. A variety of nodal and member load types should be accounted for, as well as external effects due to temperature change and support settlement. Upon completing a design run, the output data should include the section designations for the members, the analysis results for all load cases and, importantly, a clause-by-clause evaluation of member strengths and structure stiffness in accordance with the provisions of the governing steel design standard. Finally, the software system should provide on-line help messages during data input, and on-line error diagnosis during design synthesis and data output.

2.2 Computer-Based Structural Optimization

The computer-based structural optimization technique seeks a least-weight design of the structure while satisfying all strength and stiffness performance requirements simultaneously [16]. For a structure having $i = 1, 2, ..., n$ members, and taking the cross-section area a_i as the sizing variable for each member i, the general form of the design problem may be stated as:

$$\text{Minimize:} \quad \sum_{i=1}^{n} w_i a_i \tag{1}$$

$$\text{Subject to:} \quad \underline{\delta}_j \leq \delta_j \leq \overline{\delta}_j \qquad (j = 1, 2, ..., d) \tag{2}$$

$$\underline{\sigma}_k \leq \sigma_k \leq \overline{\sigma}_k \qquad (k = 1, 2, ..., s) \tag{3}$$

$$a_i \in A_i \qquad (i = 1, 2, ..., n) \tag{4}$$

Equation (1) defines the weight of the structure (w_i = weight coefficient for member i = material density × member length); Eqs. (2) define d constraints on displacements δ_j (under- and super-scored quantities denote specified lower and upper bounds, respectively); Eqs. (3) define s constraints on stresses σ_k; Eqs. (4) require each cross-section area a_i to belong to the discrete set of areas $A_i \equiv \{a_1, a_2, ...\}_i$ prevailing for the section profile specified for member i.

In their present form, the displacement and stress constraint Eqs. (2) and (3) are *implicit-nonlinear* functions of the cross-section areas a_i. To facilitate computer solution, first-order Taylor's series approximations are employed to reformulate these constraints as 'good-quality' *explicit-linear* functions of the 'reciprocal' sizing variables

$$x_i = \frac{1}{a_i}, \qquad (i = 1, 2, \ldots, n) \tag{5}$$

as follows, for displacements,

$$\delta_j = \delta_j^0 + \sum_{i=1}^{n} \left(\frac{\partial \delta_j}{\partial x_i} \right)^0 (x_i - x_i^0) \tag{6}$$

and, for stresses,

$$\sigma_k = \sigma_k^0 + \sum_{i=1}^{n} \left(\frac{\partial \sigma_k}{\partial x_i} \right)^0 (x_i - x_i^0) \tag{7}$$

where the superscript (0) indicates known quantities for the current design stage (e.g., the initial 'trial' design). The quantities $(\partial \delta_j / \partial x_i)^0$ and $(\partial \sigma_k / \partial x_i)^0$ are displacement and stress gradients referenced to the current design, while the x_i are the sizing variables for the next weight optimization. It is readily shown that [17]

$$\delta_j^0 = \sum_{i=1}^{n} \left(\frac{\partial \delta_j}{\partial x_i} \right)^0 x_i^0, \tag{8}$$

$$\sigma_k^0 = \sum_{i=1}^{n} \left(\frac{\partial \sigma_k}{\partial x_i} \right)^0 x_i^0. \tag{9}$$

Therefore, from Eqs. (1) to (9) the statement of the weight optimization problem then becomes [5]:

$$\text{Minimize:} \qquad \sum_{i=1}^{n} w_i / x_i \tag{10}$$

$$\text{Subject to:} \qquad \underline{\delta}_j \le \sum_{i=1}^{n} \left(\frac{\partial \delta_j}{\partial x_i} \right)^0 x_i \le \bar{\delta}_j \qquad (j = 1, 2, \ldots, d) \tag{11}$$

$$\underline{\sigma}_k \le \sum_{i=1}^{n} \left(\frac{\partial \sigma_k}{\partial x_i} \right)^0 x_i \le \bar{\sigma}_k \qquad (k = 1, 2, \ldots, s) \tag{12}$$

$$x_i \in X_i \qquad (i = 1, 2, \ldots, n) \tag{13}$$

where the components of each discrete set $X_i = \{x_1, x_2, \ldots\}_i$ in Eqs. (13) are the reciprocals of the cross-section areas comprising the corresponding discrete set A_i in Eqs. (4).

The displacement and stress gradients in Eqs. (11) and (12) are evaluated using a 'virtual-load' sensitivity analysis technique that accounts for both first-order elastic stiffness properties and second-order geometric stiffness properties. Each individual nodal displacement δ_j of concern to the design is related to the overall vector of nodal displacements $\underline{u}$ for the structure as

$$\delta_j = \underline{b}_j^T \underline{u} \tag{14}$$

where $\underline{b}_j$ is a constant vector and the displacements $\underline{u}$ are found from second-order analysis of the current design. Now, for a unit virtual load associated with δ_j in Eq. (14), it follows from virtual work principles that $\underline{b}_j$ can be viewed as a virtual load vector such that, from the Displacement Method of analysis,

$$\underline{u}_j = \underline{K}^{-1}\underline{b}_j \tag{15}$$

where, for $K =$ the structure stiffness matrix, $\underline{u}_j$ is the corresponding vector of virtual nodal displacements. It is then readily shown that each displacement gradient is given by [17]

$$\left(\frac{\partial \delta_j}{\partial x_i}\right)^0 = \frac{1}{x_i^0}\left(\underline{u}_j^T \underline{K}_i \underline{u}\right)^0 \tag{16}$$

where $\underline{K}_i =$ the global-axis stiffness matrix for member i. Similarly, each individual member stress σ_k of concern to the design is related to the overall vector of nodal displacments $\underline{u}$ for the structure as

$$\sigma_k = \underline{t}_k \underline{u} \tag{17}$$

where the vector $\underline{t}_k$ is constant for axial truss members, but depends on the cross-section neutral-axis position for flexural frame members [3]. Following the same development as that used to find Eq. (16), the gradient of the stress σ_k with respect to each reciprocal sizing variable x_i is readily found to be given by

$$\left(\frac{\partial \sigma_k}{\partial x_i}\right)^0 = \frac{1}{x_i^0}\left(\underline{u}_k^T \underline{K}_i \underline{u}\right)^0 \tag{18}$$

where, for $\underline{t}_k$ taken as a virtual load vector, $\underline{u}_k = \underline{K}^{-1}\underline{t}_k$ is the corresponding vector of virtual nodal displacements.

By virtue of the approximate nature of the performance constraint Eqs. (11) and (12), the synthesis is conducted through an iterative process that involves solving the weight optimization problem Eqs. (10)–(13) during each design cycle. Prior to solving at each design stage, the data for Eqs. (10)–(13) is updated to reflect the current state of the design. The displacement and stress sensitivity coefficients are calculated for the current member sizes through Eqs. (16) and (18). The displacement bounds $\bar{\delta}_j$ and $\underline{\delta}_j$ (e.g., limitations on sway left and right) are not updated but remain constant throughout the synthesis process. On the other hand, the allowable tensile and compressive stresses $\bar{\sigma}_k$ and $\underline{\sigma}_k$ are updated in accordance with the current design state and the requirements of the governing steel design standard: the tensile stress bound $\bar{\sigma}_k$ accounts for the type of member connection (bolted, etc.); the compressive stress bound $\underline{\sigma}_k$ depends on the stress state (axial, flexural or combined) and accounts for both local section buckling and overall member buckling. Each discrete set of sizing variables X_i is updated from the database of steel sections to reflect only those sections having adequate strength/stability and stiffness properties for the stress and displacement constraints currently prevailing for the design; these sections must also satisfy shear requirements and any user-specified depth limitations. The total number of sections in each set X_i is limited so as to restrict the movement in the design space and, thereby, preserve the integrity of the sensitivity coefficients when solving the optimization problem Eqs. (10)–(13).

2.2.1 Synthesis Process The iterative synthesis process is as follows [18]:

1. For the initial 'trial' design, or for the current design, structural analysis is performed to determine member stresses and structure displacements for all load cases.

2. Having the stresses and displacements from the previous step, the database of standard steel sections is searched, member-by-member, to find the least-weight sections that satisfy both the strength/stability requirements of the steel design standard and the currently most active user-specified displacement constraint. (This activity is referred to hereafter as the 'member-by-member' optimization procedure.)

3. For the member section sizes from the previous step, structural analysis is performed to determine new member stresses and structure displacements.

4. Having the stresses and displacements from the previous step, the current member section sizes are verified in accordance with the strength/stability and stiffness requirements for the design. If no constraint violations are detected, the synthesis process goes to step 6. Otherwise, the design is infeasible and the synthesis process goes to step 5.

5. If the current infeasible design is as a consequence of two successive applications of step 2, the task of restoring feasibility by the member-by-member optimization procedure is abandoned and the synthesis process goes to step 6. Otherwise, the synthesis process returns to step 2.

6. The current least-weight design of the structure is compared, member-by-member,with all previously recorded least-weight designs. If any two such designs are found to be identical, convergence has occurred and the synthesis process goes to step 12. (The synthesis process also goes to step 12 if divergent behaviour is detected in that the structure weight is observed to increase for two consecutive design stages.) Otherwise, the current least-weight design is recorded and the synthesis process goes to step 7 to commence the 'formal' weight optimization, Eqs. (10)–(13), for the next design cycle.

7. For the current section size for each member taken as the 'key,' and for the current stresses and displacements, the subset of standard sections available for the design of each member during the formal weight optimization is established for the current design stage (i.e., the discrete set X_i in Eqs. (13) is established for each member i; see previous discussion).

8. With a view to computational efficiency, those displacement and stress constraints from among Eqs. (11) and (12) that are relatively inactive for the current design are deleted from the active constraint set for the current design stage. (Deleted constraints are continuously monitored and added to the constraint set if and when they subsequently become active for the design).

9. For the current member stresses and section sizes, the allowable tensile and compressive stress bounds for the active stress constraints are established for the current design stage (i.e., limiting stresses $\bar{\sigma}_k$ and $\underline{\sigma}_k$ in Eqs. (12) are established; see previous discussion).

10. For the current member sections sizes, sensitivity analysis is conducted to establish the gradients of the active displacement and stress constraints for the current design stage (i.e., the sensitivity coefficients in Eqs. (11) and (12) are established through Eqs. (16) and (18)).

11. Having established the discrete subset of sections available for design, and the gradients and bounds for the constraints, the formal weight optimization problem Eqs. (10)–(13) is formulated and solved to find a lower weight design of the structure. The synthesis process then returns to step 1.

12. The synthesis process terminates with the history of all least-weight design solutions recorded for step 6, including the relative structure weight and the degree of feasibility or infeasibility for each design (recall from step 5 that an infeasible design may be recorded as part of the design history). The user is given the option to select the final design of the structure from among those recorded in the design history, or to continue the synthesis process for one or more design cycles (the latter option, for example, allows the user to further track any divergent behaviour of the design history that may have been detected in step 6).

The coordinated use of 'member-by-member' optimization (step 2) and 'formal' optimization (step 11) within the synthesis process gives rise to several computational advantages. The member-by-member optimization allows the database of standard steel sections to be widely addressed, which the formal optimization cannot do if it is to maintain the integrity of its approximate constraint functions. On the other hand, the formal optimization allows for a system-wide redistribution of member section sizes while simultaneously accounting for all strength/stability and stiffness provisions for the design, which the member-by-member optimization cannot do since it is not cognizant of the connectivity of the members in the assembled structure. Sometimes, because of the approximate nature of its constraint functions, the formal optimization can result in an infeasible design. The member-by-member optimization restores feasibility before completing the design cycle. Taken together, the two optimization techniques provide for a comprehensive and robust procedure for least-weight design.

A disadvantage of using formal optimization in the synthesis process is that the design problem is somewhat restricted in size in order to accommodate the computer memory requirements of the optimizer algorithm.

2.3   Mill Crane Framework

The planar steel framework with trussed roof and pinned supports in Fig. 1 is part of a mill crane building [12,18]. It is to be designed in accordance with user-specified displacement constraints and the strength/stability provisions of the Canadian Limit-States-Design steel standard [14]. Member sections are to be selected from the Canadian database of standard steel sections.

The crane framework is subject to fourteen design load cases. As indicated in Fig. 1, each load case is a particular combination of dead, live, wind and crane loading, coupled with a thermal effect caused by a +50°C temperature change for the bottom chord members of the roof truss (due to elevated temperatures within the building enclosure). Load cases 1 to 6 = 1.25 dead + 1.5 live + temperature + 1.5 crane load at nodes 27 to 32; load cases 7 to 12 = 1.25 dead + 1.5 wind + temperature + 1.5 crane load at nodes 27 to 32; load case 13 = live + crane load at node 32; load case 14 = wind + crane load at node 32. Vertical displacement at node 32 is limited to 50 mm for load case 13. Horizontal displacement at node 26 is limited to 20 mm for load case 14.

Out-of-plane bracing is applied at all 56 nodes of the frameworks, including at the intermediate column nodes 51, 52 and 53 (see Fig. 1). The framework has 103 members, consisting of nine column members and ninety-four roof truss members. For the roof truss, the vertical and diagonal members are pin-connected, while the members comprising the top and bottom chords are continuous-connected over each of the two spans and pin-connected only at the exterior and interior columns. The column members and the top and bottom chord members in the roof truss are specified to have wide-flange (W) section profiles, oriented such that bending takes place about the major axis, and having compression flange bracing only at member ends. The vertical and diagonal members in the roof truss are specified to have double-angle section profiles with long legs back-to-back (⏜⏜) separated by a 10 mm thick gusset plate. To satisfy fabrication requirements, six groups of members are identified for which all members in each group are specified to have common section properties. The six fabrication groups are: exterior column

336

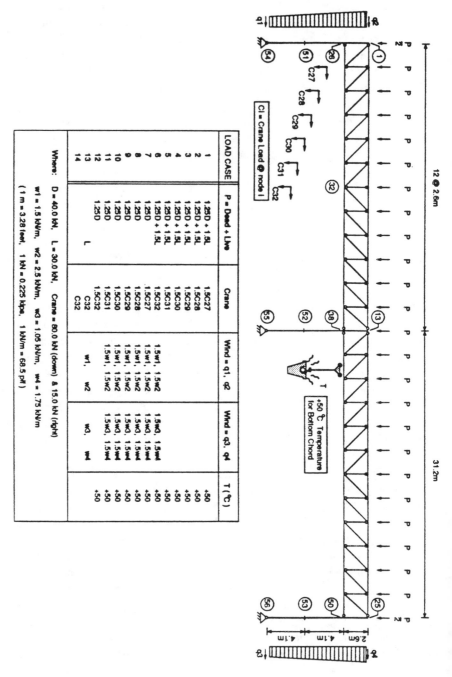

Figure 1. *Mill crane building framework.*

LOAD CASE	P = Dead + Live	Crane	Wind = q1, q2	Wind = q3, q4	T (°C)
1	1.25D + 1.5L	1.5C27	1.5w1, 1.5w2	1.5w3, 1.5w4	+50
2	1.25D + 1.5L	1.5C28	1.5w1, 1.5w2	1.5w3, 1.5w4	+50
3	1.25D + 1.5L	1.5C29	1.5w1, 1.5w2	1.5w3, 1.5w4	+50
4	1.25D + 1.5L	1.5C30	1.5w1, 1.5w2	1.5w3, 1.5w4	+50
5	1.25D + 1.5L	1.5C31	1.5w1, 1.5w2	1.5w3, 1.5w4	+50
6	1.25D + 1.5L	1.5C32	1.5w1, 1.5w2	1.5w3, 1.5w4	+50
7	1.25D	1.5C27			+50
8	1.25D	1.5C28			+50
9	1.25D	1.5C29			+50
10	1.25D	1.5C30			+50
11	1.25D	1.5C31			+50
12	1.25D	1.5C32			+50
13	L		C32		+50
14	L		C32		+50

Where: D = 40.0 kN, L = 30.0 kN, Crane = 60.0 kN (down) & 15.0 kN (right)

w1 = 1.5 kN/m, w2 = 2.5 kN/m, w3 = 1.05 kN/m, w4 = 1.75 kN/m

(1 m = 3.28 feet, 1 kN = 0.225 kips, 1 kN/m = 68.5 plf)

Ci = Crane Load @ node i

members (COLext); interior column members (COLint); top chord truss members (CHtop); bottom chord truss members (CHbot); vertical truss members (Vert); diagonal truss members (Diag).

The steel material properties for the design are: Young's modulus = 200,000 MPa; shear modulus = 77,000 MPa; yield stress = 300 MPa; ultimate stress = 450 MPa; coefficient of thermal expansion = $0.117 \times 10^{-4}/°C$.

Second-order $(P - \Delta)$ analysis is adopted as the underlying basis for the design. In keeping with the Canadian steel standard [14], the in-plane effective length factor is specified to be $K_x = 1.0$ for each roof truss member and top column member, while $K_x = 2.0$ for each lower column member. The out-of-plane effective length factor is specified to be $K_y = 1.0$ for each member (since, as previously described, out-of-plane bracing is applied at all nodes of the framework). The maximum allowable in-plane and out-of-plane slenderness ratios for each member are each specified to be $KL/r = 200$ in compression and $KL/r = 300$ in tension.

The design of the framework is conducted using the synthesis process described in Section 2.2.1 to determine a least-weight structure satisfying the user-specified displacement constraints and the strength/stability requirements of the Canadian steel standard [14]. The iterative synthesis process involves 2,886 strength and displacement constraints and is carried out using a software system developed for professional practise in North America [16]. The results for the entire design history are presented to illustrate the coordinated interaction between the member-by-member and formal optimization techniques that are jointly employed to determine the least-weight structure.

Table 1 presents the design history results from the initial design cycle 1 to the final design cycle 3 for the six member fabrication groups for the framework. The initial design given in the first column of Table 1 is so well proportioned because, actually, it is found after three consecutive continuous-variable formal optimizations for which the initial 'trial' design is defined by the largest section sizes from the specified standard section databases [5]. The least-weight design selected for the framework is given in the last column of Table 1.

In addition to the design history of the member sections for the six fabrication groups, Table 1 also gives the structure weight for each design stage along with the response ratios for the most critical strength and displacement constraints for the corresponding design. (A response ratio = actual/allowable response and, therefore, the maximum allowable response ratio for feasibility is unity).

From Table 1, the framework design found at the beginning of design cycle 1 by formal optimization weighs 19,665 kg but the response ratio = 1.292 for a strength constraint, which implies 29.2% infeasibility. (Recall that the approximate nature of the constraints for the formal optimization may result in designs that are infeasible relative to the actual constraint functions.) The member-by-member optimization restores design feasibility but, in the process, increases the structure weight to 19,725 kg. The resulting design is nearly active for strength response

DESIGN CYCLE	1 Section Designation		2 Section Designation		3 Section Designation	
	Formal Optimisation	Member-by-Member Optimisation	Formal Optimisation	Member-by-Member Optimisation	Formal Optimisation	Member-by-Member Optimisation
MEMBER GROUP:						
COLext	W610×174	W610×174	W610×113	W610×140	W610×125	W610×125
COLint	W610×195	W610×195	W530×123	W610×125	W610×125	W610×125
CHtop	W360×57	W250×49	W250×49	W250×49	W250×49	W250×49
CHbot	W410×85	W310×97	W250×89	W310×86	W310×86	W310×86
Vert	⊥⌐ 100×90×10	⊥⌐ 125×90×10	⊥⌐ 125×90×10	⊥⌐ 125×90×10	⊥⌐ 125×90×10	⊥⌐ 125×90×10
Diag	⊥⌐ 150×100×10	⊥⌐ 100×75×13	⊥⌐ 125×90×10	⊥⌐ 125×90×10	⊥⌐ 125×90×10	⊥⌐ 125×90×10
STRUCTURE WEIGHT (kg)	19,665	18,725	17,110	17,500	16,795	17,190
CRITICAL STRENGTH RESPONSE RATIO	1.292	0.978	0.994	0.996	1.048	1.000
CRITICAL DISPLACEMENT RESPONSE RATIO	0.632	0.638	1.138	0.910	0.971	0.969

Note: $Wd \times m \equiv$ wide-flange section; depth d (mm) and mass m (kg/m)
$\rceil\lfloor l_1 \times l_2 \times t \equiv$ double-angle section with long legs back-to-back; long leg length l_1 (mm), short leg length l_2 (mm), and thickness t (mm)

Table 1.

(ratio = 0.978) but somewhat inactive for displacement response (ratio = 0.638), suggesting that it may be possible to redistribute member stiffnesses to achieve further weight reduction without violating the displacement constraint.

For the feasible design from cycle 1 as its starting basis, the formal optimization for cycle 2 determines a design that weight 17,110 kg but which is 13.8% infeasible for a displacement constraint (ratio = 1.138). The member-by-member optimization restores design feasibility but, in the process, increases the structure weight to 17,500 kg. The resulting design is nearly active for both strength response (ratio = 0.996) and displacement response (ratio = 0.910), suggesting that the synthesis process is converging to the least-weight structure.

For the feasible design from cycle 2 as its starting basis, the formal optimization for cycle 3 determines a design that weighs 16,795 kg but which is 4.8% infeasible for a strength constraint (ratio = 1.048). The member-by-member optimization determines a heavier design weighing 17,190 kg but, now, the critical strength constraint is identically satisfied (ratio = 1.000) and the critical displacement constraint is nearly active (ratio = 0.969). The tight degree of satisfaction of both the strength and displacement constraints suggests that the synthesis process has converged to a least-weight feasible design of the framework. (In fact, not shown in Table 1, divergent behaviour of the design history is detected after this stage in that the weight of the structure is found to progressively increase for subsequent design cycles 4 and 5.)

Finally, it is of interest to identify the relative influence that the member-by-member and formal optimization techniques have on the determination of the least-weight design of the structure. To this end, for the design in the first column of Table 1 taken as the starting basis, the member-by-member optimization technique was alone employed and the iterative synthesis process converged after three cycles to a feasible design weighing 18,070 kg. This design represents a weight saving = 19,665 − 18,070 = 1595 kg compared to the initial design in the first column of Table 1, but is still 18,070 − 17,190 = 880 kg heavier than the final design in the last column of Table 1. In other words, while the member-by-member optimization technique can alone achieve a certain amount of the weight saving, it must be coupled with the formal optimization technique if the full weight saving for the design is to be realized. Conversely, as demonstrated in Table 1, the formal optimization technique must be coupled with the member-by-member optimization if the feasibility of the design is to be ensured.

3 Service and Ultimate Loading Conditions

This part of the lecture addresses the problem of designing structural frameworks for proper performance at a number of distinctly different loading levels. A conventional approach to design in this regard has been to separately proportion the structure for proper performance at one loading level, and to then modify the resulting design to satisfy requirements at one or more other loading levels of concern. For multistorey steel frames, for example, several procedures have been developed whereby a plastic design is initially conducted to ensure adequate safety against plastic collapse under specified ultimate loads, and then the member proportions are modified to satisfy elastic stress and displacement limitations under specified service loads. A major drawback to such an approach, however, is that design decisions at one loading level must be made in the absence of explicit information as to their consequences at the other loading levels of concern. As such, previous design gains are very often unnecessarily negated and, at best, a cumbersome iterative procedure is required to achieve a reasonably efficient design.

The iterative synthesis process described in Section 2 is readily extended to achieve a least-weight design capability whereby performance constraints are satisfied *simultaneously* at both a specified service-load level and a specified ultimate-load level [1,2,3]. Specifically, for proportional static loading, the design process ensures acceptable elastic stresses and displacements under specified service loads while, at the same time, ensuring adequate post-elastic strength reserve of the structure under specified ultimate loads. In other words, a serviceable design is found cognizant of the margin of safety against failure in a plastic mechanism mode. The design methodology represents an extension to the conventional 'limit states design' philosophy that essentially defines 'failure' as being the onset of 'first-yielding' anywhere in the structure. This latter failure definition is somewhat artificial for hyperstatic structures since it provides little information as to what the real margin of safety is against actual catastrophic failure.

Accounting for service and ultimate performance conditions together, the general form of the design problem Eqs. (1) to (4) now becomes:

Minimize: $$\sum_{i=1}^{n} w_i a_i \tag{19}$$

Subject to:
$$\underline{\delta}_j \leq \delta_j \leq \overline{\delta}_j \qquad (j = 1, 2, ..., d) \tag{20}$$

$$\underline{\sigma}_k \leq \sigma_k \leq \overline{\sigma}_k \qquad (k = 1, 2, ..., s) \tag{21}$$

$$\underline{\alpha}_m \leq \alpha_m \leq \overline{\alpha}_m \qquad (m = 1, 2, ..., p) \tag{22}$$

$$\underline{a}_k \leq a_i \leq \overline{a}_i \qquad (i = 1, 2, ..., n) \tag{23}$$

Equations (19), (20) and (21) are exactly Eqs. (1), (2) and (3), while Eqs. (22) define p ultimate-load constraints on plastic collapse-load factors α_m. Equations (23) replace Eqs. (4) since, now, it is presumed that section sizes are continuously available over a range defined by lower- and upper-bound cross-section areas $\underline{a}_i$ and $\overline{a}_i$ (i.e., the design is now a continuous-variable optimization problem).

As outlined in Section 2 for the service-load displacement and stress constraint Eqs. (20) and (21), first-order Taylor's series approximations are also employed to reformulate the ultimate-load constraint Eqs. (22) as explicit-linear functions of the reciprocal sizing variables $x_i = 1/a_i$, as follows

$$\underline{\alpha}_m - 2\alpha_m^0 \leq \sum_{i=1}^{n} \left(\frac{\partial \alpha_m}{\partial x_i} \right)^0 x_i \leq \bar{\alpha}_m - 2\alpha_m^0 . \tag{24}$$

The load factor α_m^0 in Eq. (24) defines the load level at which plastic collapse mechanism m forms for the current design, and is evaluated from plastic analysis as (superimposed 'dot' $\cdot$ denotes a timelike rate quantity)

$$\alpha_m^0 = (R^T \dot{\lambda}_m)^0 = \sum_{i=1}^{n} (R_i^T \dot{\lambda}_{im})^0 , \tag{25}$$

where R^0 is the vector of plastic capacities for all elements of the structure (R_i^0 is the subvector associated with element i), and $\dot{\lambda}_m^0$ is the vector of element plastic deformation rates associated with collapse mechanism m ($\dot{\lambda}_{im}$ is the subvector associated with element i). (An efficient 'finite-incremental' plastic analysis technique to find α_m^0 and $\dot{\lambda}_m^0$ is presented in Ref. [19]). Each of the load-factor sensitivity coefficients $(\partial \alpha_m/\partial x_i)^0$ in Eq. (24) is evaluated as [1,2,3]

$$\left(\frac{\partial \alpha_m}{\partial x_i} \right)^0 = -\frac{1}{x_i^0} (R_i^T \dot{\lambda}_m)^0 . \tag{26}$$

The ultimate-load constraint Eq. (24) is of poorer quality than the service-load constraints Eqs. (11) and (12) because collapse load factors vary directly rather than inversely with the sizing variables a_i (whereas the latter is generally the case for displacements and stresses). However, Eqs. (24) is always a 'conservative' (safe) constraint for any intermediate design stage and, as are Eqs. (11) and (12), an 'exact' constraint for the final stage that determines the least-weight design [1,2,3].

From Eqs. (5), (10), (11), (12), and (24), the least-weight design problem Eqs. (19) to (23) expressed explicitly in terms of reciprocal sizing variables x_i is

$$\text{Minimize:} \quad \sum_{i=1}^{n} w_i/x_i , \tag{27}$$

$$\text{Subject to:} \quad \underline{\delta}_j \leq \sum_{i=1}^{n} \left(\frac{\partial \delta_j}{\partial x_i} \right)^0 x_i \leq \overline{\delta}_j \qquad (j = 1, 2, ..., d) \tag{28}$$

$$\underline{\sigma}_k \leq \sum_{i=1}^{n} \left(\frac{\partial \sigma_k}{\partial x_i} \right)^0 x_i \leq \overline{\sigma}_k \qquad (k = 1, 2, ..., s) \tag{29}$$

$$\underline{\alpha}_m - 2\alpha_m^0 \leq \sum_{i=1}^{n} \left(\frac{\partial \alpha_m}{\partial x_i} \right)^0 x_i \leq \overline{\alpha}_m - 2\alpha_m^0$$

$$(m = 1, 2, \ldots, p) \quad (30)$$

$$\underline{x}_i \leq x_i \leq \overline{x}_i \qquad\qquad (i = 1, 2, \ldots, n) \quad (31)$$

where, from Eqs. (5) and (23), $\underline{x}_i = 1/\overline{a}_i$ and $\overline{x}_i = 1/\underline{a}_i$.

Equations (27)–(31) define the formal weight optimization problem for each design stage. After each weight optimization, the sensitivity coefficients are updated, if necessary the design is scaled to restore feasibility, and the weight optimization is repeated. The numbers d and s of service-load displacement and stress constraints are the same for all design stages. However, the number p of ultimate-load constraints progressively increases to account for new 'critical' plastic collapse modes as the design changes over the iteration history. Initially, for each load case, there is one such constraint corresponding to the 'critical' collapse mechanism for the given 'trial' design. For each design stage thereafter, one new ultimate-load constraint may or may not be added depending on whether or not the load factor for the currently 'critical' mechanism is strictly less than that for all mechanisms accounted for in the previous weight optimization. For each load case, the maximum possible number of ultimate-load constraints at the final design stage is equal to the number of degrees of freedom for the structure (the number is usually much less due to the influence on the design of the service-load constraints and the sizing restrictions).

3.1   Aluminum Tunnel-Ring Frame

The planar aluminum tunnel-ring frame in Fig. 2 can be viewed as a cross-section of the fuselage of a space craft [2]. The framework is discretized into 58 elements, consisting of 36 elements around the circumference of the ring, and 10 and 12 elements for the top and bottom bulkhead beams, respectively. Each element is specified to have the 'constant-shape' hollow-box shape shown in Fig. 2 such that the cross-sectional area is the sizing variable for the design.

With the view to investigate a design that has nearly the absolute minimum weight, symmetry alone is invoked to link the 58 elements together into 29 sizing groups, as indicated in Fig. 2. (While not considered herein, a more appropriate grouping for ease of fabrication is to specify only three sizing groups: ring elements; bottom beam elements; top beam elements).

As indicated in Fig. 2, the framework is subject to four static load cases involving both internal pressure and gravity loading. The design is conducted under both service-load and ultimate-load constraints. Elastic stresses for all elements are limited to $\pm172,375$ kN/m (62.5% of the material yield stress) under all four load cases, for a total of 232 stress constraints. Elastic displacements are specified

343

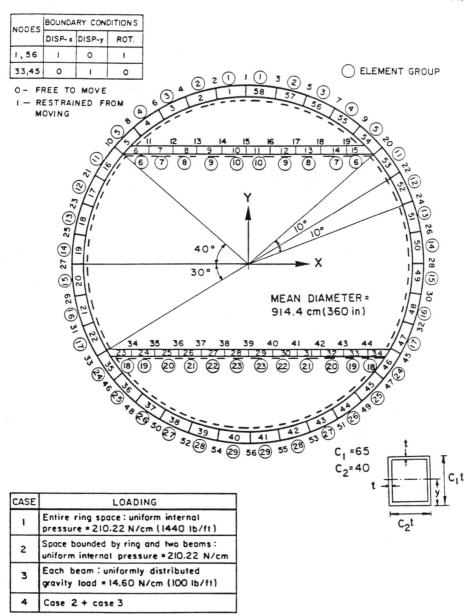

Figure 2. *Aluminium tunnel-ring frame.*

to be less than 2.5 mm horizontally at nodes 33 and 45, and less than 50 mm vertically at nodes 15 and 39, under load cases 2 and 4. Plastic-collapse constraints are
imposed for load cases 1, 2 and 4 such that the critical collapse-load factor for the
framework is not less than 2.1 (i.e., the ring frame must be capable of withstanding
an overload of 110% beyond the service-load levels indicated in Fig. 2 prior to the
formation of any plastic-collapse failure mechanism). The cross-section area for
each element is restricted to be at least 645 mm^2, but no upper bound limitation
is imposed.

The material properties for the design are: Young's modulus = 68,950 MPa;
yield stress = 275 MPa. An eight-sided piecewise linear yield condition is adopted
to govern the plastic behaviour of each element under combined axial and bending
stresses, [3].

The design of the framework is conducted using a prototype software code [2]
that carries out the iterative synthesis process while simultaneously accounting
for the elastic stress and displacement constraints under service loads and the
plastic collapse constraints under ultimate loads. The design history is indicated
in Fig. 3. For an initial 'trial' design weight of 624.6 kg, the iterative synthesis
process monotonically converged to the 'minimum' structure weight of 428.8 kg
in seven design cycles. The six plastic collapse modes indicated in Fig. 3 became
active at various stages of the iteration history (modes 1 to 5 are simple beam
mechanisms, while mode 6 also involves a buckling-type mechanism is the ring).
The active constraint conditions at the final design stage are: vertical displacement
at node 15 for load case 2 and at node 39 for load case 4; stresses in the elements
of groups 7, 11, 17, 18, 19 and 28 under one or more of the four load cases;
plastic-collapse mode 4 under load case 4. The lower-bound sizing constraints on
cross-section areas are not active at any stage of the iteration history.

4 Design under Dynamic Loads

This part of the lecture extends the synthesis process described in Section 2 to
the design of structural frameworks under dynamic loads [10,11]. In particular, to
the least-weight design of frameworks subject to dynamic displacement, dynamic
stress, natural frequency and member sizing constraints under elastic transient
dynamic loading, including simultaneous account for static loading.

In its general mathematical form, the minimum weight structural design problem Eqs. (19)–(23) becomes under elastic transient dynamic loads:

$$\text{Minimize:} \qquad W = \sum_{i=1}^{n} w_i a_i \qquad\qquad (32)$$

$$\text{Subject to:} \qquad \underline{\delta}_j \leq \delta_j(t) \leq \overline{\delta}_j \,, \qquad (j = 1, 2, \ldots, d), \ (t \in T) \qquad (33)$$

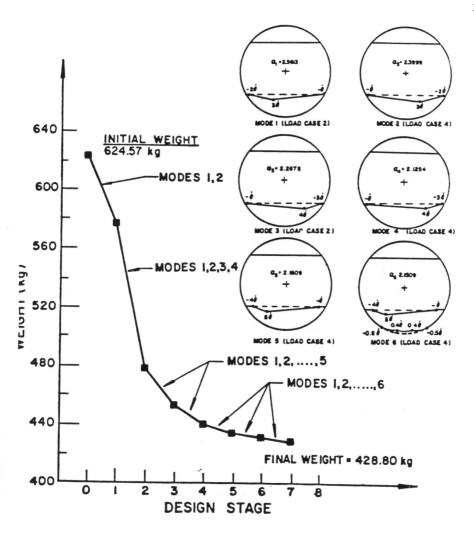

Figure 3. *Design history for tunnel-ring frame.*

$$\underline{\sigma}_k \le \sigma_k(t) \le \overline{\sigma}_k , \qquad (k = 1, 2, \ldots, s), \ (t \in T) \qquad (34)$$

$$\underline{\omega}_l \le \omega_l , \qquad (l = 1, 2, \ldots, f), \qquad (35)$$

$$\underline{a}_i \le a_i \le \overline{a}_i , \qquad (i = 1, 2, \ldots, n) \qquad (36)$$

Equation (32) is identically Eq. (19). Eqs. (33) are similar to Eqs. (20) only, now, they define d constraints on elastic time-dependent displacements $\delta_j(t)$ due to the applied dynamic loading where t is time and T is the time domain over which the response is determined. Eqs. (34) are similar to Eqs. (21) only, now, they define s constraints on elastic time-dependent stress $\sigma_k(t)$. Equations (35) replace Eqs. (22) in the design formulation and define f constraints on the natural frequencies ω_l of the structure. Equations (36) are identically Eqs. (23).

In order to obtain a solution to the implicit design problem defined by Eqs. (32)–(36), explicit constraints based on response sensitivities are achieved by formulating them as first-order Taylor series approximations in terms of repicrocal sizing variables $x_i = \frac{1}{a_i}$, i.e., each of Eqs.(33), (34) and (35) is replaced by

$$\underline{g}_m \le g_m(t)^0 + \sum_{i=1}^{n} \left(\frac{\partial g_m(t)}{\partial x_i} \right)^0 (x_i - x_i^0) \le \overline{g}_m \qquad (37)$$

where $g_m(t)^0$ is the value of the displacement, stress, or frequency (constant over time t) for the current design, $\left(\frac{\partial g_m(t)}{\partial x_i} \right)^0$ is the sensitivity of the response to design change, x_i^0 is the value of the reciprocal design variable for element i for the current design stage, and x_i is the variable to the weight optimization for the next design stage. The appropriate lower and upper bounds on the responses are represented by $\underline{g}_m$ and $\overline{g}_m$, respectively.

To facilitate the formulation of the explicit-linear dynamic constraint Eqs. (37), a modal-superposition method for elastic dynamic analysis is performed to compute dynamic displacements for each design stage as

$$\Delta(t) = \sum_{r=1}^{N} \phi_r u_r(t) \qquad (38)$$

where the ϕ_r are the mode shapes associated with free vibration of the structure (the number of modes considered, N, is generally limited to a few of the lower frequencies), and the $u(t)$ are the corresponding time-variant generalized co-ordinates. Elastic dynamic stresses $\sigma_k(t)$ are then found through the elastic dynamic displacements as

$$\sigma_k(t) = S_{lk} \Delta_l(t) \qquad (39)$$

where S_{lk} is the row of the stress matrix for member l that relates to the k-th stress constraint, and $\Delta_l(t)$ is the displacement subvector associated with the degrees of freedom pertaining to member l.

The sensitivity of each dynamic displacement $\delta_j(t)$ to change in each reciprocal sizing variable x_i is found by differentiating the relation

$$\delta_j(t) = b_j \Delta(t) , \tag{40}$$

where b_j is a vector that identifies the j-th degree of freedom from within the vector of nodal dynamic displacements $\Delta(t)$, to get

$$\frac{\partial \delta_j(t)}{\partial x_i} = \sum_{i=1}^{N} \left\{ \frac{\partial \phi_r}{\partial x_i} u_r(t) + \phi_r \frac{\partial u_r(t)}{\partial x_i} \right\} \tag{41}$$

Then, differentiation of Eq. (39) yields the required sensitivity of the dynamic stress

$$\frac{\partial \sigma_k(t)}{\partial x_i} = S_{lk} \frac{\partial \Delta_l(t)}{\partial x_i} . \tag{42}$$

where the derivatives $\frac{\partial \Delta_l(t)}{\partial x_i}$ correspond to the subset of the displacement vector that pertains to the member l associated with stress $\sigma_k(t)$.

It has been shown that a time-parametric dynamic displacement or stress constraint can be reasonably formulated as a non-parametric constraint by evaluating the corresponding Taylor series approximation at the time of occurrence of the maximum extremum response value [10]. To this end, dynamic analysis is conducted at each design stage to find the extremum displacements $\delta_j(t_{j,\text{ext}})$ and stresses $\sigma_k(t_{k,\text{ext}})$. The appropriate response sensitivities are then evaluated at the times of occurrence of the extrema, and the time-parametric displacement and stress responses are replaced by a non-parametric form. The non-parametric least-weight design problem expressed in terms of reciprocal sizing variables x_i then becomes, from Eqs. (32)–(37),

Minimize:
$$W = \sum_{i=1}^{n} \frac{w_i}{x_i}, \tag{43}$$

Subject to:
$$\underline{\delta}_j \leq \delta_j(t_{j,\text{ext}})^o + \sum_{i=1}^{n} \left(\frac{\partial \delta_j(t_{j,\text{ext}})}{\partial x_i} \right)^o (x_i - x_i^o) \leq \bar{\delta}_j,$$

$$(j = 1, 2, \ldots, d), \ (t \in T) \tag{44}$$

$$\underline{\sigma}_k \leq \sigma_k(t_{k,\text{ext}})^o + \sum_{i=1}^{n} \left(\frac{\partial \sigma_k(t_{k,\text{ext}})}{\partial x_i} \right)^o (x_i - x_i^o) \leq \bar{\sigma}_k,$$

$$(k = 1, 2, \ldots, s), \ (t \in T) \tag{45}$$

$$\underline{\omega}_l^2 \leq (\omega_l^2)^o + \sum_{i=1}^{n} \left(\frac{\partial \omega_l^2)}{\partial x_i}\right)^o (x_i - x_i^o),$$

$$(l = 1, 2, \ldots, f), \qquad (46)$$

$$\underline{x}_i \leq x_i \leq \bar{x}_i, \qquad (i = 1, 2, \ldots, n). \qquad (47)$$

where the frequency constraint Eqs. (46) are now formulated in terms of squared natural frequencies ω_ℓ^2.

As previously described in Section 2, the design problem defined by Eqs. (43)–(47) is solved iteratively through a sequence of design stages involving structural (re)analysis, design (re)formulation, optimization, and design updating. The synthesis process begins with an initial 'trial' design, and ends with a 'least-weight' structure when weight convergence occurs after a number of design cycles.

4.1 Combined Dynamic and Static Loading

Static and dynamic elastic displacements and elastic stresses can be directly summed to give,

$$\delta_{\text{sd}j}(t) = \delta_{\text{s}j} + \delta_{\text{d}j}(t) \; ; \quad \sigma_{\text{sd}k}(t) = \sigma_{\text{s}k} + \sigma_{\text{d}k}(t) \qquad (48)$$

where $\delta_{\text{s}j}$ is the static displacement at node j of the structure, $\delta_{\text{d}j}(t)$ is the dynamic displacement of that node at time t, and $\delta_{\text{sd}j}(t)$ is the combined static plus dynamic displacement at time t; and $\sigma_{\text{s}k}$, $\sigma_{\text{d}k}(t)$ and $\sigma_{\text{sd}k}(t)$ are similar static, dynamic and combined stresses at node k, respectively. The extremum responses are

$$\delta_{\text{sd}j,\text{ext}} = \delta_{\text{s}j} + \delta_{\text{d}j}(t_{\max}) \; ; \quad \sigma_{\text{sd}k,\text{ext}} = \sigma_{\text{s}k} + \sigma_{\text{d}k}(t_{\max}) \qquad (49)$$

where $\delta_{\text{d}j}(t_{\max})$ is the dynamic displacement at node j of the structure at the time $t_{\max}$ when the combined static and dynamic displacement $\delta_{\text{sd}j,\text{ext}}$ is at an extremum, and $\sigma_{\text{d}k}(t_{\max})$ and $\sigma_{\text{sd}k,\text{ext}}$ are similar stress values at node k.

Constraint functions are formulated using Taylor series approximations in a manner similar to that outlined earlier for dynamic loading alone. The combined static and dynamic displacement response is found as

$$\delta_{\text{sd}j,\text{ext}} = (\delta_{\text{s}j})^o + (\delta_{\text{d}j}(t_{\max}))^o$$

$$+ \sum_{i=1}^{n} \left\{ \left(\frac{\partial \delta_{\text{s}j}}{\partial x_i}\right)^o + \left(\frac{\partial \delta_{\text{d}j}(t_{\max})}{\partial x_i}\right)^o \right\} (x_i - x_i^o) . \qquad (50)$$

where $(\delta_{\text{s}j})^o$ is the static displacement at node j for the current design stage and $\left(\frac{\partial \delta_{\text{s}j}}{\partial x_i}\right)^o$ is the corresponding sensitivity (from Eqs. (16)), and $\delta_{\text{d}j}(t_{\max})^o$ is the dynamic displacement at node j at the time $t_{\max}$ and $\left(\frac{\partial \delta_{\text{d}j}(t_{\max})}{\partial x_i}\right)^o$ is the corresponding sensitivity (from Eq. (41)). Similarly, the combined static and dynamic stress response at node k is found as

$$\sigma_{sdk,\text{ext}} = (\sigma_{sk})^o + (\sigma_{dk}(t_{\max}))^o$$

$$+ \sum_{i=1}^{n} \left\{ \left(\frac{\partial \sigma_{sk}}{\partial x_i} \right)^o + \left(\frac{\partial \sigma_{dk}(t_{\max})}{\partial x_i} \right)^o \right\} (x_i - x_i^o) . \tag{51}$$

where the static and dynamic stress sensitivities are given by Eqs. (18) and (42), respectively. The corresponding displacement and stress design constraints

$$\underline{\delta}_j \leq \delta_{sdj,\text{ext}} \leq \bar{\delta}_j \text{ and } \underline{\sigma}_k \leq \sigma_{sdk,\text{ext}} \leq \bar{\sigma}_k . \tag{52}$$

would then replace Eqs. (44) and (45) in the design formulation when combined static and dynamic loads are of concern.

4.2 Trussed Arch

The trussed arch shown in Fig. 4 is to be designed under dynamic stress and displacement constraints, as well as for limits on the lowest fundamental frequency and on member sizes [10]. Six non-structural masses are applied at nodes 2, 4, 6, 8, 10 and 12. As well, nodes 2, 4, and 6 are subjected to the horizontal impulse force shown in Fig. 4.

The framework has 30 members, all of which experience axial force alone. To satisfy fabrication requirements, the top and bottom chord members are linked together in one group (members 1–14), while the web members are linked together in six different groups (members 15–16; 17–18; 19–20; 21–22; 23–24; 25–30).

The material properties for the design are: Young's modulus = 68,670 MPa; yield stress = 200 MPa. The dynamic compressive stress for each member is limited to 60 MPa, while the corresponding dynamic tensile stress limit is set at 125 MPa. The dynamic horizontal displacement of nodes 2, 4, and 6 is constrained to be less than 75 mm. The lowest fundamental frequency is required to be greater than 1.6 cps. Member cross-section areas are constrained to lie between 645 mm^2 and 64500 mm^2.

The design is conducted using the first two modes for the time-history analysis. A time step of 0.005 seconds is adopted and the integration is carried out over 200 time steps for a total time duration of 1.0 second. Damping of 5% is considered for each mode.

The design of the trussed arch is carried out using a prototype software code [10]. The design history is indicated in Fig. 5. For an initial 'trial' cross-sectional area of 32250 mm^2 for all members, the iterative synthesis process converged to the least-weight structure in five design cycles. The minimum weight cross-section areas found for the seven fabrication groups are also given in Fig. 5. The compressive stress constraint for one member of each of six fabrication groups governed

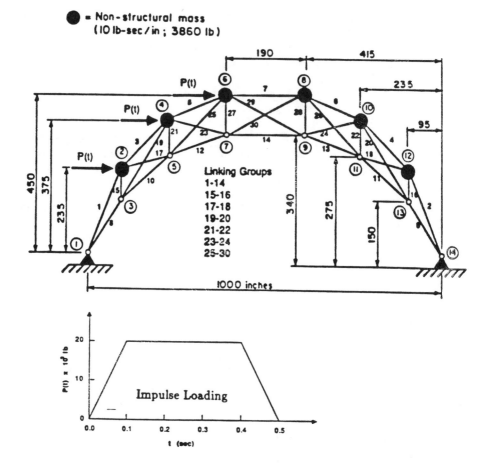

Figure 4. *Trussed arch under impulse loading.*

351

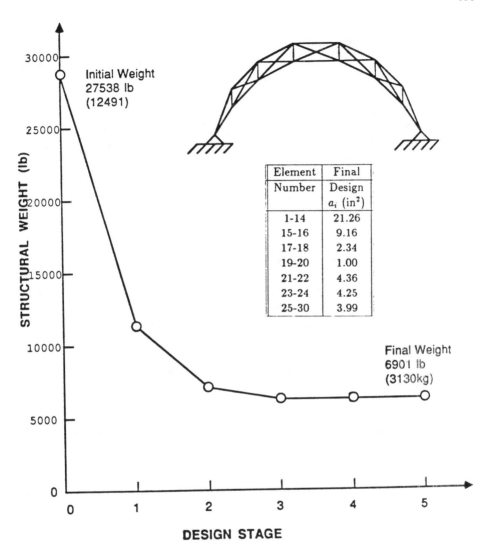

Figure 5. *Design history for trussed arch.*

the design (i.e., for members 9, 16, 17, 22, 23 and 29). The extreme horizontal deflections at nodes 2, 4 and 6 were found to be 72 mm, 64 mm and 55 mm, respectively. The frequencies of modes 1 and 2 were found to be 3.5 cps and 8.3 cps, respectively.

Acknowledgment The reported work has been sponsored by the Natural Sciences and Engineering Research Council of Canada under Grant A5306. Thanks are due to the writer's graduate assistants who conducted the underlying research at the University of Waterloo over the past number of years.

References

[1] Grierson, D. E. and Schmit, L. A., (1982), 'Synthesis under service and ultimate performance constraints,' *Computers and Structures* **15**, 4, 405–417.

[2] Chiu, T., (1982), 'Structural synthesis of skeletal frameworks under service and ultimate performance constraints,' M.A.Sc. Thesis, University of Waterloo, Department of Civil Engineering, Waterloo, Ontario, Canada.

[3] Grierson, D. E. and Chiu, T., (1984), 'Optimal synthesis of frameworks under multilevel performance constraints,' *Computers and Structures* **18**, 5, 889–898.

[4] Lee, W., (1983), 'Optimal structural synthesis of skeletal frameworks using discrete and commerically available standard sections,' M.A.Sc. Thesis, University of Waterloo, Department of Civil Engineering, Waterloo, Ontario, Canada.

[5] Grierson, D. E. and Lee, W., (1984), 'Optimal synthesis of steel frameworks using standard sections,' *Journal of Structural Mechanics* **12**, 3, 335–370.

[6] Cameron, G. E., (1984), 'Optimal structural synthesis of planar trusses subject to Canadian and American design standards using standard steel sections,' M.A.Sc. Thesis, University of Waterloo, Department of Civil Engineering, Waterloo, Ontario, Canada.

[7] Grierson, D. E. and Cameron, G. E., (1984), 'Computer-automated synthesis of building frameworks,' *Canadian Journal of Civil Engineering* **11**, 4, 863–874.

[8] Hall, S. K., (1986), 'Automated synthesis of structural frameworks under large displacements,' M.A.Sc. Thesis, University of Waterloo, Department of Civil Engineering, Waterloo, Ontario, Canada.

[9] Hall, S. K., Cameron, G. E. and Grierson, D. E., (1989) 'Least-weight design of steel frameworks accounting for $P - \Delta$ effects,' *ASCE Journal of Structural Engineering*, **115**, 6, 1463-1475.

[10] Kramer, G. J. E., (1987), 'Optimal structural synthesis of planar frames subjected to static and dynamic loads,' M.A.Sc. Thesis, University of Waterloo, Department of Civil Engineering, Waterloo, Ontario, Canada.

[11] Kramer, G. J. E. and Grierson, D. E., (1989), 'Computer-automated design of structures under dynamic loads,' *Computers and Structures*, **32**, 2, 313-325.

[12] Cameron, G. E., (1989), 'A knowledge-based expert system for structural steel design,' Ph.D. Thesis, University of Waterloo, Department of Civil Engineering, Waterloo, Ontario, Canada.

[13] Xu Lei, (1989), 'Optimal design of space structures,' Ph.D. Thesis, University of Waterloo, Department of Civil Engineering, Waterloo, Ontario, Canada (in progress).

[14] Canadian Standards Association, (1984), *CAN3-S16.1-M84 Steel structures for buildings (limit states design)*.

[15] American Institute of Steel Construction, (1986), *Load and resistance factor design specification for structural steel buildings*.

[16] _____(1987), 'SODA—Structural optimization design and analysis,' Software co-authored by D. E. Grierson and G. E. Cameron, Waterloo Engineering Software, Waterloo, Ontario, Canada.

[17] Fleury, C., (1979), 'Structural weight optimization by dual methods of convex programming,' *International Journal of Numerical Methods in Engineering* **14**, 1761–1783.

[18] Grierson, D. E., and Cameron, G. E., (1989), 'Microcomputer -based optimization of steel structures in professional practise,' *Microcomputers in Civil Engineering*, **4** 289-296.

[19] Franchi, A., (1977), 'STRUPL-Analysis: Fundamentals for a general software system,' Ph.D. Thesis, University of Waterloo, Department of Civil Engineering, Waterloo, Ontario, Canada.

Optimal Design for Static and Dynamic Loading

B. Jüdt, J. Menkenhagen and G. Thierauf
Baumechanik/Statik
Universität-GHS-Essen
Federal Republic of West Germany

Abstract A stategy for the optimization of statically and dynamically loaded structures using a finite element code and program based on the sequential quadratic programming (SQP) technique is presented. In order to reduce the number of design variables a method of decomposing the structure into substructures is shown, in which the optimization problem itself is also decomposed into one "superior" problem and a number of "inferior" problems, the later corresponding to the number of substructures.

1 Introduction

The solutions of nonlinear optimization problems in structural engineering are obtained by iterative methods. In doing so each iteration requires a complete re-analysis of the structure based on the finite element method (FEM). The numerical effort for optimization and the finite element analysis depends on the number of optimization variables and on the order of the structural system.

The computing time required to solve an optimization problem grows rapidly with the number of variables to be optimized. Therefore the number of design variables should be kept as small as possible.

In connection with the utilized optimization method, the sequential quadratic programming method (SQP), the specific reasons to reduce the number of design variables are discussed in [3]. Comprehensively it can be said that compared with the number of structural variables in the FE-analysis (more than 10000) the optimization algorithm is restricted to a relatively small number of variables (50–200). For that background the solution of optimization problems using the mentioned optimization algorithm and a finite element program [1], designed primarily for this purpose, will be discussed in the following; problems which can be solved by multi-stage optimization are considered.

B. H. V. Topping (ed.),
Optimization and Artificial Intelligence in Civil and Structural Engineering, Volume I, 355–366.
© 1992 *Kluwer Academic Publishers. Printed in the Netherlands.*

2 Optimization Method

The subsequent problems are solved using the SQP algorithm described in [2] which fulfills following requirements:

- global convergence, in order to arrive always at a permissible solution,

- few function evaluations as possible, since each necessitates a complete loop over the FE-analysis and the design,

- non-sensitivity with respect to numerical and mechanical approximations involved in the calculation of gradients.

3 Design Variables

Every finite element is allocated to a so-called cross-section group. Elements belonging to such a group all share the property of having the same cross sectional dimensions. These dimensions constitute the design variables of the optimization problem.

For example, the multi-storey framework in reinforced concrete discretizised with 207 finite elements shown in Fig. 1 is devided into 43 cross-section groups (Fig. 2), each with two design variables, one for the reinforcement and one for the cross-section dimensions of the beams.

Replacing the primary design variables

$$\underline{x}^T = (x_1, x_2, \ldots, x_{414})$$

by the new set of design variables

$$\underline{y}^T = (y_1, y_2, \ldots, y_{86})$$

and using a suitable transformation α derived from engineering judgement

$$\underline{x} = \underline{\alpha} \cdot \underline{y} \tag{3.1}$$

the number of design variables is reduced from 414 to 86.

For other problems a partitioning into mechanically stable subsystems can be chosen so that each subsystem i is connected to one design variable y_i.

4 Structures Subject to Static and Dynamic Loading

On the basis of the B&B—program for static linear and nonlinear analysis additional new modules for optimization, dynamic analysis and decomposition are

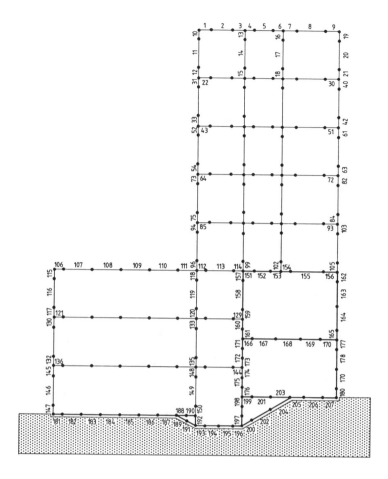

Figure 1. *FE-mesh.*

developed. The programs are based on the work done by G. Booz in the field of optimization of dynamically loaded structures using decomposition methods [4]. Displacements and forces are calculated through subspace iteration and modal analysis.

4.1 Global Problem

The problem (P) of optimal design of statically and dynamically loaded structures can be defined as follows. The objective function is formulated as the total weight of the structure:

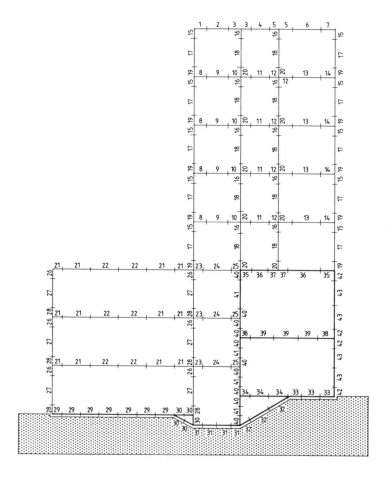

Figure 2. *Design groups (cross-section groups).*

$$\min\{Z(\underline{y}) = \sum_{i=1}^{nq} W(y_i)\}. \tag{4.1}$$

The vector of variables is

$$\underline{y} \in R^{nq}.$$

where nq is the number of cross-section groups. The total weight of the elements in group i is $W(y_i)$.

For the solution of the optimization problem the following constraints are considered:

- lower and upper bounds of the variables:

$$y_i^- \leq y_i \leq y_i^+ \quad i = 1, 2, \ldots, nq \tag{4.2}$$

- the smallest eigenfrequency ω_1 with high $(+)$ or low $(-)$ tuning: (for dynamic analysis)

$$g_1(\underline{y}) = \pm[\omega_1(\underline{y}) - \text{perm } \omega_1] \geq 0 \tag{4.3}$$

- maximum displacement r for node k

$$g_2(\underline{y}) = \text{perm } r - \max \ r(\underline{y}) \geq 0 \tag{4.4}$$

- maximum stress σ_j for each element j:

$$g_j(\underline{y}) = \text{perm } \sigma_j - \max \ \sigma_j p(\underline{y}) \geq 0 \quad j = 3, 4, \ldots, ne + 2 \tag{4.5}$$

where ne is the total number of elements.

4.2 Decomposition

A decomposition must be performed if the problem (P) has too many variables. The solution of subsystems and the grouping of design variables is detailed described in [4]. For the underlaying principales we refer to this work.

4.3 Automatic Generation of Substructures

The relationship of the discreticized finite elements of a system can be presented by its nodal connectivity in the adjacency matrix $\underline{A}^e$:

$$A_{ik}^e = \begin{cases} 1 : \text{if} I_i^n \cap I_k^n \neq \{0\} \\ 0 : \text{otherwise} \end{cases} \quad i, k = 1, \ldots, ne$$

I_i^n : set of nodes belonging to element i
I_i^n : set of nodes belonging to element i
ne : total number of elements

The symmetrical matrix $\underline{A}^e$ can be interpreted as the matrix of an undirected graph. The elements are the vertices and the nodes are the edges of the graph.

The interrelation between elements of the same cross-section group is described in the adjacency matrix $\underline{A}^q$.

$$A_{ik}^q = \begin{cases} 1 : \text{if} I_i^n \cap I_k^n \neq \{0\} \\ 0 : \text{otherwise} \end{cases} \quad i, k = 1, \ldots, nq$$

I_i^n : set of nodes belonging to element i

I_i^n : set of nodes of all elements of cross-section group i

nq : total number of cross-section groups

The number of non-zero off-diagonal elements of the symmetric matrix $\underline{A}^q$ corresponds to the nodal connection of each cross-section group. In order to mimimize the number of these non-zero off-diagonal elements a renumbering of the cross-section groups is performed. Physically the renumbering can be interpreted as a division of the cross-section groups into coherent regions. It is carried out by repeated applications of the Cuthill-McKee algorithm [5] to the adjacency matrix $\underline{A}^q$. The division is only of success in exceptional cases, since within a system any element can be associated with a definite cross-section group or design variable, respectively. One of this exceptional cases is the division of exactly one element for each cross-section group. The sum of all subsystems constitutes the global system. The sum of all elements belonging to groups within a diagonal block constitute a substructure.

4.4   Criterion for Decomposition

A criterion for the choice of cross section groups, whose elements are combined to one subsystem is the linkage of subsystems among one another. Since the possible connection of elements within a given system is determined by the geometry of the structure, the total number of connections is constant. Therefore optimal subsystems are disjoint groups of finite elements with a maximal number of interconnections, or vice-versa subsystems are disjoint groups of finite elements with a minimal number of connections between these groups.

4.5   Decomposition Algorithm

As mentioned above the renumbering of the symmetric adjacency matrix $\underline{A}^q$ is carried out by repeated applications of the Cuthill-McKee algorithm [5]. Originally the algorithm was employed for minimizing the band-width of matrices. Applications of the algorithm to the stiffness matrix $\underline{K}$ for instance leads to a reduced band-width by renumbering the nodes [6].

The application of the algorithm to the adjacency matrix $\underline{A}^q$ depends on the following conditions:

- the number of cross-section groups per substructure has to be marked out,

- the difference between the substructure with the maximum number of cross-section groups and the one with the minimum number of cross-section groups must be equal or less than one.

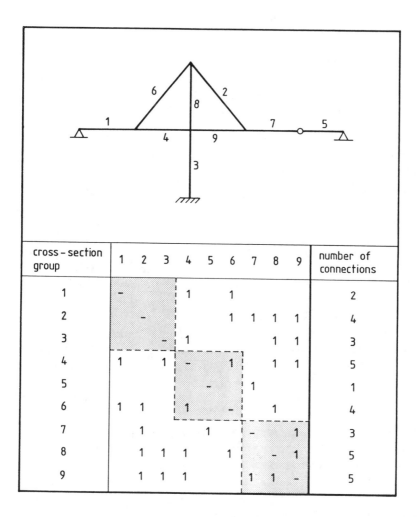

cross–section group	1	2	3	4	5	6	7	8	9	number of connections
1	–			1		1				2
2		–				1	1	1	1	4
3			–	1				1	1	3
4	1		1	–		1		1	1	5
5					–		1			1
6	1	1		1		–		1		4
7		1			1		–		1	3
8		1	1	1		1		–	1	5
9		1	1	1			1	1	–	5

Figure 3. *Framework and adjacency matrix $\underline{A}^q$ of cross-section groups.*

The algorithm starts with the original matrix $\underline{A}^q$, whose elements are determined by the original numbering of the cross-section groups. The first choosen cross-section is the one with least connections to other cross-sections. If there are more than one cross-section group with equal connections one of them will be determined arbitrary as the first one of a substructure. All other cross-section groups joined to the first are potential members of the subsystem.

For the further assignment of cross-section groups to subsystems two different cases are to be distinguished.

1. If the number of marked out cross-section groups is greater than the possible number of cross-section groups, all of these will be added to the substructure and the cross-section groups connected to them will now become potential elements of the same substructure.

2. If the number of marked out cross-section groups is less than the potential number of groups, a choice among these groups has to be made. The number of connections of each of them is counted and the cross-section group with the minimal connections is added to the substructure. This procedure will be repeated until the prescribed number of cross-section groups per substructure is reached.

Subsequently all rows and columns of the corresponding cross-section groups are deleted in the $\underline{A}^q$ matrix. The first cross-section group of the next substructure is choosen with respect to the remaining minimal number of connections.

This procedure will be repeated until all cross-section groups are assigned to a substructure. The following example is to clearify the prescribed algorithm. The framework shown in Fig. 3 is decomposed into 3 substructures, each with 3 cross-section groups. For this given arbitrary division there exist 13 connections of cross-section groups on the off-diagonal blocks of the adjacency matrix $\underline{A}^q$.

The first step is to find the least connected cross-section group and the adjoining minimal connected cross-section groups. This delivers substructure I consisting of group No. 5, 7 and 2.

After deleting the corresponding rows and columns shown in Fig. 4 cross-section group No. 1 is the first one in subsystem II, which will be accompanied by No. 4 and 6. The remaining cross-section groups 3, 8 and 9 form substructure III.

The resulting decomposition is shown in Fig. 5. A comparison of the $\underline{A}^q$ matrices points out that the connections of cross-section groups on the off-diagonal blocks, that means the connections between the substructures, are decreased from 13 to 7.

5 Optimization Problem

The decomposition of the structure into nt substructures also decomposes problem (P) into one **superior** problem $(P1)$ and nt **inferior** problems $(P2i)$, which are all on the same hierachical level.

5.1 The Superior Optimization Problem

The superior optimization problem $(P1)$ is formulated like the global problem (P), but in terms of other variables:

cross–section group	1	2	3	4	5	6	7	8	9	number of connections
1				1		1				2
2										
3			–	1				1	1	3
4	1		1	–		1		1	1	5
5										
6	1			1		–		1		3
7										
8			1	1		1		–	1	4
9			1	1			1	–		3

Figure 4. *Adjacency matrix $\underline{A}^q$ after the first subsystem determination.*

$$\min\{Z(\underline{x}) = \sum_{i=1}^{nt} x_i \sum_{j \in I_i^t} W(y_j)\} \tag{5.1}$$

$\underline{x} \in R^{nt}$: set of nodes belonging to element i
$\underline{x} \in R^{nt}$: vector of scaling factors for the subsystem,
I_i^t : set of cross-section groups belonging to subsystem i,
$W(y_i)$: objective function; the total weight of the elements in group j,
nt : number of subsystems.

All cross-section groups (design variables) of the superior optimization problem are connected. Therefore their determination can be carried out only with one variable x_i, termed as scaling-factor of all cross-section groups belonging to subsystem i. For that reason the solution of $(P1)$ the proportion of all objective functions $W(y_i)$ associated with a subsystem i, remains constant. The solution of $\underline{x}^*$ defines the corrected value by:

$$W(y_j^*) = x_j^* W(y_j) \qquad \qquad \text{for all } j \in I_i^t \tag{5.2}$$

The lower and upper bounds for the scaling factors are determined as:

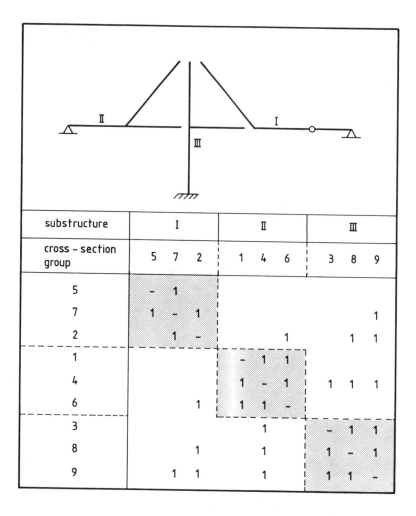

Figure 5. *Final arrangement of the adjaceny matrix $\underline{A}^q$.*

$$\min\left[W(y_j^-)/W(y_j) : j \in I_i^t\right] \le x_i \le \max\left[W(y_j^+)/W(y_j) : j \in I_i^t\right]$$

$$i = 1, \ldots, nt. \tag{5.3}$$

Displacement and eigenfrequency constraints are those of problem (P); in addition, for every subsystem one stress constraint must be fulfilled:

$$g_i^{\sigma(\underline{x})} = {}_{\text{perm}}\sigma_i - {}_{\text{max}}\sigma_i(\underline{x}) \ge 0 \qquad i = 3, \ldots, nt + 2. \tag{5.4}$$

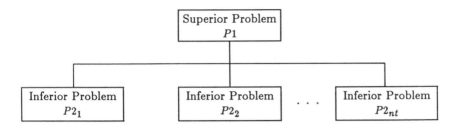

Figure 6. *Hierarchy tree of the decomposed problem (P).*

5.2   The Inferior Optimization Problem

By solving the superior problem $(P1)$ a dimensioning of the main structure is made with respect to stress, displacement and frequency constraints. Objective function and constraints for each subproblem $(P2i)$ should be formulated so that their solution represent the solution of problem $(P1)$. The objective function is the weight of each subsystem i:

$$\min\{Z^i(\underline{y}) = \sum_{j \in I_i^t} W(y_j)\}, \qquad\qquad i = 1, \ldots, nt. \qquad (5.5)$$

Assuming that the displacements of the substructure are not changed by altering the design variable y_i a simple calculation of the stresses can be carried out.

The inferior problems $(P2i)$ are constrained by:

- lower and upper bounds of variables:

$$\max\left\{y_j^-\,;y_j(1-c)\right\} \le y_j \le \min\left\{y_j^+\,;y_j(1+c)\right\},$$
$$\text{for all } j \in I_i^t \qquad (5.6)$$

  where c is a given constant (0.1)

- frequency constraints, that satisfy the displacement constraint approximately:

$$g_k(\underline{y}) = \underline{r}_k^E(\underline{y}^*)^T \underline{K}_i(\underline{y}^*)\underline{r}_k^E(\underline{y}^*) - \underline{r}_k^E(\underline{y}^*)^T \underline{K}_i(\underline{y})\underline{r}_k^E(\underline{y}^*) = 0$$
$$k = 1, \ldots, nu \qquad (5.7)$$

  where $\underline{r}_k^E(\underline{y}^*)$: eigenvector orthonormal to mass matrix $\underline{M}(\underline{y}^*)$
  $\underline{K}_i(\underline{y}^*)$: part of the global stiffness matrix belonging to subsystem i,

- stress constraints:

$$g_k(\underline{y}) = {}_{\text{perm}}\sigma_k - {}_{\text{max}}\sigma_k \geq 0, \qquad k = nu + 1, \ldots, nu + ne_i. \quad (5.8)$$

where ne_i is the number of elements in subsystem i.

6 Conclusions

Optimal design of structures with static and dynamical loadings can be formulated as multi-stage optimization problems and solved by decomposition. The decomposition is obtained by an optimal grouping of elements into substructures. The grouping is found using a bandwidth minimization technique.

References

[1] Thierauf, G. (Ed.), (1985), 'B&B—Programmsystem zur Berechnung und Bemessung allgemeiner Tragwerke,' Universität Essen, Essen.

[2] Schittkowski, K., (1982), 'Theory, implementation, and test of a nonlinear programming algorithm,' in Eschenhauer, H. and Olhoff, N. (Eds.), *Optimization methods in structural design*, Euromech–Colloquium 164, University of Siegen, FR Germany, pp. 122–132.

[3] Thierauf, G., (1987), 'Optimal Design and Optimization of Structures,' *RIL K77, Betonirakenteiden Suunnittelun Erikoiskurssi IV*, Helsinki, Finnland, pp. 146–185.

[4] Booz, G., (1986), 'Eine Dekompositionsmethode zur optimalen Bemessung von Tragwerken unter dynamischer Belastung,' Dissertation, Universität Essen.

[5] Cuthill, E. H. and McKee, J. M., (1966), 'Reducing the bandwidth of sparse symmetric matrices,' *Proceedings Conf. Assoc. for Computing Machinery*, pp. 157–166, New York.

[6] Lawo, M. and Thierauf, G., (1980), *Stabtragwerke: Matrizenmethode der Statik und Dynamik*, Teil 1: Statik, Vieweg Verlag, Braunschweig/Wiesbaden.

Optimal Design and Optimization of Structures

K. Kayvantash, U. Schilling and G. Thierauf
Baumechanik/Statik
Universität – GHS – Essen, Essen
Federal Republic of West Germany

Abstract   This paper presents a general review of some well established structural optimization techniques along with an outlook of a few particular research directions recently explored by the authors. Special attention is given to the problem of optimization under multiple load combinations in steel structures. Finally a parallel processing strategy for an optimal analysis and design of coupled field-structural problems is presented.

1   Introduction

During the last 20 years a great number of publications on structural optimization was published. Although optimal design dates back to the beginning of this century and is based partly on earlier work on variational methods, the development of high-speed computers had a great impact on the research in this field.

This contribution is not meant as a historical review, which can be found in numerous textbooks and review papers, e.g. [1], [2].

An attempt will be made to outline some important directions of research and application. Furtheron, only methods suited for optimization in structural engineering will be considered.

2   The Aims of Structural Optimization

Structural optimization can be characterized as follows:

Most engineering structures can be designed in a variety of different ways: system properties can be choosen as well as different building materials and different dimensions of the system components. These design variables must be choosen so that all functional requirements and all constraints imposed by the designing engineer in accordance with building code regulations are fulfilled.

367

B. H. V. Topping (ed.),
Optimization and Artificial Intelligence in Civil and Structural Engineering, Volume I, 367–416.
© 1992 *Kluwer Academic Publishers. Printed in the Netherlands.*

If more than one solution exists within these constraints, a choice can be made for the 'best' solution. This choice requires a measure: the simplest measure is volume or weight. In these cases we obtain minimum volume or minimum weight structures. A far more difficult measure is the cost of a structure; we all know, that cost of a structure is a most complicated quantity, depending on many factors which might even change before the design process can be finished. However, estimating the cost of a structural design is an every day task of a designing engineer and based on this experience a cost-function can be set up.

Thus 'optimality' of a structure is not the claim for the very best structure in general but only in the limited sense of the initial input. With regard to the uncertainties involved, this input should be kept in a flexible format.

In mathematical form an optimization problem is described by an objective function (merit function), the variables are the design variables, e.g. the geometric variables defining a cross-section of a beam or column or the overall geometry of a building. Integer variables can be necessary in order to describe the number of stiffeners in a steel plate.

The objective function is frequently used to distinguish structural optimization problems: weight-optimization or cost-optimization are common terms, possibly in conjunction with elastic, fully-plastic, linear or nonlinear, static or dynamic and other material or system properties.

From this we recognize that a great variety of optimization problems exists and that a proper formulation in mathematical terms is a difficult task by itself. Therefore the question what we can expect from an optimal solution compared with existing design techniques seems to be justified: experience shows, that in most cases, and even in disciplines with a long lasting development on a competitive market, a saving of 5 to 15% can be gained.

In the following some 'classical' optimization problems and solution techniques are discussed first. Then more recent methods are presented together with their application to near practical problems.

3 Some Fundamental Concepts of Structural Optimization

3.1 Michell-Trusses

An important early contribution dates back to 1904 the work of Michell [3] on beam-like trusses. Although the Michell-trusses are considered to be the beginning of structural optimization by many authors, there are only few remarks [4] on their influence on subsequent work and even on recent publications [5].

Michell's work is concerned with the optimal layout of truss-like structures (plane and space trusses) of minimal weight. Only one loading case is considered.

For demonstration we take one of Michell's optimal beams (fig. 1).

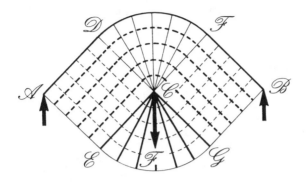

Figure 1. *Michell-truss (1904).*

A load F is applied at point C; A and B are simple supports and 'a' is the distance between A and C (or B and C). The maximal admissible stresses are Q (compression) and P (tension). The Michell-optimum is an orthogonal system of beams in full compression (thick lines) or full tension (thin lines). Dashed lines indicate that one is free to take as many truss-elements as wished; the total optimal volume is found to be

$$Fa \left(\frac{1}{2} + \frac{\pi}{4}\right) \left(\frac{1}{P} + \frac{1}{Q}\right) .$$

The Michell-truss is an ideal system and can be used only as a limiting case in design practice. Its characterization is the uniaxial elastic state of stress with upper and lower constraints. However, by applying lower and upper bound theorems for rigid plastic material [10], the same optimal design would be obtained.

On similar grounds the optimal design of shells of revolution was treated in the year 1908 by Milankovič (see [6]) and extended to rigid plastic material by Ziegler [6] in 1958.

In flexural systems the uniaxial stress-state has to be replaced by the state of constant curvature; the minimization of reinforcement in plate bending can be treated on that basis (Morley, [7] 1966). A general formulation of this class of optimal limit design problems is given by Drucker and Shield ([8], 1957).

3.2 The Prager-Shield Cost Minimization

The Prager-Shield optimality criterion ([9], 1967) is a further generalization of optimal limit design. The objective function is expressed as a 'cost-function' :

Assuming a convex yield-function $\Phi(\underline{F})$ in the generalized stress vector $\underline{F}$

$$\Phi(\underline{F}) \le \rho$$

and a specific cost function $\psi(\rho)$, which depends on the plastic potential ρ, the total cost C of a structure $\Omega(\underline{\xi})$ can be defined as follows:

$$C = \int_{\Omega} \psi(\rho) d\underline{\xi} \ .$$

The Prager-Shield optimality criterion relates the gradient $\underline{G}(\psi)$ of the specific cost–function to the generalized strain v:

A cost-optimal structure is obtained if

$$\underline{v} = \underline{G}(\psi) \ .$$

This condition is necessary and sufficient for convex functions only. The Prager-Shield condition has been applied frequently in the past (Save, [12]; Rozvany, [5], [27]) in particular for optimal design and 'layout' of beams and beam-like structures in uniaxial bending and with the further assumption, that the specific cost function can be formulated in terms of the fully plastic moment resistance. On this basis interesting 'beam–weaves' or beam–grids with constant curvature in certain regions are obtained (Fig. 2, see Prager, [11]).

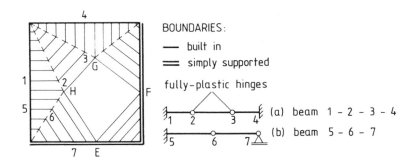

Figure 2. *Optimal beam-weave of square grillage, Prager [4].*

The equidistant lines inside the grid characterize the optimal layout of the beams. In the region EFGH the direction of the beams is arbitrary (Prager, [4]). With regard to practical applications, the ideal case of a layout is obtained. Like with the Michell-trusses the number of beams and the distance between the beams remains undefined.

3.3 Fully-Stressed Design and the Application of General Optimality Criteria

According to Melosh [13], a fully stressed design can be described by the following criterion:

The maximal admissible stress is obtained in each element of the structure at least in one of the loading cases.

For only one loading case the Michell-trusses, Morley-plates and the Prager-Rozvany beam-weaves are fully stressed designs. The associated elastic structure obtained from the Pager–Shield theory is a statically determined structure. The beam 1-2-3-4 (Fig. 2) is built in on both sides and has two fully plastic hinges; a similar beam is found in line 5-6-7 (see Fig. 2a and 2b).

Normally the fully stressed design is not related to this limit analysis. Its original form is a intuitive scaling of stress components in a linear elastic structure (Melosh, [13], Gallagher, [14]) in the following form: It is assumed, that in every element i of a structure the admissible stress $\bar{\sigma}_i$ and one design variable x_i are known. For demonstration purposes, a simple truss is considered (Fig. 3): the design variables are the cross-section areas A_i, $(i = 1, 2, 3)$.

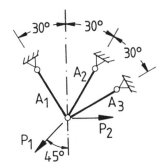

Figure 3. *Weight-optimization, 'truss,' Gallagher [14].*

The design variables are to be determined so that at least in one of the loading cases

$$(P_1 \neq 0, \qquad P_2 = 0) ,$$

$$(P_1 = 0, \qquad P_2 \neq 0)$$

a fully stressed design is obtained.

Starting from an arbitrary design vector

$$x_i^o = A_i , \quad x_i^o > 0$$

the stress in every element under each loading case is computed together with the maximum of the absolute value of the stress σ_i^o.

The iterated design variables in step $k + 1$ are found by scaling:

$$x_i^{k+1} = \frac{\sigma_i^k}{\bar{\sigma}_i} x_i^k, \qquad k = 0, 1, \ldots$$

The iteration is terminated if

$$\max_i |x_i^{k+1} - x_i^k| / x_i^k \leq \varepsilon,$$

where ε is sufficiently small.

The iteration for the cross-section area A_2 is shown in Fig. 4 (Gallagher, [14]).

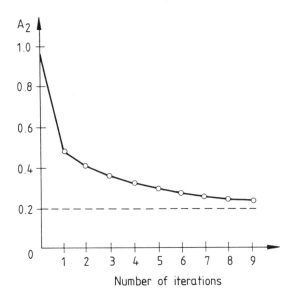

Figure 4. *Iteration, truss.*

In general this stress-scaling procedure will not end up with a minimum weight design. It can be considered, however, as the starting point for the development of more general optimization techniques based on optimality criteria (Gellatly, Berke, [15]).

These techniques are based on the principle of virtual forces, which is used to fulfill stress and displacement constraints in a similar iterative manner.

The development of these general optimality criteria was considered originally as an alternative to mathematical programming methods (see [16]). Initially it was assumed that there existed exactly one design variable per element: for this reason plane and space trusses were ideally suited for application. The 25-bar transmission tower (Fig. 5) was one of the typical examples of that time.

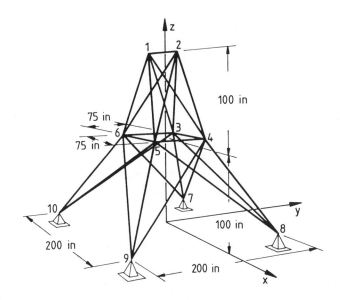

Figure 5. *Transmission-tower, 25 truss-elements.*

The fast convergence is shown in Fig. 6 (from [17]). Even for systems with a great number of variables, several loading cases and stress- and displacement constraints only ten to twenty iterations are necessary. The domgrid with 132 elements and 132 design variables (Fig. 7) required only 15 iterations, although several displacement-constraints became active (Lipp, Thierauf, [25]).

Apart from the intuitive stress-scaling design a distinction between optimality criteria and mathematical programming is hardly justified: in both cases necessary conditions for optimality are approximated by iteration.

This becomes obvious, if optimality criteria are derived by using the Lagrangian multiplier technique (Khot, [19]): We consider a general nonlinear optimization problem where the objective function $W(\underline{x})$ and the constraints $g_j(\underline{x})$, $j = 1, \ldots, m$ are continuous differentiable functions in n design variables $\underline{x}^T = (x_1, x_2, \ldots, x_n)$:

$$\left.\begin{array}{l} \text{Minimize } W(\underline{x}) \\ \text{subject to} \\ g_j(\underline{x}) \leq 0 \ . \end{array}\right\} \tag{1}$$

The constraints are transformed into equality conditions by introducing slack-variables s_j^2:

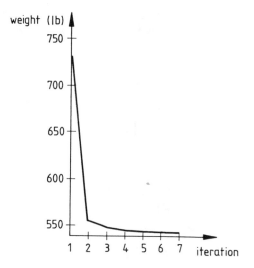

Figure 6. *Iteration, transmission-tower.*

$$g_j(\underline{x}) + s_j^2 = 0.$$

The Lagrange-function for this problem is

$$L(\underline{x}, \underline{s}, \underline{\lambda}) = W(\underline{x}) + \sum_{j=1}^{m} \lambda_j(s_j^2 + g_j),$$

$\underline{\lambda}^T = (\lambda_1, \lambda_2, \ldots, \lambda_m)$ are the Lagrange-parameters, $\lambda_j \geq 0$.
The conditions for a stationary value are:

$$\frac{\partial L}{\partial x_i} = \frac{\partial W}{\partial x_i} + \sum_{j=1}^{m} \lambda_j \frac{\partial g_j}{\partial x_i} = 0, \qquad i = 1, 2, \ldots, n \tag{2}$$

$$\frac{\partial L}{\partial s_j} = 2\lambda_j s_j = 0, \qquad j = 1, 2, \ldots, m \tag{3}$$

$$\frac{\partial L}{\partial \lambda_k} = (s_k^2 + g_k) = 0, \qquad k = 1, 2, \ldots, m \tag{4}$$

$\underline{x}^o$ must be found as the solution of $2m+n$ nonlinear equations in $2m+n$ unknowns $(\underline{x}, \underline{s}, \underline{\lambda})$.

From (3) we conclude that either $s_j = 0$ or $\lambda_j = 0$; if constraint g_j is active $s_j = 0$ and $\lambda_j \geq 0$. A passive constraint $g_k < 0$ requires that $s_k > 0$ and $\lambda_k = 0$.

Thus for all active constraints $g_j(\underline{x}) = 0$, $j \in I$ we obtain the following condition:

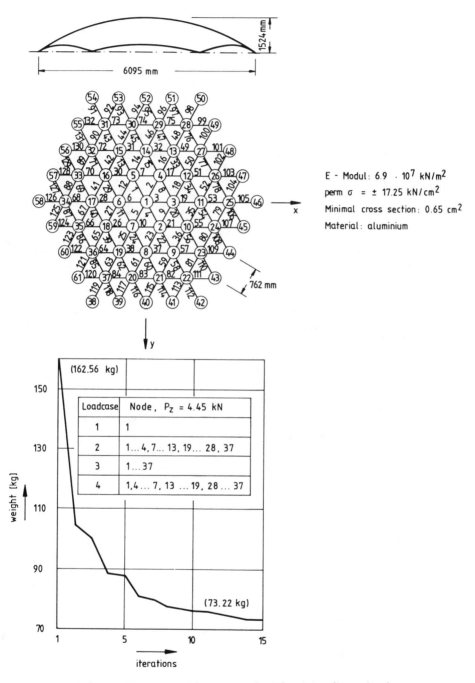

Figure 7. *Dom grid computed with optimality criteria.*

$$\left.\begin{array}{c} \dfrac{\partial W}{\partial x_i} + \sum_{j \in I} \lambda_j \dfrac{\partial g_j}{\partial x_i} = 0 \, , \qquad\qquad i = 1, 2, \ldots, n \\[3mm] \lambda_j \geq 0 \, . \end{array}\right\} \tag{5}$$

These are the (local) Kuhn-Tucker conditions for Eq (1), (Hadley, [21]). For convex functions W and g_j necessary and sufficient conditions are obtained. Apart from special cases, these functions are not convex in structural optimization. Comparing (5) with the usual formulation of optimality criteria, we find no differences.

In some cases optimality criteria are derived under the assumption that only one constraint is active, so that (5) can be expressed as a recursive relation for the design variables. Near-active constraints are treated seperately or by modifications of the step-width. Similar techniques are used in mathematical programming.

4 Integrated Analysis, Design and Optimization of Structures

4.1 Optimization of Elastic Structures Based on Finite Element Methods

In the early seventies, when structural optimization for elastic and rigid plastic material behaviour was developed in form of optimality criteria and when structural engineers began to apply mathematical programming methods, the analysis by finite elements had become a widely used tool already. For this reason the 'new' design methods were used in conjunction with finite elements and integrated finite element analysis and design programs were developed.

This concept seems to be appropriate for several reasons:

In every step ν of the iterative design an increment $\underline{d}^\nu$ of the design variables $\underline{x}$ is computed. The iterated design variables are computed from

$$\underline{x}^{\nu+1} = \underline{x}^\nu + \alpha \underline{d}^\nu \, ,$$

where α is a step-width factor. A new stiffness matrix $\underline{K}(\underline{x})$ and loading vector $\underline{R}(\underline{x})$ must be set up and the nodal displacements $\underline{r}(\underline{x})$ are computed from

$$\underline{K}\,\underline{r} = \underline{R} \, .$$

The stress components are computed subsequently.

The application of optimality criteria or of mathematical programming requires the gradient $\dfrac{\partial \underline{r}}{\partial x_i}$, which is obtained as the solution of the sensitivity equation

$$\underline{K} \frac{\partial \underline{r}}{\partial x_i} = \frac{\partial \underline{R}}{\partial x_i} - \frac{\partial \underline{K}}{\partial x_i} \, \underline{r} \, . \tag{6}$$

The nodal displacement vector and the gradient are determined by the same matrix $\underline{K}$ and can be computed simultaneously, only the right hand side has to be changed in an iterative design. Furtheron, if only one design variable is assumed per element, the element stiffness matrices can be found by scaling.

Stress constraints can be evaluated from displacements together with the computation of the stress components.

Based on the displacement method and taking into account all these computational advantages, powerful programs were developed at that time (OPTIM [20], FASTOPT [22], DESAP [23]). Advantages of a force method optimization were investigated in 1975 by the author [24] and applications in connection with optimality criteria based on mathematical programming followed ([18], [25]). Full-scale large computer programs for optimality criteria, based on the finite element force method were developed by Gellatly and Thom [26].

All integrated finite element based optimization programs can be considered as special purpose programs: problems with a great number of variables (several hundreds) can be solved in reasonable computing time and accurracy. In near optimal points convergence can become slow, depending on the particular method used.

4.2 Optimal Limit Design Based on Finite Elements

The Prager-Shield optimality criterion was applied frequently for the optimal design and layout of beam-type structures (Rozvany, [27]). A direct formulation of this theory in terms of a finite element approximation was given by the author in 1978 [28]; it can be considered as a direct link between optimality criteria methods and a sequential quadratic approximation of an optimization problem by finite element methods. For this reason the basic ideas will be outlined here.

We first assume, that only one yield-function $\Phi_i(\underline{F}_i)$ is given in every element i. $\underline{F}_i$ is the vector of linear independent forces in element i. The yield-condition is assumed to be known in form of a quadratic approximation:

$$\underline{F}_i^T A_i \underline{F}_i \leq y_i .\tag{7}$$

y_i is the plastic potential of element i.

For some element i with two force components F_1 and F_2 this approximation is shown in Fig. 8.

In accordance with Prager and Shield it is assumed that the total 'cost' W of the finite element system of m elements is a linear combination, or can be approximated by a linear form of the plastic potentials:

$$\left.\begin{aligned} W(\underline{y}) = \underline{c}^T \underline{y}, \quad \text{where} \quad & \underline{y}^T = (y_1, y_2, \ldots, y_m) \\ \text{and} \quad & \underline{c}^T = (c_1, c_2, \ldots, c_m), \\ & c_i > 0 . \end{aligned}\right\}\tag{8}$$

378

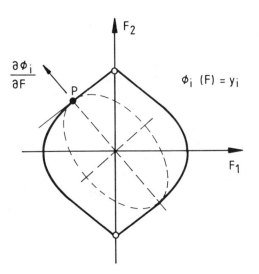

Figure 8. *Quadratic approximation of yield-surface.*

For forces $\underline{F}^T = (\underline{F}_1, \underline{F}_2, \ldots, \underline{F}_m)$ in equilibrium with the external forces $\underline{R}$, that is

$$\underline{a}^T \underline{F} = \underline{R},$$

where $\underline{a}^T$ is the equilibrium matrix, the Prager–Shield criterion can be expressed by the following optimization problem:

$$\text{minimize } W(\underline{y}) = \underline{c}^T \underline{y}, \tag{9}$$

$$\text{subject to } \underline{F}_i^T \underline{A}_i \underline{F}_i \leq y_i. \tag{10}$$

Introducing the slack–variables s_i^2, Lagrange parameters u_i and $\underline{r}$, the following Lagrange function is obtained:

$$L(\underline{y}, \underline{F}, \underline{u}, \underline{s}, \underline{r}) = \sum_{i=1}^{m}[c_i y_i + u_i(\underline{F}_i^T \underline{A}_i \underline{F}_i - y_i + s_i^2)] + \underline{r}^T(\underline{R} - \underline{a}^T \underline{F}).$$

The necessary optimality conditions are:

$$\frac{\partial L}{\partial y_i} = 0 \quad \ldots \quad c_i = u_i, \tag{11}$$

$$\frac{\partial L}{\partial s_i} = 0 \quad \ldots \quad 2\,u_i s_i = 0, \tag{12}$$

$$\frac{\partial L}{\partial F_i} = \underline{0} \quad \dots \quad 2 \operatorname{diag}(u_i \underline{A}_i) \underline{F} - \underline{a} \, \underline{r} = \underline{0} \,. \tag{13}$$

From $c_i > 0$ and from (12) we obtain $s_i = 0$. Equation (13) is the compatibility condition of an associated elastic finite element system: Introducing

$$\underline{f} = 2 \operatorname{diag}(c_i \underline{A}_i)$$

as the flexibilty matrix, we obtain the compatibility condition

$$\underline{f} \, \underline{F} = \underline{a} \, \underline{r} \,. \tag{14}$$

The Lagrange parameters $\underline{r}$ are the nodal displacements of the associated elastic structure and can be computed by the displacement method:

$$\underline{K} \, \underline{r} = \underline{R} \,,$$

where

$$\underline{K} = \underline{a}^T \, \underline{f}^{-1} \, \underline{a} \,.$$

The complementary energy $\frac{1}{2} \underline{F}^T \, \underline{f} \, \underline{F}$ of this associated system is equal to the minimum of the cost function W, which can be verified by substitution:

$$\frac{1}{2} \underline{F}^T \underline{f} \underline{F} = \underline{F}^T \operatorname{diag}(c_i \underline{A}_i) \underline{F} = \sum_{i=1}^{m} c_i y_i = W \,. \tag{15}$$

In this form the Prager–Shield criterion was formulated originally.

A finite element application on that basis together with an extension to more than one yield–condition per element is given by Pape [29]. Further applications can be found in [30].

The approximation of a Prager–Shield solution by 10 and 20 finite elements is demonstrated in Fig. 9. It can be seen that a close approximation of the exact solution (dashed curve) is obtained.

Although this method has been used only for pilot programs in optimal design, large scale optimization programs could be developed easily.

4.3 Limitations of an Integrated Finite Element Based Approach

From the foregoing results we see that the optimal design can be integrated into finite element computations by using special properties of the optimization problem. This applies for the elastic and for the plastic optimization. In both cases the optimality criteria, which involve the objective function, the constraints and gradients with respect to the design variables, must be computed in an extended finite element program. It is obvious that this approach offers the advantage of greatest

380

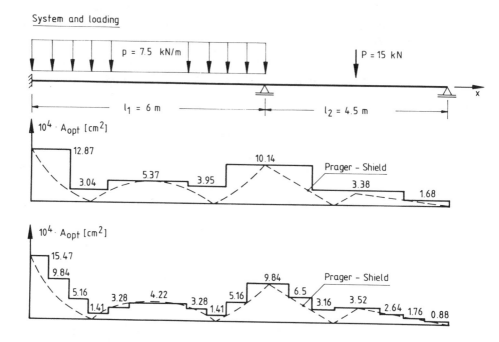

Figure 9. *Quadratic approximation of yield-condition, comparison with Prager–Shield solution.*

efficiency with respect to computation time and storage of structural data but also, that the finite element program must be specialized with respect to the objective function and the constraints. This specialization may be justified in certain fields such as aircraft design, where the minimum weight is dominant and the constraints can be standardized. As mentioned at the beginning, the situation in structural engineering is different: there is a great variety of different optimization problems together with a frequent change of the objective function and of the constraints, which are influenced by the development of design codes, construction techniques and materials. The advantages of an integrated design and analysis are outweighed by the limited capability to handle different objective functions and constraints.

An additional aspect should be mentioned: structural optimization is based on two distinct fields of research, structural analysis and mathematical programming. Great flexibility is therefore required in order to adjust for new developments in either of these fields.

5 Structural Optimization based on Modular Programming

5.1 Basic Modules and General Demands

The basic layout of a modular computer program for structural optimization is shown in Fig. 10.

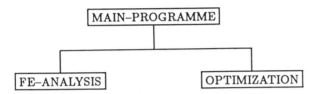

Figure 10. *Structural optimization on modular basis.*

A main program organizes the flow of data between two modules, the FE-module and the optimization module. Further, the in- and output data for both modules are prepared and transformed here to a form required by the two different modules.

The FE-module can be any available program suited for the analysis of the structure. Preferably it should have a modular structure by itself in order to adjust it to a wide class of optimization problems. The choice of the optimization module is more difficult: first it should be reliable and stable and second it should be able to solve typical engineering problems in a reasonable computing time. With regard to the connection of analysis and optimization we also have to discuss the capability to handle a great number of structural variables in analysis and an adequate number of design variables in optimization.

5.2 Selection of the Optimization Method

In section 4.3 a special quadratic approximation of constraints arising in optimal limit design has been considered. The nonlinear constraints of the special optimization problem are approximated sequentially by quadratic forms. In recent time the sequential quadratic programming approach for general nonlinear optimization problems has attained great importance, in particular in form of the NLPQL algorithm by Schittkowski [31].

The following problem is solved:

$$\min \ f(\underline{x})$$
$$g_j(\underline{x}) = 0, \qquad j = 1, \ldots, m_e$$
$$g_j(\underline{x}) \geq 0, \qquad j = m_e + 1, \ldots, m \tag{16}$$
$$(f, \ g_j : \text{continuous differentiable}).$$

Hock and Schittkowski [32] compared the NLPQL algorithm with other frequently used optimization algorithms. A sequence of 240 testing examples was chosen for comparison; the average results are shown in Table 1 (from [32]).

Table 1. *Comparision of frequently used optimization methods according to Hock and Schittkowski [32].*

Method	CPU	Function calls	Constraints	Gradients	Failure rate
Penalty method	88.1	1043	8635	957	28.5
Lagrange multipliers	42.2	158	1595	603	29.3
Generalized reduced gradients	37.7	204	2496	378	13.4
Sequential quadratic programming	14.1	18	181	64	3.3

In structural optimization function calls and gradient evaluations are normally equivalent to a FE–analysis. As the FE–analysis is the most time consuming part of the structural optimization, the NLPQL-algorithm is suited best in this sense. Further, it is far more stable than the other methods and does not require an admissible starting point like some others (e.g. inner penalty methods).

The problems tested in [32] are chosen from all fields and are not specially related to structural optimization. Vanderplaats [33] compares different optimization algorithms applied to a simple structural problem as shown in Fig. 11. A built-in beam is divided into 5 elements, each with two design variables, the width and height. This results in an optimization problem with 10 variables; stress– and displacement–constraints are considered.

The following optimization methods are compared:

(a) Sequential linear programming,

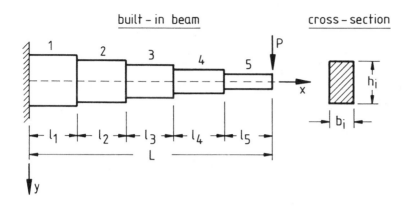

Figure 11. *Built–in beam, comparison of optimization algorithms according to Vanderplaats [33].*

(b) Exterior penalty function method,

(c) Linear extended interior penalty function method,

(d) Augmented Lagrange multiplier method,

(e) Method of feasible directions,

(f) Modified method of feasible directions,

(g) Sequential quadratic programming,

(h) Sequential convex programming.

If the product of number of iterations and number of function calls γ is chosen here as a measure of efficiency, one obtains the priority as shown in Table 2.

Table 2. *Comparison of some optimization methods for a simple built-in beam according to Vanderplaats [33].*

Method	g	a	h	f	e	c	d	b
γ	848	1100	1100	1430	1540	3936	4264	15824

As in the previous comparison, the sequential quadratic programming gives the best results, followed by sequential linear programming (23% difference).

384

A further comparison is given by Eschenauer [34]; a cylinder of a band-conveyor is optimized by using different optimization methods. The sequential quadratic approximation is not included, and the sequential linear approximation turns out as the best method. Instead of hyperspheres, which have been used by Baldur [36], constraints in form of hypercubes are imposed here on the design variables [35].

Based on these results the sequential quadratic programming method seems to be most suited for structural optimization; next best is the sequential linear programming method.

5.3   Modular Connection of FE–Analysis and Optimization

First we consider the typical structure of an optimality criteria program, e.g. OP-TIM or DESAP (Fig. 12).

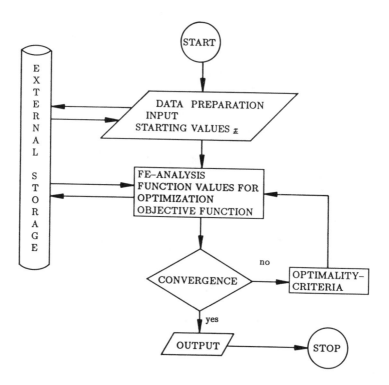

Figure 12. *Structure of integrated FE–optimization programs.*

The preparation of input data is followed by a FE–analysis incorporating all necessary computations for the subsequent application of an optimality criterion, which gives new estimates for the design variables.

A first step towards a modular structure is to take the FE–module as a submodule of the optimization (Fig. 13). The optimization module includes submodules which define the objective function, the constraints and compute the gradients. Apart from elementary applications, the gradients can not be evaluated in analytical form and must be computed by finite differences.

The main program controls the input and output and the termination criteria. At present a similar concept is frequently used in structural optimization.

Further going flexibility can be gained by a programming system [40], which consists of completely independent subprograms and a connecting network (Fig. 14).

The connecting network controls the data managment and executes the different programs as defined by the user in a high level command language (job control language). Similar concepts have been realized for strictly modular FE–systems earlier [39].

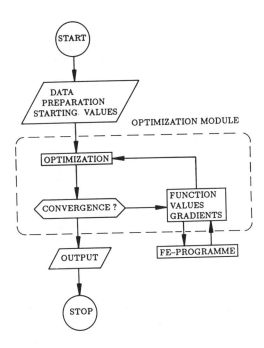

Figure 13. *Modular structure in optimal design.*

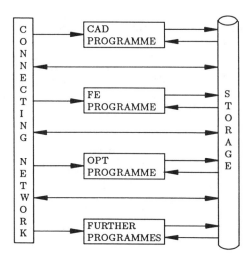

Figure 14. *Programming system, structural optimization.*

6 Reducing the Number of Design Variables

6.1 Necessity

The computing time required to solve an optimization problem grows rapidly with the number of variables; therefore the number of design variables should be kept as small as possible. In connection with the optimization methods discussed before, there are further specific reasons to reduce the number of design variables:

The sequential quadratic programming method is based on a sequence of quadratic subproblems; the quadratic term of the objective function is approximated by the Hessian matrix, the linear constraints are a linear approximation of the active (or near active) constraints. The sequential linear programming method solves a sequence of linear optimization problems and imposes constraints on the variables via hypercubes or hyperspheres. In the first case the Hessian, which normally does not have any zero elements, has to be stored. With an increasing number of variables this requires a considerable amount of storage in addition to the computing time for solving the quadratic subproblems. For the sequential linear approach the number of variables and the number of constraints determines the dimension of the linear subproblems; the solution of large linear programming problems causes similar problems ([37], [38]).

Both methods are therefore restricted to a relatively small number of variables (50 to 200) compared with the number of structural variables in a FE–analysis (more than 10.000).

6.2 Design Groups and Linking of Design Variables

An intuitive method for reducing the number of design variables will be discussed first. The design variables $\underline{x}^T = (x_1, x_2, \ldots, x_n)$ are replaced by new design variables $\underline{y}^T = (y_1, y_2, \ldots, y_m)$

$$\underline{x} = \underline{L}\,\underline{y}\,,$$

where $m < n$ and $\underline{L}$ is a suitable transformation derived from engineering judgement.

For example, a reinforced concrete plate with 128 finite elements (Fig. 15) is devided into 11 design groups, each with two variables, one for the reinforcement and another one for the thickness of the plate.

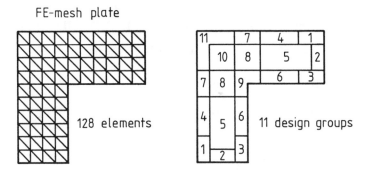

Figure 15. *FE–mesh and design groups.*

For many optimal design problems a similar grouping of the variables is indispensable for structural and economic reasons. For other problems, a partitioning into mechanically stable subsystems can be chosen so that each subsystem i is connected to one design variable y_i.

In the first step the optimization for the reduced set of design variables $\underline{y}$ is solved. The solution $\underline{y}_0$ can now be assumed as an upper limit for subsequent optimization problems on substructure level.

The selection of subsystems and the grouping of design variables can also be performed automatically but this approach requires an analysis of the structural connections and of the occurrence of design variables in constraints.

6.3 A Two–Stage Decomposition of a Structured Design Problem

For specially structured systems this decomposition can be found by inspection of the constraints: We consider the following optimization problem

$$\min f(\underline{y}, \underline{x}_1, \underline{x}_2, \ldots, \underline{x}_m)$$

subject to

$$g_1(\underline{y}, \underline{x}_1) \leq 0,$$
$$g_2(\underline{y}, \underline{x}_2) \leq 0,$$
$$\vdots$$
$$g_m(\underline{y}, \underline{x}_m) \leq 0.$$

$\underline{y}$ and $\underline{x}$ are design variables.

A similar form of an optimization problem describes the optimal design of reinforced concrete structures, which will be treated in a subsequent section.

The interconnection of the variables can be expressed by a Boolean matrix $\underline{C}$, whose elements C_{ij} are defined as follows (occurrence matrix):

$$C_{ij} = \begin{cases} 1: \text{if variable } j \text{ is contained in equation } i \\ 0: \text{otherwise} \end{cases}$$

Every column of $\underline{C}$ corresponds to one of the variables; the variables are ordered and the columns are numbered starting from 0:

$$\underline{y}, \underline{x}_1, \underline{x}_2, \ldots, \underline{x}_m.$$

Every row of $\underline{C}$ corresponds to one of the functions of the optimization problem; the rows are ordered according to

$$f = g_0, g_1, \ldots, g_m.$$

For $m = 5$ the occurence matrix is shown in Fig. 16.

$\underline{C}$ can also be expressed as a bipartite graph [41]: the nodes G_i correspond to the functions and the nodes V_i to the variables (Fig. 17).

The analysis of the connectivity of this graph can be used for the construction of an iterative solution: deleting one of the nodes V_o or G_o interrupts all cycles and thus suggests a two-stage solution, in particular, if the objective function is seperable

$$f(\underline{y}, \underline{x}_1, \ldots, \underline{x}_m) = f_0(\underline{y}) + f_1(\underline{x}_1) + \cdots + f_m(\underline{x}_m)$$

and if the variables $\underline{y}$ and $\underline{x}$ are weakly coupled, so that

$$
\underline{C} = \begin{array}{c|cccccc}
 & 0 & 1 & 2 & 3 & 4 & 5 \\
\hline
0 & 1 & 1 & 1 & 1 & 1 & 1 \\
1 & 1 & 1 & 0 & 0 & 0 & 0 \\
2 & 1 & 0 & 1 & 0 & 0 & 0 \\
3 & 1 & 0 & 0 & 1 & 0 & 0 \\
4 & 1 & 0 & 0 & 0 & 1 & 0 \\
5 & 1 & 0 & 0 & 0 & 0 & 1 \\
\end{array}
$$

Figure 16. *Occurrence matrix of a structured optimization problem.*

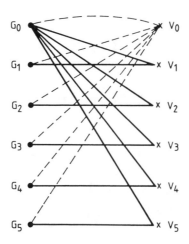

Figure 17. *Bipartite graph of occurrence matrix $\underline{C}$.*

$$\frac{\partial g_j}{\partial y_1}\frac{\partial y_1}{\partial \underline{x}_i} + \frac{\partial g_j}{\partial y_2}\frac{\partial y_2}{\partial \underline{x}_i} + \cdots + \frac{\partial g_j}{\partial y_n}\frac{\partial y_n}{\partial \underline{x}_i} \approx \underline{0}$$

The two-stage model consists of an inner iteration loop ν for the variables $\underline{x}$ and an outer loop μ for $\underline{y}$. $\underline{x}^{\nu+1}$ is computed from a stepwidth-parameter α and an increment $\underline{\xi}$:

$$\underline{x}_i^{\nu+1} = \underline{x}_i^{\nu} + \alpha\,\underline{\xi}$$

$\underline{\xi}$ is the solution of one of the m subproblems.
The iteration is started with $\nu = 0$ and $\mu = 0$ and the starting values

$$\underline{x}^{\nu} = (\underline{x}_1^{\nu}, \underline{x}_2^{\nu}, \ldots, \underline{x}_i^{\nu}, \ldots, \underline{x}_m^{\nu})\ ,\underline{y}^{\mu}$$

$$g_i(\underline{y}^{\mu}, \underline{x}_i^{\nu}) \geq 0$$

(A) Outer loop: $\underline{x}_i = \underline{x}_i^{\nu}$ constant

Solution of the optimization problem

$$\min f_0(\underline{y})$$

$$g_i(\underline{y}, \underline{x}_i) \leq 0$$

$$\text{solution } \underline{y}^{\mu+1} = \underline{y}$$

(B) Inner loop: $\underline{y} = \underline{y}^{\mu}$ constant

The following sequence of subproblems is solved:

$$\min f_i(\underline{\xi})$$

$$g_i(\underline{y}, \underline{\xi}) \leq 0$$

$$\text{solution } \underline{\xi}\ :\ \underline{x}_i^{\nu+1} = \underline{x}_i^{\nu} + \alpha\,\underline{\xi}$$

Termination:

In the outer loop the truncation error is checked, e.g. by

$$\max_i |y_i^{\mu+1} - y_i^{\mu}|/|y_i^{\mu+1}| \leq \varepsilon$$

The minimum of the original problem is obtained as the sum of the subproblems; the weak coupling of the variables $\underline{y}$ and $\underline{x}$ is checked within the iteration and, if necessary, the stepwidth is adjusted.

An admissible solution must be found as the starting value. In many cases, and in particular for problems with a sparse occurrence matrix, this solution is known from structural considerations.

In the present iteration scheme, the variables $\underline{y}$ are coordination variables for the design variables $\underline{x}$.

Depending on the structure of the occurrence matrix, multistage iteration schemes can be developed based on graph theoretic methods ([41], [42], [43]).

However, the advantages of any decomposition are gradually lost with an increasing number of non-zero elements in the occurrence matrix: speaking in mechanical terms, that means for highly connected systems a decomposition into subsystems is of no advantage compared with the solution of the complete system.

7　A Modular FE–Optimization Program

From the existing modular programs for structural optimization [44] we consider here only the programme B&B (German: 'Berechnung & Bemessung,' Analysis & Design). One reason for this restriction is the availability: B&B has been developed in the authors' department (Civil Engineering, University of Essen). Furtheron, it has been used extensively and without major drawbacks in teaching, research and in engineering practice.

Any comparison of existing programs suffers from the lack of a joint basis: most of them are specialized in some respect (objective function, constraints, loading) and despite of a modular structure any alteration requires much time.

7.1　B&B – Modular Structure and Available Versions

The smallest version of B&B is the PC–version B&B–micro for compatibles. Its main modular structure is shown in Fig. 18.

B&B–micro is a linear version containing simple types of elements (beams, plane stress, plate bending, plane 'shell' elements).

It can handle 12000 displacement components and 3000 elements in a reasonable amount of time. The PC–version itself can not be used in optimal design, however, the preprocessing of the data for the mainframe version is possible, including consistency checks. The mainframe version has an extended capability and includes design and optimzation modules. The main program versions are shown in Fig. 19.

In addition a modular connection to a sequential nonlinear programming module

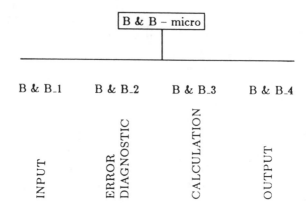

Figure 18. *Structure of B&B–micro.*

(NLPQL, [31], [32]) exists. All program versions possess the same principle structure, specific modules are added for dynamics, nonlinearities, design in reinforced concrete and optimization.

The version B&B–STAHL (steel design) will be discussed in the following.

B & B – LIN	linear elastic, small displacements
B & B – NONLIN	linear elastic, large displacements and linear stability
B & B – DYN	dynamically loaded structures
B & B – STAHL	optimization of steel structures
B & B – BETON	optimization of reinforced concrete structures
B & B – DYNOPT	optimization of dynamically loaded structures

Figure 19. *B&B program–versions.*

7.2 B&B–Stahl: Load Case Combination

The primary step in all typical civil engineering design applications is the static analysis. To this end, the Finite Element Method approach has proved itself to be a valuable tool.

In this section we introduce a design module coupled with a finite element program where the design module controles, in principle, the feasibility of the structural members.

In order to analyse general truss and/or framed structures using the finite element program the structure is discretized into elements. For the dimensioning of the structural members it is convenient to assemble the elements with identical structural function and approximately equal stress in 'groups of cross-sections.' An example for this kind of assembly into groups of cross-sections is shown in Fig. 20. In this way the number of variables of the resizing process is significantly reduced. The design module would then have to be accounted for with respect to groups of cross-sections only.

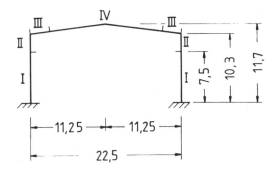

Figure 20. *Definition of various coss-section groups for a typical portal frame.*

Structural Loads

The structural loads are usually defined by service requirements of the building and the codes of practice (e.g. [55]). The codes define simple loads, the combination of which have to be considered by the designer to arrive at maximal values of stresses and deflections. Additional safety factors have to be taken into account.

All loads acting simultaneously are assembled in a load case. The designer has to define rules for the simultaneous action of different load cases. For each load case the stresses have to be evaluated at the control points of the cross-section for

394

defined locations of the structure. As a typical cross-section an I-beam is shown in Fig. 21.

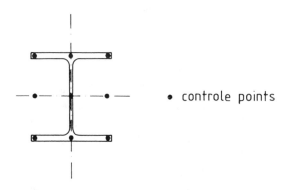

Figure 21. *Cross-section of an I–beam with control points.*

The critical combination of the load cases can be computed using an optimization algorithm. It is nessesary to consider the combination of load cases at every control point seperatly.

To determine the influence of the different load cases on the structure, a load factor α_i is defined for every load case i. The following conditions define the interdependencies of the load cases:

- upper and lower bounds

    The upper and lower bounds are defined by the safety factors and the function of the structure. In Table 3 the lower and upper bounds of the example are listed. Normaly the lower bound α_l depends on the type of load (e.g. zero for variable loads), and the upper bound α_u depends on the safety factors (see Eq. 17)

$$\alpha_l \leq \alpha_i \leq \alpha_u \tag{17}$$

    where

i : load case no. $i = 1, \ldots, n$

α : load case factor

n : no. of load cases

- compatibility condition

    The compatibility of different load cases can be defined using the condition

$$\alpha_i \alpha_j = 0 \qquad i \neq j \qquad (18)$$

where

i, j : load case no.

α : load case factor

This condition forces factor α_i to be zero for $\alpha_j \neq 0$, i.e., eliminates the load case i when j is present and vice-versa. This simulates, e.g., the dependencies of different wind load cases as defined in Fig. 22.

- condition of variable bounds

According to the German design codes the total sum of loads can be reduced when wind and snow loads act simultaniously on the structure. If the condition

$$\sum_{i=1}^{n} a_i \alpha_j + b_i \geq 0 \qquad (19)$$

where

i : load case

n : no. of load cases

α : load case factor

a, b : coefficients

is satisfied a redution in one of the load cases is permitted.

- condition of superposition

The German design codes differentiate two types or groups of loading 'H' and 'HZ' [56]. The loading type 'HZ' includes also the loads of load group 'H.' Fig. 22 and Table 3 demonstrate a typical arrangement of these two groups. The safety factors corresponding to these groups are different. The influence of these factors on the load case factors α are taken into account by the following condition:

$$-\left(\alpha_i^H - b\right)\left[\sum_{j=1}^{n}\left(a_j \alpha_j^Z - c_j\right)\right]/n \geq 0 \qquad (20)$$

where

396

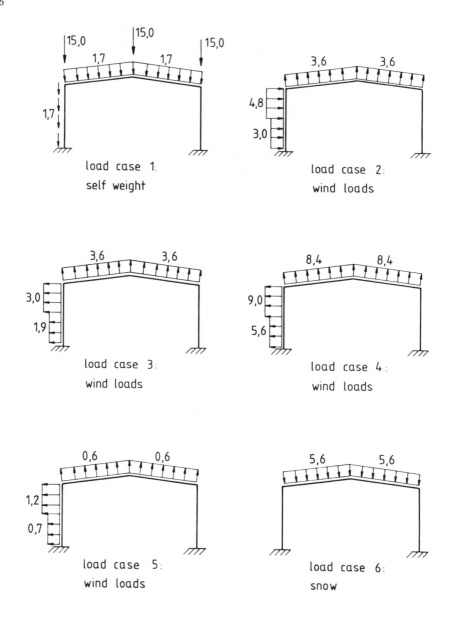

Figure 22a. *Example of structure subject to multiple loading cases, load case 1 to 6.*

397

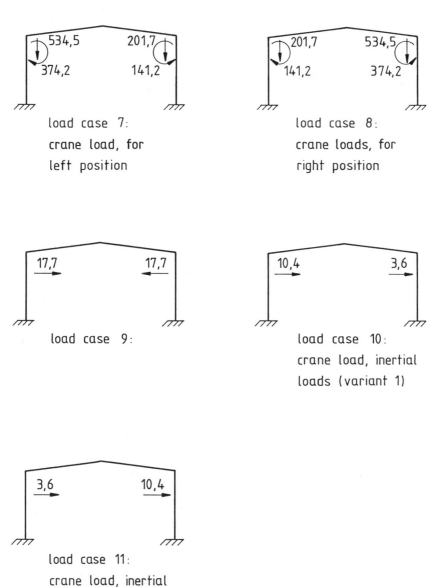

Figure 22a. *Example of structure subject to multiple loading cases, load case 7 to 11.*

Table 3. *Boundary conditions.*

No.	Equation	
1	17	$0.9 \leq \alpha_1 \leq 1.50$
2	17	$0 \leq \alpha_2 \leq 1.33$
3	17	$0 \leq \alpha_3 \leq 1.33$
4	17	$0 \leq \alpha_4 \leq 1.33$
5	17	$0 \leq \alpha_5 \leq 1.33$
6	17	$0 \leq \alpha_6 \leq 1.50$
7	17	$0 \leq \alpha_7 \leq 1.50$
8	17	$0 \leq \alpha_8 \leq 1.50$
9	17	$-1.33 \leq \alpha_8 \leq 1.33$
10	17	$-1.33 \leq \alpha_{10} \leq 1.33$
11	17	$-1.33 \leq \alpha_{11} \leq 1.33$
12	18	$\alpha_{10} \cdot \alpha_{11} = 0$
13	19	$-\alpha_7 - \alpha_8 + 1.50 \geq 0$
14	19	$-\alpha_2 - \alpha_3 - \alpha_4 - \alpha_5 + 1.33 \geq 0$
15	19	$-\alpha_2 - \alpha_3 - \alpha_4 - \alpha_5 - \alpha_6 + 2.01 \geq 0$
16	20	$-\alpha_1 \cdot (\alpha_2 + \alpha_3 + \alpha_4 + \alpha_5 + \alpha_9 + \alpha_{10} + \alpha_{11}) \geq 0$
17	20	$-\alpha_6 \cdot (\alpha_2 + \alpha_3 + \alpha_4 + \alpha_5 + \alpha_9 + \alpha_{10} + \alpha_{11}) \geq 0$
18	20	$-\alpha_7 \cdot (\alpha_2 + \alpha_3 + \alpha_4 + \alpha_5 + \alpha_9 + \alpha_{10} + \alpha_{11}) \geq 0$
19	20	$-\alpha_8 \cdot (\alpha_2 + \alpha_3 + \alpha_4 + \alpha_5 + \alpha_9 + \alpha_{10} + \alpha_{11}) \geq 0$

α_i^H : load case factor α of load case i, load group 'H'

α_j^Z : load case factor α of load case j, load group 'HZ'

a, b, c : coefficients.

The load case factor α^H is reduced to the coefficient b, when load case factor α^Z exceeds c.

The possibility to define this conditions is very flexible. It is easy to describe the bounds on the factors of the load cases.

For special applications additional conditions are nessesary.

- condition of equal sums

  The condition

$$\sum_{j=1}^{n} a_i \alpha_i + b = 0, \qquad (21)$$

where

α_i : load case factor of load case i

a, b : coefficients

forces the sum of the reactions to be constant.

- condition of equalness

  This condition fixes the load factor α of load case i

  $$\alpha_i + b = 0 \qquad (22)$$

Admissible Stress

The result of the structural analysis using the Finite Element Method are displacements and forces. The stresses are derived from the internal forces and are used to check the feasibility of the structure.

The rules for the computation of stresses are defined for many well defined types of cross-sections. Within the design module the normal stress, the shear stress and the stress invariants are evaluated at every point of interest. Fig. 21 shows the control points of a beam element with an I shaped cross section.

The shear stress and the stress invariants are treated according to their absolute values. The normal stresses are treated seperately depending on tension or compression.

The objective function of the optimization is defined as

$$\text{maximize} \quad F = \sum_{i=1}^{n} \alpha_i |S_i| \qquad (23)$$

where

α_i : load case factor of load case i

S_i : absolute value of stress (tension, compression, shear, or invariants)

n : no. of load cases

The objective function (23) sets to maximize the sum of stresses. These value depend on the load case factors α. The conditions (17) to (22) have to be satisfied.

The number of load combinations is very large, therefor it is advantageous to use methods of mathematical programming to find the critical load case. The linear objective function and the nonlinear conditions are charactaristics of this problem.

For the solution the algorithm NLPQL [31] is used. The gradients, necessary for this algorithm, are derived from the Eqs (17) to (23) with respect to the variables of optimization, i.e., the load case factors α.

The large number of control points increases the effort involved in finding the best combination of the load case factors α. It is advantageous to determine the load case combination only for regions of high stresses. A 'pre-optimization' provides the maximal stress F_v with respect to the conditions discussed. Eq. (24) shows the mathematical formulation of the pre-optimization process.

$$\left. \begin{array}{ll} \text{maximize} \quad F_v = \sum_{i=1}^{n} \alpha_i |S_i| & \\ \qquad \alpha_i = \alpha_{li} & \text{if} \quad S_i \leq 0 \\ \qquad \alpha_i = \alpha_{ui} & \text{if} \quad S_i > 0 \end{array} \right\} \qquad (24)$$

where

α_i : load bound of factor case i

α_{li} : lower bound of factor α_i

α_{ui} : upper bound of factor α_i

S_i : stress component of load case i

n : no. of load cases

The maximal stress F_v is an upper bound of the objective function (23). F_v are evaluated for every control point and sorted with respect to their values. The load case factors are determined sequentially starting with the maximal stress F_v. In this way control points with equal stresses in every load case could be treated one after another. It is then sufficient to determine the load case factors only once. If the value of the actual maximal stress F_v for a control point is less than the absolute maximum of the objective function F for this cross-section group and for this stress component, the computation can be stopped.

Figs. 20 to 23 show an application fo the design module discussed. The structure and its corresponding cross-section groups are shown in Fig. 20. Fig. 22 demonstrates the structural loading. The design coefficients defining the boundary conditions are summarized in Table 3. The results of the stress design module are presented in Table 4 and Fig. 23.

8   Thermal/Structural Interaction

A great number of practical engineering problems involve an interaction of two (or more) coupled fields requiring a simultaneous solution. This coupling inflicts a natural complexity on the nature of the solution process and is a drawback exceeding

○ Tension

× Compression

Figure 23. *Position of critical stress values.*

Table 4. *Results.*

Cross-section group	Point	Stress all stress	Load case factor 1	2	3	4	5	6	7	8	9	10	11
1 tension	1	0.4508	1.33	0.00	1.33	0.00	0.00	0.66	1.33	0.00	−1.33	0.00	1.33
compr.	2	0.8245	1.33	0.00	0.66	0.00	0.00	1.33	0.00	1.33	−1.33	0.00	1.33
2 tension	3	0.4278	1.33	0.00	0.66	0.00	0.00	1.33	1.33	0.00	−1.33	0.00	1.33
compr.	4	0.5557	1.33	0.00	0.66	0.00	0.00	1.33	1.33	0.00	−1.33	0.00	1.33
3 tension	5	0.5049	1.33	0.00	0.66	0.00	0.00	1.33	1.33	0.00	−1.33	0.00	1.33
compr.	6	0.6248	1.33	0.00	0.66	0.00	0.00	1.33	1.33	0.00	−1.33	0.00	1.33
4 tension	7	0.2912	1.33	0.00	0.66	0.00	0.00	1.33	0.00	0.00	−1.33	−1.33	0.00
compr.	8	0.3642	1.33	0.00	0.66	0.00	0.00	1.33	1.33	0.00	−1.33	0.00	1.33

the limits of most daily engineering applications. In fact in most engineering practices this interaction is ignored or at the cost of computational complexity—and reduced reliability—realized only for limited applications. Particular disadvantages become apparent if we notice that a number of 'uncoupled' solvers exist or could be realized with much less effort and more efficiency. Added to these are computational problems ranging from local programming problems such as computer storage limitations to more essential, physical problems such as the dimensional discrepancies of the fields involved.

These difficulties which obviously discourage a unique, fully integrated treatment of such coupled problems especially for large scale applications have been our main motivations in developping the proposed strategy which we shall briefly discuss in the next paragraphs.

Design Strategies

Let us review the problem from a practical point of view. Consider the example of Fig. 24.

Two plates of varying cross-sections, supported by a portal frame carry some arbitrary loads across them. The elements of the portal frame are of different cross sections. Assume we are interested in a weight (or cost) optimal design of the elements of the structure, namely the composite steel/concrete coloumns, the beams and the reinforced concrete plate itself. The structure is to be designed under a combination of multiple static dead and imposed loads along with a consideration of fire safety design (a minimum F90 design, say, is to be accounted for). Consider now, in view of the above example, the following two design strategies which we shall shortly explore and compare.

8.1   The Sequential Strategy

For each design (or optimization) loop two seperate computations must be performed in sequence. First the physical domain is discretized into elements and the time-dependent temperature field $T(X,t)$ is determined at n time-steps. These nodal temperatures should then be transformed into equivalent loads and superimposed on the other m variations of the mechanical loads resulting in $l(=m \times n)$ loading cases to be considered. A second F.E. analysis should then be carried out to determine the displacement field $D(Y,t)$. We could then proceed with the design and eventually with the search for a better or an optimal model. In fact, unless a direct element and nodal correspondance of the two meshes is preserved, i.e., unless the two are the same (which can only be obtained via a unique 3-D, solid element discretization of the system for both analyses), huge effort need be put into the "mapping" of the nodal temperatures across the displacement mesh.

Fig. 25 demonstrates a typical approach to the design of structures where both fields are of importance. After $T(X,t)$ is obtained a controle is made. If the results

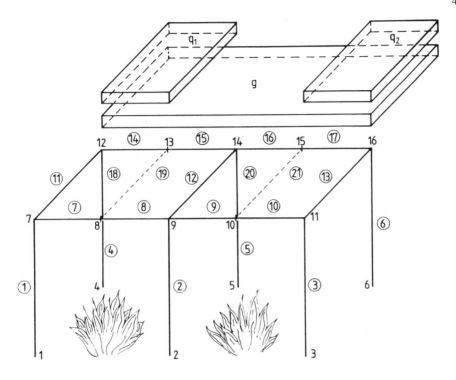

Figure 24. *Structure under combined thermal and mechanical load-*
ing with its F.E. displacement analysis discretization.

are satisfactory one can proceed with the design or optimization of the system
with respect to some criteria involving the temperature field. Having fulfilled such
criteria, we carry out the displacement analysis. Here also, after each displacement
analysis, we have the possiblity of resizing certain design variables. Each 'successful' iteration repeats the displacement F.E. analysis as an independant objective
function evaluation. An unsuccessful iteration, i.e., one which in spite of satisfying mechanical constraints, violates bounds on temperature analysis necessitates
a return to step 1 of the whole process. In fact, unless we are not interested in
an optimal shape of the structure after each iteration a return to step 1 of the
temperature analysis can not be avoided.

Although such a strategy merits from its simple data management (see Fig. 26),
it is by no means suitable for problems involving re-analysis or optimization of
structures. It is lengthy and requires a lot of CPU time and a fair amount of
computer storage. Another important disadvantage is that once we start with
the determination of an optimal solution as our goal, we are bound to proceed

404

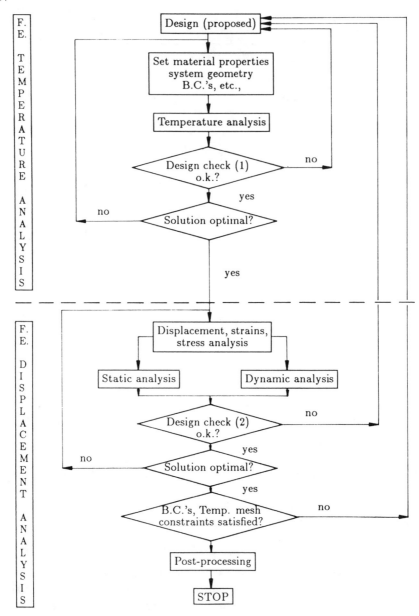

Figure 25. *Flow chart demonstrating a typical sequential strategy.*

stepwise and to repeat the whole process for the whole structure and for each iteration. We are in effect extremely limited by our strategy since it neglects our explicit knowledge of the interaction between different parts of the structure and prohibits a parallel handling of the design process and a consecutive reduction of the system still requiring modifications.

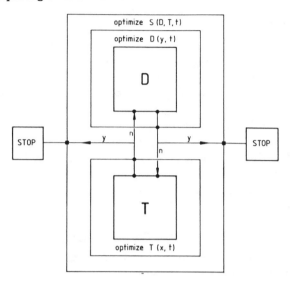

Figure 26. *Structure of data transfer in a sequential strategy.*

8.2 A Parallel Strategy

In principle two seperate computations could first be performed in parallel in order to arrive at the initial instationary temperature distribution (for the initial structure) and the initial displacement field under the combined influence of the static/dynamic loading and the vector of initial temperatures. The system discretizations could be treated completely independent of one another.

8.2.1 Sub-structuring, decomposition and the macro element concept
In this strategy the static/dynamic analysis and system design (or re-sizing) are always performed on a sub-structure level (e.g. cross-section groups could be our design variables). The initial temperature field is obtained on a micro (finite element) level where as all subsequent mapping of the two fields are carried out on a macro element level. The variations of the temperature mesh and adjustments of its boundary conditions as well as the determination of the updated temperature distribution are done on macro element level using decomposition algorithms wherever the macros share a boundary (iterating alternate Neuman-Dirichlet boundary

conditions until convergence is achieved) [54]. The following three paragraphs serve clarifying our definitions:

8.2.1.1. *Sub-structuring* refers principally to the partitioning of the whole structure into smaller parts, upon which we investigate the same global properties (functions) as those of the whole structure but to which we can also impose our explicit knowledge of their behaviour. A finite element is therefore the lowest level sub-structure conceived in our terminology.

8.2.1.2. *Decomposition* permits the reduction of the search for a global solution (function) across the whole domain of interest to a search for a set of admissible solutions across subdomains. They are admissible in the sense that they fulfill the general requirements of completeness and compatibility of the solution respecting also the boundary conditions. It might be argued that this definition is the same as that of the sub-structure. In a sense this is not completely mistaken. However, we stress that they are not exactly the same since the latter permits a parallel, i.e, an independent processing of the sub-problems where as the first requires a sequential consideration. Moreover, sub-structuring is an explicit process, where as having satisfied certain rqeuirements a structure could be implicitly decomposed often in an optimal manner.

8.2.1.3. *Macro elements* are implicit sub-structures. They are sub-regions of the system over which, after some preliminary manipulation, physical or geometrical similarities, implicit rules, or functions of space and time, can be logically or numerically deduced. These functions can then be used again and again, up to some degree of tolerance or sensitivity, for further analysis. They can be regenerated at any time, stored on some data bank and restored when necessary. They can be updated if new information are at hand (note that the macro elements' geometry could vary but not its micro element arrangements). The macros are never explicitly evaluated in terms of polynomials or else but simply evaluated and stored as time-history of earlier stages of analysis. After each iteration we shall observe a reduction in the number of macros still needing a thorough analysis or modifications since each iteration or updating provides additional data which render the interpolations more precise and the macros more independent of the neighbouring ones.

8.2.2 Parallel processing procedure The above three definitions, a computer (preferably a multi-processeor), two reliable, interacting, compatible and modular F.E. solvers [45], [51] accompanied by various design [46], pre- and postprocessor modules [49], [50] and an optimizer [31] have proved to be sufficient tools for developing and exploring our strategy. Keeping the example of Fig. 24 in mind we are now in positition to give a more thorough explanation of the major two steps mentioned above (see Figs. 27, 28).

8.2.2.1. *Step (1) – Temperature analysis* having proposed a reasonable design, we first guess our sub-structures, i.e., we define regions of the system where we

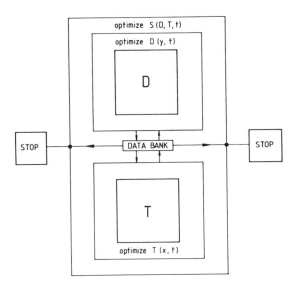

Figure 27. *Structure of data transfer in a parallel strategy.*

expect similarities to occur, e.g., beams of similar cross sections, plates of equal thickness/width ratio, etc. (see Fig. 29).

They may also correspond to the same structural element differently dimensioned for two proposed designs to choose from (or to derive derivatives or sensitivities from) in the subsequent search for the best design. Then we discretize our temperature field. This can be done completely independent of the displacement field discretization. In most cases a 2-D analysis of our sub-structure is sufficient (c.f. the 3-D displacement analysis). To each sub-structure, we then associate an appropriate number of different time-dependent loadings (multiple possible temperature variations, heat sources, convection, etc.). Note that a sub-structure considered to be subjected to two different loading cases might in reality correspond to two different parts of the structure under the same loading conditions but dimensioned differently.

Next an instationary F.E. analysis of all sub-structures, under all loading cases is performed simultaneously. If the results are satisfactory the whole system and the results are implicitly further divided into temperature-space-time macro elements.

A macro may include information concerning different parts of different sub-structures at different times. If we now require an optimal temperature distribution with respect to temperature domain variables (this is optional but advantageous as we shall see later) where multiple function evaluations are necessary all we need

408

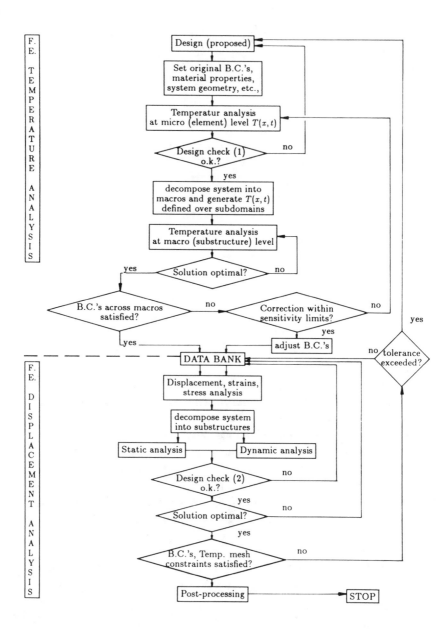

Figure 28. *Flow chart demonstrating the proposed parallel strategy.*

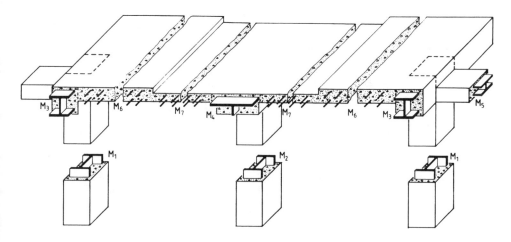

Figure 29. *2-D Macro element arrangements of the section to be discretized for the F.E. temperature analysis.*

to do is to go back to our macro elements and interpolate or extrapolate the information we need without having to perform a complete new F.E. analysis (since we have also gathered sensitivity information via our different sub-structures relating to the same structural element). We have also overcome the problem of continuity since the macros are implicit continuous functions of space and time (c.f. piecewise function defined over the micro elements) and need no more correspond to the nodes of the original discretization or be strictly limited by the its boundaries. Otherwise we would have to handle discrete sub-structures at discrete instances of time which would have severe disadvantages for the optimization process.

8.2.2.2. Step (2) – Stress Analysis we then proceed to the next step and attempt to determine the displacement field. For static problems it is again suitable to define new sub-structures. Neither the new sub-structures nor the F.E. mesh itself need to correspond (or map) exactly to the temperature field discretization. For example if we intend to find a minimum weight design of the structure (which corresponds to an optimal shape design of the temperature field), we may find it convenient to define our sub-structures as regions of equal cross section or reinforcement, etc. (which could also be defined as our design variables). We may do so ignoring temporarily any correspondance with the temperature analysis. We may also drop certain temperature field sub-structures or macros which were only essential to our temperature analysis like parts which were twice dimensioned to derive sensitivities or parts which only affect the temperature field and have no mechanical functions (isolating material, etc.). Only at the time of definition of structural loading, and further, only for thermally induced loads need we consider a "mapping" of both meshes. Here, our temperature macro elements exhibit a secondary function. They become, to our advantage, our so-called "mapping functions" extending firstly across the desired sub-structure (or cross section group), and mapping the geometry of both meshes, and secondly defining temperature functions of space (for various positions defined at a particular load case) and of time (for various load cases defined at a particular instance) within a cross section group (see Table 5).

Table 5. *Correspondance between the nodes of the displacement mesh, the mechanical loading and the thermal loading (temperature macros).*

Nodes	Load Vector Contribution
1, 4	M_1
2, 5	M_2
3, 6	M_1
7, 12	$M_1, M_3, M_5, M_6, q_1, g$
8, 13	M_6, M_7, q_1, g
9, 14	M_2, M_4, M_7, g
10, 15	M_6, M_7, q_2, g
11, 16	$M_1, M_3, M_5, M_6, q_2, g$

For dynamic problems we perform, prior to substructuring, a decomposition of the problem [47] and then proceed in a similar manner allowing the mapping functions to extend still further as temperature functions of space and of time (for various load cases defined over a period of time at a frequency equal to the frequency of the temperature variations).

Our next step is to determine the displacement field. We controle the displacements (or strains or stresses). If an optimal solution is required we call again our optimizer and iterate inside the displacement analysis loop. Having reached an optimal solution, we controle if any violation of the temperature field has occured. If so, we have the possibility to interpolate and to update, naturally within the sensitivity limits, the information we need and carry out the optimization procedure again. Only if these limits are severely exceeded need we return to step 1.

In practice we often observe that only certain constraints, i.e., certain macros remain troublesome and not the temperture field as a whole. This would have specially been the case, had we obtained an optimal temperature distribution field since this has the effect of 'smoothing' our mapping functions and stabilizing any steep variations of temperature across the macros, making them less sensitive to variations. If we should return to step 1 we could use the information we have gathered so far and sequentially reduce the size of our analysis. We would have in effect a reduced temperature analysis to fulfill in the sense that we can formulate now the macros as sub-problems, condense those which have been satisfactory out of the equation system and threat them as imposed domain and boundary conditions. These would then act as loading for the parts still requiring modification and/or evaluation wich we treat using decomposition algorithms mainly due to their effectiveness in conjunction with parallel processing.

A more detailed presentation of the above strategy could be find in [52], [53] where practical applications ranging from the optimal design of reinforced concrete structures subject to atmospheric temperature variations to design of structural composite members under fire are discussed. We find it particularly advantageous especially where optimal solutions of large scale structures of major importance. Various design criteria could also be tested with only modifications to the concepts proposed.

9 Success of Structural Optimization and Possibilities of Future Applications

The success of structural optimization is hard to quantify and there are few attempts in that direction. The question what can be gained compared with a conventional design can only be answered for such types of structures, which have been frequently designed by various engineers or for serial products with a long design tradition. In both cases it can be assumed that a 'good' design exists for comparison.

A most interesting result in this field is given by Baldur [36]: A great number of existing overhead travelling cranes was redesigned by optimization techniques. The existing cranes were all sold on a competitive market and therefore a reasonable good design can be assumed. The redesign was based on the same structural principles (box girder) so that the cost of fabrication could be expressed by a weight function. The result of this comparison confirms the initial remarks on possible savings (Fig. 30).

An average saving of 15% was obtained and in some cases even reductions of 35%.

With regard to serial products it can be expected that the saving is less dramatic. In an optimization of thin–walled steel sections (trapezoidal roof and wall panels, purlins), which was carried out by one of the authors [48], an average saving of 5% was obtained. However, with regard to the enormous amount of steel used for these products, this saving is quite substantial. Applications of this kind sometimes require techniques not discussed here, e.g. parametric solutions, sensitivity

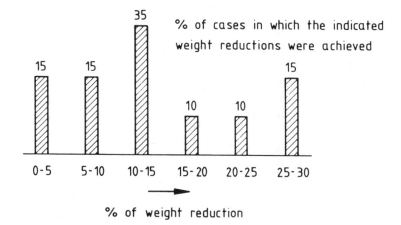

Figure 30. *Results of a systematic study of overhead traveling cranes according to Baldur [36].*

studies and Pareto–optimization [34].

Without any doubt the structural optimization will become a widely used instrument in the future; the authors believe that this development will impose high demands on practicing engineers: a profound understanding of the theoretical background paired with good knowledge of the structural principles and a responsible use will be necessary in order to avoid severe drawbacks in design practice.

References

[1] Morris, A. J. (Ed.), (1982), *Foundations of structural optimization: A unified approach*, John Wiley and Sons, Chichester.

[2] Gellatly, R., (1973), 'Survey of the state-of-the-art of optimization technology within NATO countries,' *AGARD Conference Proceedings*, AGARD–CP–123, Milan.

[3] Michell, A. G. M., (1904), 'The limits of economy of material in frame-structures,' *Phil. Mag.* **8**, 47, 6, (November).

[4] Prager, W., (1981), 'Unexpected results in structural optimization,' *J. Struct. Mech.* **9**, 1, 71–90.

[5] Rozvany, G. I. N., (1976), *Optimal design of flexural systems*, Pergamon Press, Oxford.

[6] Ziegler, H. (1958), 'Kuppeln gleicher Festigkeit,' Ingenieur-Archiv, Band XXVI.

[7] Morley, C. T. (1966), 'The minimum reinforcement of concrete slabs,' *Int. J. Mech. Sci.* **8**, 305–319.

[8] Drucker, D. C. and Shield, R. T., (1957), 'Bounds on minimum weight design,' *Quart. Appl. Math.*, XV, 3, 269-281.

[9] Prager, W. and Shield, R. T., (1967), 'A general theory of optimal plastic design,' *Trans. ASME*, March, 184–186.

[10] Massonnet, C. H., Olszak, E. and Phillips, A., (1979), *Plasticity in structural engineering, fundamentals and applications*, Springer, CISM Udine.

[11] Prager, W. and Rozvany, G. I. N., (1977), 'Optimal layout of grillages,' *J. Struct. Mech.* **5**, 1.

[12] Save, M. A., (1975), 'A general criterion for optimal structural design,' *JOTA* **15**, 1, 119–129, (January).

[13] Melosh, R. J., (1969), 'Convergence in fully–stressed designing,' *AGARD Conf. Proc. No. 36*, AGARD–CP–36–70, Istanbul, (October).

[14] Gallagher, R. H., (1973),''Fully stressed design,' in *Optimum structural design*, R. H. Gallagher, and O. C. Zienkiewicz (Eds), John Wiley and Sons, London.

[15] Gellatly, R. A. and Berke, L., 'Optimality–criterion–based algorithms,' ibid.

[16] Artek, E., et al. (Eds.), (1984), *New directions in optimal structural design*, John Wiley and Sons, Chichester.

[17] Venkayya, V. B., (1971), 'Design of optimum structures,' *Computers and Structures* **1**, 1/2, 265–309.

[18] Lipp, W., (1980), 'Ein Verfahren zur optimalen Dimensionierung allgemeiner Fachwerkkonstruktionen und ebener Rahmentragwerke,' Techn. wiss. Mitteilungen Nr. 80–3, Ruhr–Universität Bochum, Bochum.

[19] Khot, N. S., (1981), 'Algorithms based on optimality criteria to design minimum weight structures,' *Eng. Optimization* **5**, 73–90.

[20] Gellataly, R. A. et al., (1974), 'OPTIM II, A MAGIC–compatible large scale automated minimum weight design program,' AFFDL–TR–74–97.

414

[21] Hadley, G., (1964), *Nonlinear and dynamic programming*, Addison–Weslex, Reading, Mass.

[22] Wilkinson, J. *et al.*, (1977), 'FASTOPT–A flutter and strength optimization program for lifting–surface structures,' *J. Aircraft* **14**, 6, 581–587, (June).

[23] Kiusallas, J. and Reddy, G. B., (1977), 'DESAP – a structural design program with stress and displacement constraints,' NASA CR–2794.

[24] Thierauf, G. and Topcu, A., (1975), 'Structural optimization using the force method,' *World Congress FEM Struct. Mech.*, Bournemouth, October.

[25] Lipp, W. and Thierauf, G., (Sept. 1976), 'The role of the force- and displacement–method for the optimization of structures with the Langrangian–multipier–technique,' *IASBE 10th Congress*, Tokyo, September.

[26] Gellatly, R. A. and Thom, R. D., (1979), *Force method optmization*, Rep. No. D2530–953005, Bell Aerospace Textron, (December).

[27] Rozvany, G. I. N., (1984), 'Structural layout theory—the present state of knowledge,' in *New directions in optimum structural design*, E. Atrekt (Ed.), John Wiley and Sons, Chichester.

[28] Thierauf, G., (1978), 'A method for optimal limit design of structures with alternative loads,' *Comp. Meth. Appl. Mech. Eng.* **16**, 135–149.

[29] Pape, G., (1979), 'Eine quadratische Approximation des Bemessungsproblems idealplastischer Tragwerke,' Dissertation, Universität Essen.

[30] Pape, G. and Thierauf, G., (1981), 'A quadratic approximation of a nonlinear structural design problmen,' in *Physical non–linearities in structural design analysis*, J. Hult and J. Lemaitre (Eds.), Springer.

[31] Schittkowski, K., (1983), 'On the convergence of a sequential quadratic programming method with an augmented Lagrangian line search function,' *Math. Operationsforschung und Statistik, Ser. Optimization* **14**, 2, 197–216.

[32] Hock, W. and Schittkowski, K., (1983), 'A comparative evaluation of 27 nonlinear programming codes,' *Computing* **30**, 335–358.

[33] Vanderplaats, G. N., (1986), 'Numerical optimization techniques,' *Nato–ASI, Computer Aided Optimal Design–Structureal and Mechanical Systems*, M. Soares (Ed.), Troia, Portugal.

[34] Eschenauer, H. (Ed.), (1984), *Rechnerische und experimentelle Untersuchung zur Strukturoptimierung von Bauweisen*, Institut für Mech. und Regelungstechn., Siegen.

[35] Kneppe, G., 'Methode der sequentiellen Linearisierung,' SEQLI, ibid.

[36] Baldur, R., (1972), 'Structural optimization by inscribed hyperspheres,' *Proc. Am. Soc. Civ. Eng., J. Eng. Mech. Div.*, EM3, 503–518, (June).

[37] Fox, R. L., (1971), *Optimization methods for engineering design*, Addison-Wesley, Reading, Mass.

[38] Collatz, L. and Wetterling, W., (1966), *Optimierungsaufgaben*, Springer, Berlin.

[39] Schrader, K.-H., (1978), 'MESY – Einführung in das Konzept und Benutzeranleitung für das Programmsystem MESY–MINI,' Techn. wiss. Mitteilungen Nr. 78–11, Ruhr-Unversität-Bochum, Bochum.

[40] Morris, A. J., (1984), 'Structural optimization systems,' in *New directions in Optimum structural design*, E. Atrek (Ed.), John Wiley and Sons, Chichester.

[41] Harary, F., (1972), *Graph theory*, Addision-Wesley, Reading, Mass.

[42] Kirsch, U., (1981), *Optimum structural design*, McGraw Hill, New York.

[43] Steward, D. V., (1965), 'Partitioning and tearing systems of equations,' *J. Siam Num. Anal., Ser. B* **2**, 2, 345–365.

[44] Hörnlein, H. R. E. M., (1986), 'Take–off in optimum structural design,' *NATO ASI, Techn. Univ. Lisbon*, Troia, Portugal, 1, 205–234, (June).

[45] Thierauf, G. (Ed.), (1985), B&B—Programmsystem zur Berechnung und Bemessung allgemeiner Tragwerke, Universität Essen, Essen.

[46] Booz, W., (1984), 'Zweistufenoptimierung von Stahlbetontragwerken mit Hilfe der sequentiellen quadratischen Programmierung,' Dissertation, Universität Essen.

[47] Booz, G., (1986), 'Eine Dekompositionsmethode zur optimalen Bemessung von Tragwerken unter dynamischer Belastung,' Dissertation, Universität Essen.

[48] Thierauf, G., 'Optimierung von Stahltrapezprofilen,' *unveröffentlicht*.

[49] Jüdt, B., (1989), 'PREFEM – A General Purpose F.E. Pre-processor,' Universität–GHS–Essen, Fachbereich Bauwesen, Baumechanik/Statik.

[50] Jüdt, B. and Kayvantash, K., (1989), 'POSTFEM—A General Purpose F.E. Post-processor,' Universität–GHS–Essen, Fachbereich Bauwesen, Baumechanik/Statik.

[51] Kayvantash, K., (1989), 'B&B_T – Program For the Analysis and Design of Structures under Temperature Loading,' Universität–GHS–Essen, Fachbereich Bauwesen, Baumechanik/Statik.

[52] Kayvantash, K., (1989), 'Optimal Design of Structures Subject to Simultaneous Mechanical and Thermal Loading,' Doctorate Thesis, Universität–GHS–Essen, Fachbereich Bauwesen, Baumechanik/Statik, to be published.

[53] Kayvantash, K. and Zacharias, A., (1988), 'B&B/B&B_T: A Program for the Analysis, Design and Optimization of Structures,' *Proceeding of 2nd Cryogenic Technology Review Meeting held at DFVLR Köln–Porz.*

[54] Schwarz, H. A., (1869), 'Über einige Abbildungsaufgaben,' *Ges. Math. Abh.* **11**.

[55] DIN 1055—Design loads for buildings, Beuth-Verlag, Köln.

[56] DIN 18800, Part 1 (1981), Steelstructures—Design and construction, Beuth-Verlag, Köln.

Two-Stage Optimal Design in Concrete

W. Booz and G. Thierauf

Baumechanik/Statik
Universität-GHS-Essen
Essen
Federal Republic of Germany

Abstract A computer program for the analysis and optimal design of reinforced concrete structures is presented. The program is based on a nonlinear optimization (SQP) along with a FE-solver for linear and nonlinear analysis of structures. In order to achieve at a cost minimal design two optimization problems are formulated: one on a global (structural) level and another one on a local (cross-sectional) level. Two formulations of the above problems are presented.

1 Introduction

The solution of optimization problems using an optimization algorithm and a finite element program for linear and nonlinear analysis of structures [1], designed primarily for this purpose will be discussed in the following. In this program every finite element is assigned to a so-called cross-section group. Elements belonging to such a group all share the property of having the same cross-sectional dimensions and the same reinforcement in concrete design.

These dimensions and reinforcement areas constitute the design variables of the optimization problem.

2 Global Problem (PG1)

For the design of reinforced concrete structures it is preferable to minimize the total costs [2] (or an ideal volume) instead of the total weight as in steel structures since self weight is here of major importance.

The variables are then the concrete section dimensions $\underline{y}^c \in R^{nc}$ and the reinforcement areas $\underline{y}^s \in R^{ns}$, which can be combined in a vector $\underline{y} \in R^{nc+ns}$. Assuming

B. H. V. Topping (ed.),
Optimization and Artificial Intelligence in Civil and Structural Engineering, Volume I, 417–435.
© 1992 *Kluwer Academic Publishers. Printed in the Netherlands.*

418

for every cross-section group Q_i the ratio k_i^1 of the specific cost of steel to the specific cost of concrete and k_i^2 the ratio of the cost of formwork to the specific cost of concrete [3], the objective function can be formulated in the following manner:

$$\min\left\{ Z(\underline{y}) = \sum_{i=1}^{nq} \sum_{j\in I_i^e} \left[V_j(\underline{y}) + (k_i^1 - 1)V_j^s(\underline{y}) + k_i^2 A_j^f(\underline{y})\right]\right\} \qquad (2.1)$$

where

$\quad nq$: number of cross-section groups,
$\quad I_i^e$: set of elements belonging to cross-section group Q_i,
$\quad V_j$: volume of element j,
$\quad V_j^s$: volume of steel of element j,
$\quad A_j^f$: area of formwork of element j.

The following constraints can be considered:

- lower and upper bounds of the variables

$$y_i^- \le y_i \le y_i^+ \qquad\qquad i = 1, \ldots, nc + ns \qquad (2.2)$$

- constructive or logical constraints

$$g_j^y(\underline{y}) = \sum_k c_k \prod_l y_l + c_j \gtreqless 0 \quad \text{for } l \in \{1; 2; \ldots; n_c + n_s\} \qquad (2.3)$$

Equations (2.3) include the bounds of the reinforcement ratios.

- constraints on the displacements $\underline{r}$

$$g_j^r(\underline{y}) = {}_{\text{perm}}r_j - r_j \ge 0 \qquad j = 1, \ldots, nr \qquad (2.4)$$

- constraints on the shear stresses $\underline{\tau}$

$$g_j^\tau(\underline{y}) = {}_{\text{perm}}\tau_j - \tau_j \ge 0 \qquad j = 1, \ldots, nd * nl \qquad (2.5)$$

- constraints on failure under bending and axial force in linear and nonlinear analysis (w.r.t. German design codes [4])

$$g_j^f(\underline{y}) = z^* - 1 \ge 0 \qquad\qquad j = 1, \ldots, nd * nl \qquad (2.6)$$

$\quad$ where $\quad nd$: number of control points,
$\qquad\qquad\quad nl$: number of loading cases.

One gets z^* as solution of an optimization problem (PL1) on cross-section level, i.e.,

$$\max \left\{ z(\underline{v}_1) = \frac{M^f(\underline{v}_1)}{f(\underline{v}_1)M^a} \right\} \tag{2.7}$$

where $\quad \underline{v}_1^T = [e_{c1}; e_{s2}; \delta] \quad$: are three parameters necessary for a unique description of the plane of strains (Fig. 1)

with e_{c1} : minimum concrete strain,

e_{s2} : maximum steel strain,

δ : angle between the neutral axis and th applied moment's axis,

M^f : moment at failure,

$f(\underline{v}_1)$: safety factor depending on maximum steel strain in linear analysis (see [4]) with $1.75 \leq f(\underline{v}_1) \leq 2.1$,

M^a : applied moment.

The lower and upper bounds of the strains result from the gouverning material laws of steel and of concrete:

$$\left. \begin{array}{l} e_{c1}^- \leq e_{c1} \leq e_{c1}^+ \,, \\ e_{s2}^- \leq e_{s2} \leq e_{s2}^+ \,, \\ -\pi \leq \delta \leq +\pi \,. \end{array} \right\} \tag{2.8}$$

For example in linear analysis the bounds are prescribed by [4]:

$$e_{c1}^- = -0.0035 \,,$$
$$e_{s2}^+ = 0.0050 \,.$$

For mechanical reasons—a constant stress distribution must accompany a constant strain—two additional inequality constraints $g_1(\underline{v}_1)$ and $g_2(\underline{v}_1)$ result from the material laws. These constraints become active at low moments and high axial force level.

In addition, two equilibrium conditions in form of equality constraints are to be satisfied:

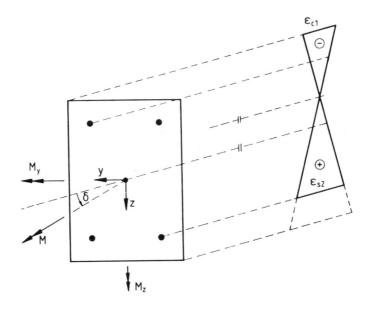

Figure 1. *Moments and strains.*

$$g_3(\underline{v}_1) = z(\underline{v}_1)N^f(\underline{v}_1) - f(\underline{v}_1)N^a = 0 \qquad \text{equilibrium of axial forces,}$$

$$g_4(\underline{v}_1) = \qquad M_\perp^f(\underline{v}_1) \qquad = 0 \qquad \text{equilibrium of moments about the axis perpendicular to the applied moment.} \Bigg\} (2.9)$$

A possible solution of (PL1) is shown in Fig. 2. The relation between problem (PG1) and problem (PL1) is shown in Fig. 3.

3 Reduced Problem (PG2)

The global optimization problem (PG1) can be reduced

- by searching minimum of the constraints for all loading cases and for all control points of each cross-section group, although here the constraints might become undifferentiable,

- by an active set strategy [5], where during each iteration step only the active constraints are considered,

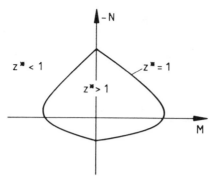

Figure 2. *Example of failure surface under bending and axial force.*

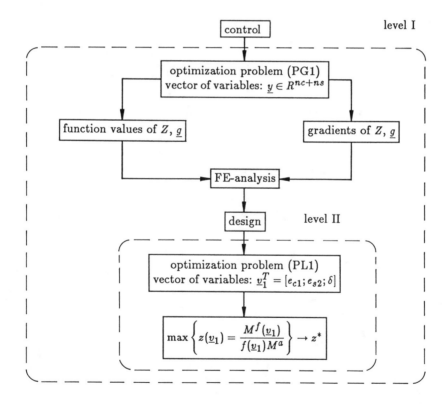

Figure 3. *Problems (PG1) and (PL1).*

- by separating the variables into independent (concrete section dimensions) and dependent ones (reinforcement). This is an approximation only in physically nonlinear analysis.

Thus the minimization of the steel volume can be reduced to a set of independent inferior problems (PL2) with same hierarchie at cross-section level where the dimensions of the concrete sections remain constant (see [6]).

The reduced problem (PG2) on structural level would then become:

$$\min\left\{Z(\underline{y}) = \sum_{i=1}^{nq}\sum_{j\in I_i^c}\left[V_j(\underline{y}) + (k_i^1 - 1)V_j^s(\underline{y}) + k_i^2 A_j^f(\underline{y})\right]\right\} \tag{3.1}$$

where $\quad \underline{y} = \underline{y}^c \in R^{nc}$

subject to

- lower and upper bounds of the variables (see (2.2))

$$y_k^- \le y_k \le y_k^+ \qquad\qquad k = 1, \ldots, nc = \sum_{i=1}^{nq} m_i \tag{3.2}$$

where $\quad m_i$: number of variable concrete section dimensions in group Q_i (Fig. 4)

- constructive or logical constraints (see (2.3))

$$g_j^y(\underline{y}) = \sum_k c_k \prod_l y_l + c_j \overset{\ge}{\underset{=}{}} 0 \qquad \text{for } l \in \{1; 2; \ldots; nc\} \tag{3.3}$$

- constraints on the displacements $\underline{r}$ (see (2.4))

$$g_j^r(\underline{y}) = {}_{\text{perm}}r_j - r_j \ge 0 \qquad\qquad j = 1, \ldots, nr \tag{3.4}$$

- constraints on the shear stresses $\underline{\tau}$ (see (2.5))

$$g_i^\tau(\underline{y}) = {}_{\text{perm}}\tau_i - {}_{\text{max}}\tau_i \ge 0 \qquad\qquad i = 1, \ldots, nq \tag{3.5}$$

- constraints on the reinforcement ratios $\underline{\mu}$ (see (2.3))

$$g_i^\mu(\underline{y}) = {}_{\text{perm}}\mu_i - {}_{\text{max}}\mu_i \ge 0 \qquad\qquad i = 1, \ldots, nq \tag{3.6}$$

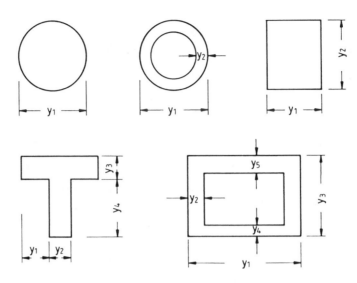

Figure 4. *Examples of cross-sections of one-dimensional elements with possible variables.*

where the maxima are searched for all loading cases and all control points in group Q_i.

On cross-section level the problem (PL2) is to minimize the steel area:

$$\min \left\{ z(\underline{v}_2) = \sum_{j=1}^{n} A_j^s \right\} \tag{3.7}$$

where $\underline{v}_2^T = \left[\underline{v}_1^T; A_1^s; A_2^s; \ldots; A_n^s \right] \in R^{3+n}$

The variable steel areas A_j^s are bounded from below by the available A_j^s, i.e. $A_j^{s-} \leq A_j^s$ for $j = 1, \ldots, n$. The upper bounds are defined by the reinforcement ratios of the superior problem (PG2). The three parameters defining the plane of strains are the same as those of problem (PL1). In addition to the two inequality constraints $g_1(\underline{v}_2)$ and $g_2(\underline{v}_2)$ (see problem (PL1)), three equilibrium conditions are to be fulfilled (see Fig. 1):

$$\left. \begin{array}{ll} g_3(\underline{v}_2) = N(\underline{v}_2) - f(\underline{v}_1)N^a & = 0 \quad \text{equilibrium of axial forces,} \\ g_4(\underline{v}_2) = M_y(\underline{v}_2) - f(\underline{v}_1)M_y^a \\ g_5(\underline{v}_2) = M_z(\underline{v}_2) - f(\underline{v}_1)M_z^a \end{array} \right\} = 0 \quad \text{equilibrium of moments,} \right\} \tag{3.8}$$

424

where $N(\underline{v}_2), M_y(\underline{v}_2), M_z(\underline{v}_2)$ are inner forces which result from integrating the stresses of concrete $\sigma^c(\underline{v}_1)$ and those of steel $\sigma^s(\underline{v}_1)$ over the cross-section.

Other constraints on the reinforcement (see (2.3)) can also be included. For example the constraint

$$g_6(\underline{v}_2) = -\sum_{j\in I_c^s} A_j^s + \sum_{k\in I_t^s} A_k^s \geq 0 \tag{3.9}$$

where I_c^s : set of indices of reinforcement in compression,

$\quad\quad I_t^s$: set of indices of reinforcement in tension,

requires the compression steel not to exceed the tension steel.

Comparing Fig. 3 and Fig. 5 the differences between problem (PG1) with inferior problem (PL1) and reduced problem (PG2) with inferior problem (PL2) become evident.

4 Solution of the Optimization Problems

In the solution of problems discussed before the following requirements should be fulfilled:

- global convergence, in order to arrive always at a permissible solution,

- few function evaluations as possible, since each necessiates a complete loop over the FE-analysis and the design in (PG1) or (PG2) respectively,

- non-sensitivity with respect to numerical and mechanical approximations involved in the calculation of gradients.

A SQP algorithm [5] satisfies all the above. Therefore all problems discussed here are solved using this algorithm.

5 Calculation of Gradients

5.1 Linear Analysis

The basic equations of displacement method are:

$$\underline{K}_e(\underline{y}^c)\underline{r} = \underline{R} - \underline{J}\underline{S}_R(\underline{y}^c) \tag{5.1}$$

where $\underline{K}_e(\underline{y}^c)$: elastic stiffness matrix,

$\quad\quad \underline{r}$: nodal displacements,

$\quad\quad \underline{R}$: nodal forces,

$\quad\quad \underline{J}$: incidence matrix,

$\quad\quad \underline{S}_R(\underline{y}^c)$: nodal equivalent loads (e.g. self weight).

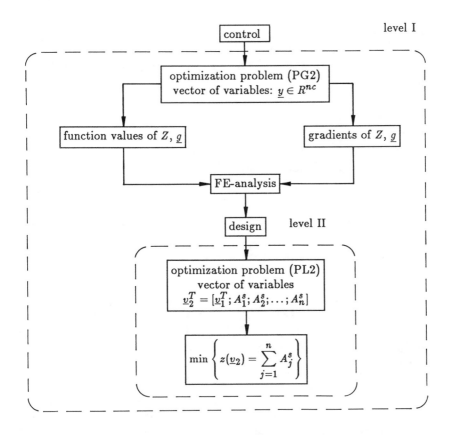

Figure 5. *Problems (PG2) and (PL2).*

One gets the stiffness matrix by formal multiplication

$$\underline{K}_e(\underline{y}^c) = \underline{J} \text{ diag } \left\{ \underline{k}_e(\underline{y}^c) \right\} \underline{J}^T \tag{5.2}$$

where diag $\{\underline{k}_e(\underline{y}^c)\}$ denotes the hyper diagonal matrix of element stiffness matrices. By partial differentiation of (5.1) with respect to $\underline{y}^c$ one obtains

$$\frac{\partial}{\partial \underline{y}^c} \left[\underline{K}_e(\underline{y}^c)\underline{r} - \underline{R} + \underline{J}\,\underline{S}_R(\underline{y}^c) \right] = \frac{\partial}{\partial \underline{y}^c}\underline{K}_e\underline{r} + \underline{K}_e\frac{\partial}{\partial \underline{y}^c}\underline{r} + \underline{J}\frac{\partial}{\partial \underline{y}^c}\underline{S}_R = \underline{0} \tag{5.3}$$

and after transformation and substitution by (5.2)

$$\underline{K}_e\frac{\partial}{\partial \underline{y}^c}\underline{r} = -\underline{J}\left[\frac{\partial}{\partial \underline{y}^c} \text{ diag } \{\underline{k}_e\}\underline{J}^T\underline{r} + \frac{\partial}{\partial \underline{y}^c}\underline{S}_R \right] \tag{5.4}$$

which can be considered as additional loading cases and which can be solved with already factorized $\underline{K}_e(\underline{y}^c)$.

Using this direct method we can derive analytically the gradients of the displacement constraints $\underline{g}^r$ (see (2.4) and (3.5)):

$$\frac{\partial}{\partial \underline{y}^c} \underline{g}^r = \underline{K}_e^{-1} \underline{J} \left[\frac{\partial}{\partial \underline{y}^c} \operatorname{diag} \{\underline{k}_e\} \underline{J}^T \underline{r} + \frac{\partial}{\partial \underline{y}^c} \underline{S}_R \right]. \tag{5.5}$$

Here the adjoint method (e.g. [7]) could be used alternativly.

With solution $\underline{r}$ of (5.1) the element nodal forces can be written as

$$\underline{S}(\underline{y}^c) = \operatorname{diag} \{\underline{k}_e(\underline{y}^c)\} \underline{J}^T \underline{r} + \underline{S}_R(\underline{y}^c). \tag{5.6}$$

Partial derivation of (5.6) yields

$$\frac{\partial}{\partial \underline{y}^c} \underline{S} = \frac{\partial}{\partial \underline{y}^c} \operatorname{diag} \{\underline{k}_e\} \underline{J}^T \underline{r} + \frac{\partial}{\partial \underline{y}^c} \underline{S}_R + \operatorname{diag} \{\underline{k}_e(\underline{y}^c)\} \underline{J}^T \frac{\partial}{\partial \underline{y}^c} \underline{r}. \tag{5.7}$$

Thus we get an analytical form of $\frac{\partial}{\partial \underline{y}^c} \underline{S}$ and of the derivatives of the element forces.

To calculate the values of $\underline{g}^f$ (see (2.6)) or those of $\underline{g}^\mu$ (see (3.6)) we have to solve the nonlinear optimization problems (PL1) or (PL2), respectively and so the gradients must be calculated numerically, i.e. semi-analytically.

To that purpose we solve problems (PL1) or (PL2) with $\underline{S} + \frac{\partial}{\partial \underline{y}^c} \underline{S} \Delta \underline{y}^c$ to obtain gradients of $\underline{g}^f$ or $\underline{g}^\mu$ by a difference scheme (e.g. [8]).

In German codes of concrete design [4] the shear stresses $\underline{\tau}$ depend on the solutions of (PL1) or (PL2), which means that the gradients of $\underline{g}^r$ (see (2.5) or (3.5)) must be calculated semi-analytically.

5.2 Nonlinear Analysis

Geometrically nonlinear analysis is carried out using an updated Lagrangian formulation (e.g. [9]) considering a multiple of loading cases simultaneously [13].

For calculation of gradients see for example [10].

Physical nonlinearities are taken care of as follows:

- when solving (PL2) by a nonlinear programming algorithm which uses gradients (see Section 4) one gets the cross-section stiffness $\underline{B}$ depending on concrete and steel dimensions (see (3.8)):

$$\underline{B} = \frac{\partial}{\partial \underline{v}_1} [g_3; g_4; g_5] * \frac{\partial \underline{v}_1}{\partial \underline{e}} \tag{5.8}$$

where $\underline{e}^T = [e_a; \kappa_y; \kappa_z]$ and e_a : axial strain,

κ_y, κ_z : curvatures.

and $\underline{k}_e(\underline{y}^c)$ by integrating these stiffnesses over the element's length or area,

- when solving (PL1) we get the stiffnesses in a similar way.

Therefore proceeding with the nonlinear analysis in this way, function values and gradients must be calculated numerically.

The calculation of the gradients can be shortened if only the last load increment is considered.

6   Examples

The two examples shown, demonstrate good results in conjunction with this kind of automatic design. The first example is a slab system with eccentrically connected beams [11] where only a linear analysis is carried out.

The second example is a three story plane frame (see [12]) combined with a simultaneous nonlinear analysis of two loading cases (see [13]).

6.1   Slab System

The system is shown in Fig. 6.

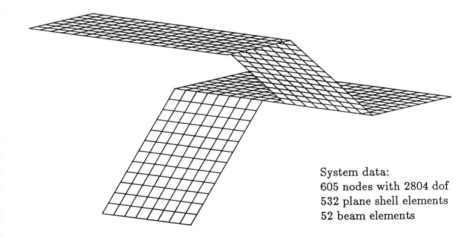

System data:
605 nodes with 2804 dof
532 plane shell elements
52 beam elements

Figure 6. *System and its FE-discretization.*

Due to eccentrical connections the stiffness matrices of the beam elements depend not only on their own dimensions but also on the variable thickness of the corresponding shell elements (see Fig. 7). Fig. 8 shows the different cross-section groups.

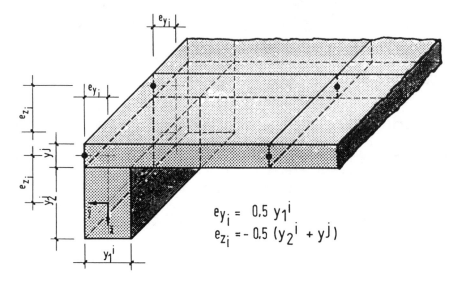

$$ey_i = 0.5\, y_1^{\,i}$$
$$ez_i = -0.5\,(y_2^{\,i} + y^j)$$

Figure 7. *Shell element and eccentrically connected beam element with rectangular cross-section.*

The following loads are applied:

• self weight

$$g_i = \begin{cases} y^i \cdot 25.0 \text{ kN/m}^2, & i = 1, \ldots, 10 \\ y_1^i \cdot y_2^i \cdot 25.0 \text{ kN/m}, & i = 11, 12 \end{cases}$$

• dead load $\tilde{g} = 2.0$ kN/m^2

• live load $p = 5.0$ kN/m^2 at different positions

The cost ratios (see (3.1)) are

$$k_i^1 = 50, \qquad k_i^2 = 0 \qquad i = 1, \ldots, 12$$

The following constraints are considered:

• lower and uppper bounds of the variables (see (3.2))

$$0.15 \text{ m} \le y^i \le 0.30 \text{ m} \qquad\qquad i = 1, \ldots, 10$$

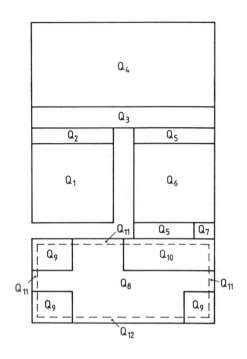

10 cross-section groups of shell elements with one variable y^i, $i = 1, \ldots, 10$

2 cross-section groups of beam elements with rectangular cross-sections with two variables y_1^i, y_2^i, $i = 11, 12$

Figure 8. *Cross-section groups.*

$$0.20 \text{ m} \leq y_1^i, y_2^i \leq 0.40 \text{ m} \qquad\qquad i = 11, 12$$

- constructive or logical constraints (see (3.3))

  These constraints are given in the first line of Table 1.

- constraints of the shear stresses (see (3.5))

  The permissible shear stresses are given in [4].

- constraints of the reinforcement ratios (see (3.6))

$$\mu_i \leq 0.02 \qquad\qquad i = 1, \ldots, 10$$

$$\mu_i \leq 0.03 \qquad\qquad i = 11, 12$$

The iteration is done with two starting vectors:

$$\underline{y}^1 = \underline{y}^-$$

and

$$\underline{y}^2 = \underline{y}^+ .$$

The iteration history is shown in Fig. 9 and the optimal solution in Table 1.

Table 1. *Variables with lower and upper bounds.*

	$y^1 = y^2$	$y^3 = y^4$	$y^5 = y^6 = y^7$	$y^8 = y^9 = y^{10}$	$y_1^{11} = y_1^{12}$	$y_2^{11} = y_2^{12}$
$\underline{y}^1 \; [m]$	0.150	0.150	0.150	0.150	0.200	0.200
$\underline{y}^* \; [m]$	0.150	0.150	0.207	0.150	0.200	0.386
$\underline{y}^2 \; [m]$	0.300	0.300	0.300	0.300	0.400	0.400

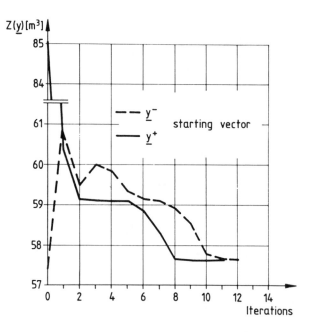

Figure 9. *Iteration history.*

The two starting points (lower and upper bounds of the variables) yield the same solution, which seems to be a global optimum.

6.2 Plane Frame

Six cross-section groups Q_i, $i = 1, \ldots, 6$ (see Fig. 10) are to be dealt with. Each group possesses two variable concrete dimensions (see Fig. 11), each element has three control points and each control point has two variable steel dimensions A_1^s and A_2^s (see Fig. 11).

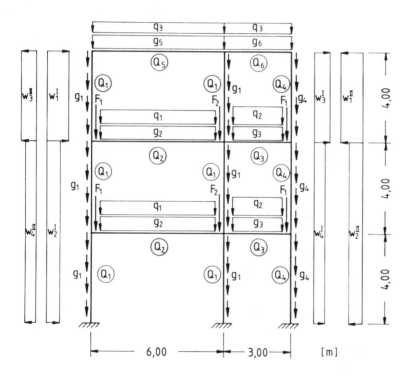

Figure 10. *System with cross-section groups, boundary conditions and loading.*

The following loads are applied (see Fig. 10):

self weight: $g_i = y_1^i \cdot y_2^i \cdot 25.0$ kN/m $i = 1, \ldots, 6$

$F_1 = 30.0$ kN

$F_2 = 67.0$ kN

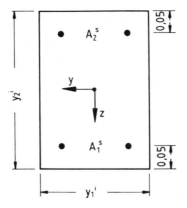

Figure 11. *Cross-section Q_i.*

$$q_1 = 42.0 \text{ kN/m}$$
$$q_2 = 57.0 \text{ kN/m}$$
$$q_3 = 9.0 \text{ kN/m}$$

$$w_1^I = w_1^{II} = 9.6 \text{ kN/m}$$
$$w_2^I = w_2^{II} = 6.0 \text{ kN/m}$$
$$w_3^I = w_3^{II} = 4.8 \text{ kN/m}$$
$$w_4^I = w_4^{II} = 3.0 \text{ kN/m}$$

The reinforcement of the columns (cross-section groups Q_1 and Q_4) does not vary across the element's length.

The cost ratios (see (3.1)) are

$$k_i^1 = 100, \qquad k_i^2 = 0 \qquad i = 1, \ldots, 6$$

The following constraints are considered:

- lower and uppper bounds of the variables (see (3.2))

$$0.20 \text{ m} \leq y_1^i, y_2^i \leq 0.70 \text{ m}$$

- constructive or logical constraints (see (3.3))

$$y_1^i = y_1^1 \qquad i = 2, \ldots, 6$$
$$y_1^i \leq y_2^i \qquad i = 1, \ldots, 6$$

- constraints on the shear stresses (see (3.5))

  The permissible shear stresses are given in [4].

- constraints on the reinforcement ratios (see (3.6))

$$\mu_i \leq 0.09 \qquad i = 1, \ldots, 6$$

These bounds of the reinforcement ratios are given in [4].

To show the influence of minimum reinforcement A_i^{s-} the iteration is done

- without minimum reinforcement, i.e. $A_1^{s-} = A_2^{s-} = 0$,

- with $A_1^{s-} = A_2^{s-} = 3.2\ cm^2$ for the columns and $A_1^{s-} = A_2^{s-} = 1.6\ cm^2$ for the beams.

During nonlinear analysis the loads are multiplied by 1.75 and the two loading cases are treated simultaneously [13].

The starting vector is

$$y_1^i = y_2^i = 0.40 \text{ m} \qquad i = 1, \ldots, 6.$$

The optimal solution is:

$$\text{width: } y_1^i = 0.20 \text{ m} = y_1^- \qquad i = 1, \ldots, 6$$

and the optimal depths are shown in Table 2.

7   Conclusion

For concrete structures it is usually necessary, because of the dimensions of the problems involved, to apply the hereby discussed separation of variables into dependent and independent ones. Especially in nonlinear analysis the calculation of gradients is very time consuming, but it can be shortened by an approximation based on mechanical arguments.

References

[1] Booz, G. et al., (1985), 'B&B—A Program for Analysis and Automatic Design of Structures—Theoretical manual and user's guide,' Forschungsberichte aus dem Fachgebiet Bauwesen, University of Essen.

[2] Kirsch, U., (1981), Optimum Structural Design, McGraw Hill, New York.

Table 2. *Objective and variables.*

	$A_1^{s-} = A_2^{s-} = 0$		c: $A_1^{s-} = A_2^{s-} = 3.2$ b: $A_1^{s-} = A_2^{s-} = 1.6$	
	linear	nonlinear	linear	nonlinear
Z° [m^3]	18.00	22.63	19.81	23.45
Z^* [m^3]	11.51	12.62	14.45	14.98
y_2^1 [m]	0.689	0.667	0.572	0.593
y_2^2 [m]	0.700	0.700	0.700	0.675
y_2^3 [m]	0.455	0.672	0.545	0.694
y_2^4 [m]	0.373	0.639	0.399	0.668
y_2^5 [m]	0.364	0.373	0.255	0.315
y_2^6 [m]	0.223	0.354	0.334	0.239

[3] Booz, W., (1985), 'Zweistufenoptimierung von Stahlbetontragwerken mit Hilfe der sequentiellen quadratischen Programmierung,' Dissertation, University of Essen.

[4] DIN 1045, (1988), *Beton und Stahlbeton—Bemessung und Ausführung*, Beuth-Verlag, Berlin, Juli.

[5] Schittkowski, K., (1982), 'Theory, implementation, and test of a nonlinear programming algorithm,' in H. Eschenhauer and N. Olhoff (Eds.), *Optimization methods in structural design*, Euromech-Colloquium 164, University of Siegen, FR Germany, pp. 122–132.

[6] Kirsch, U., (1982), 'Multilevel optimal design of reinforced concrete structures', in H. Eschenhauer and N. Olhoff (Eds.), *Optimization methods in structural design*, Euromech–Colloquium 164, University of Siegen, FR Germany, pp. 157–161.

[7] Haftka, R. T., (1986), 'Finite Elements in Optimum Structural Design', *ASI Computer Aided Optimal Design*, Troia, Portugal, Vol. 1, pp. 270–298.

[8] Gendong, C. and Yingwei, L., (1987), 'A new computation scheme for sensitivity analysis,' *Engineering Optimization*, Vol. 12, pp. 219–234.

[9] Bathe, K.-J., (1986), *Finite-Elemente-Methoden*, Springer-Verlag, Berlin/Heidelberg/New York/Tokyo.

[10] Arora, J. S. and Wu, C. C., (1986), 'Design sensitivity analysis of nonlinear structures,' ASI Computer Aided Optimal Design, Troia, Portugal, Vol. 2, pp. 228–246.

[11] Booz, G., Booz, W. and Thierauf, G., (1988), 'Multi-stage optimization of structures subject to static and dynamic loading', in H. Eschenauer and G. Thierauf (Eds.), *Discretization Methods and Structural Optimization—Procedures and Applications, Proceedings of a GAMM—Seminar*, Siegen, FR Germany, pp. 79–86.

[12] Lawo, M., (1987), *Optimierung im Konstruktiven Ingenieurbau*, Vieweg-Verlag, Braunschweig/Wiesbaden.

[13] Legewie, G., (1986), 'Geometrisch nichtlineare Berechnung und lineare Stabilitätsuntersuchung unter Berücksichtigung mehrerer Lastfälle,' Dissertation, University of Essen.

Layout Optimization in Structural Design

G. I. N. Rozvany, W. Gollub and M. Zhou
Essen University
Essen
Federal Republic of Germany

Abstract After reviewing some fundamental aspects of layout optimization, the lecture covers in detail two important techniques, *viz.* (a) iterative continuum-based optimality criteria (COC) methods for approximate layout optimization of large systems with a given grid of potential members and (b) the 'layout theory' developed by Prager and the first author for the exact optimization of the structural topology. This theory is based on the concepts of continuum-based optimality criteria (COC) and the 'structural universe' which is the union of all potential or 'candidate' members. Both 'classical' and 'advanced' layout theories are discussed: in the former, low density systems (e.g. trusses, grillages and cable nets) are considered, in which the effect of intersections on cost, stiffness and strength are neglected. Applications of advanced layout theory include 'generalized' plates (plates with a dense system of ribs) as well as perforated and composite systems. In comparing the results of iterative approximate and exact layout optimization on particular examples, a 12 digit agreement is found.

1 Introduction

It was remarked by Prager (Prager and Rozvany, 1977a) that 'structural optimization becomes a much more challenging problem when the layout as well as the cross-sectional dimensions are at the choice of the designer.' At the same time, layout optimization results in a much higher material saving than cross-sectional optimization (see, for example Rozvany, 1989, pp. 16–17). As mentioned in the first lecture by this speaker, the unknown quantities in a layout problem are: (a) the topology or configuration, describing the spatial sequence or connectivity of members and joints; (b) the location of the joints and the shape of the centroidal axes or middle surfaces between them; and (c) the cross-sectional dimensions of the members. In discretized formulations (e.g. Kirsch, 1989), these three problem solving tasks are referred to as (a) 'topological' and (b) 'geometrical' optimization and (c) 'sizing.'

Exact layout optimization is a particularly complex problem because one must

B. H. V. Topping (ed.),
Optimization and Artificial Intelligence in Civil and Structural Engineering, Volume I, 437–484.
© 1992 *Kluwer Academic Publishers. Printed in the Netherlands.*

438

consider an infinite number of potential topologies which are very difficult to classify and quantify. Moreover, at each point of the feasible space an infinite number of potential member directions, together with the cross-sectional dimensions of non-vanishing members, must be considered. Before discussing the exact layout theory of Prager and the first author in Section 3, applications of the iterative COC method to layout optimization will be considered in Section 2.

Earlier formulations of the layout theory for structural design were introduced by Prager and the first author (e.g. Prager, 1974; Rozvany, 1976; Prager and Rozvany, 1977a) and more up-to-date summaries were offered by the first author in principal lectures at NATO ASI's (Rozvany, 1981; Rozvany and Ong, 1987), in a book chapter (Rozvany, 1984) and, in particular, in a recent book (Rozvany, 1989). Valuable applications of advanced layout theory in shape optimization are also discussed by Bendsøe (e.g. Bendsøe, 1989).

2 Iterative Continuum-Based Optimality Criteria (COC) Methods for Layout Optimization

Most of the literature on layout optimization (e.g. Kirsch, 1989) is concerned with (a) *optimization of the 'geometry' for given 'topology'* or (b) *optimization of the 'topology' for given 'geometry'*. In case (a) above, the topology or spatial sequence of the members and joints is given but the coordinates of the joints are variable and are to be optimized. In case (b), the location of the joints is fixed but, out of all straight members connecting them, the non-optimal ones are to be eliminated.

The iterative COC procedure, which was explained in the first lecture by this speaker, is based on approach (b) above. However, this procedure is capable of handling a very large number of potential members which enables us to approximate very closely *both optimal topology and geometry* or even to attain them, as can be seen from the examples in Fig. 1. In the latter, potential members are shown in thin line and the optimal members in thick line. It will be shown subsequently using the exact layout theory (see Section 6), that the above designs represent the exact *optimal* solution.

In the problems in Fig. 1, we impose a constraint on the vertical deflection at the point where the load P is applied. In Fig. 1a, we consider a three bar suspension for which the obvious optimal topology consists of a single vertical bar and the sloping members take on a zero cross-section. Fig. 2b shows the iteration history for this problem. The polygon for unscaled weight is given by the iterative COC procedure. The 'weight' $\tilde{\Phi}$ here is nondimensional, $\tilde{\Phi} = \Phi Ed/(PL^2\gamma)$, where Φ is the total weight of the structure, E is Young's modulus, d is the prescribed deflection, P is the point load, L is the dimension shown in Fig. 2a and γ is the specific weight of

439

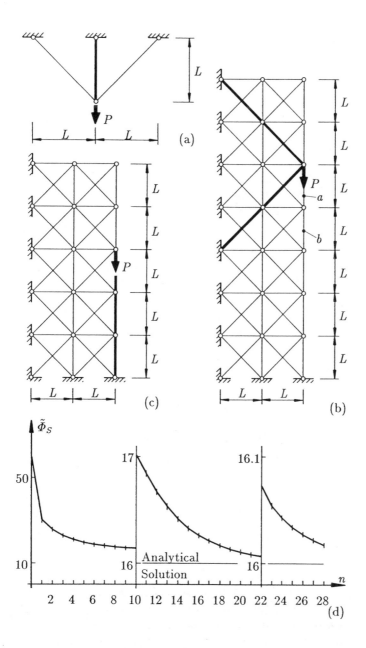

Figure 1. *(a)–(c) Various simple layout optimization problems solved by the iterative COC methods, and (d) nondimensional scaled structural weight $\tilde{\Phi}_s$ vs. number of iterations n for the problem under (b).*

440

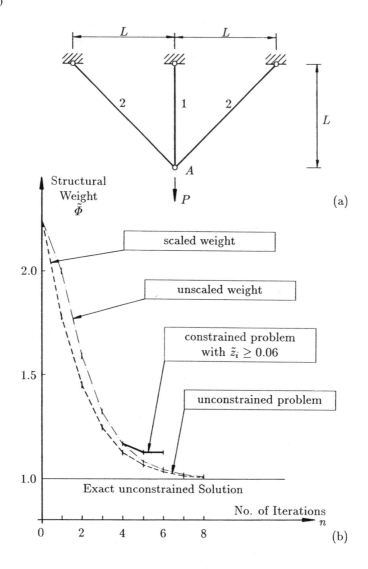

Figure 2. *Elementary layout problem with a structural universe consisting of three bars: (a) geometry and loading, and (b) convergence history for scaled and unscaled weight with and without a minimum cross-sectional area constraint.*

the material used. The nondimensional weight values yielded by the iterative COC procedure are given under 'unscaled weight' in Fig. 2b. Since the design obtained in each iterative step represents a deflection which differs slightly from the prescribed one ($u_A \neq d$), the cross-sectional areas were scaled to obtain the correct deflection after each iteration. The corresponding weight values are shown under 'scaled weight' in Fig. 2b. In a practical application, it is not necessary to calculate the scaled weights because the deflection of the unscaled solution converges rapidly to the prescribed value after a few iterations (see Fig. 2b in which the two polygons differ very little after 8 iterations). Fig. 2b shows also an alternative problem in which a prescribed minimum value (0.06) of the nondimensional cross-sectional area $\tilde{z}_i = z_i\, Ed/(PL)$ is specified. The results in Fig. 2b were obtained both by analytical hand calculations and on the computer using COC software (Rozvany, Zhou et al., 1989).

More complicated layout problems are shown in Figs. 1b and c. For these problems the COC method was combined with either (a) a finite element (FE) program developed by M. Zhou or (b) ANSYS. Both FE programs yielded identical results. Whilst for the simple example in Fig. 1a this was not necessary, in more complex problems the COC procedure for layout problems requires the specification of a small prescribed minimum cross-sectional area (z_a). For the considered problems the nondimensional value of the latter was $\tilde{z}_a = 10^{-12}$. The variation of the scaled nondimensional weight $\tilde{\Phi}_S$ in dependence on the iteration number is shown in Fig. 1d for the 56-bar truss (Fig. 1b). Using the nondimensionalization given above for $\tilde{\Phi}$, the exact optimal weight for the 56-bar truss is $\tilde{\Phi} = 16$ and the iterative COC procedure yielded after 126 iterations a weight of $\tilde{\Phi}_S = 16.000000000048$ which represents an agreement of twelve significant digits. In the COC solution, all non-optimal members took on the prescribed minimum cross-section ($\tilde{z}_a = 10^{-12}$) except members a and b (Fig. 1b) which had a cross-sectional area of $\tilde{z}_i = 2.88 \cdot 10^{-12}$ and $\tilde{z}_i = 1.74 \cdot 10^{-12}$. This indicates that all non-optimal members vanish when $\tilde{z}_a \to 0$, as in the analytical solution. The cross-sections of the optimal members (thick lines in Fig. 1b) agreed with the analytical solution for the first 12 significant digits.

In the 40-bar problem (Fig. 1c), all non-optimal members took on the value of $\tilde{z}_a = 10^{-12}$ in the iterative COC solution and the cross-sectional area for the optimal members agreed again with the analytical solution.

It can be shown that the layouts in Fig. 1 (thick lines) are also optimal for stress or natural frequency constraints. The method outlined in this section is suitable for handling a very large number (many thousand) of potential members and hence it can provide a very close approximation of both optimal 'topology' and 'geometry.'

3 The Theory of Optimal Layouts

The theory of optimal layouts for structures, developed in the late seventies by Prager and the first author, constitutes a generalization of a rather brilliant concept proposed around the turn of the century by Michell (1904) who considered the weight-minimization of pin-jointed frames. On the basis of the above theory, *both plastic and elastic design problems* can be treated. The following procedure can be used for establishing an optimal layout in *plastic design* (see Section 2 of the first lecture by this speaker):

(a) Set up a 'structural universe' consisting of all potential members.

(b) Determine the specific cost function for all members of the structural universe (see Section 2 of the first lecture).

(c) Derive continuum-based optimality criteria which are usually expressed as generalized strain-stress relations (see Fig. 8 of the first lecture). The latter, in general, give an inequality condition for vanishing cross-sections (i.e. for a zero value of the generalized stress).

(d) Construct an 'adjoint' displacement field which satisfies all kinematic boundary conditions.

(e) Calculate the strains along non-optimal (vanishing) members and check if they satisfy the optimality condition for zero generalized stresses.

(f) Determine the strains along optimal members and check if the corresponding generalized stresses equilibrate the specified external loads.

The above procedure was already demonstrated on a simple example in the first lecture (Section 3.3 and Fig. 12). Two further examples are given herein.

3.1 First Elementary Example: Pin-Jointed Frame or Truss

Consider the structural universe for a truss in Fig. 3a. Let the specific cost function for the members be (Fig. 4a)

$$\psi = k|N| , \tag{1}$$

where k is a given constant and N is the member force. The above specific cost function would apply to trusses with the same permissible stress σ_p for both tension and compression. In the considered problem, $k = \gamma/\sigma_p$ where γ is the specific weight of the truss material.

Using the Prager-Shield condition [(15) and Fig. 8 in the first lecture] , we have the following optimality conditions (Figs. 4a and b):

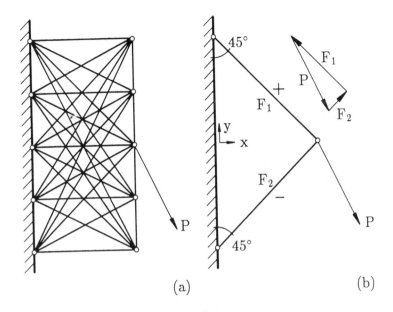

Figure 3. *Example of a structural universe: pin-jointed frame.*

$$\text{(for } N > 0) \quad \bar{\epsilon} = k \,, \tag{2}$$

$$\text{(for } N = 0) \quad -k \leq \bar{\epsilon} \leq k \,, \tag{3}$$

$$\text{(for } N < 0) \quad \bar{\epsilon} = -k \,, \tag{4}$$

where $\bar{\epsilon}$ is the axial strain in the adjoint field. The above optimality conditions refer *to plastic design for a strength condition* ($\psi \geq k|N|$) but it can be shown that the same optimality conditions apply for *elastic design with either a compliance or a natural frequency constraint*. In this case, 'compliance' means a prescribed value for the product of the point load and the displacement at the location and in the direction of the point load. In the above elastic problems, the real displacement field is proportional to the adjoint field (i.e. we are dealing with a self-adjoint problem).

It will be shown that the adjoint field satisfying (2)–(4) can be characterized by the following components:

$$\bar{u}(x,y) = 0 \,, \quad \bar{v}(x,y) = -2kx \,, \tag{5}$$

444

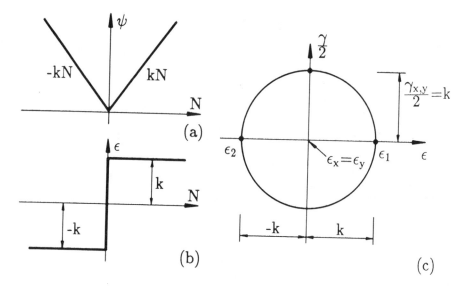

Figure 4. *Optimality conditions and the Mohr circle for the structural universe in Fig. 3.*

where $\bar{u}$ and $\bar{v}$ are displacements in the x and y directions, respectively. The above displacements satisfy the boundary conditions $\bar{u} \equiv \bar{v} \equiv 0$ along the rigid support ($x = 0$) and imply the kinematically admissible strains $[\bar{q}^K$ in (15) of the first lecture]:

$$\bar{\epsilon}_x = \frac{\partial \bar{u}}{\partial x} = 0 , \quad \bar{\epsilon}_y = \frac{\partial \bar{v}}{\partial y} = 0 , \quad \bar{\gamma}_{xy} = \frac{\partial \bar{u}}{\partial y} + \frac{\partial \bar{v}}{\partial x} = -2k . \tag{6}$$

It can be seen from the Mohr-circle in Fig. 4c that in this case the principal directions are at an angle of 45° to the x–axis. The principal strains take on the values

$$\bar{\epsilon}_1 = k , \quad \bar{\epsilon}_2 = -k . \tag{7}$$

It follows from the optimality conditions in (2) and (4) that positive (tensile) forces may only be optimal in a direction at 45° to the x–axis (sloping downwards from left to right) and negative (compressive) forces only at 45° to the x–axis (sloping upwards from left to right).

The optimal members in Fig. 3b satisfy the above conditions and provide a statically admissible set of member forces (see the vector diagram in Fig. 3b; the direction of the point load can take on any angle α with $-45° < \alpha < 45°$ to the vertical). It is still necessary to show that all non-optimal (or vanishing) members (with $N = 0$) in Fig. 3a satisfy the inequality condition in (3). That conclusion can be readily reached by an inspection of the Mohr-circle in Fig. 4c, which shows that the strains in our adjoint field have the extreme values $-k$ and k. This means that the solution in Fig. 3b and the corresponding adjoint field in (5) and (6) satisfy all optimality criteria. The latter, due to convexity of the specific cost function in Fig. 4a, are necessary and sufficient conditions of optimality and hence our solution represents a global optimum.

The above proof shows also that the considered solution would still be optimal if the structural universe consisted of all possible members in the entire x-y half-plane to the right of the support in Fig. 3a.

The total cost (weight) of the system can again be calculated from either primal or dual formulation. Assuming that the point load in Fig. 3 is vertical and its distance from the supports is L, we have $F_1 = F_2 = P/\sqrt{2}$ and the total cost becomes

$$\Phi_{opt} = \int_D \psi ds = 2k(P/\sqrt{2})(\sqrt{2}L) = 2kPL , \tag{8}$$

where s is the coordinate along the members. The same value can be obtained by dual formulation [(21) with $\hat{\psi} = 0$ in the first lecture] :

$$\Psi_{opt} = \int_D \mathbf{p} \cdot \overline{\mathbf{u}} ds = P\overline{u}_0 , \tag{9}$$

where $\overline{u}_0$ is the vertical displacement at the load. By (5) the latter becomes

$$\overline{u}_0 = 2kL , \tag{10}$$

confirming the optimal cost value in (8).

3.2 Second Elementary Example: Beam System or Grillage (Structural Universe with an Infinite Number of Members)

A square, simply supported, horizontal domain ABCD (Fig. 5a) contains two vertical point loads P of given location. These loads are to be transmitted to the supports by a system of intersecting beams of given depth having a specific cost function $\psi = k|M|$, where k is a given constant and M is the bending moment (for the derivation of this cost function, see Section 3.1 of the first lecture). In this problem, the structural universe consists of an infinite number of potential members in all possible locations and directions, covering the entire area of the domain ABCD. On the basis of the Prager-Shield condition [(15) and Fig. 8 in the first lecture], the optimality conditions for the beams are similar to those in (2)–(4):

$$\text{(for } M > 0) \qquad \overline{\kappa} = k , \tag{11}$$

$$\text{(for } M = 0) \qquad -k \leq \overline{\kappa} \leq k \,, \tag{12}$$

$$\text{(for } M < 0) \qquad \overline{\kappa} = -k \,, \tag{13}$$

$$|\overline{\kappa}|_{\max} = k \,, \tag{14}$$

In which κ is the curvature with $\kappa = \partial^2 \overline{u}/\partial s^2$, where $\overline{u}$ is the adjoint deflection and s is the coordinate along a beam. As the above optimality conditions imply that the directional maximum absolute value of the curvature i. It follows that such maximum curvatures, and by (11) and (13) also all non-vanishing, i.e. optimal beams, must have the same orientation as the principal directions of the adjoint displacement field $\overline{u}(x, y)$ with a curvature $\kappa_1 = k$ or $\kappa_2 = -k$. Moreover, the sign of the bending moments in such beams must match the sign of such principal curvatures.

This means that in order to admit potential beams over the entire beam system, the adjoint displacement field $\overline{u}(x, y)$ must consist of regions which are smoothly jointed, satisfy the kinematic boundary conditions ($\overline{u} = 0$ along the support) and must have at least one principal curvature with an absolute value of k. Such an adjoint displacement field is shown in Fig. 5b and consists of five regions (u_1 to u_5). It can be checked easily that the kinematic boundary condition $\overline{u} = 0$ is satisfied along the edges and the field $\overline{u}(x, y)$ is continuous and slope continuous along all region boundaries. The above solution is valid for plastic design as well as for elastic design with a compliance or a natural frequency constraint, and hence in the latter case Fig. 5b represents both the real and adjoint displacement fields. This is the reason for omitting overbars in the displacement functions $u_1 - u_5$ in Fig. 5b.

In the corner regions of Fig. 5b, one principal curvature has the value of $\overline{\kappa}_1 = k$ and the other one $\overline{\kappa}_2 = -k$. Such regions will be termed subsequently T-regions (Fig. 5c). In the central region (termed subsequently and in Fig. 5c an S^+-region), the curvature takes on a value of $\kappa = k$ in all directions. It follows from the optimality condition (11) that in S^+-regions beams may run in any arbitrary direction, so long as they are subject to positive bending. In the corner region, beams may be oriented only in two directions: in the direction of the diagonal passing through the considered corner the beams must be in negative bending and in the direction normal to that diagonal they must be in positive bending. This means that the optimality conditions admit an infinite number of optimal beam layouts of equal structural weight. One of these, with only positive moments in all beams, is shown in Fig. 5d, and another one, with beams in both positive (continuous line) and negative (broken line) bending in Fig. 5e. An oblique view of the adjoint field is given in Fig. 5f.

The above non-unique optimal solution was derived independently, in the context of reinforced slabs, by both Morley (1966) and the first author (Rozvany, 1966) but the latter employed only a primal formulation without making use of the

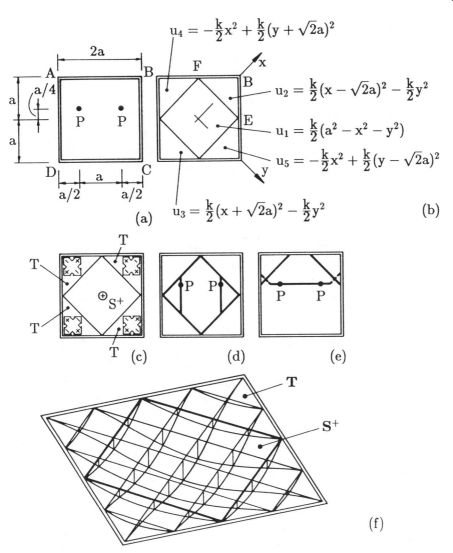

$$u_4 = -\frac{k}{2}x^2 + \frac{k}{2}(y + \sqrt{2}a)^2$$

$$u_2 = \frac{k}{2}(x - \sqrt{2}a)^2 - \frac{k}{2}y^2$$

$$u_1 = \frac{k}{2}(a^2 - x^2 - y^2)$$

$$u_5 = -\frac{k}{2}x^2 + \frac{k}{2}(y - \sqrt{2}a)^2$$

$$u_3 = \frac{k}{2}(x + \sqrt{2}a)^2 - \frac{k}{2}y^2$$

(a)

(b)

(c)

(d)

(e)

(f)

Figure 5. *Example of a structural universe consisting of an infinite number of potential members: square, simply supported grillage.*

448

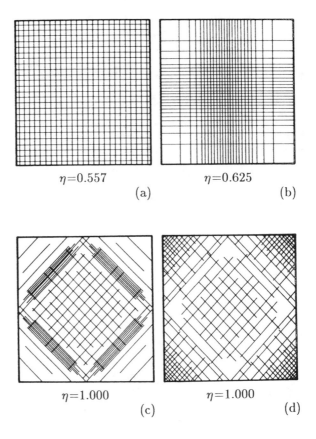

$\eta=0.557$ (a) $\eta=0.625$ (b)

$\eta=1.000$ (c) $\eta=1.000$ (d)

Figure 6. *Prager-efficiency (η) of various types of reinforcement used in destructive tests on simply supported slabs.*

adjoint field. The corresponding designs were used in full-scale destructive tests on uniformly loaded reinforced concrete slabs (Figs. 6c and d) and the reinforcement volume of the latter was compared with that of uniform mesh reinforcement (Fig. 6a) and of a reinforcement based on Hillerborg's (1956) strip method. The above experiments were carried out at Monash University, Melbourne, Australia, and confirmed the expected plastic (ultimate) load capacity of the above designs. Fig. 6 also shows the relative efficiency η of various designs; it can be seen that the conventional reinforcement (Fig. 6a) has an efficiency of only slightly over 50 per cent. In the above comparison, the so-called *Prager-efficiency* is used, which is the weight of the theoretical optimal solution divided by the weight of the considered design for the same boundary conditions and loading.

Before discussing further applications of the above technique, the difference between the two main branches of optimal layout theory will be explained.

3.3 Classical and Advanced Layout Theories

The introductory examples in Sections 3.1 and 3.2 (and in Section 3.3 of the first lecture) were based on the so-called 'classical' layout theory, a generalization of Michell's (1904) theorem. This theory has been used for the optimization of 'low-density' structural systems whose structural material occupies only a small proportion of the feasible space. Classical layout theory has two fundamental features: (a) at any point of the structural domain potential members may run in any number of directions (Fig. 7a), but (b) the effect of the member intersections on both the cost and strength (or stiffness) is neglected. It follows that the specific cost function ψ is the sum of several terms, each of which depends on a stiffness (or stress resultant) value s_i in a member direction,

$$\psi = \psi_1(s_1) + \psi_2(s_2) + \cdots + \psi_n(s_n) . \tag{15}$$

'Advanced' layout theory is used for 'high density' structures in which material occupies a high proportion of the feasible space or the structure consists of several materials whose interfaces are to be optimized. In this case, the *microstructure* of a perforated or composite structure is first optimized locally by minimizing, for given stiffnesses or stress resultants in the principal directions, the specific cost ψ (e.g. material volume per unit area or volume of the structural domain) for *perforated structures* and some factored combination of the material volumes per unit area or volume of the structural domain for *composite structures*. This means that the specific cost function, e.g. $\psi(s_1, s_2)$ in Fig. 7b, is in general a non-separable function of the principal stiffnesses or stress resultants.

Advanced layout theory results in substantial extra savings for 'high density' structures, but the optimal solutions given by this theory tend to those of classical layout theory if the material volume/feasible volume ratio approaches zero.

In the next two sections, we review briefly applications of both classical and advanced layout theories, ranging from Michell's classical findings to the latest developments in optimal elastic plate design.

4 Applications of the Classical Layout Theory

4.1 Michell-Frames or Least-Weight Trusses

This class of optimal layouts was pioneered around the turn of the century by a versatile Australian scientist, A.G.M. Michell (1904). The specific cost function for these structures was given in (1) and Fig. 4a and the optimality criteria in (2)–(4)

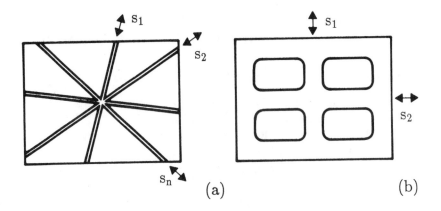

Figure 7. *Classical and advanced layout theories.*

and Fig. 4b. It follows that non-vanishing Michell frames *in the plane* may consist of the following types of regions:

$$\left.\begin{array}{llll} R^+: & N_1 > 0, & N_2 = 0, & \epsilon_1 = k, \ |\epsilon_2| \leq k, \\ R^-: & N_1 < 0, & N_2 = 0, & \epsilon_1 = -k, \ |\epsilon_2| \leq k, \\ S^+: & N_1 > 0, & N_2 > 0, & \epsilon_1 = \epsilon_2 = k, \\ S^-: & N_1 < 0, & N_2 < 0, & \epsilon_1 = \epsilon_2 = -k, \\ T: & N_1 > 0, & N_2 < 0, & \epsilon_1 = -\epsilon_2 = k, \end{array}\right\} \quad (|\epsilon_1| \geq |\epsilon_2|), \qquad (16)$$

where the subscripts 1 and 2 denote *principal* strains or forces.

In spite of a prolonged international research effort, Michell layouts have only been determined for a few simple loading conditions, most of which are summarized in an outstanding book by Hemp (1973). Hemp (1974) also derived a Michell field for a distributed load in between two point supports; his associate Chan (1975) found later that the considered solution is valid for a certain range of non-uniformly distributed loads. A few years later, Rozvany and Hill (1978a) found that certain superposition principles enable one to derive the optimal Michell layout for four alternative load conditions. The geometrical properties of Michell frames and optimal grillages have been compared by Prager and Rozvany (1977a). A Michell frame consisting of a single T-region is shown in Fig. 8, in which $3\pi/4 > \alpha > \pi/4$.

For Michell frames subject to applied loads plus *selfweight*, (30) in the first lecture yields optimality conditions similar to (2)–(4) and (16) but the quantity k is replaced everywhere by $k(1 + \bar{u})$, where $\bar{u}$ is the vertical adjoint deflection. Strang and Kohn (1983) considered the optimization of Michell frames with an *upper constraint* $|N| < \bar{N}$ *on the member forces*. Their optimality conditions take the form:

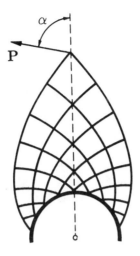

Figure 8. *Example of a Michell structure.*

$$|\epsilon| \leq k \text{ (for } N = 0), \qquad \epsilon = k \operatorname{sgn} N \quad \text{(for } 0 < |N| < \overline{N}) ,$$

$$\epsilon = ka \operatorname{sgn} N, \quad 1 \leq a \leq \infty \qquad \qquad \text{(for } |N| = \overline{N}) , \qquad (17)$$

and can be derived readily from (15) in the first lecture. The consequence of (17) is that in some optimal regions the principal *strains* have a constant absolute value $(|\epsilon| = k)$ and in other regions the principal generalized *stresses* are constant $(|N| = \overline{N})$ while the strains follow the same pattern as usual plastic slip-lines (Hencky-Prandtl nets) in plane stress or plane strain problems (Strang and Kohn, 1983).

The most recent work on Michell structures is due to Lagache (1980, 1981, 1983). Both Prager (1977, 1978a and b) and Lagache investigated the problem of 'nearly optimal' design for plane trusses.

As mentioned above, there is *no exact Michell-solution available to date for the majority of simple boundary and loading conditions.* Fig. 9a, for example, shows a domain consisting of two supported (ABC) and two free edges (ADC). The adjoint field is also shown in Fig. 9a, having $u \equiv v \equiv 0$ in Region 1 and a T-region with $\bar{\epsilon}_1 \equiv \bar{\epsilon}_x \equiv k$ and $\bar{\epsilon}_2 \equiv \bar{\epsilon}_y \equiv -k$ in the Region 2 giving $\epsilon \equiv 0$ along the boundary between Regions 1 and 2 [Region 1 is not listed in (16) because the latter only refers to regions containing non-vanishing members] . However, if we changed the proportions of the square domain in Fig. 9 into a rectangular one, then the solution would be much more difficult to obtain.

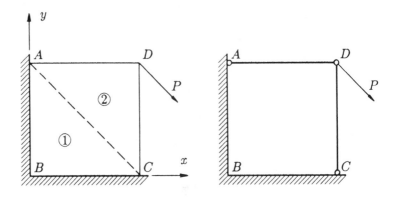

Figure 9. *Example of simple boundary and loading conditions for which a Michell-solution can be found.*

As shall be seen in the next section, the theory of least-weight grillages has been considerably more successful in locating solutions. Prager rightly pointed out (Prager and Rozvany, 1977b) that this relatively recent theory has advanced much further than that of minimum-weight trusses. Considering the complexity of the topology of these closed form, exact solutions, they constitute a rather remarkable development, representing an almost unique intellectual achievement in structural optimization. It is also very satisfying that a special class of Michell frames (see Section 4.3) has also been solved analytically for most loading conditions, even when the effect of selfweight is taken into consideration.

4.2 Least-Weight Grillages or Beam Layouts of Given Depth

The specific cost function for this problem is $\psi = k|M|$ and the optimality conditions were given in (11)–(13). The latter imply that the following types of regions may occur in the optimal solution:

$$\left.\begin{array}{llll} R^+: & M_1 > 0, & M_2 = 0, & \kappa_1 = k, \ |\kappa_2| \leq k, \\ R^-: & M_1 < 0, & M_2 = 0, & \kappa_1 = -k, \ |\kappa_2| \leq k, \\ S^+: & M_1 > 0, & M_2 > 0, & \kappa_1 = \kappa_2 = k, \\ S^-: & M_1 < 0, & M_2 < 0, & \kappa_1 = \kappa_2 = -k, \\ T: & M_1 > 0, & M_2 < 0, & \kappa_1 = -\kappa_2 = k, \end{array}\right\} \quad (|\kappa_1| \geq |\kappa_2|) . \quad (18)$$

The above optimal regions are shown in Fig. 10 in which continuous and broken lines, respectively, indicate optimal beams with positive and negative bending moments. Arrows indicate the directions of principal curvatures. In the case of circles enclosing a sign, all directions are equally optimal (the curvature in all directions is the same).

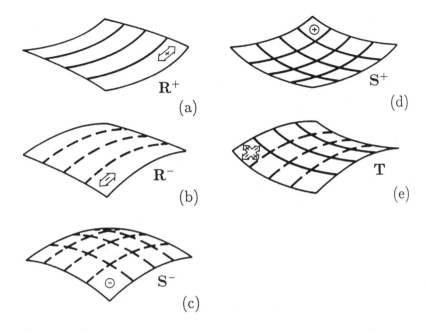

Figure 10. *Types of optimal regions for least-weight grillages.*

The *theory of optimal beam layouts* (grillages) was first explored by the author and his associates in the early seventies (for summaries of this early development see Prager 1974; Rozvany and Hill, 1976; Rozvany, 1976; Save and Prager, 1985), and later discussed extensively by Prager and Rozvany (1977a and b). The mathematical analogous problem of *optimal reinforcement in concrete plates* was introduced by Morley (1966) who used the optimality criteria in (11)–(13) and his work was extended by Lowe and Melchers (e.g. 1972). The latest developments in grillage optimization are summarized in publications by Rozvany (1981, 1984), Rozvany and Ong (1986a and b) and Rozvany and Wang (1983). For a detailed review of the literature in this field, the reader is referred to the first author's recent book (Rozvany, 1989, Section 8.5b).

As can be seen from the quotation at the end of Section 4.1, Prager regarded the grillage optimization problem as particularly important because of the following unique features:

(a) Grillages constitute the first class of truly two-dimensional structural optimization problems for which closed form analytical solutions are available for most boundary and loading conditions.

454

(b) Optimal grillages are more practical than Michell structures because the latter are subject to instability which is ignored in the formulation.

(c) The optimal rib layout of least-weight plates has been found similar to that of minimum weight grillages (see Fig. 16 in Section 5.1).

(d) A computer algorithm is available for generating analytically and plotting optimal beam layouts for a wide range of boundary conditions (Rozvany and Hill, 1978b; Hill and Rozvany, 1985); see, for example, Fig. 12 below.

(e) It has been shown that the same grillage layout is optimal for plastic design and elastic design with a stress, compliance or natural frequency constraint (Rozvany, 1976; Olhoff and Rozvany, 1982).

(f) The optimal grillage layout is independent of the (non-negative) load distribution if no internal simple supports are present.

(g) The adjoint displacement field can be readily generated and it provides an *influence surface* for any (non-negative) loading (the total structural weight equals the integral of the product of loads and deflections).

(h) A number of additional refinements have been added to the optimal grillage theory. Analytical solutions are now available for clamped and simply supported boundaries, internal simple supports, free edges, beam supported edges and corners (cusps) in the boundary. Further extensions of the theory include *plastic grillages* with up to four alternate loading conditions, non-uniform depth, partial discretization, allowance for cost of supports, bending and shear dependent cost, upper constraint on the beam density, the effect of selfweight (also for bending- and shear-dependent cost), as well as for *elastic* grillages with deflection constraints or with deflection and stress constraints (for references, see the recent book by Rozvany, 1989, Section 8.5b).

To show the complexity of the topologies involved, Fig. 11 shows the optimal beam layout for various rhomboidal domains with simple supports. In unshaded areas all beam directions are optimal, provided that the sign of the beam moments and beam curvatures is the one indicated for that region (see circled signs).

Figure 12 shows a minimum weight beam layout for a more complicated boundary shape where shading indicates a clamped edge and double lines simple supports. This exact solution was produced on a small computer in a few seconds in the seventies (Rozvany and Hill, 1978b; Hill and Rozvany, 1985). The computer program derives and plots the optimal layout for any combination of simply supported and clamped edges as well as internal supports automatically using only analytical operations.

Figure 13 shows the total weight of some optimal and non-optimal solutions for circular, simply supported grillages with allowance for selfweight as well as the

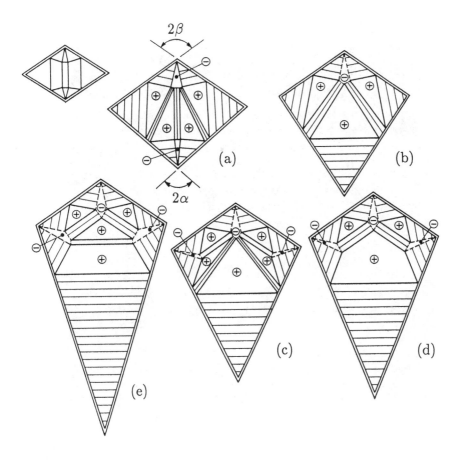

Figure 11. *Examples of optimal grillage layouts: simply supported rhomboidal domains.*

effect of both bending and shear on the specific cost. As this type of cost function is typical for long-span truss-grids, the chords and web-members are shown for such a system in Fig. 13. The parameter 'c' represents the relative magnitude of the cost of shear in comparison to the cost of bending and R is the nondimensional radius of the boundary. It will be seen that a purely circumferential chord-system is optimal up to a critical radius of the boundary. At longer spans, there are circumferential chords in an inner region, radial chords in the outer region and there is again a pair of 'concentrated' circumferential chords along the simple supports (resisting the radial moments). To give the reader some idea of the mathematical complexity

456

Figure 12. *Computer-generated optimal beam layout.*

of this solution, the radial moments M_r in the outer region are represented by the function (Yep, Sandler and Rozvany, 1986):

$$rM_r = \mathrm{e}^{\alpha r}[A\cosh(r\beta) + B\sinh(r\beta)] - cR + r \; , \tag{19}$$

with

$$A = -\sinh(a\beta)\mathrm{e}^{a(\alpha + a/2)}[\mathrm{e}^{2\alpha^2}\sqrt{\frac{\pi}{2}}2\alpha\left(\mathrm{erf}\frac{a + cR}{\sqrt{2}} - \mathrm{erf}\frac{cR}{\sqrt{2}}\right) - 1]/\beta$$

$$+ (cR - a)\mathrm{e}^{-\alpha a}[\cosh(a\beta) + \alpha\sinh(a\beta)]/\beta \; , \tag{20}$$

$$B = \cosh(a\beta)\left\{\sqrt{\frac{\pi}{2}}2\alpha\mathrm{e}^{[(a^2/2)+2\alpha^2+\alpha a]}\left(\mathrm{erf}\frac{a + cR}{\sqrt{2}} - \mathrm{erf}\frac{cR}{\sqrt{2}}\right)\right.$$

$$\left. - \mathrm{e}^{[a^2/2+\alpha a]} - (cR - a)\mathrm{e}^{-\alpha a}[\beta\tanh(a\beta) + \alpha]\right\}/\beta \; , \tag{21}$$

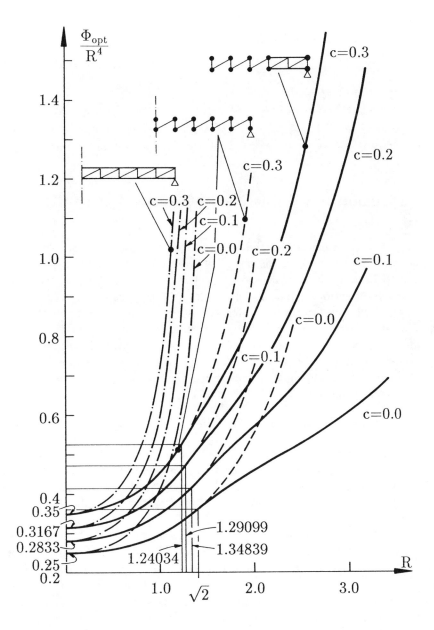

Figure 13. *Weight of optimal and non-optimal solutions for long-span circular truss grids.*

$$\alpha = cR/2, \quad \beta = (1 + \alpha^2)^{1/2} , \tag{22}$$

where

$$\text{erf} \, (r) = \frac{2}{\sqrt{\pi}} \int_0^r e^{-t^2} dt \tag{23}$$

is the 'error function.' It can be seen from Fig. 13 that there is often a several hundred per cent difference between the structural weight of optimal and non-optimal solutions.

4.3 Archgrids and Cable Nets of Optimal Layout (Prager Structures)

A Prager structure can be defined as a surface structure consisting of intersecting arches or cables for which the shape of the middle surface and the member layout are to be optimized. Moreover, the (usually vertical) loads are movable along their line of action. Alternatively, a Prager structure can be regarded as a special class of Michell frames for which (a) either the compressive or the tensile permissible stress tends to zero and (b) the position of (usually vertical) loads is unspecified and to be optimized. This special class of Michell structures has been shown to reduce always to a surface structure in 3D space (or a line structure in plane). On the basis of (15) in the first lecture, the following optimality conditions apply to Prager structures:

$$\epsilon = k \; (\text{for } N > 0), \quad -\infty < \epsilon \leq k \; (\text{for } N = 0), \quad N \geq 0, \quad \epsilon_{\text{vertical}} \equiv 0 \, . \tag{24}$$

The last condition is due to the fact that loads are 'movable' vertically, which is equivalent to having weightless (i.e. costless) members in that direction. Closed form analytical solutions are now available for any vertical axisymmetric load in three-dimensional space and for any vertical load system in a plane and also for additional selfweight [in which case, on the basis of (30) in the first lecture, k is replaced by $k(1+u)$ in (24)] . Moreover, the above solutions have been extended to 'quasi-axisymmetric' loads (concentrated loads distributed in the circumferential direction at equal angular intervals) and axisymmetric support conditions. Fig. 14a shows the general form of the optimal solution for plane Prager structures with two supports at the same level and at different levels (at an angle β), both with or without selfweight. It can be seen that for two supports at the same level the optimal form of the Prager structure is given by a funicular such that *the mean square slope is unity*. The above optimality criteria are also valid for several supports and for axisymmetric and quasi-symmetric systems.

Moreover, Fig. 14b gives optimal cost influence lines u_y for various types of plane Prager structures such that the total cost for any vertical load system is given by $\int_D p u_y \, dx$.

	WITHOUT SELFWEIGHT	WITH SELFWEIGHT
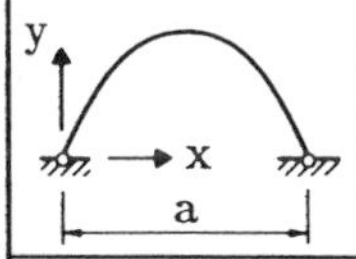	$\dfrac{\int_0^a (dy/dx)^2 dx}{a} = 1$	$\dfrac{\int_0^a (dy/dx)^2 e^{2ky} dx}{\int_0^a e^{2ky} dx} = 1$
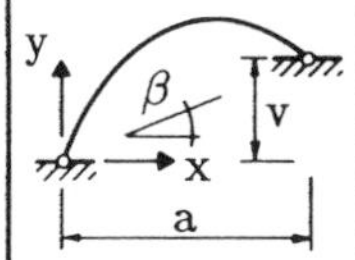	$\dfrac{\int_0^a (dy/dx)^2 dx}{a}$ $= 1 + 2\tan^2 \beta$	$\int_0^a ke^{2ky}\left[1 - (dy/dx)^2\right]\left[1 - C\int_0^x e^{-2ky}dx\right]dx - Cv = 0$ where $C = (e^{2kv} - 1)/(e^{2kv}\int_0^a e^{-2ky}dx)$

| | Optimal Geometry | (a) |

	WITHOUT SELFWEIGHT	WITH SELFWEIGHT
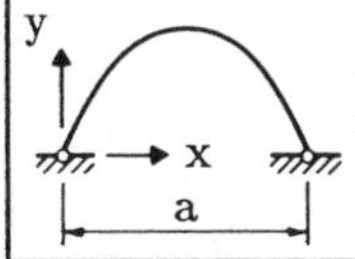	$u_y = 2ky$	$u_y = e^{2ky} - 1$
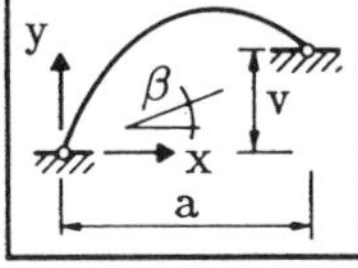	$u_y = 2k(y - x\tan\beta)$	$u_y = e^{2ky}\left[1 - \dfrac{(e^{2kv} - 1)\int_0^x e^{-2ky}dx}{e^{2kv}\int_0^a e^{-2ky}dx}\right] - 1$

| | Influence Lines | (b) |

Figure 14. *Optimal geometry and influence lines for plane Prager structures.*

It can be seen from Fig. 14b that for the simplest case (two supports at equal elevation, no selfweight) the optimal cost (structural weight) is simply the *sum* (or integral) *of the products of loads $p(x)$ and their elevation $y(x)$ multiplied by a constant $(2k)$. Using Maxwell's (1872) theorem, an alternative optimality condition for that simple case can be stated as follows: *the sum of the products of vertical forces and their elevation must equal the product of the horizontal reaction component and the distance between the two supports.*

Fig. 15a shows Prager structures *without selfweight* for two supports at different levels in the vertical plane and a linearly varying load plus point load, while Fig. 15b gives solutions *with selfweight* for a uniform external load and supports at the same level. The parameter α is the nondimensional span $(\alpha = kL$ where k is the cost factor and L is the span). For a review of the literature on Prager structures, the reader is referred to book chapters by the first author (Rozvany, 1984; 1989, Section 8.5c).

5 Applications of the Advanced Layout Theory

5.1 Optimal Plastic Design of Solid Plates

It was established already in the late sixties (e.g. Kozłowski and Mróz, 1969) and confirmed more rigorously later (Rozvany, Olhoff, Cheng and Taylor, 1982) that the weight of solid plates can be reduced to an arbitrarily small value by employing a system of sufficiently high and thin ribs. Naturally, the above solution is of purely theoretical interest because it ignores the lateral instability of such ribs. A finite value for the structural weight can be ensured, however, by introducing an *upper constraint* on the plate thickness. Cheng and Olhoff (1981) have discovered through numerical solutions that stiffener-like formations occur in the optimal solution for such systems (see Fig. 16a and b). Prager pointed out shortly before his death in 1980 that the *layout* of such ribs is similar to that of an earlier solution (Fig. 16c), derived originally by Rozvany and Adidam (1972) for grillages and for fibre-reinforced plates by Lowe and Melchers (1972). Complete solutions

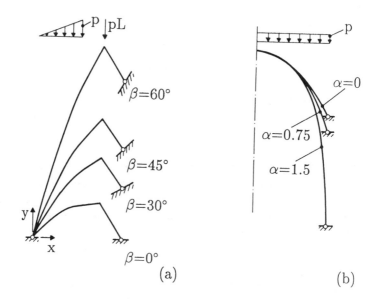

Figure 15. *Examples of plane Prager structures.*

were obtained for plastically designed solid Tresca-plates with an upper limit on the plate thickness by Rozvany, Olhoff, Cheng and Taylor (1982) and extended to other boundary conditions by Wang, Rozvany and Olhoff (1984). For simply supported circular plates, for example, the optimal solutions are compared with some non-optimal ones in Fig. 17 for various levels of the nondimensional load ν. The above solutions were obtained by establishing a specific cost function $\psi(M_1, M_2)$ through local optimization of the rib/plate configuration for given values of the principal moments (M_1, M_2) and then employing the Prager-Shield condition [(15) in the first lecture] together with principles of the (advanced) optimal layout theory (Section 3.3). The nondimensionalized specific cost function τ as a function of the nondimensional moments (μ_1, μ_2) is represented graphically in Fig. 18. All least-weight solutions have been found to consist of the following types of regions:

(α): solid plate of non-maximum thickness with $M_1 \equiv M_2$;

(β): ribs of maximum depth in one principal direction only, infinitesimal plate thickness in between ribs;

(γ) and (δ): solid plates of maximum feasible thickness.

It has also been found that at very low levels of the nondimensional load (ν in Fig. 17), the solution tends to the *optimal grillage layout* with type (β) regions only. This confirms Prager's intuitive insight that the rib layout for optimal solid plates is similar to the layout of least-weight grillages.

5.2 Optimal Plastic Design of Perforated Plates

A 'perforated plate' may only have two thicknesses, a prescribed (maximum) thickness or zero thickness. The latter occurs over 'perforations' whose in-plane dimensions are assumed to be sufficiently small so that the load over areas of zero thickness can be transmitted to the adjacent plate segments by some secondary systems of negligibly small volume. In minimizing the total material volume of perforated plates, it was assumed (e.g. Rozvany and Ong, 1986a and b; Ong, 1987) that the plate material obeys Tresca's yield condition. In perforated regions, the optimal microstructure can be shown to consist of ribs running in the directions of the principal moments (M_1, M_2). As stresses of the same sign do not influence the yield value of the major stress in Tresca's yield condition, a saving can be achieved at rib intersections with sgn $M_1 =$ sgn M_2 and the specific cost ψ becomes

$$\psi \equiv |M_1| + |M_2| - M_1 M_2 . \tag{25}$$

However, such a saving is not possible if the stresses are of opposite sign (sgn $M_1 \neq$ sgn M_2):

$$\psi = |M_1| + |M_2| . \tag{26}$$

462

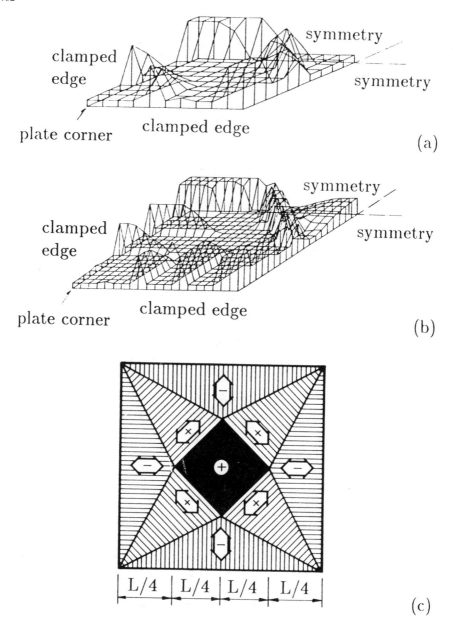

Figure 16. *(a and b) Numerical solutions by Cheng and Olhoff (1981) showing rib-like formations and (c) optimal grillage of similar layout (Rozvany and Adidam, 1972).*

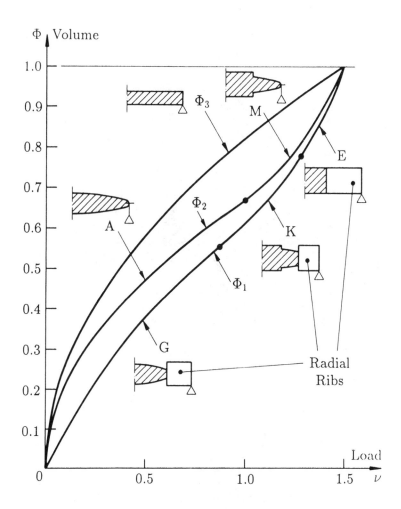

Figure 17. *Plastic plate design with a prescribed maximum thickness: a comparison of the weight of the absolute optimal (ribbed) solution Φ_1 with that of smooth Φ_2 and constant thickness Φ_3 optimal solutions.*

464

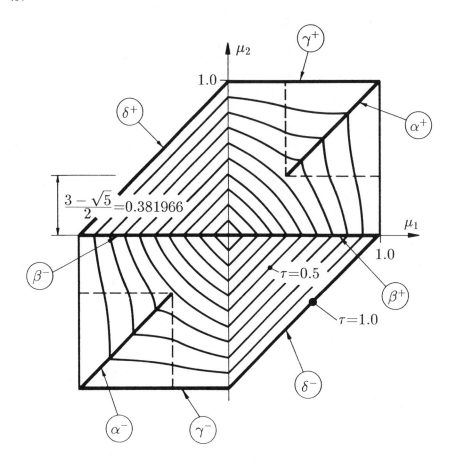

Figure 18. *Specific cost functions in terms of principal moments for solid plastic plates.*

The above specific cost function (ψ) is represented graphically in Fig. 19. Making use of the Prager-Shield condition, Ong (1987) has shown that for plastic axisymmetric perforated plates the least-weight solution may only consist of:

(α) unperforated regions (stress regimes α and γ in Fig. 19); and

(β) ribs in the radial direction ($M_\theta \equiv 0$, stress regime β in Fig. 19).

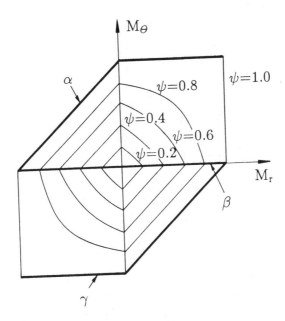

Figure 19. *Specific cost function for perforated plastic plates.*

Introducing the nondimensional notation $r = \bar{r}/\bar{R}$, $p = \bar{p}\bar{R}^2/\overline{M}$, $M_i = \overline{M}_i/\overline{M}$ ($i = r$, θ) where $\bar{r}$ is the radial coordinate, $\bar{R}$ is the plate radius, $\bar{p}$ is the load intensity and $\overline{M}$ is the maximum feasible moment capacity, the optimal solution for simply supported uniformly loaded circular plates turns out to be the following:

$$(0 \le r \le p/6) \quad M_\theta = 1, \ M_r = 1 - pr^2/6 \qquad \text{(Regime } \alpha\text{)},$$

$$(p/6 \le r \le 1) \quad M_\theta = 0, \ M_r = p(1 - r^3)/6r \qquad \text{(Regime } \beta\text{)}. \qquad (27)$$

To demonstrate the validity of the above conclusion, the volume of various intuitively selected designs is compared in Fig. 20. Design A consists of circumferential ribs only ($M_r \equiv 0$, the shear transmission is assumed to be costless), Design B of

466

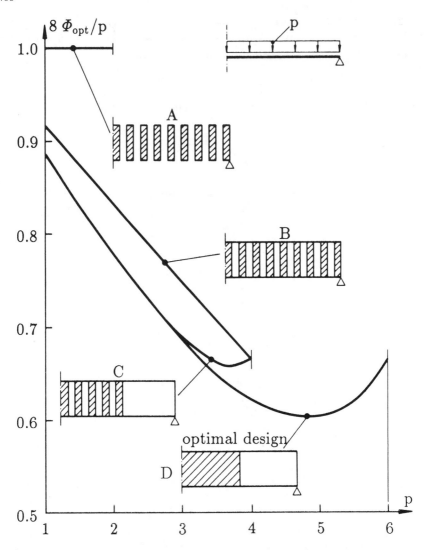

Figure 20. *A comparison of the volume of optimal and non-optimal solutions for plastic perforated circular plates.*

radial and circumferential ribs of equal width $(M_\theta \equiv M_r)$ throughout, Design C with $M_r \equiv M_\theta$ in an inner region and $M_\theta \equiv 0$ (only radial ribs) in the outer region and Design D (optimal design) with an unperforated plate in an inner region and $M_\theta \equiv 0$ (only radial ribs) in the outer region. Although even at small values of the nondimensional load p the optimal Design D gives a lower volume than Design C, the difference is too small for showing it graphically in Fig. 20.

5.3 Optimal Elastic Design of Perforated Plates with a Compliance Constraint

It was shown in recent papers by Murat and Tartar (1985), Lurie, Fedorov and Cherkaev (e.g. 1984), as well as Kohn and Strang (e.g. 1986) that optimal solutions in plane systems often contain regions with two sets of intersecting ribs (strips of material) at right angles: one such set has a first order infinitesimal spacing [of $O(\delta)$ with $\delta \to 0$] and the other one a second order infinitesimal spacing [of $O(\delta^2)$]. The implications of these results for plate optimization were investigated by Rozvany, Ong, Olhoff, Bendsøe, Szeto and Sandler (1987) who arrived at the following conclusions:

(A) Given a *horizontal system of intersecting first and second order ribs* whose depth is significantly smaller than its span, it is reasonable to assume that *under a distributed vertical load* the normal stresses in the ribs are proportional to the distance from the middle surface. As a consequence of St. Venant's principle (and a detailed finite difference/finite element analysis), any horizontal slice of a rib of second order infinitesimal width is subjected on its interior to stresses in the direction of the rib middle plane only. The same conclusion can be obtained for a *first/second order rib system in plane stress*. Hence *second order ribs do not contribute to the stiffness in the direction normal to their middle plane*. In this respect, the formulation by Rozvany et al. (1987) differs from some of the mathematical studies of the authors listed above.

(B) In minimizing the weight of a perforated plate for a *given compliance*, it was shown by Rozvany et al. (1987) on the basis of an intuitive argument that at *low rib-densities* the material consumption is smaller for a first/second order ribbed microstructure than for a first/first order system. Moreover, detailed finite element/finite difference analyses at Monash and Essen Universities by Ong, Szeto, Booz, Menkenhagen and Spengemann have shown that the *first/second order microstructure is more economical than a prismatic first/first order one at all rib densities* (see Fig. 21 in which ψ_{FF} and ψ_{FS} denote, respectively, the volume of first/first and first/second order microstructures for given equal principal stiffnesses (s) and Δ is the percentage difference between the volumes of the two microstructures.

(C) Assuming a first/second order microstructure at all rib densities, a *specific cost function* was derived for perforated plates in bending or plane stress, using the simplifying feature mentioned under (A). *For a zero value of Poisson's ratio*, the latter gives a relationship between the stiffnesses (s_1, s_2) in the principal directions and the material volume ψ per unit area of the middle surface (Fig. 22a)

$$\psi = \frac{s_1 - s_1 s_2 + s_2}{1 - s_1 s_2} \tag{28}$$

(D) The above specific cost function was then used for examining the design of *least-weight axisymmetric transversely loaded elastic perforated plates* of given compliance.

(E) It was found that the optimal design for the above problem reduces to that of *grillages* if the average rib density approaches zero (i.e. at very low load or high compliance levels).

(F) *Continuum-based optimality criteria* were derived from a variational analysis, using the proposed microstructure.

(G) The above conditions, together with static/kinematic admissibility, indicated that for transversely loaded axially symmetric plates only the following *two types of regions* may occur in loaded segments of the optimal solution:

 (i) unperforated regions (R_0);
 (ii) regions consisting of radial ribs only (R_r).

(H) On the basis of the foregoing findings, optimal solutions were derived for simply supported and clamped circular plates with uniformly distributed full and partial loading as well as a central point load and for simple supported plates with a uniform radial moment applied along their edge.

(I) The above results were confirmed by optimizing a number of intuitively selected designs with respect to their free geometrical parameters and also by independent numerical solutions by W. Booz (Essen University). In Fig. 23, for example, the volume of various partially optimized intuitive designs is compared as a function of the reciprocal compliance $1/C$. As predicted by the optimality criteria method, Design D is optimal for all $1/C$ values.

(J) More recently (Ong, Rozvany and Szeto, 1988), the above results were extended to plates with a *non-zero Poisson's ratio* $(\nu \neq 0)$. Whereas for $\nu = 0$ the specific compliance c was given by $c = M_1^2/s_1 + M_2^2/s_2$ where M_1 and M_2 are principal moments and s_1 and s_2 principal stiffnesses, for $\nu \neq 0$ the

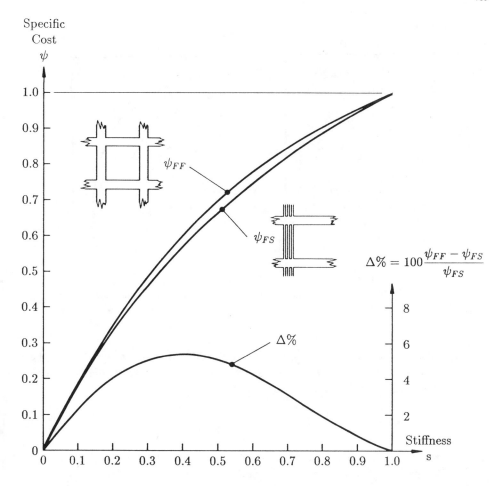

Figure 21. *Comparison of the economy of first/first and first/second order microstructures for given equal stiffnesses.*

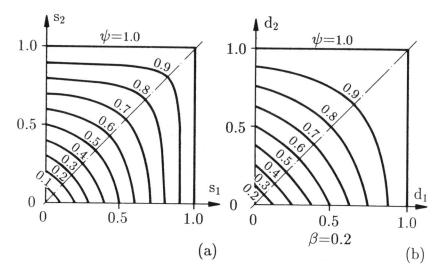

Figure 22. *Specific cost functions for (a) perforated and (b) composite plates.*

specific compliance can be expressed as $c = M_1^2/s_1 + M_2^2/s_2 - \nu M_1 M_2$ where s_1 and s_2 are no longer the principal stiffnesses, but have the same relations to the rib densities (d_1, d_2) as for $\nu = 0$:

$$s_1 = d_1, \quad s_2 = d_2/(1 + d_1 d_2 - d_1) . \tag{29}$$

Then the specific cost function can be shown to be the one in (28). Optimal solutions have been derived for various axisymmetric loading and boundary conditions. The least-weight solution for clamped circular perforated plates may consist of either one or two unperforated regions and a region with radial ribs. The range of validity of these solutions is shown in Fig. 24.

(K) Further extensions of the above approach concerned *composite plates* in which the perforations are filled with an isotropic material of lower stiffness and lower specific cost. The specific cost function for the stiffness and cost ratios $\alpha = \beta = 0.2$ is shown in Fig. 22b. It has been found that the optimal solution for *axisymmetric* composite plates may consist of:

  (i) regions consisting entirely of the stiffer material (cross-hatched in Fig. 25);

  (ii) regions consisting entirely of the less stiff material (hatched in Fig. 25); and

471

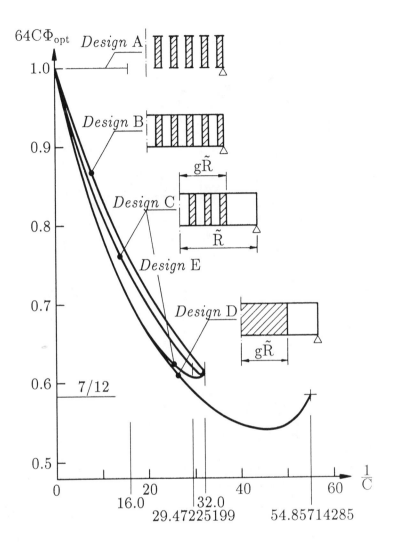

Figure 23. *A comparison of the volume of optimal and non-optimal solutions for elastic perforated plates.*

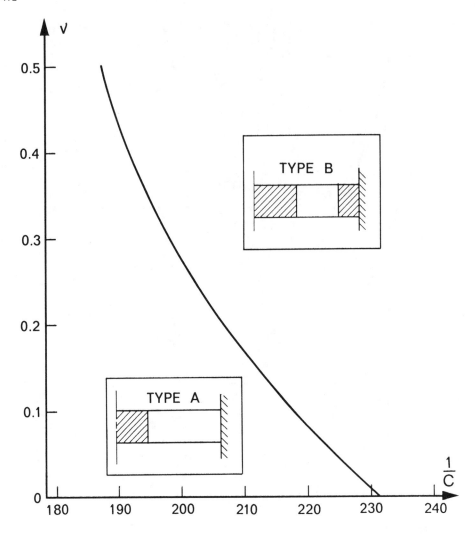

Figure 24. *Range of validity of various types of solutions; clamped perforated plates with $\nu \neq 0$.*

(iii) regions with radial ribs made out of the stiffer material and the gaps filled with the less stiff material (unhatched in Fig. 25 which gives the optimal solution for simply supported, uniformly loaded circular plates with $\alpha = \beta = 0.2$).

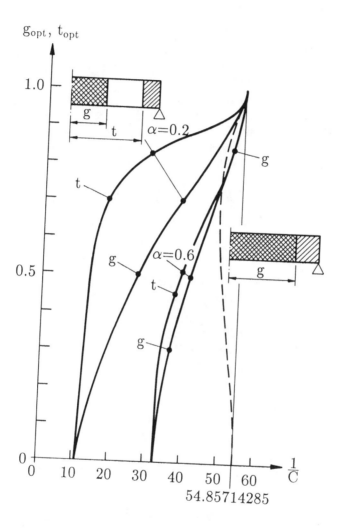

Figure 25. *Various types of solutions for composite, simply supported circular plates.*

(L) The above investigation has also been extended to

(i) plates with a prescribed deflection at a given location; and

(ii) allowance for the effect of shear on the compliance.

5.4 Applications of the Advanced Layout Theory in Shape Optimization

It was pointed out by Kohn and Strang (1983) that most shape optimization problems involving plastic or elastic continua reduce to a layout optimization problem because unconstrained shape optimization problems produce an infinite number of internal boundaries or 'holes' in the solution. Perforated continua with such cavities involve an optimal microstructure and then the layout of the latter can be optimized using 'advanced' layout theory for the corresponding specific cost function. An example of this was given by Kohn and Strang (1983) who considered the optimal plastic design of a cross-section for given torsional moment (Fig. 26). A geometrical constraint requires the cross-section to be constrained in a given square area. It can be seen from Fig. 26 that the optimal solution contains three types of regions: (a) those filled with material; (b) empty ones; and (c) those having a dense system of internal boundaries.

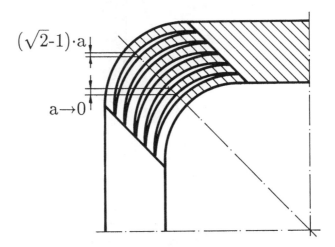

Figure 26. *Example of an optimal solution with an infinite number of internal boundaries (by Kohn and Strang).*

The same idea was used extensively by Bendsøe (e.g. 1989) for generating the optimal topology in shape optimization. In this very important development, Bendsøe used various microstructures, including those with square holes (for isotropic solutions) and first-second order rib systems (both are shown in Fig. 21 herein). Some optimal designs by Bendsøe are reproduced in Fig. 27.

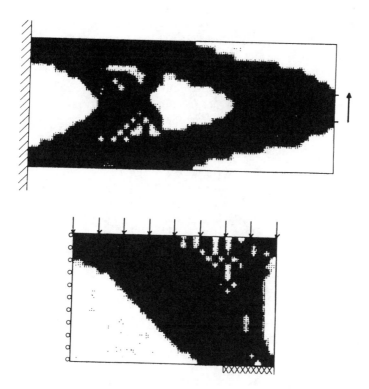

Figure 27. *Optimal shapes obtained by Bendsøe using 'advanced' layout theory.*

6 Verification of the Iterative COC Solutions in Section 2 (Fig. 1) Using Layout Theory

The global optimality of the layout in Fig. 1a can be shown quite easily by using the layout theory introduced in Section 3. The correct adjoint displacement field for this problem is (with x in the horizontal and y in the vertical direction and the origin at the top of the vertical bar):

$$\bar{u} \equiv 0 \, , \quad \bar{v} = ky \, , \quad \bar{\epsilon}_2 = \bar{\epsilon}_x \equiv 0 \, , \quad \bar{\epsilon}_1 = \bar{\epsilon}_y = k \, , \quad \gamma_{xy} \equiv 0 \, . \tag{30}$$

476

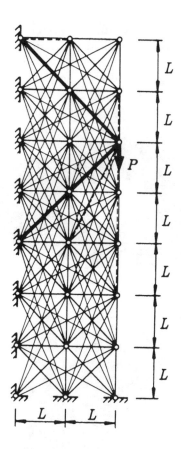

Figure 28. *Iterative COC example with a more complex structural universe.*

The above adjoint field satisfies the kinematic boundary conditions and by (2)–(4) admits non-vanishing members only in the vertical (y) direction. It also proves that the considered solution would also be optimal if the structural universe consisted of all possible members in the x-y plane.

The same does not necessarily apply to the solutions in Figs. 1b and c. For these problems the iterative COC method has also yielded absolute strain values of k, i.e. ($|\bar{\epsilon}| = k$), for all optimal members (and for members a and b in Fig. 15b) and a smaller absolute value ($|\bar{\epsilon}| < k$) for all other members. These are the correct adjoint fields for the considered universe but the same solutions would not necessarily be valid if we extended the structural universe to all possible members of the x-y plane. Iterative COC solutions considering more complex structural universes for the same boundary conditions and loading are being computed at present. One such example is shown in Fig. 28. The optimal members (with a cross-sectional area $\tilde{z}_i = 2\sqrt{2}$) are shown again by thick lines and the members having a cross-sectional area only slightly above the prescribed minimum value ($10^{-11} > \tilde{z}_i > 10^{-12}$) by broken lines. All other members have taken on the prescribed minimum cross-sectional area ($\tilde{z}_i = 10^{-12}$).

7 Concluding Remarks—Practical Aspects of Layout Optimization

These issues have been raised by the participants in discussions after the lecture; if the latter are representative of the civil engineering profession then they certainly deserve further clarifications by the authors.

It was mentioned in the lecture that the Prager-school not only anticipated, but also influenced, a number of more recent mathematical concepts now termed 'regularization,' 'relaxation,' 'homogenization,' 'smear-out process' and 'G-closure,' etc. In order to illustrate the last one, a problem was shown (see Fig. 29 left) in which a horizontal beam layout is to be optimized for simple supports along edges AB and BC, free edges AD and DC and a vertical point load at D. It was proved rigorously by Prager and Rozvany (1977b) by generating the correct adjoint for this problem that the least-weight solution is the limiting case of the sequence of solutions given in Fig. 29, that is, it (theoretically) consists of an infinite number of beams of finite length running in the AC direction and having a positive moment (continuous lines in Fig. 29), as well as an infinite number of beams of infinitesimal length running in the BD direction and subject to negative moment (broken lines). Prager termed this unusual type of singularity a 'beam-weave.' The above solution was based on a specific cost function $\psi = k|M|$ (and a similar stiffness function for elastic systems). This means that the strength condition contained only the beam moment (without the shear force) and, in the elastic problem, only flexural deformations were considered. It was claimed in a discussion that in the above solution infinite shear forces arise and hence the solution must be uneconomical, even when approximated with a finite number of beams. The speaker explained in his reply that the shear forces are actually finite even in the limiting solution but, indeed, a different solution is obtained if the specific cost function includes both the shear force and the bending moment. In fact, a more refined grillage theory was developed for that case (e.g. Rozvany, 1979) and some solutions with allowance for shear-costs were given in the lecture (see Fig. 13). The aim of Fig. 29 above was

478

to illustrate some unusual aspects of exact optimal solutions, and not to suggest a practical design, although a design with a finite number of beams having the above layout can be highly economical at very long spans, provided that mostly flexural stresses govern the design. A practical layout optimization study, considering long-span steel roof-structures and based on actual cost studies of a large construction company in Melbourne, was published already in the seventies (see Rozvany, 1976, pp. 259-262). This study revealed that the theoretical optimal layouts for beam systems are only economical at long spans.

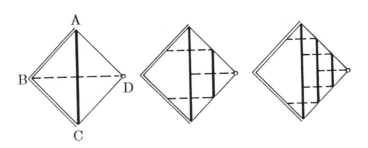

Figure 29. *Example of an optimal stress field represented by a generalized function: a 'beam-weave' (Prager).*

Much more emotional discussions were generated, however, by the conclusions that (a) a least-weight solid plate with prescribed maximum thickness contains an *infinite number of ribs* in the theoretical optimal solution and (b) a least-weight perforated plate has an *infinite number of holes* with *first and second order infinitesimal dimensions*.

It should be clarified first that for some years the above solutions have constituted *common knowledge* amongst leading mathematicians in this field (e.g. Strang and Kohn in the USA, Tartar *et al.* in France, Lurie *et al.* in the USSR, etc.) as well as amongst continuum-based researchers in structural optimization (e.g. Olhoff and Bendsøe in Denmark, Taylor in the USA, Cheng in the PR China, Armand in France, etc.). For the Prager school, similar findings have been customary since the early seventies (see Fig. 29).

The authors regard these theoretical least-weight solutions as being extremely important for the following reasons:

■ They provide a basis of comparison for assessing the relative efficiency of practical designs (see η in Fig. 6).

■ They explain as to why the optimal weight in these problems is elementation-dependent, i.e. the cost keeps on decreasing with an increase in the number of elements.

■ Optimal layouts revealed some intrinsic features of least-weight solutions and influenced pure mathematicians in developing some important new concepts (see above) as well as new function spaces (e.g. Strang and Kohn, 1983).

■ Practical designs can be obtained by using the theoretical optimal layout but with a finite (and relatively small) number of ribs, reinforcing bars or holes. Actual studies show that this way one can easily obtain designs within one per cent of the theoretical limit of economy. Even full scale tests on such 'discretized' continuum-solutions have been carried out (Fig. 6). First-second order microstructures can be replaced, without much loss of economy, by two sets of ribs having a width ratio of, say, 1 : 10.

■ Applications of the 'advanced' layout theory enable us (e.g. Bendsøe, 1989) to generate *optimal topologies* in practical shape optimization of continua because the perforated regions are usually not very extensive (see Fig. 27). At present, no other methods are available for this class of problems.

■ Finally, it happens far too often in research papers on structural optimization that the author formulates a class of problems with considerable rigour and then illustrates his general formulation with a trivial or near-trivial analytical solution only. Other authors produce only numerical solutions, the accuracy of which is not verified by a comparison with exact analytical solutions. Both types of publications can be useful, and also unavoidable, although they may fail to reveal certain pitfalls in the methodology used; however, they are some-times symptomatic of a certain intellectual laziness. One may criticize the Prager-school for being too basic research oriented; however, Prager and his associates deserve some recognition for their determination and perseverance in developing methods for deriving systematically the *exact analytical opti-mal solution for almost all possible boundary and loading conditions within certain classes of structures, in spite of the considerable complexity of the topologies and mathematical expressions involved* [see Figs. 11 and 12 and Eqs. (20)–(22) in Section 4.2].

The above discussions also seem to indicate a certain reluctance to accept new and more abstract ideas, which is more apparent in the civil engineering profession than, for example, amongst aerospace or electrical engineers and may partially account for the delay in the implementation of optimization methods by practicing civil engineers.

Acknowledgment The authors are indebted to the Deutsche Forschungsgemeinschaft (DFG) for financial support (Project No. Ro 744/1-1); to Mrs. S. Rozvany for editing the text; to Mrs. A. Fischer for processing the manuscript; and to Mrs. E. Becker for plotting and processing the diagrams.

References

Bendsøe, M. P., (1989), 'Optimal shape design as a material distribution problem,' *Struct. Optim.* **1**, 4, 193–202.

Chan, H. S. Y., (1975), 'Symmetric plane frameworks of least weight,' In Sawczuk and Mróz (Eds.), *Optimization in structural design*, (Proc. IUTAM symp. held in Warsaw, Aug. 1973), pp. 313–326. Springer-Verlag, Berlin.

Cheng, K. T. and Olhoff, N., (1981), 'An investigation concerning optimal design of solid elastic plates,' *Int. J. Solids Struct.* **17**, 3, 305–323.

Hemp, W. S., (1973), *Optimum structures*, Clarendon, Oxford.

Hemp, W. S., (1974), 'Michell framework for uniform load between fixed supports,' *Eng. Optimiz.* **1**, 61–69, Sept.

Hill, R. H. and Rozvany, G. I. N., (1985), 'Prager's layout theory: A nonnumeric computer method for generating optimal structural configurations and weight-influence surfaces,' *Comp. Meth. Appl. Mech. Engrg.* **49**, 1, 131–148, May.

Hillerborg, A., (1956), 'Theory of equilibrium for reinforced concrete slabs' (in Swedish), *Betong*, **41** 4, 171–182.

Kirsch, U., (1989), 'On the relationship between optimal structural topologies and geometries,' in C. A. Brebbia and S. Hernandez (Eds.), *Computer aided optimum design of structures: Recent advances*, (Proc. 1st Int. Conf. held in Southampton, UK, June 1989), pp. 243–253. Springer, Berlin.

Kohn, R. V. and Strang, G., (1983), 'Optimal design for torsional rigidity,' in Atluri, Gallagher *et al.*, (Eds.), *Hybrid and mixed finite element methods*, (Proc. Conf. held in Atlanta, 1981), pp. 281–288. Wiley and Sons, Chichester, England.

Kohn, R. V. and Strang, G., (1986), 'Optimal design and relaxation of variational problems,' I, II and III, *Comm. Pure Appl. Math.* **39**, 113–137, 139–182, 353–377, Jan.–March.

Kozlowski, W. and Mróz, Z., (1969), 'Optimal design of solid plates,' *Int. J. Solids Struct.* **5**, 8, 781–794, Aug.

Lagache, J.-M., (1980), 'A geometrical procedure to design trusses in a given area,' *Eng. Opt.* **5**, 1, 1–12.

Lagache, J.-M., (1981), 'Developments in Michell theory,' in Atrek and Gallagher (Eds.), *Proc. Int. Symp. on Optimum Structural Design*, (Held in Tucson, Oct. 1981), pp. 4.9–4.16. University of Arizona, Tucson.

Lagache, J.-M., (1983), 'Abstract convolution and optimum layout,' in Eschenauer and Olhoff (Eds.), *Optimization methods in structural design.*, (Proc. Euromech. Colloquium held in Siegen, Oct. 1982), pp. 340–345. Wissenschaftsverlag, Mannheim.

Lowe, P. G. and Melchers, R. E., (1972), 'On the theory of optimal constant thickness, fibre-reinforced plates,' I, *Int. J. Mech. Sci.* **14**, 5, 311–324, May.

Lurie, K. A., Fedorov, A. V. and Cherkaev, A. V., (1984), 'On the existence of solutions for some problems of optimal design for bars and plates,' *J. Optimiz. Theory Appl.* **42**, 2, 247–281, Feb.

Maxwell, J. C., (1872), 'On reciprocal figures, frames, and diagrams of force, *Trans. Roy. Soc. Edinb.* **26**, 1, Also in *Scientific Papers*, **2**, (W. D. Niven (Ed.), 1890), University Press, Cambridge, 174–177.

Michell, A. G. M., (1904), 'The limits of economy of material in frame-structures,' *Phil. Mag.* **8**, 47, 589–597, Nov.

Morley, C. T., (1966), 'The minimum reinforcement of concrete slabs,' *Int. J. Mech. Sci.* **8**, 305–319, April.

Murat, F. and Tartar, L., (1985), 'Calcul des variations et homogénéisation,' in *Les méthodes de l'homogénéisation: théorie et applications en physique.*, Coll. de la Dir. des Études et recherches de Élec. de France, Eyrolles, Paris, pp. 319–370.

Olhoff, N. and Rozvany, G. I. N., (1982), 'Optimal grillage layout for given natural frequency,' *J. Engrg. Mech. ASCE* **108**, EM5, 971–975, Oct.

Ong, T. G., (1987), *Structural optimization via static-kinematic optimality criteria*, Ph.D. thesis, Monash Univ. Melbourne, Australia.

Ong, T. G., Rozvany, G. I. N. and Szeto, W. T., (1988), 'Least-weight design of perforated elastic plates for given compliance: Nonzero Poisson's ratio,' *Comp. Meth. Appl. Mech. Eng.* **66**, 301–322.

Prager, W., (1974), *Introduction to structural optimization*, (Course held in the Int. Centre for Mech. Sci. Udine, CISM **212**), Springer-Verlag, Vienna.

482

Prager, W., (1977), 'Optimal layout of cantilever trusses,' *J. Optimiz. Theory Appl.* **23**, 1, 111–117, Sept.

Prager, W., (1978a), 'Nearly optimal design of trusses,' *Comp. Struc.* **8**, 4, 451–454, May.

Prager, W., (1978b), 'Optimal layout of trusses with finite numbers of joints,' *J. Mech. Phys. Solids* **26**, 4, 241–250, Aug.

Prager, W. and Rozvany, G. I. N., (1977a), 'Optimization of structural geometry,' in Bednarek and Cesari (Eds.), *Dynamical systems*, (Proc. Int. Symp. held in Gainsville, Florida, March, 1976), pp. 265–293. Academic Press, New York.

Prager, W. and Rozvany, G. I. N., (1977b), 'Optimal layout of grillages,' *J. Struct. Mech.* **5**, 1, 1–18.

Prager, W. and Shield, R. T., (1967), 'A general theory of optimal plastic design,' *J. Appl. Mech.* **34**, 1, 184–186, March.

Rozvany, G. I. N., (1966), 'Analysis versus synthesis in structural engineering,' *Civ. Eng. Trans. Inst. Engrs. Aust.* **CE8**, 2, 158–166, Oct., also *Proc. Conf. Inst. Engrs. Aust.*, March.

Rozvany, G. I. N., (1976), *Optimal design of flexural systems.*, Pergamon Press, Oxford, Russian translation: Stroiizdat, Moscow, 1980.

Rozvany, G. I. N., (1979), 'Optimum beam layouts: allowance for cost of shear,' *Comp. Meth. Appl. Mech. Engrg.* **19**, 1, 49–58.

Rozvany, G. I. N., (1981), 'Optimal criteria for grids, shells and arches,' in Haug and Cea (Eds.), *Optimization of distributed parameter structures,,* pp. 112–151, (Proc. NATO ASI held in Iowa City), Sijthoff and Noordhoff, Alphen aan der Rijn, The Netherlands.

Rozvany, G. I. N., (1984), 'Structural layout theory—the present state of knowledge,' in Atrek, Gallagher *et al.* (Eds.), *New directions in optimum structural design*, pp. 167–195. Wiley and Sons, Chichester, England.

Rozvany, G. I. N., (1989), *Structural design via optimality criteria.*, Kluwer Acad. Publ., Dordrecht.

Rozvany, G I N. and Adidam, S. R., (1972), 'Rectangular grillages of least weight,' *J. Eng. Mech. Div. ASCE* **98**, EM6, 1337–1352, Dec.

Rozvany, G. I. N. and Hill, R. H., (1976), 'General theory of optimal force transmission by flexure,' *Advances in Appl. Mech.* **16**, 184–308.

Rozvany, G. I. N. and Hill, R. H., (1978a), 'Optimal plastic design: superposition principles and bounds on the minimum cost,' *Comp. Meth. Appl. Mech. Engrg.* **13**, 2, 151–173, Feb.

Rozvany, G. I. N. and Hill, R. H., (1978b), 'A computer algorithm for deriving analytically and plotting optimal structural layout,' in Noor and McComb (Eds.), *Trends in computerized analysis and synthesis*, pp. 295–300. (Proc. NASA/ASCE Symp. held in Washington DC, Oct. 1978), Pergamon Press, Oxford. Also *Comp. and Struct.* **10**, 1, 295–300, April 1979.

Rozvany, G. I. N., Olhoff, N., Cheng, K. T. and Taylor, J. E., (1982), 'On the solid plate paradox in structural optimization,' *DCAMM Report* **212**, June 1981. And *J. Struct. Mech.* **10**, 1, 1–32.

Rozvany, G. I. N. and Ong, T.G., (1986a), 'Optimal plastic design of plates, shells and shellgrids,' in Bevilaqua, Feijóo, *et al.* (Eds.), *Inelastic behaviour of plates and shells*, pp. 357–384, (Proc. IUTAM Symp. held in Rio de Janeiro, August 1985), Springer-Verlag, Berlin.

Rozvany, G. I. N. and Ong, T. G., (1986b), 'Update to "Analytical methods in structural optimization".' in Steele and Springer (Eds.), *Applied mechanics update*, pp. 289–302. ASME, New York.

Rozvany, G. I. N. and Ong, T.G., (1987), 'Minimum-weight plate design via Prager's layout theory (Prager Memorial Lecture),' in Mota Soares (Ed.), *Computer aided optimal design: structural and mechanical systems*, (Proc. NATO ASI held in Troia, Portugal, 1986), pp. 165–179, Springer-Verlag, Berlin.

Rozvany, G. I. N., Ong, T.G., Sandler, R., Szeto, W. T., Olhoff, N and Bendsøe, M. P., (1987), 'Least-weight design of perforated elastic plates,' I and II, *Int. J. of Solids Struct.* **23**, 4, 521–536, 537–550.

Rozvany, G. I. N. and Wang, C. M., (1983), 'Extensions of Prager's layout theory,' in Eschenauer and Olhoff (Eds.), *Optimization methods in structural design*, (Proc. Euromech. Colloquium held in Siegen, Oct. 1982), pp. 103–110. Wissenschaftsverlag, Mannheim.

Rozvany, G. I. N., Zhou, M., Rotthaus, M., Gollub, W. and Spengemann, F., (1989), 'Continuum-type optimality criteria methods for large finite element systems with a displacement constraint,' Part I, *Struct. Optim.* **1**, 1, 47–72.

Save, M. and Prager, W., (1985), *Structural optimization—Vol. 1, Optimality criteria*, W. H. Warner (Ed.), Plenum Press, New York.

484

Strang, G. and Kohn, R. V., (1983), 'Hencky-Prandtl nets and constrained Michell trusses,' *Comp. Meth. Appl. Engrg.* **36**, 207–222.

Wang, C. M., Rozvany, G. I. N. and Olhoff, N., (1984), 'Optimal plastic design of axisymmetric solid plates with a maximum thickness constraint,' *Comp. and Struct.* **18**, 4, 653–665.

Yep, K. M., Sandler, R. and Rozvany, G. I. N., (1986), 'Optimal layout of long-span truss-grids II,' *Int. J. of Solids Struct.* **22**, 2, 225–238.

Sequential Optimisation for Structural Networks and Water Supply Systems Based on Sensitivity and Reliability

L. Gründig
Institut für Geodäsie und Photogrammetrie
Technische Universität Berlin
Berlin
Germany

Abstract Parameters reflecting reliability and sensitivity are of valuable help for an evaluation of structures. They are derived here for general network calculation problems and they are the basis of a strategy for the improvement of the network design—for an optimisation strategy. The strategy of optimisation presented here is based on a sequential reduction of a structure taking into account restrictions of deflections of the structure and reliability aspects. Applications are presented with respect to pin-joint bar structures, cable networks and membrane structures as well as water supply systems.

1 Description of a Typical Network Analysis Problem, a Hydraulic System

Since water supply systems are no longer open channels in which water flows in the direction given by the gravity, but are pipe networks, hydraulic calculations to ensure proper performance of the system have to be made. This has to be done when designing a water supply system and when (the more common case today) extending an existing network. The pressure heads at every node (tap) have to be in a given range and the flow in the different pipes is limited by material parameters. A water supply system has also to be checked to supply enough water in case of a fire emergency.

Due to history, three different types of networks are distinguished. The simplest form of a pipe network is a tree structure. It is easy to calculate because every node is reached only once and the direction of the flow of the water is therefore fixed. In case of leakage of a pipe, when a part of the net has to be shut down, all

485

B. H. V. Topping (ed.),
Optimization and Artificial Intelligence in Civil and Structural Engineering, Volume I, 485–501.
© 1992 *Kluwer Academic Publishers. Printed in the Netherlands.*

486

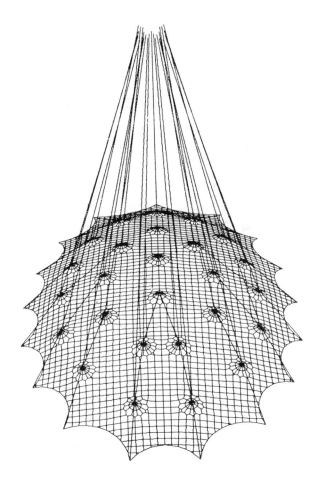

Figure 1. *Montreal Olympic roof structure as a figure of equilibrium based on the force-densities approach [13].*

subsequent outlets cannot be supplied with water.

To overcome this major drawback the principle of ring-shaped pipe networks was introduced. In these networks some major pipes are circular and so, if a shutdown of a single pipe occurs in the network it is still possible to obtain water at a lot of nodes.

Extending that principle and connecting not only major pipes (with a high flow of water) to circles, but also interconnecting smaller pipes, the security against a loss of supply of outlets due to the leakage of a single pipe is even higher. This built-up redundancy in the network, which makes it more secure against damages makes it difficult to calculate the pressure heads in every node.

In a simple, tree-shaped network all flows are easily determined after all input/output of the system is known and the coefficients of roughness for the different pipes are given. Although a unique solution exists in the more complex, redundant networks, it is not that simple to get it, because of the different ways the water can flow.

A well known way to determine the parameters of a pipe network is the Hardy-Cross method. The Hardy-Cross method makes use of the fact, that every redundant pipe in a network leads to a mesh in the system. After stating a mesh for each pipe the calculation process starts. First either the parameters 'pressure head' or the parameters 'flow' in the pipes have to be calculated using approximate values for the other type of unknown parameters. In order to achieve consistency between the two groups of parameters in one mesh of the structure, corrections to the unknowns are determined. This calculation of corrections is carried out until the corrections are below a predefined boundary value. Every mesh is treated separately and the iteration of that process may lead to an overall consistent structure with compatible parameters.

One problem here is to find correct meshes. Not every set of meshes will lead to convergence in the Hardy-Cross method. It is also not possible to solve for other parameters than pressure heads such as the coefficients of roughness. Another major drawback is that the Hardy-Cross method works only on a single mesh every step. Not the whole network is viewed and when a new iteration is made, only a single mesh is recalculated at a time—this brings naturally inconsistency into the data set during the process.

In other words, the Hardy-Cross method solves a system of equations built out of Kirchhoff's law expressed as function of the pressure heads. The inflow/outflow is given, the single flows in the pipes have to be calculated. This is identical to the calculation of the pressure heads in every node. The solution is found iterativly starting from approximate values. The rather simple relationship between the pressure loss and the flow in a pipe allows an easy calculation of each loop and leads to a simple correction. But it is not possible to obtain further information about the solved system applying the Hardy-Cross method, because only the system of equations is solved for.

A more general and nevertheless more robust strategy may be developed by starting from a global analysis of the network calculation problem.

488

2 A Systematic View to Networks Derived from Different Areas of Application

For completely different computational tasks of network problems a common approach may be found by catagorising and analysing the network properties in general. Each catagory of problems should be evaluated separately. The solution of the individual task is achieved just by comparison, thus finding the proper connection to the individual network problem afterwards. In addition to the advantage of getting more obvious views of the problem, solution strategies from different areas of application can be brought together and can be evaluated. Using such a view of the network problem it was possible to find the method of force densities for cable net structures [2],[3],[5], allowing for an easy calculation of most complex geometrical structures purely as force-structure which can be materialized by any material after achieving the figure of equilibrium.

The method of force densities for cable networks i.e. is described in the following way: Choosing the ratio of the expected bar force divided by the bar length as given input data together with the coordinates of fixed positions. Then the resulting equations of equilibrium of forces of any net structure are a linear system of equations for the coordinates of the nodes of the structure.

Starting from a systematic strategy of treating network type calculation problems the method of least squares is incorporated into that strategy in order to deduct as much information as possible from the network data.

3 Analysis of Networks in General

A network is understood as an assembly of legs or branches—with individual physical or mechanical or stochastical properties—which are connected with one another.

The points of connection of the individual branches are called nodes of the network. The coordinates of the points provide the geometry of the network. Each network may be defined topologically by its branch-node matrix C—an incidence matrix—of type (m, n) with m as the number of branches and n as the number of nodes. According to [1] C is given by:

$$C(j,i) = \begin{cases} 1 \text{ if branch } j \text{ starts in } i \\ -1 \text{ if branch } j \text{ ends in } i \\ 0 \text{ in all other cases} \end{cases}$$

C only contains the values 1 and -1 once in every row.

Using the branch-node matrix for the description of networks, certain properties of the network can already be derived. It is shown in [3] that

$$\text{rank } C = n - 1$$

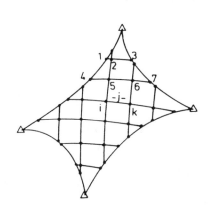
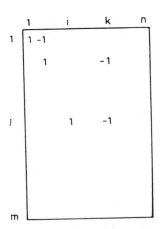

Figure 2. *Network topology and branch-node matrix* C.

This means fixing of one node is sufficient to eliminate the rank defect of the columns of C.

Several other characteristics are valid, see [11]. Using C a matrix $C^T C$ may be generated which is the node-node matrix. The rank defect of $C^T C$ equals the rank defect of C. In the same way a matrix CC^T may be formed which also has the rank $n - 1$.

In order to arrive at a complete description of a network, coordinates x must be assigned to the nodes and physical properties like the force t must be assigned to the branches. The coordinates may be arranged in a one dimensional matrix, a vector x, and the branch properties in a vector t. Using these definitions formulas may be derived which are valid for every type of network. For the differences d of the point coordinates between points which are connected by a branch, we can write:

$$d = Cx \tag{1}$$

The summation vector s of all branch variables starting in any node of the network yields:

$$s = C^T t \tag{2}$$

In the case of one dimensional coordinates and branches, (2) might be regarded as the equation for the equilibrium between internal forces and external forces s, acting in the same direction as t.

In order to complete the network description it is necessary to formulate the dependency between coordinate differences and the branch variables. This dependency might be regarded as the physical view of the problem. In general it is given by

$$t = f(d) \tag{3}$$

substituting t in (2) by (3) we get

$$s = C^T f(d) \tag{4}$$

(4) is the basic equation of equilibrium of forces in every node of the network and it will be the starting point for the derivations of formula for calculating cable networks and bar structures in general.

4 The Adjustment Approach for Networks

The equations (4) may in addition be interpreted as the basic law of a number of network problems. In geodetic network adjustment they are called the nonlinear normal equations. For statical networks they are the equations of equilibrium of forces. In electrical networks they are Kirchhoffs law and in hydraulic networks they are the basic condition that inflow and outflow add to zero is formulated. $t = f(d)$ may be chosen arbitrarily depending on the individual problem to be solved. But due to this the general approach to solve for coordinates x can be derived already here. Using suitable approximate values for t, t can be developed according to Taylor series with respect to changes of x to:

$$t_1 = t_0 + \frac{\partial f}{\partial d} \frac{\partial d}{\partial x} \Delta x \tag{5}$$

According to (2) $\frac{\partial d}{\partial x} = C$, so (4) results in:

$$\left(C^T \frac{\partial f}{\partial d} C \right) \Delta x = s - C^T t_0 \tag{6}$$

$C^T \frac{\partial f}{\partial d} C$ is the matrix of normal equations for geodetic applications. It is the stiffness matrix for structural networks. The inverse $\left(C^T \frac{\partial f}{\partial d} C \right)^{-1}$ is the cofactor matrix of the parameters x and it expresses the sensitivity of the parameters x. For structural networks it is the flexibility matrix which shows the influence of a change of the external load on the deflections of the nodes.

To describe the influence of changes on the branch variables t, Δx is substituted according to (6) into (5) which yields:

$$t_1 = t_0 + \frac{\partial f}{\partial d} C \left(C^T \frac{\partial f}{\partial d} C \right)^{-1} \left(s - C^T t_0 \right)$$

or

$$t_1 = \left(I - \frac{\partial f}{\partial d}C\left(C^T\frac{\partial f}{\partial d}C\right)^{-1}C^T\right)t_0 + \frac{\partial f}{\partial d}C\left(C^T\frac{\partial f}{\partial d}C\right)^{-1}s \qquad (5*)$$

As shown in [14]

$$C_0 = \frac{\partial f}{\partial d}C\left(C^T\frac{\partial f}{\partial d}C\right)^{-1}C^T \qquad \text{and} \qquad (I - C_0)$$

are idempotent matrices with the rank of C_0 being $n - 1$, equal to the rank of the branch node matrix C, and $(I - C_0)$ being $m - n + 1$. The trace of $(I - C_0)$ describes the degrees of freedom of the structure.

The corrections to t which the branches are possible to assume can be completely described by C_0. If only one branch variable will be modified by ∇t, the diagonal element r of $(I - C_0)$ gives the factor of t which will show up in the corrections Δt to t after the network calculation.

The following equation is valid:

$$\Delta t_k = -r_k \nabla t_k$$

r_k is called the redundancy factor of the k-th observation in geodetic adjustments and it is only defined between 0 and 1 for positive 'weights.' If r_k assumes 0 there will be no redundancy in the structure and the change Δt will in any case be zero. The structure will collapse if the branch k will be removed from the network. If r assumes 1 any change of t will have no influence on the structure, it will be completely stay local and show up locally. The redundancy factor therefore shows the essential influence of a branch element on the structure. In geodetic networks where the branches are observations, the redundancy numbers allow to identify blunders in the observations. They also allow one to show the influence of a blunder—the change of a branch on the structure (or rather its coordinates)—which is called external reliablity.

The effect on the coordinates results from:

$$\nabla x = \left(C^T\frac{\partial f}{\partial d}C\right)^{-1}C^T\nabla t_k$$

or with respect to $\nabla t_k = -\frac{\Delta t_k}{r_k}$ and $\left(\left(\frac{\partial f}{\partial d}\right)C\nabla x\right)_k$

$$\left(\frac{\partial f}{\partial d}C\nabla x\right)_k = \frac{1 - r_k}{r_k}\Delta t_k = (1 - r_k)\nabla t_k$$

The change of the structure due to changes in any branch variables can be expressed by these formulae. Such changes may be temperature loads on a structure, or imperfections of the cutting pattern etc.. Also changes of external node loads will be distributed within the structure onto the branches proportionally to the reliability numbers of the branches and proportional to $(1 - r)$ onto the nodal parameters. If a branch variable is fully redundant it will take all the additional imposed loads, if it has no redundancy all the load leads to a change of the structure, to the deflections of the structure.

As shown in [12] the redundacy factor can also be used as basis for a strategy of a sequential optimisation process of networks. A strategy based upon the cancellation of the branch with the largest redundancy factor is locally optimal with respect to precision and reliability.

For the trace of the changes in the flexibiltiy matrix caused by an elimination of branch variable k from the structure the following equation is valid:

$$\text{tr}\left(\left(\Delta\left(C^T\frac{\partial f}{\partial d}C\right)^{-1}\right)\cdot C^T\frac{\partial f}{\partial d}C\right) = \frac{1 - r_k}{r_k}$$

5 Generalisation with Respect to Spatial Networks

In order to describe real spatial networks a certain dimension must be assigned to the nodes and to the properties of the individual branches. The coordinates of point i may be arranged in

$$x_i = \begin{bmatrix} x_{i1} \\ \vdots \\ x_{ik} \end{bmatrix}$$

and the properties—projected forces—of branch j in

$$t_j = \begin{bmatrix} t_{j1} \\ \vdots \\ t_{jk} \end{bmatrix}$$

In order to assemble all nodes and branches we define the hyper vectors

$$\overline{x} = \begin{bmatrix} x_1 \\ \vdots \\ x_n \end{bmatrix} \quad \text{and} \quad \overline{t} = \begin{bmatrix} t_1 \\ \vdots \\ t_m \end{bmatrix} \tag{7}$$

consisting of the vectors x_i, $\forall i = 1, n$ and t_j, $\forall j = 1, m$

Substituting the numbers 1 by unit matrices I according to the dimension of x_i and t_j the branch-node matrix becomes

$$\overline{C} = \begin{bmatrix} I_k \ldots - I_k \ldots \\ \end{bmatrix}$$

and the equations (1) to (4) are valid again with d, s being structured according to (6); i.e.

$$\bar{s} = \overline{C}^T f(\bar{d}) \tag{4*}$$

For the calculation of the shape of a cable network equation (4*) may be the starting equation. The problem consists of the determination of the coordinates and forces of the node and bars of the structure.

In order to solve for unknown coordinates, according to (4*), generally an iterative method like Newton's method is applied. Starting from x_o we solve for $\Delta\bar{x}$

$$\bar{x} = \bar{x}_0 + \Delta\bar{x} \tag{8}$$

and get

$$C^T f(\bar{d})_0 + \frac{\partial \overline{C}^T f(\bar{d})}{\partial \bar{x}} \Delta\bar{x} = \bar{s} \tag{9}$$

with $\frac{\partial \bar{d}}{\partial \bar{x}} = C$ according to (1) we get

$$\frac{\partial f(\bar{d})}{\partial \bar{x}} = \frac{\partial f(\bar{d})}{\partial \bar{d}} \cdot \overline{C} \tag{10}$$

and

$$\overline{C}^T \left(\frac{\partial f(\bar{d})}{\partial \bar{d}} \right) \overline{C}\Delta\bar{x} = \bar{s} - \overline{C}^T f(\bar{d})_0 \tag{11}$$

It is shown in [5] and [11] that the strategy derived from (4*) may be regarded as a method of calculating displacements, but the method described is more general. Some statements might be made regarding (11):

- The necessary requirement for applying (11) in a solution process is that the Jacobi-Matrix $\frac{\partial f(\bar{d})}{\partial d}$ must exist. If it is positive definite, the system (11) will most likely converge to a solution of the non-linear problem.

- If $f(d)$ is a linear function of d, $\frac{\partial f}{\partial d}$ will be constant and the solution $\overline{\Delta x}$ will be achieved in one step. In fact Δx may replace $\bar{x}$ may replace that case.

Regarding cable network, $f(d)$ are—according to (3)—the forces in the individual branches connecting the nodes. It is elementary that

$$\bar{t}_j = F_j \begin{bmatrix} a_{xj} \\ a_{yj} \\ a_{zj} \end{bmatrix} = F_j \bar{a}_j \tag{12}$$

494

with force F_j and $\bar{a}_j$ representing the vector of projection with respect to the coordinate axes. (12) can easily be written as a function of coordinate differences according to (3)

$$f_j(\bar{d}) = \bar{s}_j = \frac{F_j}{t_j}\begin{bmatrix} d_{xj} \\ d_{yj} \\ d_{zj} \end{bmatrix} = \frac{F_j}{l_j}d_j \tag{12a}$$

with $l_j = \sqrt{d_{xj}^2 + d_{yj}^2 + d_{zj}^2}$ as the length of branch j derived from the coordinates of its end points. In general $f(d)$ according to (10a) is a nonlinear function of d because l is nonlinear in $\bar{d}$. In order to solve for coordinate changes according to (11) the matrix $\frac{\partial f}{\partial d}$ has to be formed whatever function $F(\bar{d})$ exists.

Applying the chain rule of differentiation to (12a) we get:

$$\frac{\partial f_j}{\partial d_j} = \frac{F_j}{l_j}I - \frac{F_j}{l_j^2}\bar{d}_j\frac{\partial l_j}{\partial d_j} + \frac{\bar{d}_j}{l_j}\frac{\partial F_j}{\partial d_j} \tag{13}$$

with $\frac{\partial l_j}{\partial d_j} = \frac{1}{l_j}d_j^T$ we get

$$\frac{\partial f_j}{\partial d_j} = \frac{F_j}{l_j}\left(I - \frac{1}{l_j^2}\bar{d}_j\bar{d}_j^T + \frac{1}{l_j}\frac{\partial F_j}{\partial d_j}\right) \tag{13a}$$

Regarding $\frac{\partial f_j}{\partial d_i} = 0, \forall = i+j$, we get $\frac{\partial f}{\partial d} = Q$ as a block diagonal matrix consisting of (3,3) submatrices and (11) reads:

$$\overline{C}^T\overline{Q}\,\overline{C}\Delta\bar{x} = \bar{s} - \overline{C}^T f(\bar{d})_0 \tag{14}$$

(14) corresponds to the system of equations for cable networks according to the theory of 2nd order where the deformed situation is taken into account while setting up the stiffness matrix formulation. (14) is very sensitive with respect to good initial values x when $F(d) = \frac{EA(l_j-l_{0j})}{l_{0j}}$ is derived from small elongations of the material, for example, when steel cables are used. In the case of poor initial values $F(d)$ might be positive or negative and Q might not be positive definite. The iterative calculation process—steps (14), (8)—then tends to diverge.

6 The Method of Force Densities

In order to arrive at a numerically stable solution for coordinate displacements according to (14), we may start again with formulas (10a) and ask for the conditions which let $f(u)$ become a linear function in u. Then the coordinates themselves can be solved for instead of solving for the displacements. It is evident that (13a) will be linear in u when $\frac{F_i}{l_i} = \text{const} = q$.

In [3] q are called 'force densities.' For a given constant vector

$$\bar{q}^T = [q_1, \ldots, q_m]$$

(14) becomes a linear system of equations for calculating the coordinates of all nodes of a statical network. It reads

$$\overline{C}^T \overline{Q}\, \overline{C}\overline{x} = \overline{s} \qquad (14a)$$

with $\overline{Q}$ as diagonal matrix of $\bar{q}$.

(14a) is the basic formulation of the 'force density' method and it is perfectly suited to be used for formfinding of prestressed cable networks or of membrane structures because it does not need any initial coordinates of the shape. As shown in the formfinding example in Fig. 1, estimates for the force densities may be derived from rough assumptions with regard to the expected forces and the corresponding lengths of the branches.

A more general application might be obtained if (12a) is regarded in the following way

$$\bar{s}_j = \bar{q}_j \begin{bmatrix} d_{xj} \\ d_{yj} \\ d_{zj} \end{bmatrix} = f_j(q, \bar{d}) \qquad (15)$$

Substituting (15) in (4∗) we get an underdetermined system of equations

$$\overline{f}(\bar{d}, q) = \overline{C}^T \overline{Q}\, \bar{d} = \bar{s} \qquad (16)$$

or in its linearized form:

$$\overline{C}^T \overline{QC}\Delta\bar{x} + \overline{C}^T \overline{D}\Delta\bar{q} = \bar{s} - \overline{C}^T (\overline{Qd})_0 \qquad (17)$$

(16) may be treated in a least squares way, minimizing $\Delta x^T P_x \Delta x + \Delta q^T P_q \Delta q$. Depending on the chosen weights P_x, P_q a solution can be obtained which is close to a given geometry or close to given force densities. In [3] and [5] it is shown how the least squares problem may be extended by adding functions between x and q to (16).

A third point of view may be taken into account with respect to force densities: in some problems the behaviour of a structure has to be judged with respect to its capability of carrying forces simply as normal forces. Starting from (16) we see that again a linear system of equations arises.

$$\overline{C}^T \overline{Q}\, \bar{d} = \overline{C}^T \overline{D}\, \bar{q} = \bar{s} \qquad (18)$$

If the shape is described by its topological structure and its coordinates, only the force densities are unknown. In a quadrangular spatial network more equations than free parameters q exist. We get a best fitting solution of (17) by minimizing the sum of the quadratic deflections q. The remaining misclosures show which partition of internal or external forces can not be transported by the network structure. Those misclosures cause bending moments.

All three strategies were simply derived from some basic equations valid for any type of network which only have to be interpreted in a special way. The basic strategy of calculating reliability values for judging the result may be applied in the same way as shown in Section 3.

7 Hydraulics

In the same way a hydraulic network system may be modelled and analysed [15]. Defining x as pressure heads in the nodes of such a network, t as pipe flow rate, d as pressure differences (loss of pressure), and the external flow s in the pressure heads the basic equations outflow = inflow (Kirchhoff's law) is described by

$$C^T t = s \qquad (19)$$

Since in hydraulic systems not a static case is examined, but a dynamic case is viewed, the pipes have a certain resistance to the water. That means, that the water looses energy, when it flows through the pipe, so the pressure head at the end of a pipe is—when looking only at a single pipe—lower because of the loss of energy in that pipe, than it is in the beginning of that pipe.

The relationship between the amount of flow and the loss of energy (loss of pressure) is very complex and it is not possible to state this in an explicit formula. Fortunately, there are approximate formulas, which describe the nature well enough to be used for the calculation.

In this report the formula with flow t, loss of pressure d and roughness coefficients R we assume the following 'physical property' of the branch:

$$t^\omega = \frac{d}{R} \qquad (20)$$

(19) describes the simplified Darcy's law, but this law could easily be exchanged against another one without influencing the following approach. R is the coefficient of roughness, the parameter which describes the friction between the water and the pipe. Usually ω is either 2 or 1.85, depending on the set of coefficients of roughness. Hazen-Williams coefficients use $\omega = 1.85$.

The linearisation of formula (19) with respect to the pressure loss) yields:

$$C^T t_0 + C^T \frac{\partial t}{\partial d} \Delta d = s \qquad (21)$$

The linearisation of (20) with respect to d may be done implicitly yielding to:

$$\omega \cdot t^{(\omega-1)} \Delta t = \frac{1}{R} \Delta d$$

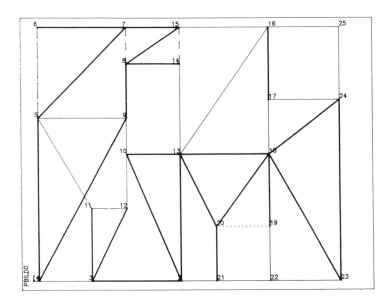

Figure 3. *Pipe network with reliability factors for the flow rate parameters. The thick lines show high redundancy (>50%), the redundancy value decreases with thin lines (50–30%), dashed lines (30–20%), dash-dotted lines (20–10%) and is close to zero (<10%) on dotted lines.*

Because we are not interested in the difference of the pressure head values, but in the pressure heads themselves, the chain rule is applied observing $\frac{\partial d}{\partial x} = C$ and we find the final formula to be:

$$C^T \omega^{-1} T^{(1-\omega)} R^{-1} C \, \Delta x = s - C^T t_0$$

This system of equations can be used for the calculation of the unknown pressure heads. It gives the same result as Hardy-Cross method. In the same way as described in Section 3 reliabiltiy values may be calculated monitoring the quality of the network.

The calculation according to (21), however, is not flexible enough. It is only feasible to fix the external loads on the system, the roughness parameters have to remain unchanged, the flow in the pipe cannot be influenced at all etc. Following the idea of Section 5 a more general solution is achieved by the extension to a least squares approach. Reading the formula (21) as observation equations in a least squares sense we obtain:

$$\left(C^T \omega^{-1} T^{(1-\omega)} R^{-1} C\right)^T P \left(C^T \omega^{-1} T^{(1-\omega)} R^{-1} C\right) \Delta x =$$

498

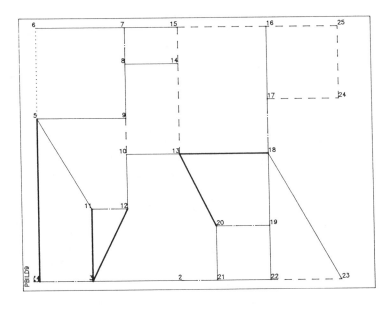

Figure 4. *The same network as in Fig. 3, the most redundant pipes were removed step by step.*

$$\left(C^T \omega^{-1} T^{(1-\omega)} R^{-1} C\right) P \left(s - C^T t_0\right) \tag{22}$$

Extending the system by additional parameters and appending additional equations to the system of (22), thus introducing them into the least squares method the reality can be modelled in a better way.

If—for instance—a flow in one of the pipes should be fixed, it is only necessary to add an equation to the system which fixes the difference in the pressure heads resulting from Darcy's law, when the flow is given.

Another problem, which occurs especially when extending an existing network is the determination of the coefficients of roughness. Due to the fact, that the pipes are getting older and that the chemical properties of the water change the roughness, they cannot always be introduced as constant (fixed) values. What can be done, is to guess that neighbouring pipes with same age have a similar roughness. With this assumption and some additional observations e.g. pressure heads at some nodes, coefficients of roughness can be determined by extending the mathematical model:

$$\Delta t = -\frac{d}{w R^2} t^{(1-w)} \cdot \Delta R \tag{23}$$

8 Redundancy and Reliability

As described earlier, redundancy describes the formation, which is not needed for the determination of the unknown parameters, but which can be used to ensure that the results of the calculation are correct and only minor influenced by an inaccurate modelling of the structure.

The redundancy value r for every observation shows to which percentage this parameter is controlled by the other paramters of observational type, $(1 - r)$ is the part of the parameter being used to determine the unknowns.

In geodetic networks the observations are formulated as functions of the variables. Regarding the (*inflow, outflow*) as an 'observation,' every such observation consists out of a sum of 2 and more edge variables. We are not interested in their redundancy because the system has to supply a certain amount of water at every outlet.

However the flow information might be subject to modifications and for practical applications the following question should be answered: What is the influence of a single pipe flow rate on the network's behaviour? The answer to this question can be found from the redundancy factors of the flow rate parameters just by rearranging the matrices of the system of equations. The formula for the redundancy factor r_k for the k-th flow rate is:

$$r_k = I - (TC)\left((TC)^T\left(CPC^T\right)(TC)\right)^{-1}(TC)^{-1}\left(CPC^T\right) \qquad (24)$$

9 Conclusion

Different types of network problems can be dealt with by using the same general network analysis. It proved to be very useful to separate the topological and physical properties of the elements of the network. After this has been done, the same strategy can generally be applied for any network type problem. This leads to a better understanding of the network problems and of their behaviour. In particular the redundancy values for each connection in a network, which describe its influence on the whole system, can be calculated. The redundancy values represent the paramaters to influence the network and therefore build the basis for an optimisation with respect to any target function.

References

[1] Fenves, S. and Branin, F., (1963), 'Network-Topological Formulation of Structural Analysis,' *Proc. of the ASCE, Journal of the Struct. Division*, August, 485–513

500

[2] Linkwitz, K. and Schek, H.-J., (1971), 'Einige Bemerkungen zur Berechnung von vorgespannten Seilnetzkonstruktionen,' *Ingenieur-Archiv* **40**, 145–158,

[3] Schek, H.-J., (1974), 'The force density methods for form finding and computation of general networks,' *Computer methods in applied mechanics and engineering* **3**, 115–134

[4] Linkwitz, K., Schek, H.-J. and Gründig, L., (1974), 'Die Gleichgewichtsberechnung von Seilnetzen unter Zusatzbedingungen,' *Ingenieur-Archiv* **43**.

[5] Gründig, L., (1975), 'Die Berechnung von vorgespannten Seilnetzen und Hängenetzen unter Berücksichtigung ihrer topologischen und physikalischen Eigenschaften und der Ausgleichungsrechnung,' Dissertation am IAGB, Universität Stuttgart, DGK-Reihe C No. 216, (1976), und SFB 64—Mitteilungen 34, (1976).

[6] Hangleiter, U., Gründig, L. and Löw, K, (1979), 'Program libray FASNET applied to formfinding of a large cable net structure,' *Pres. paper to the 2nd Intern. Symposium Weitgespannte Flächentragwerke*, Mai, Stuttgart.

[7] Stark, W., (1983), 'Untersuchungen zur Lösung und Inversion schwachbesetzter großer geodätischer Normalgleichungen,' Dissertation an der Universität Stuttgart, DGK-Reihe C, No. 301.

[8] Ströbel, D., (1984), 'Entwicklung von Hyperspartechniken zur Glechungslˇsun für Probleme in der Ausgleichungsrechnung und der Statik,' Diplomarbeit am IAGB, Universität Stuttgart.

[9] Linkwitz, K., Gründig, L., Hangleiter, U. and Bahndorf, J., (1984), 'Mathematisch-numerische Methoden der Netzberechnung,' *Abschlußbericht Teilprojekt F2, Mitteilungen des SFB 64*, No. 72.

[10] Gründig, L. and Bahndorf, J., (1984), 'Evalution of accuracy and reliability in geodetic network adjustments—Program system OPTUN,' *Journal of Surveying Engineering* **110**, 2, (August).

[11] Gründig, L., (1985), 'The "FORCE-DENSITY"—Approach and Numerical Methods for the Calculation of Networks,' *Proc. of 3rd Intern. Symposium 'Weitgespannte Flächentragwerke'*, Stuttgart, März.

[12] Gründig, L. and Bahndorf, J., (1985), 'Sequential Optimization of Geodetic Networks with Respect to Accuracy and Reliability,' *Invited Paper to the 7th Intern. Symposium on Geodetic Computations*, Krakau, Juni.

[13] Linkwitz, K., Gründig, L., Bahndorf, J., Neureither, M. and Ströbel, D., (1988), 'Geodätische Methoden als Mittel zur Form- und Zuschnittsberechnung on Flächentragwerken, Beispiel Olympiastadion Montreal, X,' *Internationaler Kurs für Ingenieurvermessung*, München, Proceedings, Kap.

[14] Bahndorf, J., (1989), 'Zur Systematisierung der Seilnetzberechnung und zur Optimierung von Seilnetzen,' Dissertation an der Universität Stuttgart, (eingereicht).

[15] Bäuerle, J., (1988), 'Berechnung von vermaschten Rohrnetzen,' Diplomarbeit am Institut für Anwendungen der Geodäsie im Bauwesen, Universität Stuttgart.

The Geometrical Analysis of Deformations Based on an Optimal Network Design and Proper Data Management Tools

L. Gründig
Institut für Geodäsie und Photogrammetrie
Technische Universität Berlin
Berlin
Germany

Abstract   The theoretical background of a geometrical analysis of deformations based on statistical tests of the deformation parameters. In addition problems of management of survey data are analysed with respect to deriving deformations of power dams, slides and buildings. A data model has been developed allowing for a comprehensive treatment of those data in a global data management system. Applying that global strategy of deformation analysis, data from different sources like photogrammetry, global positioning system GPS or ordinary terrestrial surveying can be treated properly.

1   The Concept of Detecting Significant Geometrical Deformations

Depending on the quality of geometrical data describing an object, deflections of the object can be derived from repeatedly gathered measurements of the object by a direct comparison of the measurements. The differences of the measurement values with respect to the former measurements of the objects are always partly due to random influences on the measurements and to the other part due to real deformations. The problem of deformation analysis here is to find out the significant deflections and to separate them from the miscellaneous random effects. This task is not only existent for the comparison of direct observations like repeatedly performed measurements but also for derived observations like the coordinates describing the shape of an object. In this case the arrangements of the measurements does not have to be the same for the repetition of the measurements on the object. The quality of the comparison may be improved by selecting coordinates as

B. H. V. Topping (ed.),
Optimization and Artificial Intelligence in Civil and Structural Engineering, Volume I, 503–524.
© 1992 *Kluwer Academic Publishers. Printed in the Netherlands.*

parameters for the comparison instead of direct observations, because the derived coordinates are often resulting from more observations than necessary, i.e. form an adjustment process which allows for deriving parameters of not only high accuracy but also of sufficient reliability.

The philosophy of the deformation analysis which is realised in the program systems DEFAN 'deformation analysis of networks' (Gründig, Neureither) is based on the coordinates and the variance-covariance-matrix of the points representing the state of the considered object which has to be computed for each observed epoch. The calculation of the coordinates is done by a least squares adjustment and analysis of the observations with the program system NEPTAN 'network planning and technical analysis' (Gründig, Bahndorf). In the second step using program system DEFAN the different epochs are compared with one another and deformation parameters are computed. According to our strategy the original observations are substituted by coordinates in the first step of the analysis. These coordinates serve as input data for the main part of the analysis—the determination of the deformations.

The underlying concept of the deformation analysis is a general deformation model which may be expressed by the following system of equations:

$$B^T (x + v) + Cg = w. \tag{1}$$

In this equation x represents the vector of coordinates of the epochs to be compared. Normally these coordinates result from preceeding least squares adjustments. B is the functional matrix of the differences of the coordinates between the epochs with:

$$B = [E - E], \tag{2}$$

where E is the unit matrix. The unknown deformation parameters are the vector g and C is the corresponding functional matrix of the deformation parameters. The vector v is the vector of residuals of the pseudo-observations x and w is the vector of misclosures of the system of equations (1).

(1) is a standard model of the adjustment calculus and the formula for solving this standard type called 'conditions between observations with unknown parameters' may also be applied to this system. As solution formula for the unknown parameters we get the following equation:

$$g = \left(C^T \left(B^T Q_{xx} B \right)^{-1} C \right)^{-1} C^T \left(B^T Q_{xx} B \right)^{-1} w, \tag{3}$$

where Q_{xx} is the variance-covariance matrix of the coordinates of the corresponding epochal adjustments with the original observations. Dependent on what deformation model was chosen, g can contain displacements, rotations or scale factors with respect any subgroup of points.

Two types of models are given below:

- single point movement in x, y, z

$$C_k = \begin{bmatrix} 0 & 0 & 0 \\ \cdot & \cdot & \cdot \\ 1 & 0 & 0 \\ 0 & 1 & 0 \\ 0 & 0 & 1 \\ 0 & 0 & 0 \\ \cdot & \cdot & \cdot \end{bmatrix} \left.\vphantom{\begin{matrix}1\\0\\0\end{matrix}}\right\} \text{point } k$$

$$g_k = \begin{bmatrix} g_{xk} \\ g_{yk} \\ g_{zk} \end{bmatrix}$$

- displacement, rotation and scale factor for a group l of points with respect to x, y

$$C_l = \begin{bmatrix} 0 & 0 & 0 & 0 \\ \cdot & \cdot & \cdot & \cdot \\ 1 & 0 & x_{l1} & -y_{l1} \\ 0 & 1 & y_{l1} & x_{l1} \\ \cdot & \cdot & \cdot & \cdot \\ 1 & 0 & x_{ln} & -y_{ln} \\ 0 & 1 & y_{ln} & x_{ln} \\ 0 & 0 & 0 & 0 \\ \cdot & \cdot & \cdot & \cdot \end{bmatrix} \left.\vphantom{\begin{matrix}1\\0\\0\\1\\0\end{matrix}}\right\} \text{group } l$$

$$g_l = \begin{bmatrix} g_{xl} \\ g_{yl} \\ g_{\theta l} \\ g_{scl} \end{bmatrix}$$

With the general form of equation (1) it is possible to set up any type of geometrical or physical deformation model.

After the solution for the deformation parameters these parameters are tested to verify whether they are significant or not. For this a Fisher-test is used, which allows for the comparison of two independent variances derived from independent data sets with different degrees of freedom. The test quantity is computed as a quotient of two quadratic forms the first one $(\hat{\sigma}_0^2 g)$ represents the decrease of the weighted sum of residuals of system (1) due to the deformation parameters being tested; the second one $(\hat{\sigma}_0^2 ij)$ is the weighted sum of residuals from the single adjustments. The decrease is computed as

$$\hat{\sigma}_{0g}^2 = \frac{g^T P_{gg} g}{m_g} = \frac{g^T \left(C^T \left(B^T Q_{xx} B \right)^{-1} C \right) g}{m_g}, \tag{4}$$

where P is the weight matrix and m the number of the parameters. If the test quantity

$$T = \frac{\hat{\sigma}_{0g}^2}{\hat{\sigma}_{0ij}^2} \tag{5}$$

exceeds the boundary value set by the Fisher-distribution $F_{1-\alpha, r_i + r_j, m_s}$, where the sum of redundancies from the single adjustments $r_i + r_j$ and m_g are the degrees of freedom and $1 - \alpha$ is the significance level, then the deformation parameters are significant.

2 Problems of Data Management

Each technical computer program system consists of both algorithmic and data management tools. Both kinds of tools are important since the flexibility of the program system depends not only on the algorithmic and numerical solution but also on the management of data.

Some problems and requests for a good data management should briefly be stated:

Automatic data transfer An automated data transfer is important if the data are processed by a complex program system. The components of the system should read the data from a common data base or at least from a standardized data mangement file. Manual operations of the user should only be allowed for correcting false data if these corrections are well controlled by a special program module enforcing the integrity of the data after the data manipulation.

Standardization of the data; use of certain defined data structures The data should be well structured, so that access to the data and operations with the data may easily be performed by certain software tools. Furthermore the data structuring achieves a standardization of the data, such that data from different sources can be treated with the programs that use those data structures.

Logical and internal structure of the data To achieve a high performance the internal physical data structure has to be adapted for the operations that are undertaken with the data. The user on the other hand should only have access to the data in a logical way. So from the user's viewpoint the logical data structure must be independent from the physical one.

Consistency of the data The consistency of the data must be guaranteed. This should be performed by the programs or data operators.

Problem of dependency between algorithm and data As far as possible the algorithm of the program should be independent from the managing operations on the data. By this it is possible only to change the data management routines and not the algorithm if any data structures have to be changed. On the other hand improvements of the algorithm do not effect an adaption of the data.

Data management operators for flexibility in application Some data management operators have to be created so that the programs may be used in a flexible way. Flexibility can be achieved by general concepts of the algorithm as well as by comprehensive facilities of the data management operators.

3 Data Management Concept and Data Structure

In our program system DEFAN for deformation analysis we realized a data management concept which is demonstrated in Fig. 1. This concept should meet the requests as they are given in Section 2.

The central unit of the system is a data file. On this data file the source data as well as results of the computations can be stored. All programs of the system have access to this file via two routines for obtaining information from and putting information to the file (GETDB and PUTDB). By this the essential data are absolutely transferred by the system itself between its components. The only task of the user is to give a minimum of information to control transfer of and access to the data.

The data file itself consists of certain data tables, which differ according to the physical structure and the logical meaning of the data they contain. The access to the tables is controlled by a directory where net name and epoch number are stored (see Fig. 2). This directory indicates the tables belonging to a certain net and data type. The following data tables exist:

Statistical Parameters global statistical information about the network (epoch), e.g. dimension of the network, kind of the network (given by type of observations and number and type of rank defects), number of points, global redundancy, quadratic sum of residuals, additional parameters

Points and Coordinates name and type of the points, coordinates

Variances-Covariances variance-covariance-matrix of the coordinates of the points

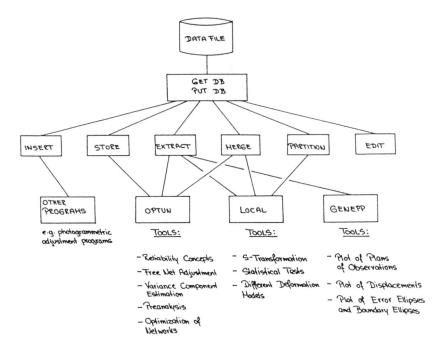

Figure 1.

Observations type of observation, value of observation, precision of observation

Deformations deformations between corresponding points of different epochs

Ellipses error ellipses, boundary ellipses of detectability of deformations

This data structure allows the inclusion of most important network type survey data. Furthermore with the INSERT operator it is possible to adapt data from other sources to the data structure given above. So it is possible also to analyse data which request special software for evaluation (e.g. special photogrammetric data etc.) with the program DEFAN.

4 Data Management Operators

The data management in the programs is achieved by several operators. These operators are written as routines, which are independent of the algorithms in the program. The operators control as far as it is possible the consistency of the

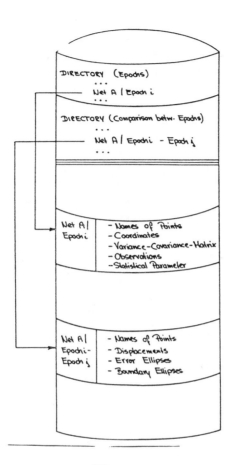

Figure 2.

data. Error messages are printed out, if the consistency between the data is not guaranteed. The following data management operators are available:

EDIT The operator EDIT creates and modifies data for the network and deformation analysis. It is intended to realise this operator also as an interactive graphic routine.

STORE With the operator STORE the data and results of a least squares network adjustment performed by NEPTAN or PICTRAN (photogrammetric bundle adjustment program) are stored on the central data file.

EXTRACT (Fig. 3) This operator allows the extraction of certain parts of a network or even of certain types of coordinates (e.g. x or y-coordinates). The user defines the part of the network in a logical way by indicating the names of the points belonging to this partial network and/or the type of coordinates. The data concerning this part are automatically made available by the system, so that the consistency between the data is guaranteed. Afterwards the computations of the program are applied to the extracted data. EXTRACT may be used in all programs.

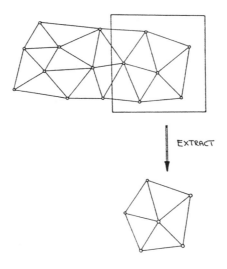

Figure 3.

PARTITION (Fig. 4) This is a special operator for the deformation analysis program DEFAN. DEFAN distinguishes between reference points and object points. So all related information (e.g. variance-covariance-elements) has to be ordered by this criterium. PARTITION is invoked if the user preselects a reference system and if the reference system changes after a reference point cannot be regarded as to be stable.

MERGE (Fig. 5) MERGE is used to form a common epoch out of a number of single epochs. This is useful if one wants to perform a common adjustment of some epochs (e.g. to compute the reliability of deformations between two epochs). Another reason may be to exploit the information of several measured epochs for future deformation analyses. With MERGE the unstable points between the epochs are introduced as different points, where the stable points are identical in the considered epochs.

These operators maintain a high flexibility in the use of the programs without the necessity of changing the underlying algorithms.

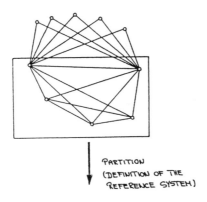

PARTITION
(DEFINITION OF THE
REFERENCE SYSTEM)

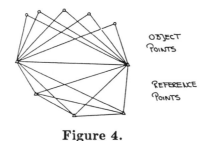

OBJECT
POINTS

REFERENCE
POINTS

Figure 4.

5 Datum Transformation

In the last two sections it was shown that the data standardisation has positive effects on the realisation of data management operators which then can be used in all programs using this data structure. With the substitution of the original observation data by coordinates for the purpose of deformation analysis standardisation of data is also achieved. This enables the technique of deformation analysis to be applied to the data regardless of type of network and observations. However, a problem which occurs with coordinates is the datum-dependency. Coordinates and their variances and covariances are always related to the datum they are based on. The coordinates and error ellipses computed by an adjustment are different whether the adjustment is performed as a free network adjustment or as an adjustment with some coordinates fixed. An example for three different results of the same network are shown in Fig. 6.

512

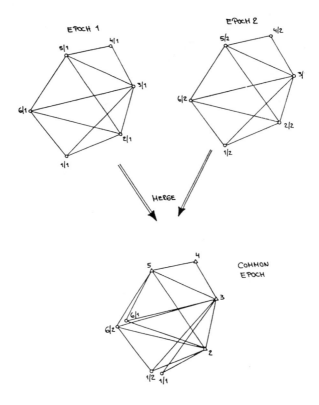

Figure 5.

The transformation from one result to another (each called S-system) can be done without performing a new adjustment. The algorithmic instrument for this is called an S-transformation. The formula of the S-transformation from system 1 to system 2 are the following:
for the transformation of the coordinates

$$x_2 = S_2 x_1 \tag{6}$$

for the transformation of the variance-covariance matrix

$$Q_2 = S_2 Q_1 S_2^T \tag{7}$$

where S_2 is the transformation matrix

$$S_2 = E - G \left(G_2^T G \right)^{-1} G_2^T \tag{8}$$

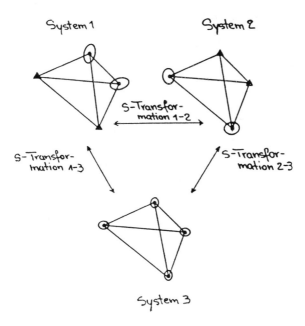

Figure 6.

(E is the unit matrix; G is a matrix containing conditions for a free network—dependent on its type—to allow for the computation of the coordinates; G_2 is the part of G containing only the rows of the datum points)

This process of S-transformation is of essential meaning for a deformation analysis with coordinates. Of course, the deformation parameters (e.g. the displacements as differences of coordinates) are also datum-dependent. A comparison of two epochs with different datum leads to unplausible deformations (see Fig. 7). So it has to be guaranteed that a common and proper computational base is choosen to be able to compute correct deformations (Fig. 8).

This computational base consists of the points of the reference system which can be preselected by the user and which is modified by the system if any reference point has moved. The choice of the datum for the network adjustment therefore is independent from the computational base in the deformation analysis. All concerned data are transformed to this new datum by the means of an S-transformation before the deformation analysis starts. With this tool it is also possible to compare different types of networks (e.g. GPS and terrestrial networks) by controlling the datum transformation by the number and kind of datum conditions of the matrix G. Furthermore, the influences of different datum pa-

514

Net A / System 1

Net B / System 2

Deformations

Figure 7.

Net A / System 3

Net B / System 3

Deformations

Figure 8.

515

rameters may be separated from the deformation parameters by performing the S-transformation with additional free datum parameters (e.g. a free scale between two epochs with distance measurements).

Thus the concept of datum transformation allows a flexible treatment of different types of networks, regardless of their type, their datum or their configuration.

6 Examples of Applications of the Deformation Analysis

Finally, some application examples are presented. The first example is the network ARGEN which was established to detect slides and movements of a net area near a highway construction in Southern Germany. Sixteen epochs have already been measured, each epoch with a slightly different network design (see Figs. 9, 10), resulting from the visibility between points due to the constructions on the site. Because of the flexible data management and the datum transformations, which are processed automatically without any intervention of the user, the analyses of the different epochs do not cause any problems. The results of a comparison is given in Fig. 11. The used deformation model for this comparisons was a single point movement model, because of the different topography of the area.

A second example which is described in detail in (Fraser, Gründig) is the Turtle Mountain in Alberta/Canada. The input data for the deformation analysis came from a photogrammetric data acquisition and evaluation. By an interface programm these data were put to the above described data file. The results of the analysis with DEFAN have been of good quality which could be approved by simulated deformations between the epochs.

The third example is the network HOLLISTER in California/USA. In order to determine the crustal movements near the faults San Andreas, Calaveras and Sargent a high precision network was installed where slope distances with a Geodolite were measured. Because of the poor information of the z-coordinates (exact levelling data were missing) only the x, y-deformations could be analysed. This was achieved by an extraction of the x and y-information using the operator EXTRACT. The result of the example in Fig. 12 (comparison epoch 1 to epoch 6) consisted of different deformation parameters for the group of points being separated by the faults.

An example for applying the deformation analysis for the monitoring of buildings is the use of NEPTAN and DEFAN for the computation of deformations for the Lausanne Cathedral (Gründig, Neureither). During repair work people discovered cracks in the stonework of the cathedral. No explanation could be given for this effect. Therefore it was decided to establish a geodetic monitoring network consisting of 69 objects at five different levels on the pillars where the cracks have been observed (principal plan of the situation of the points given in Fig. 13).

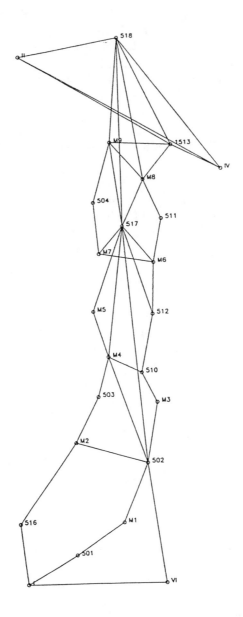

Figure 9.

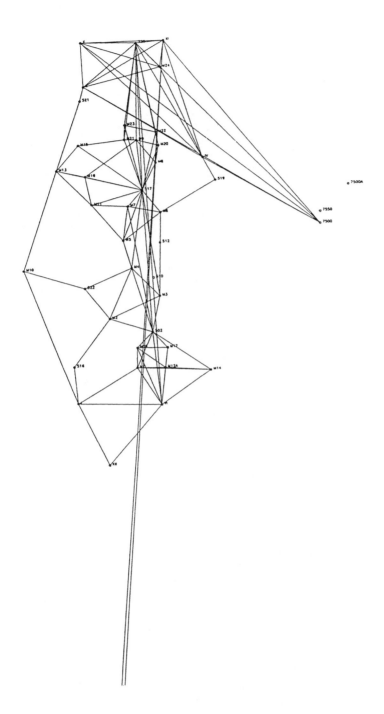

517

Figure 10.

518

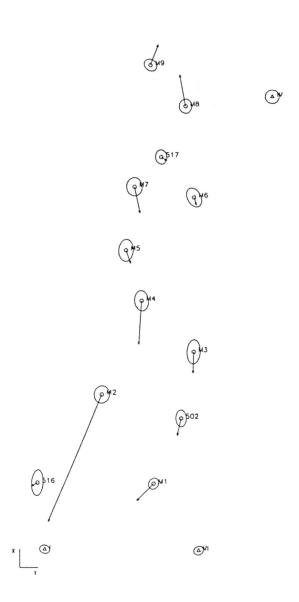

Figure 11.

519

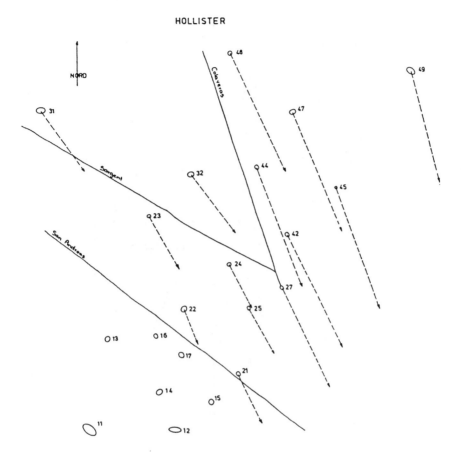

Figure 12.

520

Figure 13.

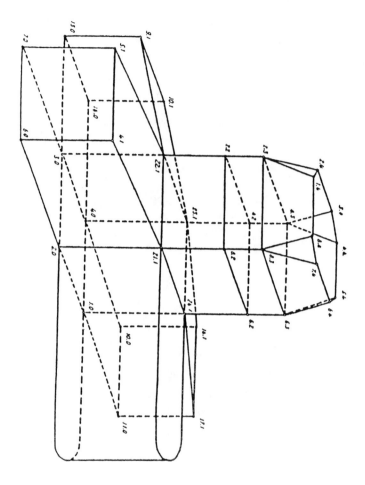

The measurements were taken in fixed time intervals and to assess the effects of repair work. The use of NEPTAN/DEFAN for the analysis of two representative epochs brought very practical results, mainly by exploiting the facility of the data management operators.

A first request was the visualization of certain parts of the observation plan (given in Fig. 14) to check the completeness and the configuration of the network. The operator EXTRACT was used for this purpose, so that graphical outputs for single parts of the whole network could be easily produced (Fig. 15). EXTRACT was also applied for the representation of deformation vectors of special groups of points, e.g. the points situated on the second level as given in Fig. 16. As it was found by the results of DEFAN, deformations in the height were mostly not significant. So EXTRACT was used to separate the x, y-information from the z-information to achieve more sensible results for the movements in the x, y-plane.

The operator PARTITION was necessary to initially define and during the localization process of deformations to redefine the reference system of stable points. By the use of this operator in this operator in the combination with the algorithmic tool 'S-transformation' the datum definition for the adjustment for the single epochs can be chosen completely independently from the datum definition used for deformation analysis. The computations of the single epochs were performed as free network adjustments by fixing the center of all approximate coordinates. The datum for the deformation analysis, however, has been defined only by the points situated on the floor of the cathedral, because it was assumed that these points would not participate in the movements of the stonework. This hypothesis was tested using DEFAN and confirmed with the exception of two single points which had moved significantly. To check the reliability of the computed deformations the data management operator MERGE combines the data of two different epochs and creates a new common epoch. An adjustment of this common epoch—where non-identical points are renamed—is necessary to compute the possible effects of non-detectable small blunders of observations on the deformations (external reliability of deformations). In the case of the Lausanne Cathedral the reliability of the deformations in the outer parts of the network was bad because of the poor configuration of the network in these parts.

The main result of the Lausanne Cathedral deformation analysis was the detection of a slight tilt of the tower in the middle of the cathedral (Fig. 16), which has effected the cracks in the pillars. This important global information as well as further detailed information can be obtained by the flexible use of data management operators realized in NEPTAN and DEFAN.

522

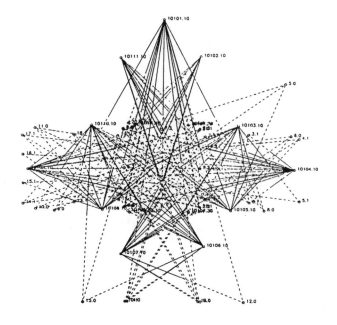

Figure 14.

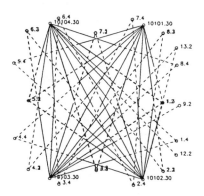

Figure 15.

Figure 16.

7 Conclusion

Combination and interaction of algorithmic and data management tools is essential for a high quality computer program system. In the case of the deformation analysis some operators have been described. It has been shown that the implementation of these operators serves for more flexibility in applications.

References

[1] Gründig, L. and Bahndorf J., (1984), 'Accuracy and Reliability in Geodetic Networks—Program System OPTUN,' *Journal Surveying Engineering* **110**, 2, (August).

[2] Gründig, L., Neureither, M. and Bahndorf, J., (1985), 'Detection and Localization of Geometrical Movements,' *Journal Surveying Engineering* **111**, 2, (August).

[3] Fraser, C. S. and Gründig L., (1985), 'The Analysis of Photogrammetric Deformation Measurements on Turtle Mountain,' *Photogrammetric Engineering and Remote Sensing* **51**, 2, (February).

[4] Gründig, L., Bahndorf, J. and Neureither, M., (1983), 'Zur Auswirkung von Beobachtungsfehlern auf Deformationen oder Deformationen als Beobachtungsfehler gedeutet,' *Schriftenreihe Vermessungswesen HSBW München* **9**, Munich.

[5] Strang van Hees, G. L., (1982), 'Variance-Covariance-Transformations of Geodetic Networks,' *Manuscripta Geodaetica* **7**, 1–20.

[6] Gründig, L. and Neureither, M., (1988), 'Data Management Tools with Respect to the Geometrical Analysis of Deformations,' *Computers and Structures* **30**, 3, 623–635.

Interaction of Criteria, Accuracy and Reliability and its Implication for the Design of Engineering Survey Control Networks

L. Gründig
Institut für Geodäsie und Photogrammetrie
Technische Universität Berlin
Berlin
Germany

Abstract Accuracy and reliability aspects play an important role in describing the quality of a survey network design. In this contribution the interaction of these quality criteria is illuminated. It is shown that a network is more sensitive with respect to criteria of reliability than with respect to accuracy. A strategy for optimizing the design of networks was therefore developed based on preserving sufficient reliability and still enforcing a given accuracy level.

1 Introduction and Motivation

Engineering survey control networks are established for setting out bridges, tunnels, power dams and for monitoring those artificial constructions or natural ones like slides, subsidence areas etc. Mostly the network design is done according to the experience of the surveying engineer. Taking into account the situation at the site, control points are established in areas which are regarded to be stable and from which points on the object to be surveyed can be determined relative to the control points via geodetic means. However, it may not be guaranteed that the control points remain stable. In general they have to be observed and evaluated relative to one another and to the object points in a more complex geodetic network.

Objective criteria for network design which should be considered are those of precision, sensivity, reliability and economy. The precision can be derived from the stochastic properties of the observations applying the law of propagation of errors. In the case of adjustment, the variance covariance matrix of the parameters to be determined describes the precision behaviour. The same is true for the sensivity with regard to an expected deformation behaviour. It is based on the

B. H. V. Topping (ed.),
Optimization and Artificial Intelligence in Civil and Structural Engineering, Volume I, 525–544.
© 1992 *Kluwer Academic Publishers. Printed in the Netherlands.*

error distribution of the observations and on the geometry of determination of the parameters. The reliability of a network design is mainly based not upon the propagation of errors but rather on the degree to which the observations control each other. If they don't control each other within a network or a part of it there is no internal reliability, and a blunder in an observation will remain undetected and might falsify the result. The falsification of the network parameters due to blunders of any kind is described by the external reliability.

Although reliabilty aspects in geodetic networks are mentioned often in scientific publications, the realization of the theory and its efficiency for application in practice are not adequately confirmed. Most program system in use still only provide parameters and their precision in a network adjustment; possibly due to this fact the method of least squares is often regarded as a strategy where observational errors might remain undetected and falsify the results.

In the following, the concept of reliability analysis is described starting from the formulations of parametric adjustments. It is shown that the tools of statistical test for outliers, internal reliability and variance component estimation can be derived from basic formulas and that they act as powerful means for network design and analysis.

2 The Mathematical Model of Accuracy and Reliability

In the adjustment of geodetic networks, the following functional model is set up:

$$\underline{v} = f(\underline{x}) - \underline{b} \tag{1}$$

where $\underline{b}$ are the observables, $\underline{v}$ the residuals and $f(\underline{x})$ the (non-linear) functions of the unknowns. Using approximate unknowns $\underline{x}_0$, we get the linearized observation equations

$$\underline{v} = \underline{A}\,\underline{\Delta x} - \underline{l} \quad \text{with} \quad \underline{A} = \left.\frac{\partial f}{\partial x}\right|_0 \quad \text{and} \quad \underline{l} = \underline{b} - f(\underline{x}_0) \tag{1a}$$

and the least squares solution observing $\underline{v}^T \underline{P}\,\underline{v} = \min$:

$$\underline{\Delta x} = (\underline{A}^T \underline{P}\,\underline{A})^{-1} \underline{A}^T \underline{P}\,\underline{l} \tag{2}$$

It is well known that an estimable quantity of any adjustment is

$$\hat{\sigma}_0^2 = \frac{\underline{v}^T \underline{P}\,\underline{v}}{r} \tag{3}$$

with r the number of redundant observations in the adjustment.

The quotient

$$\frac{\hat{\sigma}_0^2}{\sigma_0^2} \quad (\sigma_0 = \text{a \textit{priori} variance of unit weight}) \tag{3a}$$

may be tested by applying the F-test.

The test determines whether or not the assumed functional or stochastical model has to be rejected with a given statistical probability. Because many observations are included in the test, it might not be sensitive enough for single observations, i.e., there might be errors present in the observations without the test indicating a re- jection of the model.

It was shown by Baarda (1968) that the F-test not only can be applied to the global test of rejecting the functional or stochastical model, but it may also be favorably applied to test whether or not every single observation might be an outlier.

Substituting Δx according to (2) into (1a), we get the desired equations for the internal reliability of the observations. Starting from

$$v = \left((A^T P A)^{-1} A^T P - I\right) l \tag{4}$$

with unit matrix I, and using

$$Q_{vv} = P^{-1} - A(A^T P A)^{-1} A^T \tag{5}$$

we get

$$v = -Q_{vv} P l. \tag{6}$$

It is well known that the redundancy of the adjustment problem can be calculated according to

$$r = \text{tr}(Q_{vv} P) \tag{7}$$

If we define r_j as the jth diagonal element of $Q_{vv} P$, we get σ_0^2 according to (3) derived from a single observation l_j, as:

$$\hat{\sigma}_{0j} = \sqrt{\frac{v_j^2 P_j}{r_j}} \tag{8}$$

2.1 Statistical Tests for Outliers

The testable quantity for a single outlier is:

$$w_j = \sqrt{\frac{\hat{\sigma}_{0j}^2}{\sigma_0^2}} \tag{9}$$

Assuming a diagonal matrix P, we get the 'data snooping' value

$$w_j = \frac{|v_j|}{\sigma_{vj}} \tag{10}$$

For the application of the F-test, a probability α of a correct decision—there is an outlier—will be chosen. With $\alpha = 99\%$ level of significance we get a critical value $k = \sqrt{F(\alpha, 1, \infty)} = 2.53$ indicating the limit for rejecting the null hypothesis.

In order to decide that the test correctly concludes that there is an outlier, a probability β—the power of the test—has to be chosen as well. Thus we test w_j against the boundary value $\sqrt{\lambda} = \sqrt{F(\alpha, \beta, 1, \infty)}$ of the non-central F-distributation. With $\alpha = 99.9\%$ and $\beta = 80\%$ we get the boundary value $\sqrt{\lambda} = \sqrt{F(\alpha, \beta, 1, \infty)}$ of the non-central F-distributation. With $\alpha = 99.9\%$ and $\beta = 80\%$ we get the boundary value $\sqrt{\lambda} = 4.1$.

For the practical application $w_j < k$ is used as a first check, and if w_j exceeds $\sqrt{\lambda}$ the observable l_j is flagged for rejection.

It has to be emphasized that the hypothesis being tested is a single outliers in the adjustment problem. Numerous applications of this test to actual adjustments show, however, that even in the case of multiple outliers which are not clustered the test is very sensitive and correctly identifies erroneous observations.

2.2 Blunder Detection; International and External Reliability

Förstner (1979) showed that it is possible to derive the size of a single blunder after the adjustment. If we assume only l_j being an outlier, then v_j is only caused by $\underline{\nabla l_j}$. With

$$\underline{\nabla l_j} = \begin{vmatrix} 0 \\ 0 \\ \nabla l_j \\ 0 \\ 0 \\ 0 \end{vmatrix} \tag{11}$$

we get according to (6)

$$v_j = -r_j \nabla l_j \tag{12}$$

or

$$\nabla l_j = -\frac{v_j}{r_j} \tag{13}$$

where ∇l_j is now the blunder calculated from the actual residual v_j which was assigned to the observation in the adjustment. Substituting v_j according to (8) assuming $w_j = \sqrt{\lambda}$, we get a boundary value ∇l_{0j} for a detectable blunder

$$\nabla l_{0j} = \sqrt{\frac{\lambda}{r_j P_j}} \qquad (14)$$

∇l_{0j} is a measure of internal reliability of the adjustment. It may be calculated for any observation. It is important because it indicates the controllability of the observations.

Because the variables ∇x are used for further applications, instead of the adjustment observations it is essential to know which influence a possible blunder might have on the coordinates. Substituting l by ∇l_{0j} according to (11) into (2) we get

$$\nabla x_{0j} = -(A^T \underline{P} A)^{-1} A^T \underline{P} \nabla l_{0j} \qquad (15)$$

∇x_{0j} is the change of x caused by a just detectable blunder ∇l_{0j}. ∇x_{0j} is dependent on the chosen datum (constraints of the adjustment but there exist functions of x which are datum independent, namely adjusted observables.

Applying (15) to the pre-analysis of a deformation monitoring network, the dependency of ∇x_{0j} on the datum is not essential if for each individual epoch the same outer constraints are used (fixed stations or minimum constraints). ∇x_{0j} indicates the deformation which could be caused by a blunder ∇l_{0j}.

A scalar function of ∇x being datum independent may be derived starting from (4) and arriving at the following equivalence

$$\underline{v}^T \underline{P} \underline{v} = \underline{l}^T \underline{P} \underline{l} - \Delta x^T A^T \underline{P} \underline{l} = \underline{l}^T \underline{P} \underline{l} - \Delta x^T A^T \underline{P} A \Delta x \qquad (16)$$

Because $\underline{v}^T \underline{P} \underline{v}$ and $\underline{l}^T \underline{P} \underline{l}$ both are datum independent, $\Delta x^T A^T \underline{P} A \Delta x$ also is datum independent. Substituting Δx by Δx_{0j} we arrive at

$$\delta_j = \nabla x_{0j}^T A^T \underline{P} A \nabla x_{0j} \qquad (17)$$

Using (10) we get

$$\delta_j = \nabla l_{0j}^T \underline{P} A \left(A^T \underline{P} A \right)^{-1} A^T \underline{P} \nabla l_{0j} \qquad (17a)$$

and with (14) and (12):

$$\delta_j = (1 - r_j)\, p_j \nabla l_{0j}^2 = \left(\frac{1 - r_j}{r_j} \right) \lambda \qquad (17b)$$

δ_j indicates the influence of a blunder ∇l_{0j} on the variables of the adjustment, i.e. an increase of the quadratic form (17). Using the terminology of Baarda (1968), it is called external reliability.

In geodetic networks often only the influence of an error on a subset of unknowns is of interest. As shown in (4) the influence on nuisance parameters like orientation unknowns can be separated from the influence on coordinate parameters.

The importance of blunder detection and internal and external reliability is that they utilize all the information included in the observations.

Fig. 1 shows a network where a blunder of 50 cm has been introduced in distance observation 100–200. The statistical analysis indicates an error (boundary value > 4.1). The largest value w indicates the most probable error. The corresponding value ∇l gives the size of that blunder which is equal to the blunder introduced in the adjustment.

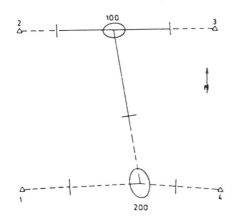

directions

from	to	observ.	residuals	$\nabla l (")$	w
100	2	300.0010	5.8	−50.0	4.9 **
100	3	100.0004	5.4	−41.3	4.6 **
100	200	193.3930	−11.3	23.2	4.8 **

distances

from	to	observ.	residuals	$\nabla l (cm)$	w
100	2	500.007	2.0	−2.9	2.4
100	3	499.995	−2.2	3.1	2.6 **
100	200	481.700	.5	−47.3	5.2 **
200	1	550.350	−.5	.8	.6
200	4	450.500	3.6	−5.3	4.3 **

Figure 1. *Result of the adjustment, indicating the largest data snooping value $w = 5.2$ in observation 100–200.*

2.3 Effects of a Blunder on Critical Parts of a Network

While equation (17b) gives an estimation of the worst influence of undetected blunders on the variables, equation (15) may be used to check the influence of blunders on critical variables.

For engineering applications the influence of an error in observation l_m on the relative positioning of points i, k (see Fig. 2) is often required. The relative positionings may be the breakthrough error in tunnel networks (see Fig. 3) or the critical parts in control networks. Applying (15) to $\underline{\nabla x}$ due to l_m, we get the variable ∇x_i, ∇x_k, ∇y_i, ∇y_k. If we refer these coordinate shifts as along (∇d_{ki}) and perpendicular (∇p_{ki}) to the connection $\vec{ik}$, we get

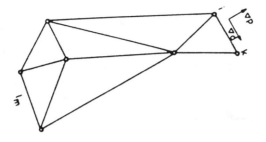

Figure 2. *Influence of a blunder in l_m on the relative positioning of points i, k.*

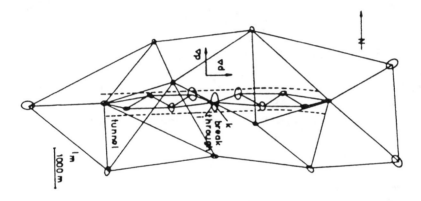

Figure 3. *Influence of a blunder in l_m on the breakthrough error.*

$$\nabla d_{ki} = (\nabla x_k - \nabla x_i)\cos\phi_{ik} + (\nabla y_k - \nabla y_i)\sin\phi_{ik} \qquad (18a)$$

$$\nabla p_{ki} = (\nabla x_k - \nabla x_i)\sin\phi_{ik} + (\nabla y_k - \nabla y_i)\cos\phi_{ik} \qquad (18b)$$

with ϕ_{ik} the orientation of $\vec{ik}$.

∇d_{ki} and ∇p_{ki} may be calculated for each individual observation, thus indicating the worst possible influence of an undetectable error on the desired function of the coordinates.

2.4 Variance Component Estimation

From the adjustment, the variance of unit weight may be calculated. It gives an estimation of the accuracy of an observation of unit weight.

A problem arises when heterogeneous geodetic observations are to be used together in an adjustment model, i.e. the distances and directions in a geodetic network. The problem is to estimate the weight relation between the various observations. As shown in (10), an estimation of variance of unit weight may be derived from single observations taking into account the individual redundancy number assigned to the observation in the adjustment.

Förstner (1979) showed that it may be also calculated for a group of k observations according to

$$\overline{\sigma}_{0k}^2 = \frac{\left(\underline{v}^T \underline{P}\, \underline{v}\right)_k}{\sum_{j=1}^k r_j} \tag{19}$$

In order to be a correct estimation of the variance of unit weight, the individual groups of observations must generate equal or nearly equal values. This property may be favorably applied to establish an iterative procedure for an estimation of variance components of observational groups.

If the group values σ_{0k}^2 differ from σ_0^2, the respective weights are premultiplied by σ_0^2/σ_{0k}^2 in every iteration. This procedure converges within a reasonable number of iterations. Experimental tests by the author confirmed the convergence if the observational groups have reasonable redundancy ($\sum r_j > 20$ from empirical tests). Convergence was achieved in less than 6 iterations.

It must be emphasized that the variance component estimation, although it is a useful check of the proper choice of the relative weights of observational groups, is only applicable in the absence of gross errors because it presumes the errors are normally distributed.

If gross errors are still in the adjustment model, the variance components of the respective group may get an unreasonably large value causing a down weighting of that group and the data snooping technique being applied afterwards might indicate no significant error due to the down weighting.

Because the application of data snooping presumes a proper variance for the observations entering the adjustment while the application of variance component estimations presumes the absence of gross errors, data snooping must be applied as the first step of checking of the observations using given variances.

2.5 Combination of Analysis and Preanalysis

The following algorithm for combination of measured, and of planned observations has been derived and applied. Using m for measured and p for planned observations, we get the following error equations:

$$\underline{v}_m = \underline{A}_m \Delta\underline{x} - \underline{l}_m$$

$$\underline{v}_p = \underline{A}_p \Delta\underline{x} \tag{20}$$

Minimizing $\underline{v}_m^T \underline{P}_m \underline{v}_m + \underline{v}_p^T \underline{P}_p \underline{v}_p$ leads to the normal equations

$$\left(\underline{A}^T \underline{P}_m \underline{A}_m + \underline{A}_p^T \underline{P}_p \underline{A}_p\right) \Delta \underline{x} = \underline{A}^T \underline{P}_m \underline{l}_m \tag{21}$$

and to the solution

$$\Delta \underline{x} = \left(\underline{A}^T \underline{P}_m \underline{A}_m + \underline{A}_p^T \underline{P}_p \underline{A}_p\right)^{-1} \underline{A}^T \underline{P}_m \underline{l}_m \tag{22}$$

After calculating $\underline{X}_{k+1} = x_j + \Delta \underline{x}$, an iterative process of adjustment has to be repeated until it converges to $\Delta \underline{x} = 0$.

From a numeral point of view, systems (21) and (22) may be regarded as a damped solution of nonlinear equations with the damping factors $\underline{A}_p^T \underline{P}_p \underline{A}_p$. Because the matrix of damping factors is at least positive semi-definite, it can be regarded as a type of Levenberg damping (Levenberg 1944) which is known to be numerically stable. The strategy shown above is especially useful in checking the influence of additional observations on the precision and reliability of adjustment results.

3 Design Strategy with Respect to Precision and Reliability

Fig. 4 shows a control network which was planned with distance and angular measurements and which was well designed with respect to the precision of the points and to the sensitivity of detecting point movements. In the analysis of the observations a poor reliability showed up. Gross observational errors might remain undetected in this design and might cause 'artificial' movements. The dashed arrows show the significant movements determined from a deformation analysis. The solid part of the arrows show the maximum influence of possible just undetectable blunders in the directions of the movements. It becomes apparent that the movements might at least partly be due to observational blunders in one or both measurement epochs.

This example shows that reliability aspects in deformation control networks are very essential in order to ensure the result given the possibility of blunders. They are also essential in order to ensure the correct result given the possibility of a systematic falsification of the parameters, i.e. due to the influence of scale of distance measurements. A good reliability is guaranteed if observations control each other and if the influence of a blunder on the parameters of the network remains sufficiently small.

For an evaluation of the global falsification of the parameters with respect to a possibly undetected blunder, the equation (17b) which is independent of the reference system can be used.

A mutual interaction of precision and internal reliability can be derived when the influence of the elimination of an observation on the precision of the parameters is evaluated. The cofactor matrix $Q_{xx} = \left(A^T P A\right)^{-1}$ is enlarged by ΔQ_{xx} if

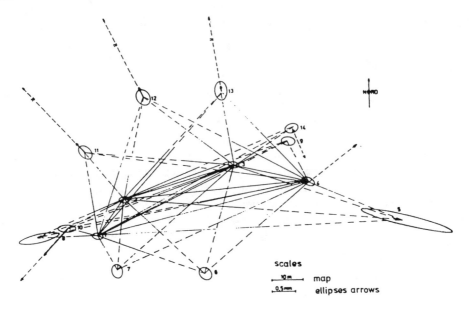

Figure 4. *Influence of observational errors on point movements.*

the observation showing redundancy factor r_i is cancelled. According to Gründig (1986) the following equation holds:

$$\text{tr}\left(\Delta Q_{xx_i} A^T P A\right) = \frac{1 - r_i}{r_i} \tag{23}$$

Because $A^T P A$ is a constant matrix for a given observational scheme tr $\left(\Delta Q_{xx} A^T P A\right)$ is small if

$$\frac{1 - r_i}{r_i}$$

is small enough, i.e. r is large, because

$$\frac{1 - r_i}{r_i}$$

is a monotonic function of r_i.

The smallest enlargement of precision with respect to $\text{tr}(\Delta Q_{xx i} A^T P A)$ by eliminating an observation is achieved when the most redundant observation r_i is removed. In the formulation

$$\text{tr}\left(\Delta Q_{xx i}\right) = \frac{p_i a_i^T Q_{xx}^2 a_i}{r_i} \tag{24}$$

which describes the increase of the variance covariance matrix the same statement becomes apparent.

A strategy based upon the cancellation of the observation with largest redundancy factor r_i is locally optimal with respect to precision and reliability.

3.1   Example of the Improvements of Network Designs

The improvement of the design is done not by changing the observational weights, but by modifying the observational scheme. Those observations which have no significant contribution to the target function of precision and reliability are removed step by step. The proposal strategy is therefore a process of reduction starting from an allowable design and ending in a design which is just sufficient according to the imposed restriction.

Figures 5 and 6 show the application of the reduction strategy applied to a special purpose network, when only a deformation in position 4 should be detectable.

The influence of a blunder on a function of the variables was locally minimized also.

In order to visualize the effect of cancelling one observation at a time several quality criteria of a network are calculated within the individual steps applied in the proposed strategy.

These criteria are:

$r_{\min}$	minimum redundancy factor of an observation
$r_{\max}$	maximum redundancy factor of an observation
∇l_0	boundary value for a non-detectable blunder
$\nabla x_{\max}$	falsification of the parameters
σ_p	average point standard deviation
$a_{\max}$	maximum diameter of the point error ellipses
$\text{cond}_{\max}$	maximum condition number as quotient of largest divided by the smallest eigenvalue
$a_{\max}/a_{\min}$	ratio of maximum versus minimum diameter of the point error ellipses
a_{emax}	maximum diameter of relative error ellipses

536

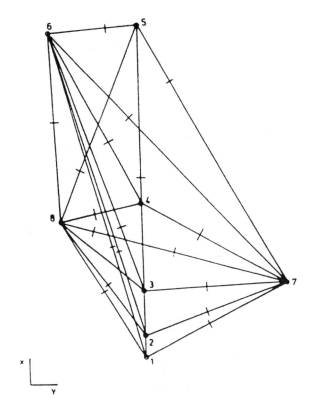

Figure 5. *Starting situation.*

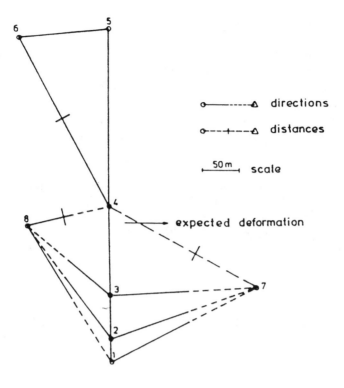

Figure 6. *Reduced observational scheme.*

a_{emax}/a_{emin} ratio of maximum versus minimum of relative error ellipses.

The design strategy may be characterized as a gradient strategy because in each individual step the target function is minimized in the direction of the local gradient. This direction happens to be described by the maximal redundancy of an observation. The corresponding observation may be superfluous with respect to reliability and accuracy as well. In order to generalize the approach and to make it more economical an evolution strategy of optimization or a dynamic strategy may also be satisfactory. These strategies might allow one to regard more than a single cancellation at a time; also in this type of approach the sensitivity analysis evaluated here would play an important role.

4 NEPTAN Program Package

A program system called NEPTAN—network planing and technical analysis—has been developed which allows for the adjustment of one, two and three dimensional geodetic networks, as well as the computation of accuracy and reliability values. The main features of the system are directed toward the task of enabling the user to judge and make optimal use of all information contained in the observations; the system thus provides a powerful tool for geodetic network adjustment and blunder detection.

The main features of the program system are data management of input and output, external reliability with respect to critical parts of networks, free network adjustment. The examples and numerous network adjustments confirm that the system is capable of fulfulling the imposed requirements. The flexibility was achieved by means of a high modularization of the program and a powerful data management scheme.

4.1 NEPTAN - Features

Tools for quality control of networks, such as accuracy and reliability checking and devices for variance component estimation and for free net adjustment have been implemented in the FORTRAN programming package NEPTAN.

At this time the following observation types are included in the adjustment system:

- directions or angles
- azimuths
- horizontal distances
- slope distances
- scaled distances
- zenith angles
- height differences
- observed coordinates
- observed correlated coordinates differences.

The directions may be given in grads or degrees. Angles are presently represented by two directions. Because the nuisance parameters associated with the directions are eliminated, they don't enlarge the system of equations to be solved.

538

An appropriate weight may be attached to any individual observation or to any subgroup of observations. For each group of observations, the variance component of unit weight σ_{0k} is calculated according to Section 2. σ_{0k} provides the information about a correct relative weighting of groups of observations in the absence of blunders.

In order to be able to solve large systems of equations as well as to solve systems of equations originating from spezial purpose applications such as deformation networks, different techniques for solving linear equations were evaluated and implemented in the NEPTAN software.

If a complete variance covariance matrix of the parameters is not required, i.e. for large control networks, the profile of the matrix of normal equations is minimized according to a modified Banker's strategy (Snay 1976) and the inverse within the profile is calculated. If deformation analysis is required after running NEPTAN, the task 'store the result on data management file' is combined with a control parameter causing the complete inverse to be calculated applying Stiefel's exchange algorithm (Stiefel 1976). Thus the program system automatically chooses the algorithm most appropriate for the type of problem to be solved.

In addition, special features are implemented like calculating the eigenvalues of the matrix of normal equations in order to evaluate the global accuracy and the sensitivity of the network with respect to critical movements.

The results of the adjustment—the coordinates and their covariance matrix—are, upon request stored on a data management file for further use by program system DEFAN which localizes movements between different measurement epochs.

In order to apply NEPTAN for a preanalysis of planned observations, only initial coordinates and the planned observational data types with their assumed a priori variances are provided. The planned observations are described by their from-to (topological) information only.

The preanalysis includes accuracy calculations and reliability checking. Choosing specific points and their critical directions of movement, the influences of all marginally non detectable errors of the observations on any chosen direction are calculated. This enables the designer to check the critical parts of the network and allows him to optimize a network from a reliability point of view according to the strategy described in Section 3. Step-by-step, that observation can be eliminated which shows the least effect on the desired function of the unknowns.

As stopping criteria the minimum redundancy value was limited to 0.33 and the average of the parameters was limited to 0.2 cm.

In order to check the accuracy between neighbouring points, the connections between neighbouring points are automatically determined applying a triangularization of the adjusted points. The corresponding relative accuracy between points is calculated using the covariance matrix and applying the law of error propagation.

1	−5.	−337.	
2	−18.	335.	
11	−90.	−282.	
21	−66.	324.	
14	−23.	−79.	
24	−24.	171.	
12	−227.	−44.	Approximate
22	−199.	159.	coordinates of
13	−171.	16.	positions
23	−318.	145.	
10	229.	−325.	
20	295.	281.	
.	.	.	
.	.	.	
.	.	.	
10	14	7.1	
	24	7.1	
	21	7.1	
	2	7.1	
	20	7.1	
11	21	11.9	
	2	11.9	Planned
	24	11.9	observations
	14	11.9	with standard
	1	11.9	deviations
12	23	9.8	
	22	9.8	
	21	9.8	
.	.	.	
.	.	.	
.	.	.	

Figure 7. *The data input for a typical preanalysis.*

4.2 Combining Preanalysis and Adjustment

Besides allowing preanalysis and adjustment (including analysis of the results), NEPTAN enables the designer to combine planned observations with real observations in one computer run. This device is especially valuable if the original observational plan has to be modified due to changed external conditions. The use of real observables allows one to check this data; introducing the additional planned observables allows one to check the reliability of the network due to the

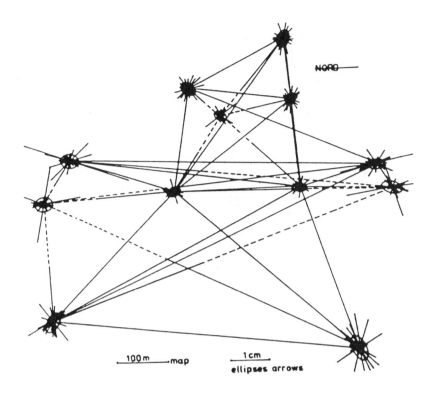

Figure 8. *The influence of all non detectable observational errors on the coordinates of a free network. (The stars represent these effects at any point) (Gründig et al., 1983)*

changed conditions.

4.3 Realization of NEPTAN

The flow diagram showing the structure of program system NEPTAN is shown in Fig. 9.

5 Conclusion

The basic idea when designing program system NEPTAN was to combine the evalution of accuracy and reliability in a common approach for two and three dimensional networks. NEPTAN was originally meant for engineering applications but its application is not restricted to those problems. The general approach used makes it possible to include deflections of the vertical, special refraction models, any

542

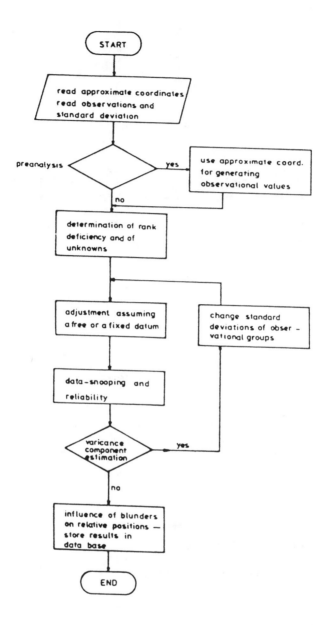

Figure 9. *NEPTAN flow chart.*

Table 1. *Result of the optimization process.*

Free Network 12 Stations	Characteristic numbers of the original network	Characteristic numbers of the network 'optimized'
Number of observations	120	73
Total redundancy	87	40
Minimum redundancy value r_i	0.39	0.33
Maximum effect of a blunder on coordinates	0.78 cm	0.94 cm
Maximum coordinate variance	$(0.2 \text{ cm})^2$	$(0.21 \text{ cm})^2$
Average variance of coordinates	$(0.15 \text{ cm})^2$	$(0.20 \text{ cm})^2$

in advance or iteratively performed centering calculations, correlated coordinate differences (GPS data), etc..

The interaction of precision and reliability of an observational scheme is reported and applied in a sequential optimization procedure. After justifying the sequential procedure because of their locally optimal properties with respect to precision and reliability, the sensitivity of the approach is described with respect to different design criteria. Based on the sensitivity analysis, it was found adequate to optimize the design stepwise with regard to the redundancy number of each individual observation if an overall precise and reliable design is desired.

It has been shown that the concept of free net adjustment and the automated calculation of relative accuracy and reliability with regard to special functions of the unknowns, provide an enormous flexibility of the system. Major advantages of the system are the combination of preanalysis and adjustment in one computer run and the possible optimization of reliability with regard to the assumed functions. The methods reported in this paper were effectively applied because of the incorporation of data management strategies and a modular design.

References

[1] Baarda, W., (1973), 'S-Transformations and Criterion Matrices,' *Netherlands Geodetic Commission* **5**, 1, (Delft).

[2] Baarda, W., (1968), 'A Testing Procedure for Use in Geodetic Networks,' *Netherlands Geodetic Commission* **2**, 5, (Delft).

544

[3] Bahndorf, J., (1982), 'OPTUN—Ein Programmsystem zur Planung und Analyse von geodätischen Netzen,' Technische Akademie Esslingen, November.

[4] Förstner, W., (1979), 'Das Programm TRINA zur Ausgleichung und Gütebeurteilung geodätischer Lagenetze,' *Zeitschrift für Vermessungswesen* **104**, Heft 2.

[5] Gründig, L., (1986), 'Interaction of Criteria of Accuracy and Reliability and its Implication for the Design of Engineering Survey Control Networks,' *Proceedings of the XVII FIG Congress*, Toronto.

[6] Gründig, L., (1985), 'Special Adjustment Tools for Networks,' PEDS Workshop 1985, Division of Surveying Engineering, the University of Calgary, Calgary.

[7] Gründig, L., Bahndorf, J., (1984), 'Accuracy and Reliability in Geodetic Networks,' *Journal Surveying Engineering* **110**, 2, (August).

[8] Gründig, L., Bahndorf, J., (1984), 'Optimale Planung und Analyse von 2- und 3-dimensionalen geodätischen Netzen im Ingenieurbereich—Programmsystem OPTUN,' IX Internationaler Kurs für Ingenieurvermessung, Graz.

[9] Gründig, L., Neureither, M., Bahndorf, J., (1983), 'Zur Auswirkung von Beobachtungsfehlern auf Deformationen oder Deformationen als Beobachtungsfehler gedeutet,' Deformationsanalysen 1983. Schriftenreihe HBSW München, Heft 9.

[10] Levenberg, K., (1944), 'A Method for the Solution of Certain Nonlinear Problems in Least Squares,' *Quarterly Journal of Applied Mathematics*, No. 1.

[11] Snay, R. A., (1976), 'Reducing the Profile of Sparse Symmetric Matrices,' *National Geodetic Survey*, NOS NGS4.

[12] Stiefel, A., (1976), *Einführung in die numerische Mathematik*, 5. Auflage.

A Systems Approach to Real-Time Reservoir Operations

Mark H. Houck

School of Civil Engineering
Purdue University
West Lafayette, Indiana
United States of America

Abstract   The problem of real-time reservoir operations is: given uncertainty about future conditions such as precipitation, streamflows and water related demands, and given long-term operating goals or objectives, and given shorter term guidelines for operation, how much water should be released from each dam during the next hour or day? The current methods used to address this problem in the United States often given considerable discretion to the reservoir operators who generally use their experience, rules of thumb, and intuition to decide on the next period's releases. This decision process can often be well represented by an optimization model that formalizes the objectives and constraints of operation, and provides a focus for discussion on improved operations. This lecture will provide a review of the real-time reservoir operations problem; an overview of current operating methods; a survey of several optimization modeling approaches to the problem; and comments on several applications of these approaches.

1   Introduction

In the past ten years, considerable progress has been made in the study of real-time reservoir operations. In particular, a significant effort has been made to design models that mimic the decision making process or at least produce decisions that are close to those of actual reservoir operators. These models are often mathematical programs (optimization models) whose solution is a recommended operating strategy for a multiple purpose, multiple reservoir system.

To begin this study of optimization models as decision support systems, a review of real-time, short-term reservoir operation is provided in the remainder of this section. The second section contains a description of one of the first attempts at constructing an optimization model for an existing reservoir system. The model is a linear program and was constructed to enhance the current operation of the

B. H. V. Topping (ed.),
Optimization and Artificial Intelligence in Civil and Structural Engineering, Volume I, 545–565.
© 1992 *Kluwer Academic Publishers. Printed in the Netherlands.*

four multipurpose reservoirs in the Green River Basin, Kentucky. The third to fifth sections contain descriptions of different approaches for designing an optimization model. Goal programming applied to real-time, short-term reservoir operations is described in section three. Chance-constrained programming is described in section four. And a very effective balancing procedure using cumulative distribution functions of operating performance is described in section five.

1.1 Reservoir Operation

A reservoir is a storage facility for water. The simplest analogy is a bucket into which a stream flows. There are two ways for water to leave the bucket: it may spill over the top if the bucket fills to the brim and it may exit through a faucet located at the bottom of the bucket. Thus, the inflows to the bucket are stochastic and not controllable unless, of course, there is another dam and reservoir upstream. The outflows from the bucket are partially controllable as long as water is not spilling over the top. The principal decisions in operating the reservoir or bucket are when to open the faucet and how much to open it.

Of course the bucket analogy disregards some important points. For example, reservoirs can lose significant quantities of water through seepage into the groundwater system or under the dam and through evaporation. And operators of a reservoir system usually have several ways to release controlled amounts of water, instead of one faucet. Nevertheless, the reservoir operations problem is how best to use the facilities that are available, given a stochastic inflow and a limited ability to control it.

1.2 The Uses of a Reservoir

The major uses of reservoirs may be classified into three categories: flow attenuation, reservoir pool stabilization, and hydroelectric energy production. Flow attenuation involves the use of the reservoir to reduce peak flows by only releasing a portion of the inflow and storing the rest. It also involves increasing low flows by supplementing them with water already in storage.

Flow attenuation is important because flood damage downstream from the reservoir is reduced if peak flows can be reduced. If low flows can be reliably supplemented, then the amount of water available for downstream industrial, municipal, commercial and domestic use is increased, thereby facilitating additional development. Navigability of the stream is improved because there will be more days when the flow is sufficient to permit ship or boat traffic. Water quality may be improved due to the dilution of pollutants in more water. And instream recreation below the dam may be enhanced because of the increased low flow. This is not an exhaustive list but does indicate the importance of flow attenuation in the operation of a reservoir.

Pool stabilization is principally for two reasons. Recreation at the reservoir may be adversely affected by significant changes in the pool elevation (or reservoir storage). It is difficult to use a beach when there is fifty yards of muck between the end of the sand and the water; and it is difficult to use a picnic table that is under water. Significant, rapid changes in pool elevation may also result in severe bank sloughing, thereby increasing sediment loads into the reservoir and potentially decreasing the storage capacity. Any land owners abutting the reservoir will also suffer loss of property.

Hydroelectric energy production is classified separately because the amount of energy produced is a function of storage volume (actually, the depth of water (head) above the turbines) multiplied by the release. Therefore, it does not fit into any category involving storage alone or flow alone.

1.3 Difficulties in Multi-Reservoir Multipurpose Operation

One can begin to consider the difficulties by considering a single reservoir with only one purpose and an infinite capacity. If the purpose is flood control, the best policy is to keep the reservoir empty until a flood arrives. Then release only the non-damaging portion and store the rest. If the purpose is water supply, the reservoir should be kept filled until a drought arrives. Then the maximum amount of water will be available to supplement the very low natural flows. If the purpose is recreation at the reservoir, the pool elevation should be kept constant and an amount equal to the inflow is released. Clearly, this simple example demonstrates the potential conflicts that can exist in a multipurpose reservoir system.

When more than one reservoir must be considered because they substantially interact, the operations problem becomes more complex. Some means of resolving the conflicts must be found.

Cost-benefit analysis offers some help in evaluating alternatives. But so many externalities and intangibles are involved that economic evaluation is not enough. Multiobjective analysis is often essential.

1.4 Real-Time Short-Term Reservoir Operation

Thus far, two important issues have been ignored. The first is: in what environment are the operating decisions made? Real-time operation implies that the decisions are made in an uncertain and risky environment. Forecasts of future streamflows may be available to the decision makers but the forecasts are not totally reliable and they are only for a limited time horizon. Therefore, the operator decides how much water to release from the reservoir and then waits. Sometime later, the operator reevaluates the condition of the reservoir system and makes another decision. This process of making a decision with limited uncertain information, waiting, updating the available information, and making another decision is repeated time after time.

The second issue that has been ignored is: how often are decisions made? Short-term implies a time step of hours, days, or at most a week. In the work described below, the time step is always taken as one day. However, the time step could be shortened to several hours or increased to several days without any severe consequences.

2 Green River Basin: Operation Optimization

The Green River Basin is located in Kentucky and contains four multipurpose reservoirs operated by the U.S. Army Corps of Engineers. For more than a decade, this reservoir system has been studied by researchers, led by the late Professor G.H. Toebes, at Purdue University in cooperation with the Corps of Engineers. Previous investigations have identified rainfall-runoff properties of the basin and have selected appropriate river routing models. The latest work has focused on constructing an optimization model to assist in real-time, daily operation of the system.

The remainder of this section will summarize the work done constructing and testing the Green River Basin Operations Optimization Model (GRBOOM). Because a complete description of this work is contained in Yazicigil (1980) and Yazicigil *et al.* (1983), some details are omitted. The important features of the operating environment and the model are included.

2.1 Current Reservoir Operation Rules

For the reservoirs in the Green River Basin (GRB) as well as for many other reservoir systems in the US, there exists a hierarchy of rules that regulate the operation of the system. In general, these rules can be broken into three categories: the rule curve, release schedules, and operating constraints.

The rule curve is a guide to long-term operation. It defines the desired reservoir storage or elevation level for the entire year. Fig. 1 is one example of a rule curve where the reservoir elevation is maintained at one level for about half the year and at another level for the rest of the year.

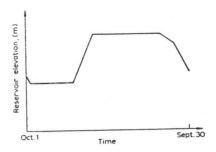

Figure 1. *Typical rule curve.*

During actual operation, it will often be impossible or undesirable to maintain the reservoir level at the rule curve. For example, if a large rainfall results in a large inflow to the reservoir, it may be desirable to store some of the inflow to mitigate flood damage downstream. As a result, the reservoir storage will be increased and may no longer equal the rule curve storage level. To accommodate this type of event, release schedules are designed. Each of these comprises a set of rules that defines how the reservoir storage is allowed to deviate from the rule curve.

An example may be helpful. Imagine the single reservoir system shown in Fig. 2. Below the dam is a tributary stream and then a control station where a streamflow gauge is located. To reduce flood damage at or near the control station, the release from the dam may be restricted by these rules:

Flow at control station (m³/s)	Maximum release from reservoir (m³/s)
0–1000	500
1001–2000	200
2000–	50

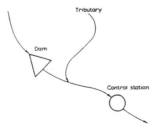

Figure 2. *Reservoir, tributary, control station example.*

The results of these rules are that as the uncontrolled tributary flow increases, the controlled release from the reservoir is reduced, and therefore the storage level in the reservoir may be increased.

If the storage level is above the rule curve, another release schedule may define how the storage level is to be reduced to the rule curve. For example, these rules would gradually reduce the storage level to the rule curve when the inflow to the reservoir is below 400 m³/s.

Reservoir elevation above rule curve (m)	Reservoir release (m³/s)
10	900
5 < 10	750
0 < 5	600

Fig. 3 illustrates the results of implementing the release schedules during a period when a heavy rainfall occurs. Because the length of time is so short, the rule curve is essentially flat. The storage begins on the rule curve. Due to the large streamflows in the tributary and into the reservoir, as a result of the rain, the release from the dam is restricted. The reservoir storage increases because water is being stored. At some point, the streamflow tapers off to a point where the mandated release from the reservoir exceeds the inflow, thereby reducing the storage level. Eventually, the storage level returns to the rule curve.

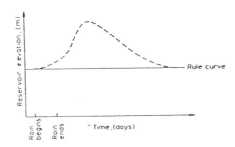

Figure 3. *Effect of release schedules on reservoir elevation.*

Operating constraints are restrictions that take precedence over other operating guidelines and rules. For example, an operating constraint may restrict the storage volume to exceed some minimum storage level. If one of the release schedules would result in the minimum storage constraint being violated, then the release schedule would not be followed. Other operating constraints may restrict releases.

A great deal of work has been done to identify all of the rule curves, release schedules, and operating constraints for the Green River Basin (Toebes and Rukvichai, 1978). Further work was done to compare historical operations to these operating rules (Yazicigil, 1980; Toebes and Rukvichai, 1978). The result was that although the combination of rule curves, release schedules, and operating constraints is fairly complete, not all operations are considered. Furthermore, in discussions with the operators, a need for a model that mimicked the decision making process that the operators go through was identified. With such a model, the operators could investigate more alternatives during actual operation and could use the model to learn how to improve the operating guidelines.

2.2 Operation Objectives

Assume that there exist targets for storage level (perhaps the rule curve), release, and any other measure of reservoir condition or operation. If the reservoir could be operated so that all of the targets are hit, it would be called ideal operation. Any deviation from ideal operation, therefore, represents non-ideal operation.

Suppose a function exists that relates the deviation of storage level from its target to an amount of penalty, measured in penalty units or dollars. A function could also relate the deviation of the release from its target to an amount of penalty. In general, any non-ideal operation could be valued in terms of the total penalty incurred. Fig. 4 illustrates a typical penalty function. The shapes of the functions are convex implying that as operations deviate farther from ideal operations (i.e. from their targets), the marginal penalty incurred does not decrease.

The obvious choice of an objective for selecting future operations is to minimize the total penalty incurred by those operations. For the Green River Basin, it was possible to adopt such an objective because the operators were able to identify targets and all the necessary penalty functions. In other words, they provided the knowledge base that permitted the development of a decision support system.

2.3 GRBOOM

The Green River Basin Operation Optimization Model (GRBOOM) is a linear program that minimizes the sum of all penalties incurred for non-ideal operations at all four reservoirs and the nine control stations in the basin over a T day time horizon (Yazicigil *et al.*, 1983). Typically, the time horizon is between one and five days, which corresponds to the streamflow forecast horizon used by the Corps of Engineers.

552

The constraints of the linear program include: mass balance equations (continuity), minimum and maximum storage restrictions, minimum and maximum release restrictions based on the release schedules and operating constraints, minimum and maximum rate of change of release restrictions, and river routing equations. All of these were identified in consultation with the Corps of Engineers.

The GRBOOM's formulation may be illustrated by considering the single reservoir example shown in Fig. 2. Equations 1–7 provide the simplest version of GR-BOOM type model. Equation 1 is the objective function which is to minimize the total penalty incurred over the T day time horizon. Equation 2 is a mass balance constraint on the contents of the reservoir. Equation 3 is a linear routing model: the tributary flows and reservoir releases are routed to the control station. Equations 4–7 limit the range of possible storages and releases.

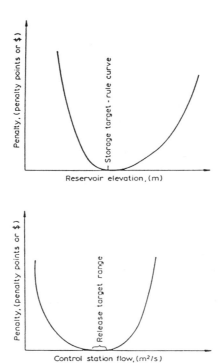

Figure 4. *Typical penalty functions.*

$$\text{minimize } PENALTY = \sum_{t=1}^{T} PQ(Q_t) + \sum_{t=1}^{T} PS(S_{t+1}) \tag{1}$$

subject to:

$$S_{t+1} + R_t - S_t = I_t \qquad\qquad t = 1, 2, \ldots, T \tag{2}$$

$$Q_t - \sum_{j=0}^{J} \alpha_j R_{t-j} - \sum_{j=0}^{J} \beta_j TR_{t-j} = 0 \qquad\qquad t = 1, 2, \ldots, T \tag{3}$$

$$S_{t+1} \leq S_{\max} \qquad\qquad t = 1, 2, \ldots, T \tag{4}$$

$$S_{t+1} \geq S_{\min} \qquad\qquad t = 1, 2, \ldots, T \tag{5}$$

$$R_t \leq R_{\max} \qquad\qquad t = 1, 2, \ldots, T \tag{6}$$

$$R_t \geq R_{\min} \qquad\qquad t = 1, 2, \ldots, T \tag{7}$$

where,

I_t	forecasted inflow to reservoir during day t (m³)
$PENALTY$	total penalty incurred over T day time horizon (penalty points)
$PQ(Q_t)$	penalty incurred due to a flow of Q_t (penalty points)
$PS(S_{t+1})$	penalty incurred due to a storage level S_{t+1} (penalty points)
Q_t	streamflow at control station during day t (m³)
R_t	release during day t (m³)
$R_{\max}$	maximum permissible daily release (m³)
$R_{\min}$	minimum permissible daily release (m³)
S_t	storage at beginning of day t (m³)
$S_{\max}$	maximum permissible storage (m³)
$S_{\min}$	minimum permissible storage (m³)
T	model and forecast horizon (days)
TR_{t-j}	forecasted $(t - j \geq 1)$ or actual $(t - j \leq 0)$ tributary flow on day $t - j$ (m³)
α_j	fraction of a day's reservoir release that arrives at control station j days later
β_j	fraction of a day's tributary flow that arrives at control station j days late

Because the penalty functions in the objective function are all convex, it is possible to approximate them in a linear program. The other typical non-linearity in such models is the river routing equation. Fortunately, in the Green River Basin, linear routing models (Equation 3) were found to be acceptable. And because the model is relatively small, with fewer than 200 constraints and 400 variables for the complete GRBOOM, it is solved as a linear program.

2.4   Use of GRBOOM in Real-Time Operations

Before the optimization model can be used in real-time operations, it must be calibrated. This means that all of the constraints and the objective function must be correct and more importantly that the operator judge to be feasible, reasonable, and good the outputs of the model (i.e. the suggested operations).

Once the model is calibrated, to use it in the actual decision making process is relatively simple. The GRBOOM is an interactive model so that it is easy to change any portion of the linear program, if the operators decide to do so. Furthermore, it is easy to input the required data for running the model, including current storages, past streamflows that are needed for the river routing equations, and forecasted streamflows. Because the model is small, it takes only seconds to obtain a solution. If the operators wish to explore the effect of changes in the input data (for example, what happens if the forecasted streamflows are 10% lower?) it is only a matter of entering the new data and resolving the model.

2.5   Using GRBOOM to Improve Operating Guidelines

It is possible to simulate the operation of the Green River Basin reservoirs over a long period of time by repeatedly solving GRBOOM while updating the forecast and past input data. If this is done, then changes in the river routing equations or the penalty functions or another part of the model can be evaluated. If changes are made systematically, then it is possible to identify those changes that will result in an improved operation.

A hypothetical example of how better guidelines may be found is shown in Fig. 5. Under one set of penalty functions, the simulated operation of a single reservoir system like that shown in Fig. 2 may be plotted as dashed lines. Under a different set of penalty functions, the simulated operation is plotted as dotted lines. If the target range for releases and the target storage (rule curve) are correct, then clearly the dotted lines are preferable to the dashed lines. Therefore, the penalty functions used to produce the dotted lines must be preferable. It may seem improbable that the best penalty functions could not be selected a priori. However, experience with this type of model indicates that this is not so (Yazicigil et al., 1983; Nzewi, 1987).

2.6 Value of Information in Reservoir Operations

Once a GRBOOM type model is calibrated and then recalibrated as described in the previous section, it is possible to estimate the value of good forecast data in reservoir operations. An example of this is given in Fig. 6. By parametrically adjusting the importance of the storage and flow penalties in the objective function, it is possible to define the relationship between these two types of penalty. Each curve in Fig. 6 represents this relationship for a particular model or forecast time horizon.

It should be noted that the curve for a three day forecast horizon is far to the left and down from the curve for the one day forecast horizon. The five day curve is to the left of the three day curve but not as much as the three day curve is away from the one day curve. Because ideal operations would result in a point on Fig. 6 at the origin—no flow penalty and no storage penalty—it is possible to evaluate the contribution that additional days of forecast data have on operation.

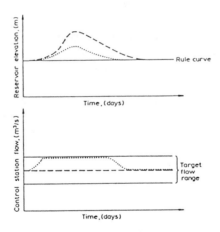

Figure 5. *Hypothetical simulated operation.*

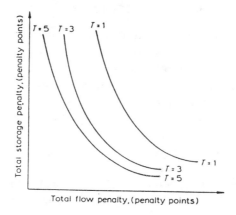

Figure 6. *Achievable storage and flow penalty totals in a season of simulated operation with different forecast horizons (T).*

Clearly, in this example, three days of forecast information is much better than only one day of forecast information. Five days, however, is not that much better than three days. And if seven days of forecast information were available, it may not result in operations any better than the operations based on five days of data. In fact, if the reliability of the forecast information deteriorates with time into the future, it is possible that a longer forecast horizon may result in worse operations.

3   Real-Time Operation By Goal Programming

The GRBOOM described in the previous section requires a set of penalty functions that relate deviations from ideal operations to an amount of penalty. Obtaining these penalty functions may be very difficult. An alternative approach to modeling the real-time reservoir operations decision making process that does not require the penalty functions is goal programming. A detailed description of the application of goal programming to real-time reservoir operations has been provided elsewhere (Can, 1982; Can *et al.*, 1982; Can and Houck, 1984). A summary of this work is given below.

3.1 Goal programs

There are several types of goal program; only one type is described here. This goal program includes a set of constraints like those in the example in Equations 2–7. The objective is expressed differently from the form of Equation 1, however. Instead of a single equation that defines the objective, a hierarchy of goals is specified. The goals are ordered by priority with the highest priority goal first and the least important goal last.

The goal program is solved by considering the goals one at a time in order. First, a solution which maximizes attainment of goal 1 is found. A constraint which ensures that goal 1 will always be satisfied to a level equal to the one just found is added to the constraint set. Goal 2 is now considered. A solution is found that maximizes the attainment of goal 2. Another constraint which ensures that goal 2 will always be satisfied to the level just found is now added to the constraint set. This process continues until all the goals have been considered.

This procedure is only appropriate if goal 1 is much more important than goal 2 and goal 2 is much more important than goal 3 and so on. It does not permit a reduction in the attainment of a higher priority goal in order to improve the attainment of a lower priority goal. For modeling real-time reservoir operations, goal programming may be appropriate.

3.2 The Goals of Real-Time Reservoir Operations

Fig. 7 is an example of the form of penalty function used in the GRBOOM. Experience with the Corps of Engineers in selecting these penalty functions for the Green River Basin indicates that the points a, b, TAR, c, d, and e are relatively easily chosen. They correspond to physical performance levels such as:

Point	Storage	Flow
a	beaches unusable	fish just surviving
b	boat ramps unusable	no rafting possible
TAR	rule curve	flow target
c	picnic tables underwater	minor flooding
d	spillway elevation	major flooding
e	top of dam	disaster

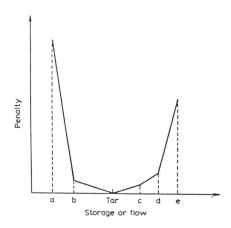

Figure 7. *Typical penalty function.*

The selection of the penalty values associated with the various storage and flow levels is more difficult. However, it is easier to list the priority of operating in different ranges. For example, operating in the range a–b may be the worst of all ranges. Operating in the d–e range may be next to worst. Operating in the range c–d may be next. One way to summarize these observations is with a hierarchy of goals such as:

    goal 1 operate in range a–e
    goal 2 operate in range b–e
    goal 3 operate in range b–d
    goal 4 operate in range b–c
    goal 5 operate at target TAR

The first goal may be interpreted as: find a feasible operating policy without regard to the goodness of the policy. The second goal eliminates those operations that are the most detrimental. The remaining goals limit the range of operations step by step, with the last goal being ideal operation.

3.3 Performance of the Goal Program

Specification of the hierarchy of goals requires less information than specification of all the penalty functions. Therefore, it may be expected that operations suggested by the goal programming model could not be better than those suggested by a GRBOOM type model that explicitly uses the penalty functions. This is not true.

If either the GRBOOM type model or the goal programming model is used to simulate operation of a reservoir system over several months, the simulated operation is the result of the solution of a whole sequence of optimization models. Furthermore, only a portion of the optimal solution of each of the models is used in the simulation: only the suggested release for the first day is implemented; the model is updated and resolved for the next day. Therefore, even if the simulated operation is evaluated in terms of the penalty functions used explicitly in the GRBOOM type model, there is no guarantee that the GRBOOM's suggested operations will be better than the goal program's suggested operations.

An example of this initially surprising result is given in Table 1. The GRBOOM was used in a simulation of the operation of the four reservoir system in the Green River Basin. Two simulations were made. Each was for 35 days and the first had a forecast horizon in the model of one day, the second had a three day forecast horizon. A goal program based on the penalty functions used in the GRBOOM was also used to simulate the system's operations. The actual cumulative penalties incurred at various times during the 35 days are listed in Table 1.

Although the goal program and GRBOOM penalties for a forecast horizon of one day are not significantly different, over the 35 day simulation, the goal program produces less penalty. For a three day forecast horizon, the GRBOOM operations are less penalized than the goal program operations. However, from days 31–33, the goal program outperforms the GRBOOM.

Thus far, no way has been found to determine when the goal program and a GRBOOM type model will yield significantly different operations. The only procedure for discovering which is the preferable approach is to test both.

Table 1. *Cumulative penalties of goal programming (GP) and GR-BOOM operations.*

Number of	Cumulative Penalty			
days of	T = 1		T = 3	
operation	GRBOOM	GP	GRBOOM	GP
1	2.13	2.13	2.13	2.13
5	8.18	8.19	8.17	8.35
10	12.26	12.27	12.26	12.73
15	13.34	13.35	13.34	14.03
20	13.35	13.35	13.34	14.04
25	13.44	13.44	13.44	14.14
30	13.81	13.80	13.86	14.71
33	14.85	14.81	14.75	15.52
35	15.28	15.26	15.23	16.28

4 Real-Time Operation by Chance Constrained Programming

It is possible to use either the goal programming approach or the linear programming approach (GRBOOM type model) described previously to consider the effects of imperfect streamflow forecasts. One brute force way to measure the effects is to solve the optimization model many times under different streamflow forecast assumptions. A much more elegant procedure that explicitly considers the uncertainties in streamflow forecasts is chance constrained programming. Datta (1981) and Datta and Houck (1984) have provided detailed discussions of this form of optimization modeling applied to real-time, short-term reservoir operation. Only a short description is included here.

4.1 Chance Constrained Programming

In problems that involve random variables, it is sometimes important to restrict the probability of occurrence of some events. For example, chance constrained programming has been used in long-term planning and operating models for reservoir systems. The random variable has been monthly or seasonal streamflow and the events that have been constrained have been storage levels, release rates, and hydroelectric power production. For example, a chance constraint on release may be:

$$Pr[\text{release} \leq R_{\text{max}}] \geq 0.9 \qquad (8)$$

or

$$Pr[\text{release} \geq R_{\text{min}}] \geq \alpha \qquad (9)$$

Because release is a function of the reservoir inflow and because inflow is a random variable, the release is a random variable. The selection of an appropriate reliability value such as 0.9 or α, however, is not necessarily easy.

Before a chance constraint can be included in an optimization model, it usually must be converted into a form accepted by the optimization algorithm. Typically, this conversion results in an equation that is called the deterministic equivalent (of the chance constraint) and that is in the proper form for inclusion in a linear program or some other optimization procedure.

4.2 Chance Constraints Applied to Real-Time Operations

One constraint and an objective function can be used to illustrate how the real-time reservoir operations problem can be cast in the form of a chance constrained program. The mass balance equation for reservoir storage from the beginning of the first day of operation to the beginning of the second day of operation is:

$$S_2 = S_1 + I_1 - R_1 \qquad (10)$$

In this equation, I denotes the forecast inflow to the reservoir during the first day.

The error in the forecast is a random variable whose distribution may be determined. Therefore, the forecast is a random variable whose distribution may be determined. And through the mass balance equation, the storage at the beginning of the second day (S_2) becomes a random variable. Chance constraints restricting the possible beginning storages for the second day may be formulated as:

$$Pr[S_2 \geq S_{\text{min}}] \geq 0.9 \qquad (11)$$

or

$$Pr[S_2 \leq S_{\text{max}}] \geq \gamma \qquad (12)$$

Appropriate objectives that correspond to these chance constraints (Equations 11 and 12) are: to maximize the value of S_{min}, which is the storage that is exceeded by at least 90% of all possible actual storages at the beginning of the second day; or to maximize the value of γ, which is the reliability of actual storage at the beginning of the second day not exceeding a specified value S_{max}.

Other performance levels of the operation can be included in either the constraints or objective of the optimization model. For example, hydroelectric power production could be restricted to be within some designated range with high probability. The probability that the storage at the end of five days is within some specified rage of the rule curve could be maximized. Or the probability of the flow at a downstream control station falling within a target range could be maximized.

4.3 Chance Constrained Models

Only limited experience testing chance constrained models for real-time, short-term reservoir operation is available. Datta (1981) has described a single reservoir system whose operation was simulated using a chance constrained model. He also argues strongly in favor of the chance constrained approach applied to real-time reservoir operations. Only with additional testing, however, will the usefulness of this optimization modeling approach be determined.

5 Balancing Real-Time Reservoir Operations

Another approach to modeling the decision making process of real-time reservoir operation balances different parts of the operation to achieve a good overall operating policy. The balancing is done with cumulative distribution functions of different measures of reservoir operation, such as storage, release, and hydroelectric energy production. This approach has been discussed in detail elsewhere (de Mansabert *et al.*, 1983; Houck, 1982; Nzewi, 1987).

5.1 Balancing Models

The only difference between the balancing model and a GRBOOM type model or a goal programming model is the form of the objective function; the constraints are identical. The objective of the balancing model is to minimize the maximum probability measure associated with any of the performance criteria over the forecast horizon. An example is the easiest way to explain what this statement of the objective means.

Suppose that it is possible to forecast perfectly the necessary streamflows for a GRBOOM type model for an infinite time horizon. Then it would be possible to find the optimal operation for the reservoir system over a long period. It would also be possible to use the historical operations over some long period if it is believed that they represent good operations. In either case, a long base period of operations is available.

These operations may be used to construct empirical cumulative distribution functions of storages, releases, energy production, and any other physical characteristic of the operation that is deemed important. Although the horizontal axes of the CDFs may have significantly different scales, the vertical axes all have a scale from zero to one. The CDFs can be used to convert any storage, release, etc. to associated values which are called probability measures.

Consider a single day's operation. For a given inflow, the ending storage and the day's release have a one to one relationship. Each unit of water that is released reduces the ending storage by one unit. One possible operation may be to release a large amount, resulting in a small ending storage. Another may be to release a small amount and end with a large storage. Still a third operation may be to release a moderate amount and end with a moderate storage. In all cases, the associated probability measure for ending storage and release could be found. Table 2 summarizes these data.

Table 2. *Possible one day operating strategies.*

Operations strategy	Ending storage	Probability measure of ending storage	Release	Probability measure of release
1	small	0.2	large	0.8
2	moderate	0.6	moderate	0.6
3	large	0.9	small	0.4

Operating strategy one results in an ending storage that is so small that only 20% of the storages in the base operations period were below it; yet 80% of the base period releases were below the release made in strategy one. Strategy three results in an ending storage larger than 90% of the base period storages and a release larger than only 40% of the base period releases. Strategy two results in an ending storage and a release that are larger than 60% of the storages and releases in the base period.

One conclusion is that strategies one and three are out of balance and strategy two is in balance. Both strategies one and three have one performance level (storage or release) that is very large and another that is very small when compared to the good operations included in the base period. It is possible to reduce the extremes of these two strategies simply by reallocating water among the storage and release. At best, both release and storage will have equal probability measures like strategy two.

The objective of the balancing model is to minimize the maximum probability measure encountered over the operating or forecast horizon. This is equivalent to maximizing the minimum probability measure in this case. And it is equivalent to balancing or equalizing the probability measures encountered.

5.2 Efficiency of Balancing Model

The most thorough testing of the balancing model thus far has been reported by Nzewi (1987). In comparisons of a GRBOOM type model, other models, and the balancing model, the balancing model performed extremely well. No model was significantly better than the balancing model, yet most were significantly worse.

6 Conclusions

In recent years, a considerable amount of work has been done on the design of optimization models to be used in real-time, short-term reservoir operation. One principal focus of this work has been modeling the Green River Basin multi-reservoir, multipurpose system. These different modeling approaches appear to have special characteristics that are very useful in real-time operation.

Acknowledgment The work described in this paper is due in large measure to the efforts of a number of scholars including Emre Can, Bithin Datta, Sharon deMonsabert, Gerrit Toebes, and Hasan Yazicigil. This paper is substantially the same as 'Designing an Expert System for Real Time Reservoir System Operation,' that appeared in *Civil Engineering Systems* **2**, 1, 30–37, March, 1985. It is reprinted here with permission of the journal.

References

Can, E. K., (1982), 'Optimization of Real-Time Operations of Reservoir Systems and the Implementation of Optimization Techniques,' Ph.D. Dissertation, Purdue University, West Lafayette, Indiana, USA.

Can, E. K., and Houck, M. H., (1984), 'Real-Time Reservoir Operations by Goal Programming,' *Journal of Water Resources Planning Management* **110**, 3, 297–309.

Can, E. K., Houck, M. H., and Toebes, G. H., (1982), 'Optimal Real-Time Reservoir Systems Operation: Innovative Objectives and Implementation Problems,' Technical Report 150, Purdue University Water Resources Research Center, West Lafayette, Indiana, USA.

Datta, B., (1981), 'Stochastic Optimization Models for Long-Term Planning and Real-Time Operation of Reservoir Systems,' Ph.D. Dissertation, Purdue University, West Lafayette, Indiana, USA.

Datta, B., and Houck, M. H., (1984), 'A Stochastic Optimization Model for Real-Time Operation of Reservoirs Using Uncertain Forecasts,' *Water Resources Research* **20**, 8, 1039–1046.

deMonsabert, S. M., Houck, M. H., and Toebes, G. H., (1983), 'Management of Existing Reservoir Systems by Interactive Optimization,' Technical Report 149, Purdue University Water Resources Research Center, West Lafayette, Indiana, USA.

Houck, M. H., (1982), 'Real-Time Daily Reservoir Operation by Mathematical Programming,' *Water Resources Research*, 1345–1351.

Nzewi, E. U., (1987), 'Real Time Multipurpose Reservoir System Management Using Probabilistic Balancing Rule Models,' Ph.D. Dissertation, Purdue University, West Lafayette, Indiana, USA.

Toebes, G. H., and Rukvichai, C., (1978), 'Reservoir System Operations Policy-Case Study,' *Journal Water Resources Planning and Management Division, ASCE* **104**, WRI, 175–191.

Yazicigil, H., (1980), 'Optimal Operation of a Reservoir System Using Forecasts,' Ph.D. Dissertation, Purdue University, West Lafayette, Indiana, USA.

Yazicigil, H., Houck, M. H., and Toebes, G. H., (1983), 'Daily Operation of a Multipurpose Reservoir System,' *Water Resources Research* **19** 1, 1–13.

Approaches to Risk Management and Operational Guidance for Municipal Water Supplies

Richard N. Palmer
Department of Civil Engineering
University of Washington
Seattle, Washington
United States of America

Abstract The proper initiation of water use restrictions for municipal water systems should depend upon the risk of system shortfall, the economic loss associated with those shortfalls, and the range of management options available. Often these considerations are overshadowed by more subjective considerations that can not be easily quantified. This paper presents the results of a risk methodology designed to integrate quantitative factors into optimal policies through the use of an operational database. This methodology evaluates the range of economic risk associated with each management option. A discussion of a response curve approach also is presented with an application to the Seattle, Washington water supply system. These issues are pursued further in additional papers found in these proceedings by the author in which the more subjective aspects of the problems are addressed.

1 Introduction

Recent climatological events have focussed attention on the potential economic, social, and environmental impacts of droughts. Municipal water authorities throughout the United States have found their sources inadequate in the face of increasing water demands and seemingly less predictable supplies (Boland *et al.*, 1980). The summer of 1989 has already been predicted to be a potential drought year for major portions of northern California (where a third consecutive year of low flows are being experienced), the metropolitan areas of Boston and New York, the Delaware River Basin, parts of Georgia and South Carolina, the Missouri River Basin, and southern Florida (EOS, 1989). Significant portions of Great Britain have also been affected during th summer of 1989. Reports of significant changes in rainfall and

567

B. H. V. Topping (ed.),
Optimization and Artificial Intelligence in Civil and Structural Engineering, Volume I, 567–576.
© 1992 *Kluwer Academic Publishers. Printed in the Netherlands.*

temperatures also have fueled public interest. Although the degree of public concern over reliable water supplies has increased, the water utility professionals have long been aware of the necessity to determine the dependable yield of their water supplies (Palmer *et al.*, 1982, Smith and Lampe, 1988; Woo, 1982).

The necessity to develop accurate estimates of system yield and water supply management strategies during droughts is ubiquitous in the water supply industry. For smaller utilities, managers often rely solely on the services of consultants in making these calculations. For larger utilities, in-house staff participate in the evaluations, although consultants may play a primary role. Unlike many aspects of the water supply industry (such as disinfection), stringent procedures and guidelines do not exist for the calculation of system yield.

The process of calculating yield estimates and drought management strategies requires a consistent set of planning and management procedures. These calculations require a significant amount of information about the physical and hydrological characteristics of the water supply and estimated costs of shortfalls and impacts of water use restrictions. Much of this information is often incomplete, known only with a significant degree of uncertainty or not known at all. The collection and interpretation of the information needed is a costly and time consuming process, but it is required for the selection of a cost-effective drought management alternative. A consistent set of procedures is needed although unique situations exist at individual water supply utilities.

2 Dependable Yield Analysis and Drought Management Policies

Dependable yield analysis has been a topic of discussion for centuries. Since the time of the Roman Empire and its expansion, water supply has been the first concern of city engineers. Perhaps the seminal paper of water supply reliability is that of Rippl (1883) in which a simple, graphical procedure was suggested to calculate the storage requirement of surface reservoir to meet a specified demand. Since that time, this topic has been a subject of active research (Yeh, 1985).

Initially, yield analysis centered on the development of 'safe yield' calculations. The safe yield of a system is defined as the maximum amount of water that can be extracted from a water supply system over the available historic hydrologic record. It should be recognized that this is a very useful concept. It allows managers to evaluate the relative reliability of their system. It can be easily calculated for simple systems having only one water supply reservoir. With more effort, computer models or spreadsheets can be developed to calculate safe yields for more complex systems with multiple sources. Currently, safe yield is the most commonly used measure of water supply system reliability and the technique is applied in much the same way (albeit with a computer rather than graph paper) as it was at the turn of the century.

Although useful, safe yield does not address many of the more necessary concepts in yield analysis and water supply management (Hashimoto, *et al.*, 1982; Palmer and Lettenmaier, 1983). Since the early sixties, research emphasis has been placed on the use of optimization techniques to model yield analysis and supply management. More recent investigations have focussed on two areas: the calculation of probabilistic yields and the characterization of supply system failure. Unfortunately, little of this research has been transferred into common professional use.

As currently practiced, the calculation of safe yield is a function of the length of the historic record available for evaluation. The longer the hydrologic record available, the more likely it is to have a significant drought. A need has been recognized to evaluate systems with similar record length. Two techniques emerged to address this problem, streamflow extension techniques and synthetic streamflow generation. Streamflow extension attempts to extend existing streamflow records by correlation with other, longer records. The longer records can be of rainfall, streamflow at other sites, or other explanatory phenomenon (tree rings). Synthetic streamflow generation attempts to characterize the statistics of a stream and generate numerous, equally likely, streamflow scenarios that can be used to evaluate reliability. The purpose of both of these approaches is to place the calculation of safe yield on a consistent probabilistic basis.

Researchers have also been concerned that safe yield addresses only the extreme event on record and no other. As managers attempt to operate systems more closely to their safe yield, more interest has been expressed in characterization of system failure. The estimation of parameters such as the average storage of systems over time, the maximum possible deficit, the average deficit, and the number of deficits in a period have become useful planning concepts.

During periods of rapid increase in demand and restrictions on system expansion, it has become obvious that the actual range of operational alternatives during droughts is small. In general, these alternatives can be characterized as a series of voluntary and mandatory restrictions in which outdoor and indoor water uses are limited to both the public and private sector (AWWA, 1980; Hresoe et al., 1986; Lord et al., 1983; Wiley, 1983). These restrictions have economic losses associated with them because they preclude the use of water in ways in which it would typically be consumed. Evaluation of management strategies requires that at any point in time, a manager must decide which of several potential restriction levels, if any, is the most appropriate, based on the range of potential events he may experience. There are a number of reasonable objective functions the manager may want to consider. The manager may select a strategy that minimizes the expected economic losses, one that is appropriate for the most extreme drought on record, or one that reflects more subtle parameters such as the perceived response of public officials. Some larger municipal water supply systems have developed water use restriction policies although most have not. Policies for many utilities are qualitative at best and based upon little numeric information. To this date, no consistent approach has developed within the industry for generating drought management strategies.

3 Risk and Multi-Objective Analysis

The term 'risk analysis' has been defined in a variety of fashions. Here, risk analysis is defined to be the process of evaluating the broad economic, social and environmental risks associated with an existing or predicted set of conditions and selecting management actions that minimize these risks in a cost effective manner. From a manager's perspective, this definition implies two activities. First, the costs and reductions in risk of an action must be evaluated. Second, these costs and reductions must be compared to the marginal reductions in risk that would occur if other actions were taken.

Drought management risk evaluation is composed of three difficult but necessary steps:

1. Identify the economic, social, and environmental impacts of droughts,

2. Quantify the impacts of these conditions for a particular setting, and

3. Select an appropriate drought management response.

For drought management, management decisions can be divided into planning and operational responses. Planning responses are those made in a time period in which significant changes to the water supply system can be made, such as the addition of new sources of water, the addition of groundwater extraction, or the change in the capacity of an element of the system. Operational responses are those that can be made in the short-term such as the call for water use restrictions, changes in water allocations (between municipal, industrial, and environmental uses), and similar real-time decisions.

Difficult decisions in water supply are always multi-objective. A utility is constantly balancing the costs of new sources of water with the increased reliability that the new sources will provide. During drought periods, utilities may balance the impacts of a long period of minor water use restrictions versus the use of very severe restrictions for a short period of time.

During the last two decades, efforts have been made to address the problem of multi-objective decision making. A primary interest in this effort has been the development of procedures for multi-objective analysis for situations in which the information is either subjective or can not be expressed quantitatively. It should be recognized that the problem of multi-objective analysis is critical in evaluating risks. Situations often arise in which the goals of comparative risk management are conflicting and can not be expressed in a single metric such as dollars. A procedure for addressing the multi-objective decision making problem must be contained in any expert system developed for drought management.

4 Restrictions

Voluntary and mandatory water use restrictions are extremely useful municipal water management options for dealing with potential water shortages. Such restrictions can delay the construction of expensive, new sources of water and can minimize the losses endured when unusual hydrologic events occur. The initiation of water use restrictions are a function of three primary parameters, 1) the probability of a shortfall, 2) the economic impact of the shortfall, and 3) the degree to which managers are willing to accept the risk of shortfalls.

For the water manager, difficulties arise in determining the timing and severity of such restrictions. Complicating factors include the uncertainty associated with predicting future streamflows and water demands, unpredictable user responses to water use restrictions, and the desire not to initiate restrictions unless they are truly required. There is a natural hesitancy to call for either voluntary or mandatory water use restrictions because such calls may be viewed by some as an indication of improper management. Although this is not necessarily true, this perception can limit the use of restrictions as a water management tool. If water use restrictions are to be used effectively, a coherent plan of action must be established that provides clear justification for the initiation of water use restrictions and that quantifies the benefit of their use. The plan must also make apparent any subjective and non-quantifiable information that is used to establish the risk adversity of the managers. The remainder of this paper describes a procedure for the establishment of water use restrictions and its application to the Seattle Water Department water supply system.

5 Water Resource Risk Evaluation

In the water resources literature, the estimation of the probability of hydrologic events is addressed in two fundamental ways, the evaluation of individual past events and the estimation of the probability distributions that are associated with past events. These approaches lead to two general techniques for evaluating the impact of events, deterministic optimization or simulation models, and implicitly (or explicitly) stochastic optimization or simulation models. Difficulties arise in both approaches. In deterministic optimization, the evaluation is limited to the hydrologic record, thus the full range of potential hydrologic events may not be considered. In stochastic approaches, assumptions must be made about the distributional forms of streamflow or rainfall data. This approach is very accurate for

estimation of average flow characteristics. However, the estimation of streamflow
cross correlations and serial correlations during extreme events is speculative at
best. For the case study described here, approximately fifty years of streamflow
record was available for analysis. The length of this record made deterministic
optimization the best approach available.

It is important to note that for municipal water supply management the range
of operational alternatives is small. In general, these can be characterized as a
series of voluntary and mandatory restrictions in which the indoor and outdoor
uses of water are limited in both the public and private sectors. These restrictions
have economic losses associated with them because they preclude the use of water
in ways in which people would typically pay to consume it. Risk evaluation for
water supply requires that at any point in time, a manager must decide which of
several potential restriction levels, if any, is the most appropriate, based on the
range of potential events he may experience. There are a number of reasonable
objective functions the manager may consider. The manager may select a strategy
that minimizes the expected economic losses, one that is appropriate for even the
most extreme droughts on record, or one that reflects more subtle parameters such
as the perceived response of public officials.

6 Problem Application

The Seattle Water Department (SWD) serves over one million people in the Seattle,
Washington metropolitan area. SWD provides direct service to approximately half
of these customers and serves as a wholesaler to the remainder. Its primary sources
of water are two surface reservoirs, located on the Tolt and Cedar Rivers. SWD
estimates the conjunctive safe yield of this system to be 169 million gallons a
day (MGD) at a 98% reliability. Current demand for water is approximately 170
MGD with a growth rate of 2 MGD per year. A small quantity of groundwater
is available for use to handle short-term growth until new sources of water are
identified, evaluated, and developed.

A water shortage response plan (WSRP) has been developed to encourage an
orderly response to any potential water shortage. Five stages of water use restrictions exist in the WSRP. The guidelines for initiating the restrictions are relatively
vague, allowing water managers a great deal of freedom. The actions required
and the responses of the five stages are estimated to be: Stage 1, eliminate non-
essential water use by the City of Seattle with an expected savings of 3 MGD;
Stage 2, call for public voluntary reduction in outdoor water use with total savings of 7.7 MGD; Stage 3, require mandatory outdoor water use reductions and
decrease instream flow requirements of fish with total savings of 20 MGD; Stage 4,
increase mandatory outdoor water use restrictions, call for voluntary indoor and
commercial restrictions and further reduce releases from reservoirs for non-water

purposes with a total savings of 22 MGD; and Stage 5, implement water rationing with a total savings of 57.4 MGD.

7 Problem Approach and Database Development

The approach used to calculate operational risk and generate potential management policies required the evaluation of a large operational database. This software for risk management of reservoir operating decisions is denoted here as DREAMER(Data base Risk Evaluation and Management for Reservoirs). To generate the database in DREAMER, a linear programming model of the water supply system was created. The objective function of this program was to minimize the economic losses experienced over a sixteen week period while meeting all other operational considerations. The economic loss function replicated the customer losses associated with the initiation of each of the five stages of water use restrictions described previously. Forty-six years of hydrologic data were used. This data included streamflows from seven sites within the watershed. Instream flow requirements and other operational considerations were incorporated into the model as constraints. The model operated on a weekly time step. The results of the model included both the total economic losses associated with a particular sixteen week hydrologic record and the optimal implementation of water use restrictions.

Operational risk was recognized to include two types of parameters: current conditions and predicted conditions. Current conditions included the month for which a policy was desired, the previous month's system demand, and the current system storage. Predicted conditions included both total system inflow and total system demand. The influence of current conditions on operating policy was evaluated by investigating each of these parameters over the hydrologic record. Storage levels were discretized into nine values and five low flow summer months were selected. Demand levels predicted for 1990, 1995 and the year 2000 were incorporated into the model. Investigating the forty-six years of record requires 6,210 runs of the linear program. Some of these configurations are of little interest since shortages do not occur and therefore water use restrictions were not initiated. This problem was reduced in size by first calculating the yield of the system for each of the nine storage levels and five starting periods for each of the forty-six years of records (2070 model runs). Then the operating policy was investigated for all situations in which the three potential demands were greater than system yield.

8 Risk Approaches in DREAMER

The database in DREAMER contains operational policies for a wide range of system conditions, including thousands of situations which have never occurred. To develop policies relevant for any situation the user must first determine the current system status (system storage, demands, and month) and then determine the range of future conditions of interest. Once these have been identified, the database contains sufficient information to suggest an operating policy associated with these conditions. This information then supplies the manager with quantitative information for developing a risk analysis by describing the frequency with which specific operating policies have proven optimal for various conditions and the economic impacts of selecting those policies.

Selecting the range of future conditions of interest is a difficult task. It is a central issue in any risk analysis and should be approached in an iterative fashion. This iterative approach is suggested for the SWD because of the relatively poor quality of streamflow predictors that are available during the summer months for this primarily rainfall driven system. Because accurate predictions are not available, the water manager must be made aware of the sensitivity of the optimal release policies relative to predicted inflows and demands. Three approaches to analyzing the database are available in DREAMER to emphasize the system sensitivity. The first approach assumes a perfect inflow forecast and therefore does not have a risk component. The second approach allows the user to select ten representative streamflow and demand patterns, assign a probability of occurrence to these flows, and determine the operating policies that minimize the expected economic loss. The third approach assigns an equal probability to all past streamflows and provides a system response surface describing the frequency with which the initiation of each restriction level was optimal for each system configuration.

In the first approach, the manager enters into DREAMER the best 'single estimate' of the inflows and demands that are to be placed on the water supply system during the following sixteen week period. These inflows and demands are translated directly into an operational policy minimizing the economic losses that would result if these inflows and demands did occur. No sensitivity analysis is provided although the manager can initiate the process again with a different best 'single estimate' to develop ad-hoc sensitivity analysis. In this method operating risk is associated with the quality of the streamflow and demand forecasts.

The second approach allows the manager to select up to ten streamflow and demand scenarios in DREAMER that are associated with past streamflow events for evaluation. The manager provides information describing the system storage (between 20 and 100% storage by increments of 10%), time of year (beginning of the months of June, July, August, September and October), and the municipal and industrial (M&I) demand (one of three values). The manager can also provide the probability to be associated with the occurrence of each of the streamflows. For each of these events, the optimal operating policy is available from the database. Expected economic losses and expected optimal operational policies are then calculated by weighting the operational strategies with their respective probabilities.

The third approach gives the manager the opportunity to evaluate the full database to determine the frequency with which the six alternative operating policies prove optimal. In this approach, it is assumed that each of the forty-six streamflow scenarios have an equally likely probability of occurring. The manager provides the same information to DREAMER as in the second approach to initiate this process. A summary is then developed indicating the frequency with which each of the policies proved optimal over the forty-six year record for the initial conditions indicated. Next, a response curve is created that indicates the sensitivity of the policies to each of the initial parameters provided by the manager. This is accomplished by calculating the rate at which the frequency of the policies change as a function of the initial storage, demand, and month. The optimal operating policies for the most severe drought on record for the specified initial storage, demand, and month is also presented.

9 Summary

An interactive methodology is suggested for providing water managers a range of operational policies that explicitly consider the risk associated with water shortages. This methodology includes the development of an operational database from which the manager can evaluate the use of restrictions on past and predicted hydrologic events and calculate their potential economic impacts. This topic is further pursued in two other papers in this proceedings by the author. In the first, the use of expert systems in drought management is explored. In the second, the use of natural language tools for database access is discussed.

References

American Water Works Association, (1980), *Water Conservation Strategies*, Denver, Colorado.

Boland, J. J., Carver, P. H. and Flynn, C. R., (1980), 'How Much Water Supply Capacity is Enough?,' *Journal of the American Water Works Association* **72**, 7, 368–374, (July).

Clark, R. M. and Males, R. M., (1986), 'Developing and Applying the Water Supply Simulation Model,' *Journal of the American Water Works Association* **78**, 8, 61–65, (August).

EOS, (1989), Vol. 70, No. 11, March 14, p. 162.

Hashimoto, T., Stedinger, J. R. and Loucks, D. P., (1982), 'Reliability, Resiliency, and Vulnerability Criteria for Water Resource System Performance Evaluation,' *Water Resources Research* **18**, 1, 14–20, (February).

Hirsch, R. M., (1979), 'Synthetic Hydrology and Water Supply Reliability,' *Water Resources Research* **15**, 6, 1603–1615, (December).

Hreso, M. S., Bridgeman, P. G. and Walker, W. R., (1986), 'Managing Droughts Through Triggering Mechanisms,' *Journal of the American Water Works Association* **78**, 6, 53–60, (August).

Lord, W.B., Chase, J.A. and Winterfield, L. A., (1983), 'Choosing the Optimal Water Conservation Policy,' *Journal of the American Water Works Association* **75**, 7, 324–329, (July).

Palmer, R. N. and Lettenmaier, D. P., (1983), 'The Use of Screening Models in Determining Water Supply Reliability,' *Civil Engineering Systems* **1**, 1, 15–22, (September).

Palmer, R. N., Smith, J. A., Cohon, J. L. and ReVelle, C. S., (1982), 'Reservoir Management in the Potomac River Basin,' *Journal of Water Resources Planning and Management, ASCE* **108**, 1, 47–66.

Rippl, W., (1883), 'The Capacity of Storage Reservoirs for Water Supply,' *Proceedings of the Institute of Civil Engineers* **71**, 270–278, London.

Smith, R. L. and Lampe, L. K., (1988), 'Dependable Yield Evaluations: How Much Water is Really There?,' *Journal of the American Water Works Association* **80**, 9, 66–70, (September).

Wiley, R. D., (1983), 'Denver's Water Conservation Program,' *Journal of the American Water Works Association* **75**, 7, 320–323, (July).

Woo, V., (1982), 'Drought Management: Expecting the Unexpected,' *Journal of the American Water Works Association* **74**, 3, 126–131, (March).

Yeh, W. W-G., (1985), 'Reservoir Management and Operating Models,' *Water Resources Research* **21**, 12, 1797–1818, (December).

Optimization Modeling in Water Resources: A Sample of Applications

Mark H. Houck
School of Civil Engineering
Purdue University
West Lafayette, Indiana
United States of America

Abstract Optimization modeling has been extensively described in the literature and less frequently applied in practice. The range of applications is enormous, spanning optimal pipe network design to optimal long term planning and operations of multiple reservoir systems. The purpose of this lecture is to describe a select set of optimization models that illustrate the range of development and use of this type of tool in water resources. The modeling developments described include: (1) the optimal design of gravity sewer networks using dynamic programming; (2) the optimal design of dual purpose detention systems or ponds to control water quantity and water quality in an urban drainage network using nonlinear programming; and (3) the long-term planning and operation of multipurpose reservoir systems using a variety of optimization techniques.

1 Introduction

There are many diverse and interesting applications of optimization modeling in water resources. Wright *et al.* (1980) provide a review of water resources models currently available and used by federal agencies in the United States. The literature is repleat with proposed model formulations, case studies, and reports of actual implementation. In this paper, three applications are considered. The first of these is an optimization approach for designing gravity sewers. This program and an application will be described extensively. Two other applications will be described more briefly. These are the optimal design of urban drainage networks, especially when water quantity and water quality constraints and objectives are important; and the long-term planning and operation of multi-purpose reservoir systems.

B. H. V. Topping (ed.),
Optimization and Artificial Intelligence in Civil and Structural Engineering, Volume I, 577–596.
© 1992 *Kluwer Academic Publishers. Printed in the Netherlands.*

2 Optimal Gravity Sewer System Design

The Optimal Gravity Sewer Design Program (OGSDP) is a Fortran computer program (Nzewi et al., 1985) that is an effective tool in the design of gravity sanitary sewers. An initial sewer system design is obtained using a heuristic called the Initial Solution Algorithm, which finds a feasible design and attempts to use the smallest available pipes (Pacheco and Gray, 1982). This design is then improved using discrete dynamic programming with successive approximations. The objective is to minimize the total cost of the system using cost information supplied by the user or default costs embedded in the program. The designs produced by OGSDP observe most of the criteria specified by the Ten State Standards (1978).

2.1 Literature Review

Generally, two distinct classes of sewer system design models have been considered in the literature. The first optimizes crown elevations, manhole depths and pipe sizes for a predetermined horizontal sewer lay-out; OGSDP is in this class of models. The second class optimizes the horizontal sewer lay-out, as well as the vertical design. Some models also consider special conditions such as lift or pumping stations.

Several sewer design models have been formulated as linear programs (LPs). Deininger (1973) proposed an LP model assuming a linear cost function for excavation and sewer costs. Dajani and Gemmell (1972) also proposed an LP-based model. Linear programming models are difficult to formulate appropriately because the sewer design problem is discontinuous (e.g., pipes only come in discrete diameters) and nonlinear (e.g., the equations for flow through pipes).

More recent work has focused on the use of dynamic programming. Meritt and Bogan (1973) developed a model with rather complex hydraulics for the design of sanitary sewers. Partially filled pipe flow was modeled explicitly. In addition, the model minimized energy variations across manholes. Minimum and maximum velocities of flow were predicted by Manning's open channel flow formula.

Walsh and Brown (1973) proposed a model which considers partially filled sewers. Determination a priori of the feasible region is necessary so that an initial solution may be found. This initial solution is then used as a starting solution for the algorithm which generates a number of optional 'optimal' designs from which the users may pick one that best satisfies their requirements. The optimal design is not determined per se by the model.

Mays and Yen (1975) have proposed a model for storm sewers with simple hydraulics. The sewers are sized using the Manning formula for gravity open channel flow. Full pipe flow is assumed at the design flow rate. An initial trajectory or feasible solution is found by assuming an average slope for each link in the network. The sewer network is divided into groups of serial (end-to-end links) subsystems and a main system. This allows a discrete differential dynamic programming (DDDP) technique to be applied to a branched sewer system. Mays and Wenzel (1976) have updated the search algorithm previously proposed by Mays and Yen (1975) introducing the concept of isondal lines which group manholes into sets based on their proximity to the outlet of the system.

Another model which applies DDDP but also provides risk-based design of storm sewer systems was presented by Tang et al. (1975). The solution algorithm is the same as the one presented by Mays and Wenzel (1976). The hydraulics of this model are simplistic, however. The design flows are calculated by the product of an average (assumed) flow and the risk level (factor) for that given link. The cost of each link comprises pipe material costs, installation costs (including excavation) and the estimated cost of damage (which depends on the risk level assumed). Water storage facilities or pump stations are not considered.

Froise and Burges (1978) have also proposed a model that determines the least-cost design of a storm sewer system. They consider water storage facilities (detention ponds) and pump stations in the design. Hydrograph attenuations and flow interactions within the system are both modeled in the optimal design process.

Han et al. (1980) developed a computer program which implemented discrete dynamic programming (DDP) to study two small storm sewer systems. Gupta et al. (1983) have also presented a computer program that determines the least-cost design of a gravity sanitary sewer system. The model comprises two algorithms: one that generates the feasible designs at each manhole and another that selects the optimal set of inverts and pipe size at junction manholes to complete the design process. The hydraulics are described by the modified Hazen-Williams formula which accounts for the variation of apparent roughness with depth of flow.

Typically, the sizes (diameters) of pipes in the sewer system are not permitted to decrease in the downstream direction. Walters and Templeman (1979) and Templeman and Walters (1980) are among the few who have developed methods using dynamic programming with two state variables (pipe crown or invert elevation at a manhole, and pipe size or diameter entering the manhole) to consider this problem optimally. They report that the inclusion of the second state variable (pipe size) in the formulation does not significantly increase the computational burden of obtaining the globally optimal solution. Nonetheless, OGSDP was designed using a single state variable formulation and hence cannot guarantee a globally optimal solution. OGSDP uses a heuristic to attempt to find a good solution to the sewer design problem when non-decreasing pipe sizes are required. OGSDP also can be used to determine an upper bound on the error of the heuristic solution, thus permitting the user to assess the overall quality of the proposed design.

2.2 Design Constraints

OGSDP designs a minimum cost sewer system based on certain assumptions and parameter values. The collection system is assumed to be a non-looping network of pipes with manholes at the upstream and downstream ends of each pipe. The user specifies the number of residence-equivalent connections to each pipe. The design flow in each pipe is based on the product of the number of upstream connections, the number of persons per connection, the average sewage flow per person and the peak factor. The program allows the user to specify the latter four parameters or to use default values. In addition, the Harmon formula may be used to determine a peak factor for each pipe based on the upstream population.

As recommended by the Ten State Standards (1978), the minimum pipe size is 8 inches. Other commercial pipe sizes that are incorporated in OGSDP are 10, 12, 15, 18, 21, 24, 27, 30 and 36 in.

The model respects the value of the minimum depth of cover, the vertical distance from the crown of any pipe to the ground surface. If a value is not given by the user, a default value of 6 ft is used. This restriction is not enforced at the outlet of the system in case other restrictions are required.

OGSDP aligns the pipes to produce acceptable flow velocities. The default minimum just-full flow velocity is 2 ft/s. This velocity is required to prevent sedimentation and eventual blockage of the system. The user is free to increase this value. The default maximum velocity is 8 ft/s. It may be increased or decreased. However, maximum velocities in excess of 10 ft/s are not advisable as erosion of the pipe may result.

OGSDP assumes the validity of the full pipe Manning equation. A recommended value (for concrete pipes) of 0.013 is used as the default for Manning's roughness coefficient, n. The user is free to input any other appropriate value. The crown elevations for the lowest incoming pipe and the pipe that drains a manhole are matched except for drop manholes. Whenever the difference in invert elevations between the highest pipe emptying into a manhole and the pipe draining the manhole exceeds 2 ft, a statement indicating the need for an external drop manhole is printed.

In practice, pipe sizes usually are not allowed to decrease in the downstream direction but this restriction is sometimes ignored (relaxed). This may result in two problems. First, as noted previously, OGSDP uses a single state variable dynamic programming formulation that theoretically guarantees the global optimal design for the relaxed problem but does not guarantee it for the original (unrelaxed) problem. Second, because OGSDP matches crown elevations, a manhole with decreasing diameters may fail to drain: the invert of the draining pipe may be higher than the invert of an incoming pipe. If this ever happens, a warning is printed identifying which manhole is affected. Both problems may be addressed by selecting options available within OGSDP.

OGSDP is organized with three options. The first option is to solve the relaxed design problem; i.e., ignore the constraint on decreasing downstream pipe sizes. The second option is to solve the unrelaxed problem by only considering non-decreasing downstream pipe sizes at each stage of the solution. A third option is to solve the unrelaxed problem and then in the neighborhood of that solution, search for a better solution to the relaxed problem. Use of these three options is very effective because the solution of the relaxed problem is a lower bound to the actual (unrelaxed) problem; therefore, the user can always tell how good the final solution is. In practice, the relaxed solution is almost always 'very close' to the practical (unrelaxed) solution because sewer size generally tends not to decrease in the downstream direction.

Nonlinear cost functions are used to compute the cost of pipe and other material as well as the costs of excavation and installation of pipes and manholes. The default cost functions were obtained from Han et $al.$ (1980). They are:

$$PCOST = \begin{cases} LP + L(1.93D + 1.688H - 12.6); \\ \quad \text{if } 0 \le H \le 20 \\ \\ LP + L(0.69D + 2.14H + 0.559HD - 13.56); \\ \quad \text{if } 20 < H \end{cases} \qquad (1)$$

where:

$COST$ = pipe material and installation costs ($)
P = unit pipe material cost ($/ft)
      Default values are used if the user does not
      specify other appropriate unit costs.
L = length of pipe, (ft)
D = diameter of pipe (in)
H = average invert depth of pipe (ft)

Manhole costs are calculated thus:

$$MHCOST = 259.4 + 56.4Y \qquad (2)$$

where:

$MHCOST$ = manhole cost ($)
Y = manhole depth (ft)

The cost functions were formulated after a review of the literature and local material costs. The equations show that unit material and installation costs for each pipe are a function of average invert depth H (ft) and pipe diameter D (in). Manhole costs are a function of depth Y (ft). All costs are in 1980 $. The user may update the unit pipe costs in the input file for OGSDP to reflect prevailing costs. With additional effort, the entire set of cost equations may be replaced with more appropriate functions; this requires the rewriting of one subroutine in OGSDP.

2.3   A Brief Description of the Model

OGSDP consists of three main interconnected algorithms:

- Flow information algorithm

- Initial solution algorithm

- Optimization algorithm

They have been listed in the sequence in which they are encountered when the model is implemented. Fig. 1 further describes how they are each used in the design process.

Flow Information Algorithm   This algorithm computes the design flow for each pipe (link) in the given sewer system. First, for each link, the population increment due to that link is calculated from the product of the number of residence-equivalent service connections for that link and the assumed number of persons per service connection (usually four). Next, the total population contributing flow to each link is calculated from the sum of the population increments for that link and those of any upstream pipes. Finally, the design flow is calculated from the product of the total population contributing flow to any link, the (assumed) daily average sewage flow *per capita*-day and the peak factor (input by the user or computed using the Harmon equation).

Intital Solution Algorithm   The function of this algorithm is to provide a starting solution for the optimization algorithm. It is a heuristic; no formal optimization techniques have been incorporated. However, it may yield the optimal design in certain cases.

It designs a system with the objective of using a combination of the smallest, but adequate pipe sizes that satisfy the restrictions on the invert elevation of the pipe at the outlet of the sewer system. The total fall from the invert of any upstream pipe to the specified invert elevation at the outlet is limited. If the upstream pipes are designed with steep slopes, downstream pipes may be unable to carry the design flow because their slopes will be too small. Therefore, in the Initial Solution Algorithm, the average slope from the upstream invert of the pipe to the outlet is not exceeded, unless the minimum depth of cover or the relevant flow velocity requirements cannot otherwise be met.

While this algorithm does not consider costs explicitly, it does so implicitly by trying to use smaller but adequate pipe sizes. This should result in lower pipe costs and reduced excavation (therefore lower excavation costs). The function of this algorithm is to provide an initial feasible design for the Optimization Algorithm. All constraints on the design of the system except one are incorporated within the

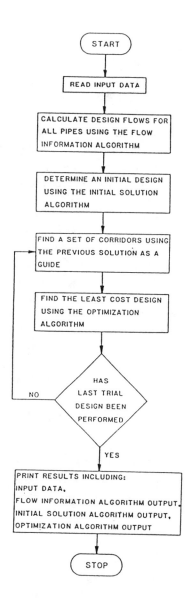

Figure 1. *Generalized flow chart of OGSDP.*

Initial Solution Algorithm. The one exception is that maximum velocity restrictions may be violated if no solution is found that satisfies them. This situation is flagged whenever it occurs. The maximum velocity restriction is enforced in the final design obtained from the Optimization Algorithm.

Optimization Algorithm Once an initial design has been determined, OGSDP uses a sequential optimization procedure to improve the design. First, a trial range of crown elevations at each manhole location is calculated. The ranges, which include the crown elevations determined in the previous solution, are searched for an improved trial design. The minimum depth of cover requirements are respected when these ranges of crown elevation (called corridors) are determined. Discrete dynamic programming is exploited to search the corridors for a better design.

The design obtained on any trial is used to set up a new array of corridors for another trial design if necessary or desired. This iterative process may be described as successive approximations. The sizes of the corridors in each succeeding trial design are reduced by a factor which may be specified by the user.

The scheme applied here has certain unique features. One is that the previous design is specifically compared with any succeeding designs. This allows for the re-use of the previous solution. Another feature is that pipe sizes may be prohibited from decreasing in the downstream direction. A heuristic is used to limit the pipes considered at each stage of the solution to those diameters that are non-decreasing in the downstream direction.

The Optimization Algorithm solves a given design problem stage by stage from the most upstream links towards the outlet of the system. At each stage, a set of least-cost combinations of downstream crown elevations are determined along with the adequate pipe size while the relevant constraints for that trial design are enforced. At the final stage, the least-cost invert elevation of the system is chosen. A more elaborate description of the computer program is given by Nzewi et al. (1983).

2.4 Example Problem

The input data required for OGSDP include: the number of pipes in the system, the desired invert elevation at the outlet, the minimum allowable depth of cover, the peak factor, the incremental depth used to define the set of possible elevations of pipes and manholes in the dynamic program, the maximum number of elevations considered in the dynamic program for any manhole location, the number of trial optimal designs desired, unit pipe costs for each pipe size (if the user desires to update these values); and, for each pipe, the pipe number, upstream and downstream manhole numbers, the length, the number of service-equivalent connections, and the upstream and downstream ground elevations at the manhole locations. A switch indicating whether pipes are allowed to decrease in diameter in the downstream direction must be turned on or off.

The solution of a sample problem with 20 pipes, which are permitted to decrease in the downstream direction, is the focus of this section. The input file is not presented but complete details on the input format are available elsewhere (Nzewi et al., 1983). Part of the output from OGSDP has been included for illustration purposes.

Table 1 contains the results of the Initial Solution Algorithm. This initial design costs $351,426. Each line of the table contains information about one pipe in the system. For example, the first line describes pipe number 1. The recommended pipe diameter is 10 in. and the recommended slope is 1.429%. The cover above the pipe at manholes 1 and 2 is 8 ft. and the invert elevations are 491.17 ft. and 486.17 ft. at manholes 1 and 2, respectively. The just-full pipe capacity is 2.63 cfs which exceeds the design flow of 1.671 cfs. The just-full flow velocity if 4.8 ft/s. The initial design violates the maximum flow velocity restriction of 8.0 ft/s in pipe 14, noted at the bottom of the table. The need for an external drop manhole at manhole 21 in order to meet the specified invert elevation at the system exit is noted at the bottom of the table also.

Table 2 is the final solution obtained by the Optimization Algorithm. Its total cost is $338,309. In this solution, all pipes have non-decreasing diameters in the downstream direction and none of the design constraints is violated.

3 Dual Purpose Urban Drainage Network Design

Stormwater detention basins are increasingly being used to control both urban flooding and nonpoint source pollution. In order to insure the effectiveness of such basins for a wide range of operating conditions, it is imperative that a regional stormwater management approach be used (McCuen, 1979; Amandes and Bedient, 1980; and Bennett, 1983). Conventionally, most detention basins are designed using a trial and error approach. Besides being time consuming, such an approach may fail to consider the interactions of flow rates and pollutant loads within a watershed. As a result, designs may be obtained that can actually aggravate potential flood hazards (McCuen, 1974; Smiley and Haan, 1976; and Abt and Grigg, 1978). In order to consider the regional impact of stormwater detention basins, some type of design algorithm or model is usually needed. Several attempts at the development of such models have been presented by Lakatos (1976), Hopkins et al. (1978), Dendrou and Delleur (1978), Mays and Bedient (1982), Ormsbee et al. (1984), and Bennett and Mays (1985). These algorithms have used either optimization, simulation, or both in the development of general stormwater management plans. The majority of these attempts have been limited to general planning considerations and have excluded any consideration of water quality impacts.

The objective of the dual purpose design problem is to determine the locations and sizes of detention basins so as to minimize the overall design costs of the system while satisfying both water quantity (flow rate) and water quality (pollutant load) constraints at specified control stations. The general problem involves two different levels of optimization: the optimal design of the individual basins, and the optimal location of the individual basins within the watershed. Ultimately, the problem of the optimal location of the basins can be imbedded into the optimal design problem. If a particular basin location yields a design with zero storage then that location is not needed.

To solve the dual purpose detention basin design problem, some type of design model is required. Because of the serial nature of the problem, dynamic programming was investigated for use in developing a general design algorithm. This technique has been applied with apparent success in the planning of stormwater detention systems by Flores et al. (1982), Mays and Bedient (1982), Bennett (1983), and Bennett and Mays (1985). In each case, the problem formulation was restricted to water quantity considerations.

The following scheme is suggested for the application of dynamic programming to the design of detention basins. First, let each stage correspond to the distance from the watershed outlet as measured by the number of detention basins along any reach. Associated with each stage i is a set of J detention basins denoted by $\{i\}$.

Associated with any detention basin may be several different state variables. Different combinations of the values of the state variables may be represented by a state vector S_{ij}. The value associated with a particular combination of state variables may be represented by S_{ijk} where K different combinations are possible $(k = 1, 2, ..., K)$. The state variables represent the conditions or components associated with a particular basin design. The exact choice of state variables depends on the nature of the dynamic program formulation. For example, for one formulation the basin base area, side slope and orifice outlet dimension could all be specified as state variables. For another formulation, the actual outflow hydrograph and pollutant load from each basin could be specified as state variables.

In addition to a set of state variables associated with each detention basin location, a set of decision variables must be specified. The decision variables represent the set of possible transitions from the state space associated with one detention basin site to the state space associated with a second downstream site. Associated with each of the transitions or decision variables is a particular cost or return. This cost or return may be quantified using a return function. In general, the return function is a function of both the decision variable and the state variable associated with a particular stage.

This dynamic programming approach has been applied to the design of an urban drainage network in Glen Ellyn, Illinois (Ormsbee et al., 1984). The objective of this application was to design the three detention basins at a minimum cost

Table 1. *Initial solution algorithm output.*

Cost of the entire system = $ 351 426 (1980 $)

Pipe No.	Manhole No. Upper	Manhole No. Lower	Pipe diam. (in)	Pipe length (ft)	Slope (ft/ft)	Cover Upper (ft)	Cover Lower (ft)	Ground elevation Upper (ft)	Ground elevation Lower (ft)	Invert elevation Upper (ft)	Invert elevation Lower (ft)	Design flow (cfs)	Pipe capacity (cfs)	Full velocity (ft/s)
1	1	2	10.0	350.00	0.01429	8.0	8.0	500.0	495.0	491.17	486.17	1.671	2.63	4.8
2	2	3	10.0	400.00	0.02000	8.0	8.0	495.0	487.0	486.17	478.17	2.340	3.11	5.7
3	3	4	12.0	350.00	0.02000	8.0	8.0	487.0	480.0	478.00	471.00	3.888	5.05	6.4
4	6	5	8.0	400.00	0.01250	8.0	8.0	490.0	485.0	481.33	476.33	0.891	1.36	3.9
5	5	4	10.0	430.00	0.01163	8.0	8.0	485.0	480.0	476.17	471.17	1.783	2.37	4.3
6	4	10	15.0	550.00	0.01818	8.0	8.0	480.0	470.0	470.75	460.75	6.785	8.74	7.1
7	7	8	8.0	500.00	0.01000	8.0	8.0	490.0	485.0	481.33	476.33	0.418	1.21	3.5
9	9	10	12.0	350.00	0.01429	8.0	8.0	475.0	470.0	466.00	461.00	2.702	4.27	5.4
10	10	11	18.0	500.00	0.01000	8.0	8.0	470.0	465.0	460.50	455.50	9.709	10.54	5.9
11	14	13	8.0	500.00	0.02000	8.0	8.0	485.0	475.0	476.33	466.33	1.263	1.71	4.9
12	13	12	12.0	350.00	0.1429	8.0	8.0	475.0	470.0	466.00	461.00	2.823	4.27	5.4
13	12	11	12.0	350.00	0.1429	8.0	8.0	470.0	465.0	461.00	456.00	2.969	4.27	5.4
14	11	15	21.0	600.00	0.1667	8.0	8.0	465.0	455.0	455.25	445.25	13.792	20.52	8.5†
15	18	17	8.0	400.00	0.01000	8.0	8.0	468.0	464.0	459.33	455.33	1.114	1.21	3.5
16	17	16	10.0	300.00	0.01333	8.0	8.0	464.0	460.0	455.17	451.17	1.560	2.54	4.6
17	16	15	18.0	350.00	0.1429	8.0	8.0	460.0	455.0	450.50	445.50	8.244	12.60	7.1
18	15	19	27.0	400.00	0.01000	8.0	8.0	455.0	451.0	444.75	440.75	23.596	31.07	7.8
19	19	20	27.0	500.00	0.00669	8.0	8.3	451.0	448.0	440.75	437.41	25.267	25.41	6.4
*20	20	21	27.0	600.00	0.00746	8.3	9.8	448.0	445.0	437.41	432.93	25.407	26.84	6.7

* Indicates that a drop connection is required at downstream manhole

† Design violates maximum velocity restrictions: maximum velocity = 8.51 ft/s

NOTE: Since the restrictions on the maximum velocity are enforced in the final design it is possible that a feasible solution may not be found by the optimization algorithm. Therefore, it may be necessary to raise the maximum allowable velocity using the value obtained here as a guide.

Table 2. *Solution 3 from the optimization algorithm.*

Cost of the entire system = $338,309 (1980 $)

Pipe No.	Manhole No. Upper	Manhole No. Lower	Pipe diam. (in)	Pipe length (ft)	Slope (ft/ft)	Cover Upper (ft)	Cover Lower (ft)	Ground elevation Upper (ft)	Ground elevation Lower (ft)	Invert elevation Upper (ft)	Invert elevation Lower (ft)	Design flow (cfs)	Pipe capacity (cfs)	Full velocity (ft/s)
1	1	2	8.0	350.00	0.01898	8.0	9.6	500.0	495.0	491.33	484.69	1.671	1.67	4.8
2	2	3	10.0	400.00	0.01589	9.6	8.0	495.0	487.0	484.52	478.17	2.340	2.77	5.1
3	3	4	12.0	350.00	0.02000	8.0	8.0	487.0	480.0	478.00	471.00	3.888	5.05	6.4
4	4	6	8.0	400.00	0.01250	8.0	8.0	490.0	485.0	481.33	476.33	0.891	1.35	3.9
5	5	4	10.0	430.00	0.01163	8.0	8.0	485.0	480.0	476.17	471.17	1.783	2.37	4.3
6	4	10	15.0	550.00	0.01818	8.0	8.0	480.0	470.0	470.75	460.75	6.785	8.73	7.1
7	7	8	8.0	500.00	0.01000	8.0	8.0	490.0	485.0	481.33	476.33	0.418	1.21	3.5
8	8	9	8.0	450.00	0.02222	8.0	8.0	485.0	475.0	476.33	466.33	1.309	1.81	5.2
9	9	10	10.0	500.00	0.01509	8.0	8.3	485.0	478.0	476.17	468.62	2.702	2.70	4.9
10	10	11	18.0	500.00	0.00944	8.3	8.0	475.0	470.0	465.50	460.78	9.709	10.23	5.8
11	11	13	8.0	500.00	0.02000	8.0	8.8	475.0	466.0	466.33	456.33	1.263	1.71	4.9
12	12	13	10.0	350.00	0.01651	8.8	8.0	471.0	464.0	461.17	455.39	2.823	2.82	5.2
13	13	11	12.0	350.00	0.01206	8.0	8.0	470.0	466.0	461.00	456.78	2.969	3.92	5.0
14	14	13	18.0	600.00	0.01711	8.0	8.3	475.0	465.0	465.50	455.23	13.792	13.78	7.8
15	15	15	8.0	400.00	0.01000	8.0	8.0	465.0	461.0	456.33	452.33	1.114	1.21	3.5
16	16	16	10.0	300.00	0.01333	8.0	8.0	464.0	460.0	455.17	451.17	1.560	2.54	4.7
17	17	17	15.0	350.00	0.01617	8.0	8.7	460.0	455.0	450.75	445.09	8.244	8.24	6.7
18	18	15	24.0	400.00	0.01080	8.7	9.0	455.0	451.0	444.34	440.02	23.596	23.58	7.5
19	19	20	27.0	500.00	0.00663	9.0	9.3	451.0	448.0	439.77	436.45	25.267	25.29	6.4
*20	20	21	27.0	600.00	0.00670	9.3	10.3	448.0	445.0	436.45	432.43	25.407	25.41	6.4

* Indicates that a drop connection is required at downstream manhole

The largest incremental depth for any iteration is now = 0.023 ft

The least incremental depth for any iteration is now = 0.020 ft

while meeting both water quantity and water quality constraints at the watershed outlet. Only one pollutant, total suspended solids, was considered. One of the state variables was the outflow hydrograph from the basins. The inclusion of these hydrographs into the overall problem can yield inseparable problem formulations. Neverless, such formulations may still be expected to yield good designs in a fast and efficient manner, and provide the basis for the development of valuable design heuristics.

4 Long-Term Multipurpose Reservoir Operations

In total, there is enough water worldwide for everyone. However, its temporal and spatial distribution is variable such that oftentimes, the water needs and water availabilities at particular locations and times are not compatible. Streamflow regulation by a reservoir partially overcomes the problem of inadequate or excessive water through alteration of inflows to the reservoir.

One major challenge to water resource planners is the development of the best strategy for scheduling releases or a detailed reservoir system operating procedure. Improved operation is an effective way of using currently available water in an existing reservoir system to provide benefits associated with a wide range of purposes. Some of the possible conflicting purposes are flood control, water supply, recreation, hydroelectric power generation, navigation, and water quality improvement. The problem of finding the best strategy for efficient operation or management of reservoir systems can be exceedingly complex. However, improvement in operational methods is an economical alternative to making major capital investments in new facilities.

In the past two decades, the difficulties encountered in finding new sources of water and the escalating cost of new development has led to an upsurge in studies on the design and operation of water resource systems. The same period of time has seen advances in mathematical programming techniques, streamflow analysis, and computer technology. These recent developments currently make possible organized and systematic analysis of reservoir system to obtain good or optimal operating rules.

Numerous techniques have been proposed and reported in the water resource literature to solve the important and practical problem of determining optimal reservoir operating rules. The foremost motivation in applying these techniques is the possible improvement of current operation because relatively minor improvements produce large benefits (Beard, 1975). In general, the techniques for finding optimal operating rules often involve several areas or disciplines such as operations, hydrology, statistics, operations research and economics. This section contains a literature review of relevant techniques that are currently available.

590

Depending on the treatment of hydrologic information, the approaches to the solution of the reservoir operation problem may be grouped into five general classes:

(a) simulation,

(b) implicitly stochastic optimization (ISO),

(c) explicitly stochastic optimization (ESO),

(d) alternative stochastic optimization (ASO), and

(e) deterministic optimization.

Roefs and Bodin (1970) identified the first three modeling approaches. Croley (1974a and 1974b) discussed ESO and ISO and, in addition, developed the ASO technique. Palmer *et al.* (1982) added the last type to complete the classification of the approaches.

Simulation is a straightforward approach that simply tests alternative operating policies that are evaluated according to a desired objective. Successful implementation of simulation in reservoir operation problems has been widespread (e.g., Sigvaldason, 1976). Simulation studies are limited, however, in the sense that detailed examination of all possible alternatives may be impractical. Furthermore, finding an optimal solution can not be guaranteed.

ISO includes generation of a certain number of synthetic inflow sequences as a first step. Subsequently, optimization program runs are made for a large number of equally likely future flow sequences to obtain a sample of optimal solutions. Finally, by using techniques of multivariate analysis, the optimal rule is deduced. The published works of Young (1967), Croley and Rao (1977, 1979), and Amborsino *et al.* (1984) are good examples of ISO. A major problem with the ISO approach concerns the criterion for choosing among the distributions of the optimum resulting from the sample (Askew *et al.*, 1971). This question was bypassed by Young (1967) who solved the mathematical program only once, with one long synthetic series.

The ESO approach has been used by many researchers (e.g., Revelle *et al.* 1969; Su and Deininger, 1974; Roefs and Guitron, 1975; Bogle and O'Sullivan, 1979; Maidment and Chow, 1981; Loucks *et al.*, 1981). In this approach, the streamflow is represented by a process from which probabilities and conditional distribution of flows are obtained. The probability distribution instead of inflow sequences is then used in the optimization over the time horizon of interest. As pointed out by Croley (1979), ESO has an advantage over ISO because ESO is based on proposed conditional probability distributions. Thus, more information is available for the choice of decisions. However, the complexity of the systems could severly limit ESO's application unless assumptions on the probability distribution (unchanging or steady-state distribution is a common assumption) are made.

ASO involves repeated optimization for each stage for many input realizations over a limited number of stages into the future. In other words, less than the entire operation horizon is considered. A distribution of relative optimal values of the objective function that is likely with the system is thus obtained.

Deterministic optimization can be used when exact knowledge of future hydrology can be assumed. There are instances when deterministic optimization may be preferable to the comparatively involved stochastic optimization (Klemes, 1977b). Deterministic optimization has been reported to result consistently in near optimal policies with great savings on computational efforts when compared to stochastic methods (Harley and Chidley, 1978). This was attributed to the inaccuracies possibly introduced when the statistical properties of the data were simplified to make them suitable as input to stochastic programming techniques. Deterministic optimization also becomes an attractive and advantageous approach if the cost of carrying out stochastic optimization procedures would exceed the value of improvement in operation control. Additionally, the deterministic assumption might be necessary to analyze more complex systems. (For instance, Yakowitz (1982) reported that the largest system solved by stochastic dynamic programming is composed of only two reservoirs compared to ten reservoirs by deterministic dynamic programming.) Furthermore, nonstochastic modeling is also useful because its result may be used to define a starting point for a more elaborate analysis by simulation.

Acknowledgment The work described in this paper is due in large measure to the efforts of a number of scholars including Rosalinda Cantiller, Jacques Delleur, Donald Gray, Emmanual Nzewi, Lindell Ormsbee, and Jeff Wright.

References

Abt, S. R. and Grigg, N. S., (1978), 'An Approximate Method for Sizing Detention Reservoirs,' *Water Resources Bulletin* **14**, 4, 959–961.

Amandes, C. B. and Bedient, P. B., (1980), 'Storm Water Detention in Developing Watersheds,' *Journal of the Environmental Engineering Division, ASCE* **106**, 2, 403–409.

Ambrosino, G., Fronza, G. and Guariso, G., (1979), 'Real-Time Predictor Versus Synthetic Hydrology for Sequential Reservoir Management,' *Water Resources Research* **15**, 4, 885–890, (August).

Bellman, R. E., (1957), *Dynamic Programming*, Princeton University Press, Princeton, New Jersey.

Bennett, A. B., (1983), 'Dynamic Programming Model for Determining Optimal Sizes and Locations for Detention Storage Facilities,' *Proceedings of the International Symposium on Urban Hydrology, Hydraulics, and Sediment Control*, University of Kentucky, Lexington, Kentucky, 461–468, July.

Bennett, M. S. and Mays, L. W., (1985), 'Optimal Design of Detention and Drainage Channel Systems,' *Journal of Water Resources Planning and Management Division, ASCE* 11, 1, 99–112.

Bogle, M. G. V. and O'Sullivan, M. J., (1979), 'Stochastic Optimization of a Water Supply System,' *Water Resources Research* 15, 4, 778–786, (August).

Cantiller, R. R. A., Houck, M. H. and Wright, J. R., (1986), 'A Nonlinear Implicitly Stochastic Method for Determining Reservoir Operating Rules,' Technical Report CE-HSE-86-12, School of Civil Engineering, Purdue University, West Lafayette, Indiana, (September).

Croley, T. E., II and Rao, K. N. R., (1977), 'Stochastic Trade-Offs for Reservoir Operation,' Iowa Institute of Hydraulic Research, No. 197, The University of Iowa, (January).

Croley, T. E., II and Rao, K. N. R., (1979), 'Multiobjective Risks in Reservoir Operation,' *Water Resources Research* 15, 4, 807–814, (August).

Croley, T. E., II, (1974b), 'Efficient Sequential Optimization in Water Resources,' *Hydrology Papers No. 69*, Colorado State University, Fort Collins, Colorado, (September).

Croley, T. E., II, (1972), 'Sequential Deterministic Optimization in Reservoir Operation,' *Journal of the Hydraulics Division, ASCE* 100, HY3, 443–459, (March).

Dajani, J. S. and Gemmell, R. S., (1972), 'Optimal Design of Wastewater Collection Networks,' *Journal of Sanitary Engineering Division, ASCE* 98, SA6, 853–867, (December).

Deininger, R. A., (1973), 'System Analysis for Water Supply and Pollution Control,' in *Natural Resources Systems Models in Decision Making*, G. H. Toebes (Ed.), Water Resources Center, Purdue University, West Lafayette, Indiana.

Dendrou, S. A. and Delleur, J. W., (1982), 'Watershed Wide Planning of Detention Basins,' *Proceedings Engineering Foundation Conference on Stormwater Detention*, Hennicker, New Hampshire, 72–85, August.

Driscoll, E. D., (1982), 'Analysis of Detention Basins in EPA NURP Program,' *Proceedings of the Conference on Stormwater Detention Facilities*, Hennicker, New Hampshire, 21–31, August.

Flores, A. C., Bedient, P. B. and Mays, W. L., (1982), 'Method for Optimizing Size and Location of Urban Detention Facilities,' *Proceedings of the International Symposium on Urban Hydrology, Hydraulics, and Sediment Control*, University of Kentucky, Lexington, Kentucky, 357–366, July.

Gupta, A., Mehndiratta, S. L. and Khanna, P., (1983), 'Gravity Wastewater Collection Systems Optimization,' *Journal of Environmental Engineering Division ASCE* **109**, 5, 1195–1208, (October).

Froise, S. and Burges, S. J., 'Least-Cost Design of Urban-Drainage Networks,' (1978), *Journal of Water Resources Planning Management Division, ASCE* **104**, WR1, 75–92, (November).

Han J., Rao, A. R. and Houck, M. H., (1980), 'Least Cost Design of Urban Drainage Systems,' Technical Report Number 138, Purdue University Water Resources Research Center, West Lafayette, Indiana, (September).

Harley, M. J. and Chidley, T. R. E., (1978), 'Deterministic Dynamic Programming for Long Term Reservoir Operating Policies,' *Engineering Optimization* **3**, 63–70.

Hopkins, L. D., (1978), 'Land Use Allocation for Flood Control,' *Journal of Water Resources Planning and Management Division ASCE* **104**, 1, 93–104.

Klemes, V., (1977a), 'Discrete Representation of Storage for Stochastic Reservoir Optimization,' *Water Resources Research* **13**, 1, 149–158, (February).

Lakatos, D. F., (1976), 'Penn State Runoff Model for the Analysis of Timing of Subwatershed Response to Storms,' *Proceedings of the International Symposium on Urban Hydrology, Hydraulics, and Sediment Control*, University of Kentucky, Lexington, Kentucky, 25–29, July.

Loucks, D. P., Stedinger, J. R. and Haith, D. A., (1981), *Water Resources Systems Planning and Analysis*, Prentice-Hall, Englewood Cliffs, New Jersey.

Maidment, D. R. and Chow, V. T., (1981), 'Stochastic State Variable Dynamic Programming for Reservoir Systems Analysis,' *Water Resources Research* **17**, 6, 1578–1584, (December).

Mays, L. W. and Bedient, P. B., (1982), 'Model for Optimal Size and Location of Detention,' *Journal of Water Resources Planning and Management Division, ASCE* **108**, 3, 270–285, (October).

594

Mays, L. W. and Wenzel, H. G., Jr., (1976), 'Optimal Design of Multilevel Branching Sewer Systems,' *Water Resources Research* **12**, 5, 913–917, (October).

Mays, L. W. and Yen, B. C., (1975), 'Optimal Cost Design of Branched Sewer Systems,' *Water Resources Research* **11**, 11, 37–47, (February).

McCuen, R. H., (1974), 'A Regional Approach to Urban Storm Water Detention,' *Geophysical Research Letters* **1**, 7, 321–322.

McCuen, R. H., (1979), 'Downstream Effects of Stormwater Management Basin,' *Journal of Hydraulics Division, ASCE* **105**, 11, 1343–1356.

Merritt, L. B. and Bogan, R. H., (1973), 'Computer-Based Optimal Design of Sewer Systems,' *Journal of Environmental Engineering Division, ASCE* **99**, EE1, 35–53, (February).

Nzewi, E. U., Gray, D. D. and Houck, M. H., (1983), 'Computer-Aided Optimal Design of Gravity Sanitary Sewerage Systems,' Technical Report Number CE-HSE-83-2, School of Civil Engineering, Purdue University, West Lafayette, Indiana, (April).

Nzewi, E. U., Gray, D. D. and Houck, M. H., (1985), 'Optimal Design Program for Gravity Sanitary Sewers,' *Civil Engineering Systems Analysis* **2**, 132–141, (September).

Ormsbee, L. E., Delleur, J. W. and Houck, M. H., (1982), 'Development of a General Planning Methodology for Stormwater Management in Urban Watersheds,' Technical Report 163, Purdue University Water Resources Research Center, West Lafayette, Indiana, (March).

Ormsbee, L. E., Houck, M. H. and Delleur, J. W., (1987), 'Design of Dual Purpose Detention Systems Using Dynamic Programming,' *Journal of Water Resources Planning and Management* **113**, 4, 471–484, (July).

Pacheco, M. A. and Gray, D. D., (1982), 'Computer Aided Design of Gravity Sanitary Sewer Systems,' Technical Report Number CE-HSE-82-1, School of Civil Engineering, Purdue University, West Lafayette, Indiana, (January).

Palmer, R. N., Smith, J. A., Cohon, J. L. and ReVelle, C. S., (1982), 'Reservoir Management in Potomac River Basin,' *Journal of Water Resources Planning and Management Division, ASCE* **108**, WR1, 47–66, (March).

ReVelle, C., Joeres, E. and Kirby, W., (1969), 'The Linear Decision Rule in Reservoir Management and Design, 1. Development of the Stochastic Model,' *Water Resources Research* **5**, 4, 767–777, (August).

Roefs, T. G. and Guitron, A. R., (1975), 'Stochastic Reservoir Models: Relative Computational Effort,' *Water Resources Research*, **11**, 6, 801–804, (December).

Roefs, T. G. and Bodin, L. D., (1970), 'Multireservoir Operation Studies,' *Water Resources Research* **6**, 2, 410–420, (April).

Sigvaldason, O. T., (1976), 'A Simulation Model for Operating a Multipurpose Multireservoir System,' *Water Resources Research* **12**, 2, 263–278, (April).

Smiley, J. and Haan, C. T., (1976), 'The Dam Problem of Urban Hydrology,' *Proceedings of the International Symposium on Urban Hydrology, Hydraulics, and Sediment Control*, University of Kentucky, Lexington, Kentucky, July.

Su, S. Y. and Deininger, R. A., (1974), 'Modeling the Regulation of Lake Superior Under Uncertainty of Future Water Supplies,' *Water Resources Research* **10**, 1, 11–25, (February).

Tang, W. H., Mays, L. W. and Yen, B. C., (1975), 'Optimal Risk-Based Design of Storm Sewer Networks,' *Journal of Environmental Engineering Division, ASCE* **101**, EE3, 381–398, (June).

Templeman, A. B. and Walters, G. A., (1979), 'Optimal Design of Storm-Water Drainage Networks for Roads,' *Proceedings of the Institute of Civil Engineers*, Part 2, 67, 573–587, (September); and 'Discussion,' 69, 875–880, (September).

Ten State Standards, (1978), Great Lake-Upper Mississippi River Board of Sanitary Engineers, *1978 Recommended Standards for Sewage Works*, Health Education Service, Albany, New York (The ten states are: Illinois, Indiana, Iowa, Michigan, Minnesota, Missouri, New York, Ohio, Pennsylvania, and Wisconsin).

Walsh, S. and Brown, L. C., (1973), 'Least Cost Method for Sewer Design,' *Journal of Environmental Engineering Division, ASCE* **99**, EE3, 333–345, (June).

Walters, G. A. and Templeman, A. B., (1979), 'Non-Optimal Dynamic Programming Algorithms in the Design of Minimum Cost Drainage systems,' *Engineering Optimization*, **4**, 3, 139–148, (December).

Wright, J., Malik, A., Steiner, R., Cohon, J. and ReVelle, C., (1980), 'Water Resources Models Currently Available and Used by Federal Agencies,' Department of Geography and Environmental Engineering, The Johns Hopkins University, Maryland, July.

Yakowitz, S., (1982), 'Dynamic Programming Applications in Water Resources,' *Water Resources Research* **18**, 4, 673–696, (August).

Young, G. K., Jr., (1967), 'Finding Reservoir Operating Rules,' *Journal of the Hydraulics Division, ASCE* **93**, HY6, 297–321, (November).

Queues, Construction Operations and Optimization

David G Carmichael
School of Civil Engineering
The University of New South Wales
Kensington
Australia

Abstract   Certain construction operations lend themselves very neatly to be modelled as queueing processes. In particular earthmoving type operations are shown to be naturally modelled as finite source queueing processes or cyclic queueing processes. Such models take account of the variabilities in the operations as well as the interaction of equipment. Using two-stage cyclic queueing processes under various probability assumptions as a basis, the paper develops optimization in the planning and management of earthmoving operations. Specific problems formulated and solved include the choice of the optimal hauling unit fleet size, optimal loading policies, optimal heterogeneity (mix of equipment), optimal operation layout, optimal choice of equipment characteristics and optimal preemptive and non-preemptive priority treatment of equipment. The objective functions are typically cost, production, cost per production or equipment utilization. Constraints may include time, production or other considerations. The use of optimal planning and management strategies is shown to provide a more rational framework than conventional earthmoving procedures. In the process of developing the optimal strategies, new queueing models are put forward to handle the idiosyncrasies of the civil engineering problems considered.

1   Introduction

Construction operations such as the one illustrated in Fig. 1 lend themselves very neatly to be modelled as queueing processes whether of the finite source type or cyclic type. The models implicitly allow for operation variabilities and equipment interaction [1].

Optimization in the planning and management of such operations is facilitated by the use of these queueing models. The objective function or criterion component of the optimization is typically cost, production, cost per production or equipment utilization. The third component, namely constraints may include time, production or other considerations.

B. H. V. Topping (ed.),
Optimization and Artificial Intelligence in Civil and Structural Engineering, Volume I, 597–616.
© 1992 *Kluwer Academic Publishers. Printed in the Netherlands.*

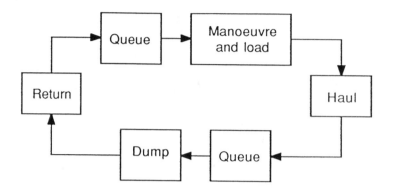

Figure 1. *Example construction operation; earthmoving.*

Specific optimization problems addressed below relate to the operation shown in Fig. 1 and include the choice of the optimal hauling unit fleet size, optimal operation layout, optimal choice of equipment characteristics and optimal preemptive and non preemptive priority treatment of equipment. Each problem is formulated in terms of its pertinent model, objective function and constraints.

Background to related problems and related solution methods may be found in [2].

2 Models

The favoured queueing models for operations of the form shown in Fig. 1 reduce the schematic to that shown in Fig. 2. Stage 1 contains a service phase (typically loading and manoeuvring) and a queueing phase. For ease of discussion the number of loaders or servers in Stage 1 is taken as one. Stage 2 may be regarded as a self-service stage, there being no queueing and as many servers as there are hauling units (customers); it is an amalgamation of the haul, queue at dump, dump and return phases. Hauling units or customers cycle repeatedly between Stage 1 and Stage 2.

Notation relating to Fig. 2 includes:

Stage 1

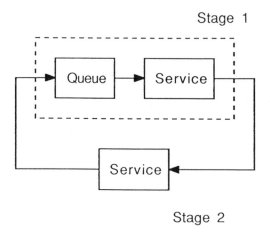

Stage 2

Figure 2. *Schematic for queueing analysis.*

$1/\mu$ average service time for Stage 1
$1/\lambda$ average service time for Stage 2
W_q average queue waiting time per customer in Stage 1
K number of customers in the operation
η Stage 1 server utilization

Where different customer types are present, these symbols are subscripted such as μ_i, λ_i, W_{qi}, K_i and η_i, $i = 1, 2, \ldots, V$, where i is the customer type and V is the total number of customer types.

Various assumptions may be made on the distribution of service times in the two stages including Erlang (denoted E_α, where α is a shape parameter), exponential and deterministic. Generally it is seen from field data that Erlang distributions reflect observations closely while exponential and deterministic distributions lead to results which bound observed behaviour.

Models may be further classified according to whether the operation is homogeneous or heterogeneous. Heterogeneity implies different customer characteristics or in terms of Fig. 2, the stage service times are different for different customer types.

The central piece of information deriving from a queueing model is the utilization of the Stage 1 server. Denoted η it is the proportion of time that the Stage 1 server is busy. From η all other relevant information on the operation may be obtained [2]. In particular,

$$\text{Production} = \mu \eta C T$$

where C is the capacity (m^3, tonnes, ...) of a customer (alternatively C_i for the heterogeneous case) and T is the time over which the production is being calculated. And,

$$\text{Average customer cycle time} = W_q + \frac{1}{\mu} + \frac{1}{\lambda} = \frac{K}{\mu\eta}$$

For the heterogeneous case the Stage 1 server utilization derives from the contributions of the different customer types,

$$\eta = \sum_{i=1}^{V} \eta_i$$

While,

$$\text{Average customer type } i \text{ cycle time} = W_{qi} + \frac{1}{\mu_i} + \frac{1}{\lambda_i} = \frac{K_i}{\mu_i \eta_i}$$

2.1 Homogeneous Case

For deterministic assumptions on the service times it is shown in [2] that the utilization (η) of the Stage 1 server is:

$$\eta = \min\left\{\frac{K/\mu}{1/\mu + 1/\lambda}, 1\right\}$$

For exponential assumptions on the service times, the state n, $n = 0, 1, \ldots, K$, is defined as the number of customers in Stage 1 [2]. Balance equations may be written and solved in terms of P_n the probability of being in state n with

$$\eta = 1 - P_0$$

For Erlang assumptions on the service times where Stage 1 follows an E_m distribution and Stage 2 follows an E_h distribution, the state is defined as (r, s) where r and s are the number of Erlang phases of service remaining in Stage 1 and Stage 2 respectively [1]; $r = 0, 1, \ldots, mK$; $s = 0, 1, \ldots, hK$. Solving the balance equations for $P(r, s)$, the probability of being in state (r, s) enables η to be calculated;

$$\eta = 1 - P(0, 0)$$

2.2 Heterogeneous Case

For deterministic assumptions on the service times, there is no closed form solution for the general case. However [3] outlines an algorithm based on a postulated order of service in Stage 1 and a redefinition of Stage 1 service times for customers appearing a multiple number of times in one postulated cycle.

For exponential assumptions on the service times, the state describes the order of customers in Stage 1. Balance equations in terms of $P_{...cba}$, the steady state probability of being in state $...cba$ may be written. Here $...cba$ implies a customer type a in service in Stage 1 and customers type b, c, ... waiting in queue for service in Stage 1. Solving the balance equations gives

$$\eta = 1 - P_{...000}$$

which is composed of η_i contributions.

Non-preemptive priority implies customers are not interrupted during their service in Stage 1. Accordingly, for exponential assumptions the state may be defined as $(n_1, n_2, .../i)$ where there are n_1 type 1, n_2 type 2, ..., n_v type V customers in Stage 1 and type i, $i = 1, 2, ..., V$ is currently in service in Stage 1. Here

$$\eta = 1 - P(0, 0, .../0)$$

where $P(n_1, n_2, .../i)$ is the steady state probability of being in state $(n_1, n_2, .../i)$. η again is composed of η_i contributions.

Preemptive priority implies customers of higher priority on their arrival in Stage 1 interrupt the Stage 1 servicing of lower priority customers. For exponential service times, the state may be written as $(n_1, n_2, ...)$ where n_i, $i = 1, 2, ..., V$ is the number of customers of type i in Stage 1. Here

$$\eta = 1 - P(0, 0, ...)$$

where $P(n_1, n_2, ...)$ is the steady state probability of being in state $(n_1, n_2, ...)$ and η has η_i contributions.

3   Objective functions

Relevant objective functions include cost, production, cost per production or equipment utilization.

Cost   Let the all-up hourly costs (owning plus operating) of the Stage 1 serving device and a hauling unit (customer) be c_0 and c_1 (or $c_i, i = 1, 2, ..., V$) respectively. The total hourly cost of the operation is then:

$$\text{Hourly cost} = c_0 + Kc_1 = c_0 + \sum_{i=1}^{V} K_i c_i$$

Production   For a time unit of hour, the hourly production of the operation is a modification of an earlier given result, namely:

$$\text{Hourly production} = \mu \eta C = \sum_{i=1}^{V} \mu_i \eta_i C_i$$

Cost per production Cost per production may be obtained from the previous two expressions:

$$\text{Cost/production} = \frac{c_0 + Kc_1}{\mu\eta C} = \frac{c_0 + \sum_{i=1}^{V} K_i c_i}{\sum_{i=1}^{V} \mu_i \eta_i C_i}$$

(In the examples that follow, constant terms in this expression are not included in the calculations; this permits the optimal value of the decision variable to be obtained but not the absolute value of the cost per production.)

Server utilization The utilization of the Stage 1 server is measured by η. This may be maximized, or alternatively the proportion of time the Stage 1 server is idle, $1 - \eta$, may be minimized.

Customer utilization In every cycle the proportion of time each customer is idle is,

$$\frac{W_q}{W_q + 1/\mu + 1/\lambda}$$

or occupied is,

$$\frac{1/\mu + 1/\lambda}{W_q + 1/\mu + 1/\lambda}$$

4 Constraints

Relevant constraints may include restrictions on time or production. Time and production are related through the earlier quoted result for production and can be included in the optimization calculations through this.

5 Optimization Problems

A number of optimization problems are formulated and solved below. The presentation is via examples as closed form solutions appear only obtainable for the most elementary assumptions.

5.1 Problem $\mathcal{A}$ - Optimal Hauling Unit Fleet Size

For a given earthmoving operation, the relevant decision variable is K, the hauling unit fleet size. Assuming an homogeneous operation the relevant models may be based on deterministic, exponential or Erlang distributions. The objective function is one of cost per production. Time or production constraints may be present.

Consider an operation where the average service time for Stage 1 is 3.25 minutes, the average service time for Stage 2 is 4.6 minutes and the capacity of the hauling units is 15m³. Table 1 and Fig. 3 illustrate the calculations for the unconstrained case.[1] The optimum K value is 2 for all distributions. The Erlang results are for a shape parameter of 5 for both service stages; shape parameters of 10 to 20 are preferred as being closer to observed values and these give results midway between the exponential and deterministic cases [5].

The problem may be turned around such that the cost ratio c_1/c_0 becomes the left hand side. In particular c_1/c_0 may be found from noting that at the transition from a fleet of K hauling units to a fleet of $K + 1$ hauling units the cost per production is one value, namely proportional to

$$\frac{1 + K\left(\frac{c_1}{c_0}\right)}{\eta_K} = \frac{1 + (K + 1)\left(\frac{c_1}{c_0}\right)}{\eta_{K+1}}$$

where the subscript on η indicates the hauling unit fleet size. That is,

$$\frac{c_1}{c_0} = \frac{\eta_{K+1} - \eta_K}{(K + 1)\eta_K - K\eta_{K+1}}$$

For example the transition between hauling unit fleet sizes is shown in Table 2. This is plotted in Fig. 4. The regions between the vertical lines indicate the optimal fleet size. The vertical lines represent transitions between fleet sizes where two fleet sizes are equally optimal. For a cost ratio of 1.1 the optimal K value is 2 as before.

The inclusion of time or production constraints is straightforward. For example Table 3 and Fig. 5 show the behaviour for a production requirement of 10000m³ in a 38 hour period. Fleet sizes less than 3 cannot fulfill this requirement.

5.2 Problem $\mathcal{B}$ - Optimal Operation Layout

The problem of the optimal layout of an operation may be interpreted as one of establishing the optimal Stage 2 service or backcycle characteristics of the operation. As such $1/\lambda$ becomes the decision variable. The model, objective function and constraints remain as in the previous problem.

For the same values as in the previous example, but now fixing K at 2, Table 4 and Fig. 6 indicate the typical calculations. A trade-off is now performed between the lost production of large Stage 2 service times versus the cost of decreasing the Stage 2 service times. A typical application of such a situation occurs in deciding when to move a mobile crusher as the pit face gradually gets further from the crusher.

---

[1]Subsequent figures and tables can be found at the end.

604

5.3 Problem C - Optimal Choice of Equipment Characteristics

The equipment used for the servicing in Stage 1 reflects the value chosen for $1/\mu$. Accordingly $1/\mu$ may be thought of as the decision variable. The model, objective function and constraints remain as before.

Table 5 and Fig. 7 indicate the optimization calculations. Again a trade-off calculation between the cost of providing additional or faster service in Stage 1 versus the improved production is required. Typically such calculations are required when selecting the size and type of a loader, shovel, excavator or dragline.

5.4 Problem D - Optimal Loading Policies

For nonlinear load-growth curves (load versus time relationship) an optimal time can be derived, above and below which the production is less and the cost per production is more [4]. Such load-growth curves apply for pusher-scraper operations. For loader-truck operations the load-growth curve is linear or stepped (Fig. 8) and irrespective of the truck fleet size it can be demonstrated that it is optimal to load the trucks to their capacity assuming the loader's bucket has been matched to the trucks.

Table 6 shows this result for the assumption of exponential distributions and four equal buckets to a truck. The calculations remain the same for nonlinear load-growth curves although then a distinct optimum load time is obtained.

5.5 Problem $\mathcal{E}$ - Optimal Heterogeneity

As an example it is assumed that two types of hauling units are available for an open cut mining operation. Data for the two hauling unit types (type 1 and type 2) are respectively as follows: Stage 1 service times (minutes) of 3.8 and 2.7; Stage 2 service times (minutes) of 4.0 and 5.2; and capacities (m^3) of 17.5 and 12.5. The ratio of hourly costs is taken as 1:1.2:1 for the *loader : hauling unit type 1 : hauling unit type 2*.

To establish the optimum mix of hauling unit, the relevant model is that for heterogeneous operations. For an unconstrained solution Tables 7a and 7b and Figs. 9a and 9b show the relevant calculations and behaviour. Such calculations indicate the influence of customer mixes versus homogeneous bench marks.

5.6 Problem $\mathcal{F}$ - Optimal Non-Preemptive Priority Treatment of Equipment

For certain operations involving heterogeneity it can be demonstrated that improved production and equipment utilization or reduced cost per production may be obtained by specifying priority treatment of some equipment over others. A non-preemptive priority classification of hauling units implies the loading of hauling units is not interrupted on the arrival of a higher priority hauling unit. Where the Stage 1 server has a choice of hauling units to load, that with the higher priority is chosen.

For the data given in the heterogeneity problem, Table 8 and Fig. 10 indicate the behaviour under non-preemptive priority and exponential assumptions. Such calculations indicate the influence of non-preemptive priorities versus no priority and also homogeneous bench marks.

5.7 Problem $\mathcal{G}$ - Optimal Preemptive Priority Treatment of Equipment

A different type of priority classification of hauling units is that of preemptive priority where the loading of hauling units is interrupted on the arrival of a higher priority hauling unit.

For the data of the heterogeneity problem, Table 9 and Fig. 11 indicate the behaviour under preemptive priority and exponential assumptions. A comparison with Figs. 9a and 10 indicates the influence of priorities.

6 Closure

Additional insight may be gained into the planning and management of many construction operations through the use of optimization as a methodology and queueing concepts for the operation model. A range of optimization problems was formulated and solved in the paper related to earthmoving type operations. Generally the optimization was carried out numerically; no closed form solutions appear possible although general trends and conclusions may be recognized. Queueing tables such as presented in [5] can facilitate the ready transfer of the approach of this paper into day-to-day industry practice.

References

[1] Carmichael, D. G., (1987), 'A Refined Queueing Model for Earthmoving Operations,' *Civil Engineering Systems* 4, pp. 153–159.

[2] Carmichael, D. G., (1987), *Engineering Queues in Construction and Mining*, Ellis Horwood Ltd (John Wiley and Sons Ltd), Chichester.

[3] Carmichael, D. G., (1989), 'Heterogeneity in Deterministic Finite Source Queues,' submitted to *Civil Engineering Systems*.

[4] Carmichael, D. G., (1987), 'Optimal Pusher-Scraper Loading Policies,' *Engineering Optimization* **10**, pp. 51–64.

[5] Carmichael, D. G., (1989), 'Production Tables for Earthmoving, Quarrying and Open-Cut Mining Operations,' in *Applied Construction Management*, D. G. Carmichael (Ed.), Unisearch Ltd. Publishers, Sydney, pp. 275–281.

Table 1. *Optimal hauling unit fleet size* ($\mu \eta C$ *in units of* $\mathrm{m^3/h}$).

K	Exponential			Erlang			Deterministic		
	η	$\mu \eta C$	$\frac{1+(1.1)K}{\eta}$	η	$\mu \eta C$	$\frac{1+(1.1)K}{\eta}$	η	$\mu \eta C$	$\frac{1+(1.1)K}{\eta}$
1	0.414	114.6	5.072	0.414	114.6	5.072	0.414	114.6	5.072
2	0.708	196.1	4.520	0.745	206.3	4.295	0.828	229.3	3.864
3	0.879	243.4	4.892	0.939	260.0	4.579	1.000	276.9	4.300
4	0.959	265.6	5.631	0.994	275.2	5.433	1.000	276.9	5.400
5	0.989	273.9	6.572	1.000	276.9	6.500	1.000	276.9	6.500

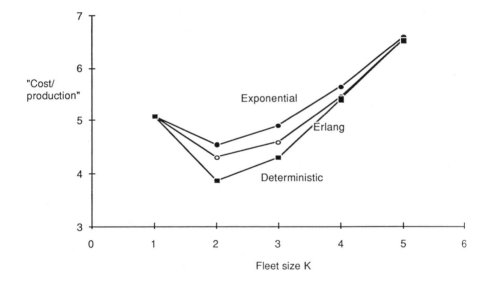

Figure 3. *Optimal hauling unit fleet size.*

Table 2. *Optimal hauling unit fleet size; size transitions.*

	c_1/c_0		
K to $K+1$	Exponential	Erlang	Deterministic
1 to 2	2.450	3.988	∞
2 to 3	0.467	0.543	0.355
3 to 4	0.125	0.071	0
4 to 5	0.036	0.006	0

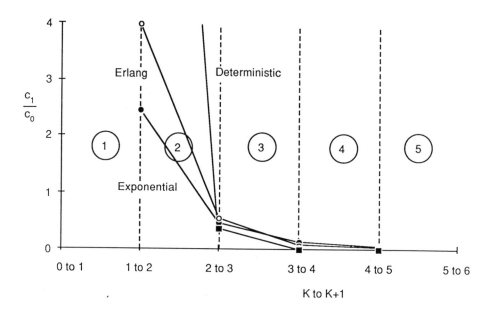

Figure 4. *Optimal fleet size transitions.*

Table 3. *Production-time constraint calculations.*

K	Production (m^3)		
	Exponential	Erlang	Deterministic
1	4356	4356	4356
2	7450	7839	8712
3	9249	9880	10522
4	10091	10459	10522
5	10406	10522	10522

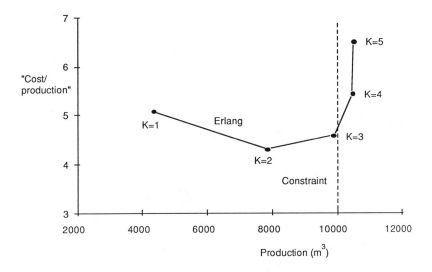

Figure 5. *Optimal fleet size, constrained.*

Table 4. *Optimal operation layout;* $1/\mu = 3.25$ *min,* $K = 2$.

1/λ (min)	ρ	Exponential η	Exponential μηC (m³/h)	Erlang η	Erlang μηC (m³/h)	Deterministic η	Deterministic μηC (m³/h)
3.25	1.000	0.800	221.5	0.852	235.9	1.000	276.9
4	0.813	0.746	206.6	0.790	218.8	0.897	248.4
5	0.650	0.681	188.6	0.716	198.3	0.788	218.2
6	0.542	0.625	173.1	0.652	180.5	0.703	194.6
7	0.464	0.576	159.5	0.596	165.0	0.634	175.6
8	0.406	0.533	147.6	0.549	152.0	0.578	160.1
9	0.361	0.495	137.1	0.508	140.7	0.531	147.0
10	0.325	0.462	127.9	0.473	131.0	0.491	136.0

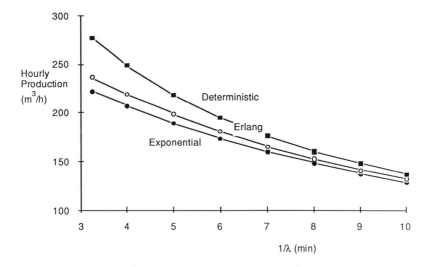

Figure 6. *Optimal operation layout.*

Table 5. *Optimal choice of equipment characteristics; $1/\lambda = 4.6$ min, $K = 2$.*

		Exponential		Erlang		Deterministic	
$1/\mu$ (min)	ρ	η	$\mu\eta C$ (m³/h)	η	$\mu\eta C$ (m³/h)	η	$\mu\eta C$ (m³/h)
1.0	0.217	0.346	311.4	0.350	315.0	0.357	321.3
1.5	0.326	0.463	277.8	0.474	284.4	0.492	295.2
2.0	0.435	0.555	249.8	0.573	257.9	0.606	272.7
2.5	0.543	0.626	225.4	0.653	235.1	0.704	253.4
3.0	0.652	0.682	204.6	0.717	215.1	0.789	236.7
3.5	0.761	0.727	186.9	0.769	197.7	0.864	222.2
4.0	0.870	0.764	171.9	0.812	182.7	0.930	209.3
4.5	0.978	0.794	158.8	0.845	169.0	0.989	197.8
4.6	1.000	0.800	156.5	0.852	166.7	1.000	195.7

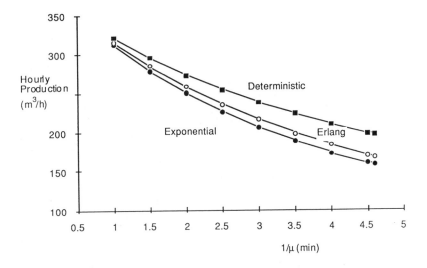

Figure 7. *Optimal equipment characteristics.*

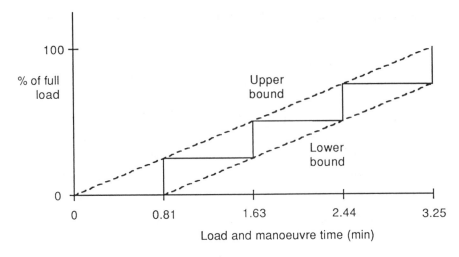

Figure 8. *Load-growth curve.*

Table 6. *Optimal loading policies.*

$1/\mu$ (min)	ρ	μC	$K = 1$		$K = 2$		$K = 3$	
			η	$\mu\eta C$ (m³/h)	η	$\mu\eta C$ (m³/h)	η	$\mu\eta C$ (m³/h)
0.00	0.000	0.0	0.000	0.0	0.000	0.0	0.000	0.0
0.81	0.176	276.9	0.150	41.5	0.293	81.1	0.426	118.0
1.63	0.354	276.9	0.262	72.5	0.489	135.4	0.675	186.9
2.44	0.530	276.9	0.347	96.1	0.618	171.1	0.807	223.5
3.25	0.707	276.9	0.414	114.6	0.707	195.8	0.878	243.1

Table 7a. *Optimal heterogeneity; exponential model.*

K	K_1	K_2	η_1	η_2	η	$\mu_1\eta_1C_1$ (m^3/h)	$\mu_2\eta_2C_2$ (m^3/h)	$\sum \mu_i\eta_iC_i$ (m^3/h)	$\frac{1+\sum K_ic_i/c_0}{\sum \mu_i\eta_iC_i}$
1	1	0	0.487	—	0.487	134.5	—	134.5	0.0164
1	0	1	—	0.342	0.342	—	94.9	94.9	0.0211
2	2	0	0.787	—	0.787	217.3	—	217.3	0.0156
2	1	1	0.432	0.279	0.711	119.3	77.4	196.7	0.0163
2	0	2	—	0.623	0.623	—	172.9	172.9	0.0174
3	3	0	0.930	—	0.930	256.8	—	256.8	0.0179
3	2	1	0.679	0.222	0.901	187.5	61.6	249.1	0.0177
3	1	2	0.373	0.486	0.859	103.0	134.9	237.9	0.0177
3	0	3	—	0.800	0.800	—	222.0	222.0	0.0180

Table 7b. *Optimal heterogeneity; deterministic model.*

K	K_1	K_2	η_1	η_2	η	$\mu_1\eta_1C_1$ (m^3/h)	$\mu_2\eta_2C_2$ (m^3/h)	$\sum \mu_i\eta_iC_i$ (m^3/h)	$\frac{1+\sum K_ic_i/c_0}{\sum \mu_i\eta_iC_i}$
1	1	0	0.487	—	0.487	134.5	—	134.5	0.0164
1	0	1	—	0.342	0.342	—	94.9	94.9	0.0211
2	2	0	0.974	—	0.974	269.0	—	269.0	0.0126
2	1	1	0.481	0.342	0.823	132.8	94.9	227.7	0.0141
2	0	2	—	0.684	0.684	—	189.8	189.8	0.0158
3	3	0	1.000	—	1.000	276.2	—	276.2	0.0167
3	2	1	0.738	0.262	1.000	203.8	72.7	276.5	0.0159
3	1	2	0.413	0.587	1.000	114.0	162.9	276.9	0.0152
3	0	3	—	1.000	1.000	—	277.5	277.5	0.0144

614

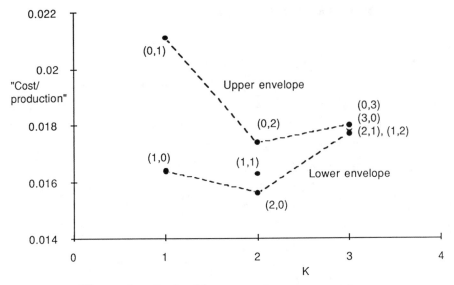

Figure 9a. *Optimal heterogeneity; exponential case.*

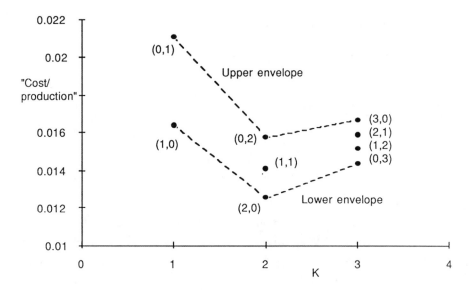

Figure 9b. *Optimal heterogeneity; deterministic case.*

Table 8. *Optimal non-preemptive priority; exponential model.*

K	K_1	K_2	η_1	η_2	η	$\mu_1\eta_1C_1$ (m³/h)	$\mu_2\eta_2C_2$ (m³/h)	$\sum\mu_i\eta_iC_i$ (m³/h)	$\frac{1+\sum K_ic_i/c_0}{\sum\mu_i\eta_iC_i}$
1	1	0	0.487	—	0.487	134.5	—	134.5	0.0164
1	0	1	—	0.342	0.342	—	94.9	94.9	0.0211
2	2	0	0.787	—	0.787	217.3	—	217.3	0.0156
†2	1	1	0.432	0.279	0.711	119.3	77.4	196.7	0.0163
*2	1	1	0.432	0.279	0.711	119.3	77.4	196.7	0.0163
2	0	2	—	0.623	0.623	—	172.9	172.9	0.0174
3	3	0	0.930	—	0.930	256.8	—	256.8	0.0179
†3	2	1	0.716	0.186	0.902	197.7	51.6	249.3	0.0176
*3	2	1	0.652	0.248	0.900	180.0	68.8	248.8	0.0177
†3	1	2	0.396	0.464	0.860	109.4	128.8	238.2	0.0176
*3	1	2	0.348	0.509	0.858	96.1	141.2	237.2	0.0177
3	0	3	—	0.800	0.800	—	222.0	222.0	0.0180

† Priority to type 1   * Priority to type 2

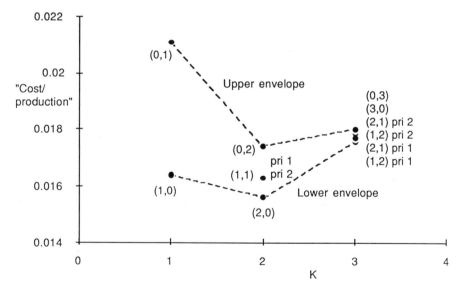

Figure 10. *Optimal non-preemptive priority.*

Table 9. *Optimal preemptive priority; exponential model.*

K	K_1	K_2	η_1	η_2	η	$\mu_1\eta_1C_1$ (m³/h)	$\mu_2\eta_2C_2$ (m³/h)	$\sum\mu_i\eta_iC_i$ (m³/h)	$\frac{1+\sum K_ic_i/c_0}{\sum\mu_i\eta_iC_i}$
1	1	0	0.487	—	0.487	134.5	—	134.5	0.0164
1	0	1	—	0.342	0.342	—	94.9	94.9	0.0211
2	2	0	0.787	—	0.787	217.3	—	217.3	0.0156
†2	1	1	0.487	0.228	0.716	134.5	63.3	197.8	0.0162
*2	1	1	0.365	0.342	0.707	100.8	94.9	195.7	0.0164
2	0	2	—	0.623	0.623	—	172.9	172.9	0.0174
3	3	0	0.930	—	0.930	256.8	—	256.8	0.0179
†3	2	1	0.788	0.117	0.905	217.6	32.5	250.1	0.0176
*3	2	1	0.555	0.342	0.897	153.3	94.9	248.2	0.0177
†3	1	2	0.241	0.612	0.853	66.6	169.8	236.4	0.0178
*3	1	2	0.487	0.377	0.864	134.5	104.6	239.1	0.0176
3	0	3	—	0.800	0.800	—	222.0	222.0	0.0180

† Priority to type 1   * Priority to type 2

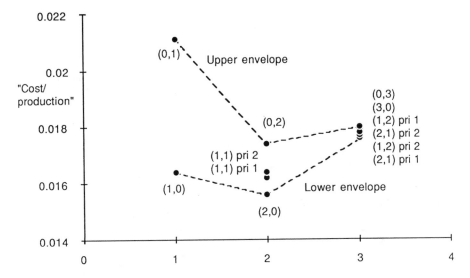

Figure 11. *Optimal preemptive priority.*

Sensitivity Analysis in Reliability-Based Shape Optimization

J. D. Sørensen and Ib Enevoldsen
Department of Building Technology and
Structural Engineering
The University of Aalborg
Aalborg
Denmark

Abstract A reliability-based shape optimization problem is formulated with the total expected cost as the objective function and requirements for the element or systems reliability measures as constraints. As design variables both sizing and shape variables are used.

It is shown how the sensitivities (gradients) for use in optimization algorithms can be obtained if the structural model is modelled by linear or non-linear finite element models with or without stochastic variables used in connection with the stiffness calculations.

Further it is shown how a sensitivity analysis of the optimal design can be performed. Finally a simple example where the techniques are utilized is considered.

1 Introduction

Application of first-order reliability methods (see Madsen, Krenk and Lind [10]) in structural design problems has attracted growing interest in recent years, see e.g. Frangopol [4], Murotsu, Kishi, Okada, Yonezawa and Taguchi [11], Thoft-Christensen and Sørensen [18] and Sørensen [16]. In probabilistically based optimal design of structural systems some of the quantities used in the modelling are modelled as stochastic variables. The stochastic variables could be related to the strength (yield stress), the load (wind, wave or earthquake loads) or the mathematical model (model uncertainty). The first order reliability measures are determinate solving some optimization problems, see Section 3.

A reliability-based shape optimization problem can be formulated with the total expected costs as objective function and some requirements for the reliability measures (element or systems reliability measures) as constraints, see Section 2. As design variables sizing (diameters, thicknesses etc.) and shape variables (ge-

617

B. H. V. Topping (ed.),
Optimization and Artificial Intelligence in Civil and Structural Engineering, Volume I, 617–637.
© 1992 *Kluwer Academic Publishers. Printed in the Netherlands.*

ometrical variables) are used in this paper. Some procedures to solve the design problems are mentioned in Section 4.

When the optimal design has been determined a sensitivity analysis can be performed. The sensitivity analysis can be related to the optimal design and/or to the reliability measures. In Sections 5 and 6 it is shown how the sensitivity analysis can be performed effectively using some of the results from the reliability analysis. Further it is shown how the sensitivity measures can be obtained if the structural model is modelled by linear or non-linear finite element models with or without stochastic variables used in connection with the stiffness calculations.

Finally, in Section 7, a simple example with a model of an offshore platform is considered.

2 Reliability-Based Shape Optimization

A number of structural shape optimization problems based on reliability measures can be formulated, see e.g. Sørensen [16]. Here the shape optimization problem is formulated with a systems reliability constraint

$$\min \qquad W\left(\bar{z}\right) \qquad\qquad\qquad (1)$$

$$\text{s.t.} \qquad \beta^S\left(\bar{z}\right) \geq \beta^{\min} \qquad\qquad\qquad (2)$$

$$z_i^{\min} \leq z_i \leq z_i^{\max}, \quad i = 1, 2, \ldots, N \qquad\qquad\qquad (3)$$

where $z_1, z_2, \ldots, z_N$ are the optimization variables and β^S is the systems reliability index determined as described below in Section 3. $\beta^{\min}$ is a minimum acceptable systems reliability index. $z_i^{\min}$ and $z_i^{\max}$ are simple lower and upper bounds of z_i. $W\left(\bar{z}\right)$ is the objective function which can e.g. be the weight or the cost (for example the total expected cost during the design lifetime) of the structural system. The optimization problem, Eqs. 1–3, is generally non-linear and non-convex.

The optimization variables $z_1, z_2, \ldots, z_N$ in the shape optimization problem are often divided into two groups, sizing variables and shape variables. Sizing variables can represent cross-sectional dimensions of the structure and shape variables can represent overall layout quantities, e.g. placing of nodes in a jacket platform. However, it will be seen later that the division is not necessary from a mathematical point of view, but it can be effective from a computational point of view.

In a more general structural optimization problem also topological and material optimization variables can be included.

3 Reliability Analysis

3.1 Single Failure Elements

In reliability-based structural analysis random variables $\overline{X} = (X_1, X_2, \ldots, X_n)$ are used to model uncertain quantities connected with the description of the strength (structural resistance), the load (wind, wave etc.) and the mathematical model (model uncertainty). The reliability model of a structural system is modelled by a number of failure elements modelling potential local failure modes of the structural system, e.g. yielding of a cross-section, stability failure or fatigue failure of a welded joint. Each failure element is described by a failure function:

$$g(\overline{x}, \overline{z}, \overline{p}) = 0 \qquad (4)$$

where $\overline{z}$ is the vector of optimization variables at a given iteration during the solution of the problem stated in Eqs. 1–3, and $\overline{p}$ is the vector of remaining parameters in the model. Realizations $\overline{x}$ of $\overline{X}$, where $g(\overline{x}, \overline{z}, \overline{p}) \leq 0$, correspond to failure states while realizations $\overline{x}$ where $g(\overline{x}, \overline{z}, \overline{p}) > 0$ correspond to safe states.

In first order reliability methods (FORM), see e.g. Madsen, Krenk and Lind [10], a measure of the reliability (the reliability index β) is defined. A transformation T of the correlated and non-normally distributed stochastic variables $\overline{X}$ into standardized and normally distributed variables $\overline{U} = (U_1, U_2, \ldots, U_n)$ is defined. Let $\overline{X} = T(\overline{U})$. In the $\overline{u}$ space the reliability index β is defined as

$$\beta = \min_{g(T(\overline{u}), \overline{z}, \overline{p}) = 0} \left(\overline{u}^T \overline{u} \right)^{\frac{1}{2}} \qquad (5)$$

The reliability index β is thus determined by solving an optimization problem with one constraint. The optimization problem is generally non-convex and non-linear and can in principle be solved using any general non-linear optimization algorithm, but the iteration algorithm developed by Rackwitz and Fiessler, see e.g. Madsen, Krenk and Lind [10], has shown to be fast and effective in FORM analysis. The Rackwitz-Fiessler algorithm solves Eq. 5 using at each iteration (for given $\overline{u}$)

- the value of the failure function : g

- the gradient of g with respect to $\overline{u}$: $\nabla_{\overline{u}} g(\overline{u}, \overline{z}, \overline{p})$

The solution point $\overline{u}^*$ of the optimization problem is called the design point and is the point on the failure surface defined by Eq. 4 closest to the origin in the $\overline{u}$-space.

In FORM the safety margin M defined by

$$M = g\left(T\left(\overline{U}\right), \overline{z}, \overline{p}\right) \qquad (6)$$

is linearized in the design point

$$M \approx -\overline{\alpha}^T \overline{U} + \beta \tag{7}$$

where the elements in the $\overline{\alpha}$-vector are given by

$$\alpha_i = \frac{u_i^*}{\beta} = \frac{-1}{|\nabla_{\overline{u}} g|} \frac{\partial g}{\partial u_i} \tag{8}$$

If the failure function is not too non-linear the probability of failure P_f can with good accuracy be determined from

$$P_f \approx \Phi(-\beta) \tag{9}$$

where $\Phi(\cdot)$ is the standard normal distribution function.

3.2 Series Systems at Level 1

If the structural system is defined as having failed if one element fails (failure is modelled at level 1) the reliability model is a series system whose elements are the potential failure modes described above. With m failure modes the systems reliability index β^S can be estimated from

$$\beta^S = -\Phi^{-1}\left(1 - \Phi_m\left(\overline{\beta}, \overline{\overline{\rho}}\right)\right) \tag{10}$$

where $\Phi_m(\cdot)$ is the m-dimensional normal distribution function. $\beta_1, \ldots, \beta_m$ are the reliability indices of the failure modes and $\overline{\overline{\rho}}$ the corresponding correlation coefficient matrix determined by

$$\rho_{ij} = \overline{\alpha}_i^T \overline{\alpha}_j \qquad i, j = 1, 2, \ldots, m \tag{11}$$

3.3 Series Systems at Higher Levels

Let failure of the structural system be defined at level L, i.e. the system is defined as having failed if L failure elements fail. The reliability model then becomes a series system of parallel systems where each parallel system contains L elements.

The significant parallel systems can be identified using e.g. the β-unzipping method, see Thoft-Christensen and Murotsu [17] or a branch-and-bound technique, see e.g. Guenard and Cornell [7].

In order to estimate the probability of failure of a given parallel system the first step is to determine the joint design point $\overline{u}^*$ by solving the following optimization problem

$$\left.\begin{array}{l} \min\limits_{\overline{u}} \ \dfrac{1}{2}\overline{u}^T\overline{u} \\[2mm] \text{s.t.} \quad g_1\left(\overline{T}(\overline{u}), \overline{z}, \overline{p}\right) \leq 0 \\ \qquad \vdots \\ \qquad g_L\left(\overline{T}(\overline{u}), \overline{z}, \overline{p}\right) \leq 0 \end{array}\right\} \tag{12}$$

where $g_1(\overline{x},\overline{z},\overline{p}),\ldots,g_2(\overline{x},\overline{z},\overline{p})$ are the failure functions of the L failure elements.

In the joint design point $\overline{u}^*$ the constraints are linearized and each of the L_A active elements is written as $M_i = \beta_i - \overline{\alpha}_i^T \overline{U}$.

The probability of failure of the parallel system is

$$P_p = P(M_1 \leq 0 \wedge \cdots \wedge M_L \leq 0)$$

$$\approx P(\beta_1 - \overline{\alpha}_1^T \overline{U} \leq 0 \wedge \cdots \wedge \beta_{L_A} - \overline{\alpha}_{L_A}^T \overline{U} \leq 0)$$

$$= \Phi_{L_A}(-\overline{\beta},\overline{\rho})$$

$$\approx P(\beta^e - \overline{\alpha}^{e^T}\overline{U} \leq 0) \tag{13}$$

where β^e and $\overline{\alpha}^e$ are the equivalent reliability-index and α-vector for an equivalent linear failure element.

β^e and $\overline{\alpha}^e$ are determined from

$$\beta^e = -\Phi^{-1}\left(\Phi_{L_A}\left(-\overline{\beta},\overline{\rho}\right)\right) \tag{14}$$

$$\overline{\alpha}^e = \frac{\overline{a}^e}{|\overline{a}^e|} \tag{15}$$

where

$$a_j^e = \frac{1}{\varphi(-\beta^e)} \sum_{i=1}^{L_A} \alpha_{ij}\varphi(\beta_i)\Phi_{L_A-1}\left(-\overline{\beta}_i^a;\overline{\rho}_i^a\right) \tag{16}$$

$$\rho_{ij} = \overline{\alpha}_i^T\overline{\alpha}_j \tag{17}$$

$\overline{\beta}_i^a$ and $\overline{\rho}_i^a$ are conditional reliability indices and correlation coefficient matrices obtained from $\overline{\beta}$ and $\overline{\overline{\rho}}$, see Sørensen [16] and [14]. $\varphi(\cdot)$ is the standard normal density function.

Based on the equivalent linear failure elements the system reliability index β^S can now be estimated using Eq. 10.

The FORM analysis can e.g. be performed with PRADSS [13].

4 Optimization Procedures

The optimization problem, Eqs. 1–3, can in principle be solved using any general non-linear optimization algorithm. In this paper the NLPQL algorithm developed by Schittkowski [12] is used in an example shown in Section 7. The NLPQL algorithm is based on the optimization method by Han, Powell and Wilson, see Gill, Murray and Wright [5]. Generally, it is a very effective method where each iteration consists of two steps. The first step is determination of the search direction by solving a quadratic optimization problem formed by a quadratic approximation of the Lagrange function of the non-linear optimization problem and a linearization of the constraints at the current design point. The second step is a line search with an augmented Lagrangian merit function.

NLPQL requires estimates of the gradients of the objective function and the constraints. If the structural weight is used as objective function the gradients of $W(\bar{z})$ are easily determined numerically. Gradients of the reliability constraint, which are generally time-consuming to estimate numerically, can be determined semi-analytically.

If failure is defined at level 1 (see Eq. 10) and a series system with m elements is considered, then

$$\frac{\partial \beta^S}{\partial z_j} = \sum_{i=1}^{m} \frac{\partial \beta^S}{\partial \beta_i} \frac{\partial \beta_i}{\partial z_j} + 2\sum_{i<k}^{m} \frac{\partial \beta^S}{\partial \rho_{ik}} \frac{\partial \rho_{ik}}{\partial z_j} \tag{18}$$

where

$$\frac{\partial \beta^S}{\partial \beta_i} = \frac{\varphi(\beta_i)}{\varphi(\beta^S)} \Phi_{m-1}\left(\overline{\beta}_i^a; \overline{\overline{\rho}}_{ik}^a\right) \tag{19}$$

$$\frac{\partial \beta^S}{\partial \rho_{ik}} = \frac{\varphi_2(\beta_i, \beta_k; \overline{\overline{\rho}}_{ik})}{\varphi(\beta^S)} \Phi_{m-2}\left(\overline{\beta}_{ik}^b; \overline{\overline{\rho}}_{ik}^b\right) \tag{20}$$

$\overline{\beta}_i^a$, $\overline{\overline{\rho}}_{ik}^a$, $\overline{\beta}_{ik}^b$ and $\overline{\overline{\rho}}_{ik}^b$ are the conditional index and correlation coefficient matrix, respectively, obtained from $\overline{\beta}$ and $\overline{\overline{\rho}}$, see Sørensen [16] and [14].
$\frac{\partial \beta_i}{\partial z_j}$ is determined by

$$\frac{\partial \beta_j}{\partial z_i} = \frac{1}{|\nabla_u g_j|} \frac{\partial g_j(\overline{u}^*, \overline{z}, \overline{p})}{\partial z_i} \tag{21}$$

where $\overline{u}^*$ is the solution of Eq. 5. $\frac{\partial \rho_{ik}}{\partial z_j}$ is determined as shown in Appendix A.

If failure is defined at a higher level $(L > 1)$ $\frac{\partial \beta^S}{\partial z_j}$ can in principle be determined from the same formulas as above, see Appendix A.

If the structural system is analyzed using a FEM-modelling then determination of the gradients becomes more difficult, see Section 6.

Due to the different nature of the variables in the reliability-based optimization problem in Eqs. 1–3 it can be appropriate to separate the variables (in sizing variables and shape variables) and to solve the optimization problem sequentially with alternately one of the two types of variables fixed. Alternatively a multi-level optimization technique can be used.

Another procedure to solve the optimization problem (Eqs. 1–3) is to generate a sequence of optimization problems where the systems reliability constraint (Eq. 2) is replaced by m element reliability constraints. Element reliability index based optimization problems can generally easily be solved because the reliability gradients can be determined exactly by the simple formula (Eq. 21). At each step in the sequence the lower bounds of the reliability indices are modified such that the systems reliability requirement at convergence is satisfied, see Sørensen [16]. The advantage of using the sequential procedure is that determination of the time-consuming approximation to the gradient of the systems reliability index can be avoided.

5 Sensitivity Analysis

With the techniques described above it is possible to find a shape optimal design which satisfies the requirement for the systems reliability. In this Section methods to estimate the sensitivities of the optimal design due to changes in parameters p_l, $l = 1, \ldots, L$ in the structural model are considered.

At optimum the following Kuhn-Tucker conditions must be satisfied (see Eqs. 1–3)

$$\frac{\partial W}{\partial z_i} + \lambda_0 \frac{\partial \beta^S}{\partial z_i} + \lambda_i - \lambda_{N+i} = 0, \qquad i = 1, 2, \ldots, N \qquad (22)$$

$$\lambda_0(\beta^S - \beta^{\min}) = 0, \qquad \lambda_0 \geq 0 \qquad (23)$$

$$\lambda_i(z_i - z_i^{\min}) = 0, \qquad \lambda_i \geq 0, \qquad i = 1, 2, \ldots, N \qquad (24)$$

$$\lambda_{N+i}(z_i^{\max} - z_i) = 0, \qquad \lambda_{N+i} \geq 0, \qquad i = 1, 2, \ldots, N \qquad (25)$$

where $\lambda_0, \lambda_1, \ldots, \lambda_{2N}$ are Lagrangian multipliers.

The sensitivity with respect to a parameter p_l in the model (typically a distribution parameter or a deterministic quantity) is found by requiring that the Kuhn-Tucker conditions must remain valid when a small change in p_l is introduced, see e.g. Vanderplaats [19]

$$\sum_{j=1}^{N} \frac{\partial^2 W}{\partial z_i \partial z_j} \frac{\partial z_j}{\partial p_l} + \frac{\partial^2 W}{\partial z_i \partial p_l} + \frac{\partial \lambda_0}{\partial p_l} \frac{\partial \beta^S}{\partial z_i} + \lambda_0 \sum_{j=1}^{N} \frac{\partial^2 \beta^S}{\partial z_i \partial z_j} \frac{\partial z_j}{\partial p_l}$$

$$+ \lambda_0 \frac{\partial^2 \beta^S}{\partial z_i \partial p_l} + \frac{\partial \lambda_i}{\partial p_l} - \frac{\partial \lambda_{N+i}}{\partial p_l} = 0, i = 1, 2, \ldots, N \qquad (26)$$

$$\frac{\partial \beta^S}{\partial p_l} + \sum_{j=1}^{N} \frac{\partial^2 \beta^S}{\partial z_i \partial z_j} \frac{\partial z_j}{\partial p_l} = 0, \qquad i = 1, 2, \ldots, N; \text{ if } \lambda_{N+i} = 0 \quad (27)$$

$$\frac{\partial z_i}{\partial p_l} = 0, \qquad i = 1, 2, \ldots, N; \text{ if } \lambda_{N+i} = 0 \quad (28)$$

$$\frac{\partial z_i}{\partial p_l} = 0, \qquad\qquad i = 1, 2, \ldots, N; \ \ \text{if } \lambda_{N+i} = 0 \quad (29)$$

It is assumed that the constraints active at optimum remain active.

The above equations can be written in following matrix form

$$
\begin{bmatrix}
\Delta W + \lambda_0 \Delta \beta^S & \left\{\frac{\partial \beta^S}{\partial z_i}\right\} & \overline{\overline{G}} \\[2mm]
\left\{\frac{\partial \beta^S}{\partial z_i}\right\}^T & 0 & \overline{\overline{0}} \\[2mm]
\overline{\overline{G}}^T & \overline{\overline{0}} & \overline{\overline{0}}
\end{bmatrix}
\begin{bmatrix}
\left\{\frac{\partial z_i}{\partial p_l}\right\} \\[2mm]
\frac{\partial \lambda_0}{\partial p_l} \\[2mm]
\left\{\frac{\partial \lambda_i}{\partial p_l}\right\}
\end{bmatrix}
$$

$$
=
\begin{bmatrix}
-\left\{\frac{\partial^2 W}{\partial p_l \partial z_i}\right\} - \lambda_0 \left\{\frac{\partial^2 \beta^S}{\partial p_l \partial z_i}\right\} \\[2mm]
-\frac{\partial \beta^S}{\partial p} \\[2mm]
\overline{0}
\end{bmatrix}
\qquad (30)
$$

where the elements in $\overline{\overline{G}}$ are 0 or 1 depending on which of the simple constraints are active. The system reliability constraint is assumed to be active.

When Eq. 30 is solved the sensitivity of the objective function can then be found

$$\frac{dW}{dp_l} = \sum_{j=1}^{N} \frac{\partial W}{\partial z_j} \frac{\partial z_j}{\partial p_l} + \frac{\partial W}{\partial p_l} \qquad (31)$$

The objective function is assumed to be relatively simple (compared to the reliability constraints) so the derivatives of W in Eq. 30 can easily be calculated numerically.

The first derivatives of β^S are calculated as described in Eqs. 18–20.

Analytical estimation of $\Delta \beta^S$ can generally be expected to be very time-consuming compared to the following semi-analytical approximation of the elements in $\Delta \beta^S$

$$\frac{\partial^2 \beta^S}{\partial z_i z_j} \simeq \frac{1}{\Delta z_j} \left(\frac{\partial \beta^S (z_i, \ldots, z_j + \Delta z_j, \ldots, z_N)}{\partial z_i} \right.$$
$$\left. - \frac{\partial \beta^S (z_i, \ldots, z_j, \ldots, z_N)}{\partial z_i} \right) \qquad (32)$$

where Δz_j is a small change in z_j.

6   Sensitivity Analysis in FEM-Models

As stated in Sections 3 and 5 an important part of the sensitivity analysis is to estimate $\frac{\partial \beta_i}{\partial z_j}$. In formulations where the structural response can be determinate directly (and quickly), e.g. statically determined structures, $\frac{\partial \beta_i}{\partial z_j}$ can be found as stated in Eq. 21 with numerical differentiation. An example of a shape optimization where $\frac{\partial \beta_i}{\partial z_j}$ is found directly with Eq. 21 is shown in Enevoldsen, Sørensen and Thoft-Christensen [3].

However, usually the structural response is calculated based on a finite element model where the equations of equilibrium can be written

$$\overline{f}(\overline{a}) - \overline{P} = \overline{0} \tag{33}$$

$\overline{f}(a)$ can be considered as the structural reaction to the external load $\overline{P}$ corresponding to the actual system of nodal degrees of freedom, $\overline{a}$ are the nodal displacements.

In a linear FEM-model Eq. 33 is written

$$\overline{\overline{K}}\overline{a} = \overline{P} \tag{34}$$

where $\overline{\overline{K}}$ is the global stiffness matrix. If the reliability-based optimization problem is solved using a gradient based optimization algorithm then estimation of $\frac{\partial \beta_i}{\partial z_j}$ by numerical differentiation becomes very time consuming because each step in the iteration process requires $N + 1$ assemblies and inversions of $\overline{\overline{K}}$.

Further, the complexity of the reliability-based optimization problems also depends highly on whether some of the quantities in the stiffness matrix are modelled by stochastic variables.

Below the following 4 cases are considered:

- A linear FEM-model without stochastic variables in the stiffness matrix $\overline{\overline{K}}$.

- A geometrical non-linear FEM-model without stochastic variables in connection with the stiffness calculations.

- A linear FEM-model with stochastic variables in $\overline{\overline{K}}$.

- A geometrical non-linear FEM-model with stochastic variables in connection with the stiffness calculations.

6.1 Calculation of $\nabla_u g$ and $\frac{d\beta_i}{dz_j}$ in a Linear FEM-Model Without Stochastic Variables in The Stiffness Matrix $\overline{\overline{K}}$

In the linear case without stochastic variables in the stiffness matrix the failure functions and the finite element equations can be written

$$g\left(\overline{u}, \overline{z}, \overline{a}(\overline{u}, \overline{z})\right) = 0 \tag{35}$$

$$\overline{\overline{K}}(\overline{z})\overline{a}(\overline{u}, \overline{z}) = \overline{P}(\overline{u}, \overline{z}) \tag{36}$$

The parameters $\bar{p}$ are omitted in Eq. 35 and in the following equations in this Section.

The elements in the gradient vector $\nabla_u g$ are determined from

$$\frac{dg}{du_i} = \frac{\partial g}{\partial u_i} + \sum_{j=1}^{J} \frac{\partial g}{\partial a_j} \frac{\partial a_j}{\partial u_i} \tag{37}$$

where J is the number of degrees of freedom. $\frac{\partial \bar{a}}{\partial u_i}$ is calculated using the pseudo-load vector technique:

$$\frac{\partial \bar{a}}{\partial u_i} = \bar{\bar{K}}^{-1} \frac{\partial \bar{P}}{\partial u_i} \tag{38}$$

Generally $\frac{\partial g}{\partial u_i}$, $\frac{\partial \bar{g}}{\partial a_j}$ and $\frac{\partial \bar{P}}{\partial u_i}$ are easily determined numerically.

$\frac{\partial \beta}{\partial z_j}$ is estimated semi-analytically using a pseudo-load vector technique, see Sørensen [15]

$$\frac{\partial \beta}{\partial z_j} = \frac{1}{|\nabla_u g|} \left[\frac{\partial g}{\partial z_j} + \sum_{k=1}^{J} \frac{\partial g}{\partial a_k} \frac{\partial a_k}{\partial z_j} \right] \tag{39}$$

where the elements in $\frac{\partial \bar{a}}{\partial z_j}$ are estimated from

$$\frac{\partial \bar{a}}{\partial z_j} = \bar{\bar{K}}^{-1} \left\{ \frac{\partial \bar{P}}{\partial z_j} - \frac{\partial \bar{\bar{K}}}{\partial z_j} \bar{a} \right\} = \bar{\bar{K}}^{-1} \bar{P}_j^s \tag{40}$$

In Eqs. 39 and 40 $\frac{\partial g}{\partial z_j}$, $\frac{\partial g}{\partial a_k}$, $\frac{\partial \bar{P}}{\partial z_j}$ and $\frac{\partial \bar{\bar{K}}}{\partial z_j}$ are estimated numerically (or analytically if possible). The number of pseudo-load vectors P_j^s defined by Eq. 40 is N.

The main advantage of estimating the gradients of the reliability index using Eqs. 39–40 instead of simple differentiation of β is, as described in the previous Section, that a very large number of reliability index calculations and stiffness matrix assemblies and inversions can be omitted, thus reducing the computer time consumption considerably.

6.2 Calculation of $\nabla_u g$ and $\frac{d\beta_i}{dz_j}$ in a Geometrical Non-Linear FEM Model Without Stochastic Variables In Connection With The Stiffness Calculations

From Section 6.1 it follows that it is generally easy to calculate $\nabla_u g$ in a linear model because $\bar{\bar{K}}$ was independent of $\bar{u}$ and then constant during a Rackwitz-Fiessler iteration (with $\bar{z}$ is fixed).

In the geometrical non-linear problem it becomes more difficult to find $\nabla_u g$ because a constant stiffness matrix $\bar{\bar{K}}$ does not exist.

In the following it is described how $\nabla_u g$ (to be used in determination of the reliability index) can be found using a pseudo-load vector technique.

The failure function and the finite element equations are written

$$g(\overline{u}, \overline{z}, \overline{a}(\overline{u}, \overline{z})) = 0 \tag{41}$$

$$\overline{f}(\overline{z}, \overline{a}(\overline{u}, \overline{z})) = \overline{P}(\overline{u}, \overline{z}) \tag{42}$$

where $\overline{f}$ is defined in Eq. 33. For given $\overline{u}$ and $\overline{z}$, $\overline{a}$ is determined from the non-linear FEM-problem in Eq. 42 with e.g. a Newton-Rapson iteration scheme, see Zienkiewicz [20].

The elements in $\nabla_u g$ are estimated from

$$\frac{dg}{du_i} = \frac{\partial g}{\partial u_i} + \sum_{j=1}^{J} \frac{\partial g}{\partial a_j} \frac{\partial a_j}{\partial u_i} \tag{43}$$

where the elements in $\frac{\partial \overline{a}}{\partial u_i}$ are determined from

$$\frac{\partial \overline{a}}{\partial u_i} = \overline{\overline{K}}_T^{-1} \frac{\partial \overline{P}}{\partial u_i} \tag{44}$$

where the elements in the tangent stiffness matrix $\overline{\overline{K}}_T$ are given by, see Zienkiewicz [20]

$$K_{T_{ij}} = \frac{\partial f_i}{\partial a_j} \tag{45}$$

As in Section 6.1 $\frac{\partial g}{\partial u_i}$, $\frac{\partial g}{\partial a_j}$ and $\frac{\partial \overline{P}}{\partial u_i}$ are calculated numerically.

The above formulas require that the exact $\overline{\overline{K}}_T^{-1}$ is known in each iteration in the Rackwitz-Fiessler algorithm. The geometrical non-linear case is thus much more computer time consuming than the linear case.

$\frac{d\beta}{dz_j}$ is estimated using Eq. 39 where $\frac{\partial \overline{a}}{\partial z_j}$ is determined from

$$\frac{\partial \overline{a}}{\partial z_j} = \overline{\overline{K}}_T^{-1} \left\{ \frac{\partial \overline{P}}{\partial z_j} - \frac{\partial \overline{f}}{\partial z_j} \right\} \tag{46}$$

6.3 Calculation of $\nabla_u g$ and $\frac{d\beta_i}{dz_j}$ in a Linear FEM Model With Stochastic Variables in The Stiffness Matrix $\overline{\overline{K}}$

The failure function and the finite element equations are written

$$g(\overline{u}, \overline{z}, \overline{a}(\overline{u}, \overline{z})) = 0 \tag{47}$$

$$\overline{\overline{K}}(\overline{u}, \overline{z}) \overline{a}(\overline{u}, \overline{z}) = \overline{P}(\overline{u}, \overline{z}) \tag{48}$$

The elements in $\nabla_u g$ are estimated using Eq. 43 where $\frac{\partial \bar{a}}{\partial u_i}$ is determined from

$$\frac{\partial \bar{a}}{\partial u_i} = \bar{\bar{K}}^{-1} \left\{ \frac{\partial \bar{P}}{\partial u_i} - \frac{\partial \bar{\bar{K}}}{\partial u_i} \bar{a} \right\} \tag{49}$$

The difference from the case considered in Section 6.1 is that it is now necessary to determine one $\bar{\bar{K}}^{-1}$ in each iteration in the Rackwitz-Fiessler algorithm one $\bar{\bar{K}}^{-1}$ calculation was adequate for each reliability index calculation in the case without stochastic variables in $\bar{\bar{K}}$.

$\frac{\partial \beta}{\partial z_j}$ is determined as in Section 6.1 (using Eqs. 39 and 40).

6.4 Calculation of $\nabla_u g$ and $\frac{d\beta_i}{dz_j}$ in a Geometrical Non-Linear FEM Model With Stochastic Variables in Connection With The Stiffness Calculations

The geometrical non-linear case with stochastic variables in the stiffness calculations is a combination of the cases in Sections 6.2 and 6.3.

The failure function and the finite element equations are written

$$g(\bar{u}, \bar{z}, \bar{a}(\bar{u}, \bar{z})) = 0 \tag{50}$$

$$\bar{f}(\bar{u}, \bar{z}, \bar{a}(\bar{u}, \bar{z})) = \bar{P}(\bar{u}, \bar{z}) \tag{51}$$

The gradient $\nabla_u g$ is calculated using Eq. 43 where $\frac{\partial \bar{a}}{\partial u_i}$ is determined from

$$\frac{\partial \bar{a}}{\partial u_i} = \bar{\bar{K}}_T^{-1} \left\{ \frac{\partial \bar{P}}{\partial u_i} - \frac{\partial \bar{f}}{\partial u_i} \right\} \tag{52}$$

$\frac{d\beta}{dz_j}$ is estimated using Eq. 39 where $\frac{\partial \bar{a}}{\partial z_j}$ is determined from

$$\frac{\partial \bar{a}}{\partial z_j} = \bar{\bar{K}}_T^{-1} \left\{ \frac{\partial \bar{P}}{\partial z_j} - \frac{\partial \bar{f}}{\partial z_j} \right\} \tag{53}$$

7 Example

7.1 Platform Model

The plane model of a tubular steel-jacket platform is shown in Fig. 1. The loads on the platform are assumed to consist of two vertical loads P_1 on the top of the legs and horizontal loads modelling the wave and wind load. P_1 is modelled as a stochastic variable. The horizontal load is assumed to be proportional with the length and the diameter of the tubular members and with the stochastic variable P_2.

The system is modelled by a linear finite element model with the tubular elements divided into the 7 groups shown in Table 1 where the initial geometrical configuration is also shown

Finite element No. 7 models the stiffness of the top side. The modulus of elasticity is 2.1×10^8 kNm^{-2}.

7.2 Shape Optimization Model

A shape optimization problem is formulated as described in Section 2 with the structural weight as the objective function.

As design variables $\bar{z}$ the tubular thicknesses of the element groups 1,2,4,5 and 7 and the three shape variables y_1, y_2 and y_3 are used, see Fig. 1, i.e. $N = 9$. All structural elements are assumed to remain straight. The lower bounds of the thicknesses are 0.02 m, and for the shape variables $(y_1, y_2$ and $y_3)$, 20.0 m, 20.0 m and 40.0 m are used. The upper bounds of the shape variables are chosen as 30.0 m, 35.0 m and 60.0 m.

The minimum acceptable systems reliability index is chosen as $\beta^{\min} = 4.5$.

7.3 Reliability Analysis

Failure of the structural system is defined at level 1 and the systems reliability index is estimated by the Hohenbichler approximation [6] with the 10 significant elements placed in a series system. The 10 elements are selected from 62 failure elements used to model yielding, stability and punching failure. The failure elements are numbered as shown in Fig. 2. The failure functions of the three types of failure elements are described in Sørensen [13].

In Table 2 the statistical characteristics of the basic stochastic variables are shown.

The stochastic variables Y_1–Y_7 model the yield stress of the seven groups of tubular beams. Y_- models the reduced yield stress used in modelling the stability failure elements. The model uncertainty variable Z_1 models uncertainty with yielding failure elements. Z_2 and Z_3 model uncertainty with stability failure elements. Z_4, Z_5 and Z_6 model uncertainty with punching failure elements. The stochastic variables $\ln Y_1, \ldots, \ln Y_7$ are assumed to be correlated with correlation coefficients $\rho = 0.3$. All other stochastic variables are assumed to be independent.

7.4 Shape Optimization Results

The shape optimization problem is solved using the NLPQL-algorithm [12] and the reliability calculations are performed using PRADSS [13].

The results are seen in Table 3 and the optimal geometrical shape is seen in Fig. 3.

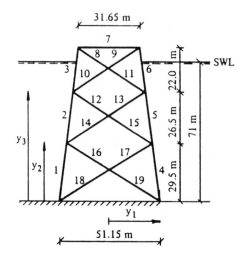

Figure 1. *Geometrical model of offshore platform.*

Table 1. *Groups of tubular elements (initial thicknesses).*

Group	Finite elements No.	Diameter (m)	Thickness (m)
1	1,4	2.01	0.068
2	2,3,5,6	1.99	0.056
3	7	1.20	1.000
4	8,9	1.02	0.052
5	10,11	1.00	0.040
6	12,13,14,15	1.10	0.056
7	16,17,18,19	1.40	0.056

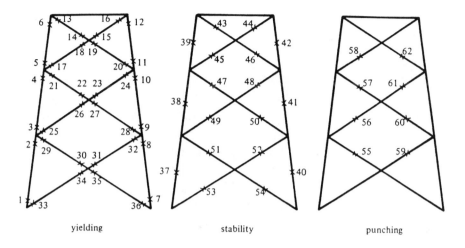

yielding   stability   punching

Figure 2. *Numbering of failure elements.*

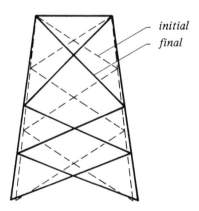

initial
final

Figure 3. *Optimal geometry.*

632

Table 2. *Statistical characteristics (EX1 : extreme type 1, N : normal, LN : log-normal).*

No.	Basic variable	Distribution	Expected value	Stand. deviation
1	P_1	N	15000.	7500.
2	P_2	EX1	1500.	525.
3	Y_1	LN	340000.	34000.
4	Y_2	LN	340000.	34000.
5	Y_3	LN	340000.	34000.
6	Y_4	LN	340000.	34000.
7	Y_5	LN	340000.	34000.
8	Y_6	LN	340000.	34000.
9	Y_7	LN	340000.	34000.
10	Z_1	N	1.0	0.01
11	Y_-	LN	34000.	3400.
12	Z_2	N	1.2	0.144
13	Z_3	N	0.5	0.019
14	Z_4	N	1.0	0.050
15	Z_5	LN	1.161	0.2113
16	Z_6	LN	1.227	0.1706

7.5 Sensitivity Analysis

In Eq. 30 it is seen that the matrix on the left-hand side is independent of parameter p_l due to which the sensitivity is determined, i.e. it is only necessary to calculate the vector on the right-hand side in Eq. 30 for each new parameter and to solve the equations to determine the sensitivities.

Here only the sensitivities due to the expected values and standard deviations of the two loads P_1 and P_2 are determined and shown in Table 4.

The third column in Table 4 show $\frac{dW}{dp_l}$ multiplied by 1% of p_l so the sensitivities are made comparable.

In Table 4 it is seen that, as expected, the sensitivities with respect to the horizontal load are higher than the sensitivities for the vertical load.

The above example demonstrates the techniques described in the previous Sections in the linear case. It must be emphasized that it is necessary to be very careful in the numerical implementation, especially in the calculation of the second derivatives because these are very sensitive to small inaccuracies in the reliability calculations.

Table 3. *Results of shape optimization.*

Opt.	Initial	Optimum
z_1	0.068	0.0492
z_2	0.056	0.0498
z_3	0.049	0.0200
z_4	0.038	0.0200
z_5	0.052	0.0201
z_6	0.052	0.0200
z_7	25.2	27.69
z_8	28.5	20.00
z_9	54.4	40.00
W	118.0	72.7
β^S	4.72	4.50

Table 4. *Results of sensitivity analysis.*

p_l	dW/dp_l	$dW/dp_l \cdot 0.01\, p_l$
$p_1 = E(P_1)$	2.95×10^{-4}	0.044
$p_2 = \sigma(P_1)$	3.63×10^{-4}	0.027
$p_3 = E(P_2)$	1.41×10^{-2}	0.21
$p_4 = \sigma(P_2)$	2.29×10^{-2}	0.12

8 Conclusions

Reliability-based shape optimization problems are formulated based on requirements for systems reliability measures. The design variables are related to sizing variables and shape variables.

It is shown how the reliability measures are determined and how sensitivity measures can be obtained with respect to the design variables. Both linear and non-linear finite element models are considered.

Further, it is shown how a sensitivity analysis of the optimal design can be performed. The analysis gives the sensitivity of the objective function (usually related to the expected cost of the structure) with respect to some quantities which are not modelled as optimization variables or as stochastic variables.

A simple example is considered where a linear plane model of an offshore steel jacket platform is shape optimized and the sensitivities due to some parameters in the model are determined.

References

[1] Bjerager, P. and Krenk, S., (1987), 'Sensitivity Measures in Structural Reliability Analysis,' *Proc. of the first IFIP WG 7.5 Working Conf.*, pp. 459–470.

[2] Bjerager, P. and Krenk, S., (1989), 'Parametric Sensitivity in First Order Reliability Analysis,' *ASCE, Journal of Engineering Mechanics.*

[3] Enevoldsen, Ib, Sørensen, J. D. and Thoft-Christensen, P., (1989), 'Shape Optimization of Mono-Tower Offshore Platform,' Presented at OPTI 89 Conference on Computer Aided Optimum Design of Structures, Southampton.

[4] Frangopol, D. M., (1985), 'Sensitivity of Reliability-Based Optimum Design,' *ASCE, Journal of Structural Engineering* **111**, 8.

[5] Gill, P. E., Murray, W. and Wright M. H., (1981), *Practical Optimization,* Academic Press.

[6] Gollwitzer, S. and Rackwitz, R., (1986), 'An Efficient Numerical Solution to the Multinormal Integral,' Berichte zur Zuverlassigkeitstheorie der Bauwerke, Heft 80, LKI, Technische Universitat München.

[7] Guenard, Y. and Cornell, C. A., (1986), 'A Method for the Reliability Analysis of Steel-jacket Offshore Platforms under Extreme Loading Condition,' *9th Advances in Reliability Technology Symposium.*

[8] Kirkegaard, P. H., Enevoldsen, Ib, Sørensen J. D. and Brincker, R., (1989), 'Reliability Analysis of a Mono-Tower Platform,' Submitted to *Journal of Offshore Mechanics and Arctic Engineering ASME.*

[9] Madsen, H. O., (1988), *Sensitivity Factors for Parallel Systems,* Technical Note, Danish Engineering Academy, Denmark.

[10] Madsen, H. O., Krenk, S. and Lind, N. C., (1986), *Methods of Structural Safety,* Prentice Hall.

[11] Murotsu, Y., Kishi, M., Okada, H., Yonezawa, M. and Taguchi, K., (1984), 'Probabilistically Optimum Design of Frame Structure,' *11th IFIP Conf. on System Modelling and Optimization,* Springer-Verlag, pp. 545–554

635

[12] Schittkowski, K., (1986), 'NLPQL, A FORTRAN Subroutine Solving Constrained Non-Linear Programming Problems,' *Annals of Operations Research*.

[13] Sørensen, J. D., (1987), 'PRADSS: Program for Reliability Analysis and Design of Structural Systems,' Structural Reliability Theory, Paper No. 36, The University of Aalborg, Denmark.

[14] Sørensen, J. D., (1986), 'Reliability-Based Optimization of Structural Elements,' Structural Reliability Theory, Paper No. 18, The University of Aalborg, Denmark.

[15] Sørensen, J. D., (1987), 'Reliability-based Optimization of Structural Systems,' *13th IFIP Conference on System Modelling and Optimization*, Tokyo, Japan.

[16] Sørensen, J. D., (1988), 'Probabilistic Design of Offshore Structural Systems,' *Proc. 5th ASCE Spec. Conf.*, pp. 189–193, Virginia.

[17] Thoft-Christensen, P. and Murotsu, Y., (1986), *Application of Structural Systems Reliability Theory*, Springer Verlag.

[18] Thoft-Christensen, P. and Sørensen J. D., (1987), 'Recent Advances in Optimal Design of Structures from a Reliability Point of View,' *Quality and Reliability Management* 4, 1, 19–31.

[19] Vanderplaats, G. N., (1984), *Numerical Optimization Techniques for Engineering Design*, McGraw-Hill.

[20] Zienkiewicz, O. C., (1977), *The Finite Element Method*, 3rd edition, McGraw-Hill.

Determination of $\dfrac{\partial \rho_{jk}}{\partial z_i}$

Estimation of the sensitivity of the correlation coefficient $\frac{\partial \rho_{jk}}{\partial z_i}$ with respect to a design parameter z_i can e.g. be performed as described in Madsen [9] or Bjerager and Krenk [2]. In this appendix it is shown how $\frac{\partial \rho_{jk}}{\partial z_i}$ can be estimated using the Kuhn-Tucker conditions. Both parallel and series systems are considered.

If the reliability index is determined from (14) (parallel systems) then

$$\frac{\partial \rho_{jk}}{\partial z_i} = \frac{\partial \overline{\alpha}_j^T}{\partial z_i} \overline{\alpha}_k + \frac{\partial \overline{\alpha}_k^T}{\partial z_i} \overline{\alpha}_j \tag{A1}$$

If $\overline{\alpha}_j = \frac{\nabla_u g_j}{|\nabla_u g_j|}$ is used then

$$\frac{\partial \overline{\alpha}_j}{\partial z_i} = \frac{-d\nabla_u S_j}{|\nabla_u g_j|} + \frac{\nabla_u g_j d\nabla_u g_j^T \nabla_u g_i}{|\nabla_u g_j|^3} \tag{A2}$$

where $d\nabla_u S_j = \frac{\partial \nabla_u g}{\partial z_i} + \Delta g_j \frac{\partial \overline{u}}{\partial z_i}$ and $\{\Delta g_j\}_{kl} = \frac{\partial^2 g_j}{\partial u_k \partial u_l}$

$\frac{\partial \overline{u}}{\partial z_i}$ can in principle be determined using the sensitivity analysis technique described in Section 5. The Kuhn-Tucker conditions for the optimization problem in Eq. 12 are formulated as in Eq. 30. The following system is then obtained when L_A constraints are active

$$\begin{bmatrix} \overline{\overline{I}} + \sum_{l=1}^{L_A} \lambda_l \Delta g_l & \overline{\overline{D}} \\ \overline{\overline{D}}^T & \overline{\overline{0}} \end{bmatrix} \begin{bmatrix} \frac{\partial \overline{u}}{\partial z_i} \\ \frac{\partial \overline{\lambda}}{\partial z_i} \end{bmatrix} = \begin{bmatrix} -\overline{\overline{E}}\,\overline{\lambda} \\ -\frac{\partial \overline{g}}{\partial z_i} \end{bmatrix} \tag{A3}$$

where $\overline{\overline{D}} = \{\nabla_u g_1, \ldots, \nabla_u g_{L_A}\}$, $\overline{\overline{E}} = \left\{ \frac{\partial \nabla_u g_1}{\partial z_i}, \ldots, \frac{\partial \nabla_u g_{L_A}}{\partial z_i} \right\}$ and $\overline{\lambda}$ is a vector containing the multipliers of the active constraints.

The solution of the system in (B4) gives $\frac{\partial u_j}{\partial z_i}$ and $\frac{\partial \rho_{jk}}{\partial z_i}$ can then be determined from (A1).

If the system reliability index is determined from (10) (series system) and the elements in the series system are also determined from the optimization problem in (12) with just one constraint ($L=1$) then (A1) with (8) can be formulated as

$$\frac{\partial \rho_{jk}}{\partial z_i} = \left(\frac{\partial \overline{u}_j}{\partial z_i} \frac{1}{\beta_j} - \frac{\overline{u}_j}{\beta_j^2} \frac{\partial \beta_j}{\partial z_i} \right)^T \frac{\overline{u}_k}{\beta_k} + \frac{\overline{u}_j^T}{\beta_j} \left(\frac{\partial \overline{u}_k}{\partial z_i} \frac{1}{\beta_k} - \frac{\overline{u}_k}{\beta_k^2} \frac{\partial \beta_k}{\partial z_i} \right)$$

$$= \frac{1}{\beta_j \beta_k} \left(\frac{\partial \overline{u}_j^T}{\partial z_i} \overline{u}_k + \frac{\partial \overline{u}_k^T}{\partial z_i} \overline{u}_j \right) - \rho_{jk} \left(\frac{1}{\beta_j} \frac{\partial \beta_j}{\partial z_i} - \frac{1}{\beta_k} \frac{\partial \beta_k}{\partial z_i} \right) \tag{A4}$$

where $\frac{\partial \bar{u}_i}{\partial z_i}$ is determined from (A3) with failure element j as the only active element. $\frac{\partial \beta_i}{\partial z_j}$ is determined as described in Section 6.

If the systems reliability index is determined from (14) with each element modelling an equivalent failure element for a parallel system (Eqs. (12)–(17)) (series system of parallel systems) then the derivatives can in principle be determined using the above formulas and (18)–(20). The resulting formulas are not written here.

Genetic Algorithms in Automated Structural Synthesis

P. Hajela
Department of Aerospace Engineering,
Mechanics and Engineering Science
University of Florida, Gainesville
United States of America

Abstract   The present paper describes an implementation of genetic search methods in the optimum synthesis of structural systems. These methods represent an adaptation of principles of natural selection and evolution to conduct a randomized search in the most promising regions of the design space. The approach requires only an evaluation of the objective function and constraints. Furthermore, the search proceeds by updating an entire population of designs, an attribute that makes the method less susceptible to convergence to a local optimum. The focus of the paper resides in examining the numerical characteristics of genetic algorithms. An adaptation of the approach in simultaneous analysis and design is also presented. The approach is combined with finite element analysis for the design of truss and frame structures.

1   Introduction

Among the simplest strategies for function optimization are the enumerative search techniques, where the function is determined at a very large number of predetermined points. Successive refinement and evaluation reveals the optimum design. An improvement on simple enumeration are methods such as random walk and random walk with direction exploitation [1]. For large problems, these methods cannot be expected to perform much better than exhaustive enumeration. Such strategies are computationally burdensome, even moreso in structural design where the behavioral response constraints typically require solutions of very large systems of algebraic equations.

Nonlinear programming algorithms have emerged as among the most generally applicable and widely used methods for the optimum design of structural systems [2–4]. These methods are extremely efficient and robust when applied to

B. H. V. Topping (ed.),
Optimization and Artificial Intelligence in Civil and Structural Engineering, Volume I, 639–653.
© 1992 *Kluwer Academic Publishers. Printed in the Netherlands.*

problems that present properly scaled and unimodal design spaces. The more attractive of this class of methods are gradient based, and require at least the first derivative of the objective function and constraints with respect to the design variables. In such methods, a design is initialized and then recursively updated along search directions computed from the gradient information. In the event that the design space is nonconvex, the selection of the initial design point becomes critical in determining the global optimum.

It is critical to examine alternative methods that are not as computationally demanding as the enumerative search or random walk methods, and are also less susceptible to pitfalls of the 'hill-climbing' nonlinear programming algorithms. Genetic algorithms are proposed to fill this gap, and their implementation in a series of structural design problems, is the subject of the present paper. Genetic algorithms are highly exploitative search techniques, based on a randomized selection from that restricted region of the design space which yields an improvement in the objective function. The presence of nonconvexities or disjointness in the design space does not affect their performance. This is largely attributed to their lack of dependence on function gradients and on the fact that design proceeds in parallel from several points, and an entire population of designs is updated over a cycle. In this regard, they are distinct from traditional optimization methods.

Genetic search derives from Darwin's theory of survival of the fittest. Analogous to a natural process where a population of a given species adapts to a natural habitat, a population of designs is created and allowed to adapt to the design space. Conditions favorable to the process of adaptation may be a minimum weight configuration with no constraint violation. The design alternatives representing a population in a given generation are allowed to reproduce and cross among themselves, with bias allocated to the most fit members of the population. Combination of the most desirable characteristics of mating members of the population results in a progeny population that is more fit than the parent population. In a situation where the initial population corresponded to infeasible designs, there would be an improvement in the constraint violation. Hence, if the measure by which fitness of a generation is indicated is also the desired objective of a design process, successive generations produce better values of the objective function.

There is a range of structural synthesis problems for which the use of genetic algorithms is particularly attractive. Design of statically loaded beam grillage structures for stress and displacement constraints yields nonconvex design spaces [5]. Likewise, harmonically loaded structures with dynamic stress or displacement constraints present nonconvex or disjoint design spaces [6,7], dependent upon the inclusion of structural damping in the analytical model. As will be shown in subsequent sections of the paper, genetic algorithms work from a representation of the design variables rather than the variables themselves. This feature allows one to handle a design space with a mix of continuous, discrete, and integer design variables without resorting to expensive branch-and-bound strategies. Lastly, since genetic search methods work by updating entire populations of design, it is possible to simultaneously determine designs corresponding to various local optima that may exist in the design space.

The parallelism involved in updating the entire population of designs, and which contributes significantly to circumventing convergence to a local optimum, is not without cost. The computational requirements are high, albeit not to the extent of exhaustive enumeration. Also the computational requirements do not increase with problem dimensionality to the same extent observed with traditional gradient based methods.

Subsequent sections of the paper present the terminology of genetic search, its principal components, and key concerns in its numerical implementation. The application of the method on a series of problems drawn from the literature serves to enumerate both the advantages and shortcomings of the proposed approach. A discussion of current work aimed at enhancing the efficiency of the method and exploring its potential applications in structural design is also included.

2   Genetic Algorithms

The central idea behind genetic algorithms was first proposed by Holland [8], where possible solutions to a given problem were coded by bit string representations, and transformations analogous to biological reproduction and evolution were used to vary and improve these coded solutions. An analogy was drawn with the natural process of reproduction in biological populations, where genetic information stored in chromosomal strings evolves over generations to adapt favorably to a changing environment. This chromosomal structure represents the generational memory, and is altered when members of the population reproduce. The three basic processes which affect the chromosomal makeup in natural evolution are inversions of chromosome strings, an occasional mutation of genetic information, and a crossover of genetic information between the reproducing parents. The last process is an exchange of genetic material between the parents, and allows for beneficial genes to be represented in the progeny. A mathematical simulation of these processes makes up the principal components of genetic algorithms. This is discussed in greater detail in [9] and is summarized here for completeness.

At the very outset, one needs a scheme or code by which to represent possible solutions to the problem. Binary coding of individual design variables, and stacking of such binary digits end-to-end, is used in the present work. The manipulation of these bit strings by transformations representing the biological reproduction process, is another important component of these algorithms. Such manipulations are typically governed by a numerical fitness factor associated with each member of the design population. The definition of this factor has some influence on the convergence characteristics of the algorithm. Other parameters which are critical

642

to the performance of genetic search are the optimal population size, lengths of chromosome strings representing a design, and the frequency and manner in which the transformation functions simulating biological reproduction are invoked. Some discussion of these problems is available in [10]. The present paper also examines the influence of these parameters in representative structural design problems. The mechanics of genetic search are discussed next in the context of a structural optimization.

2.1 The Fitness Function

In typical structural sizing problems, the structural weight is usually minimized, subject to a range of design constraints on the structure behavioral response. In the parlance of genetic search, the minimum weight structure with no constraint violation can be considered to be the most fit. The constraint functions can be appended to the structural weight as in an exterior penalty function formulation [1] to obtain the objective function. However, since the fitness of the population is maximized in an evolutionary process, an inverse of the objective function is defined as the fitness function. Another possibility is to prescribe a constant which is larger than the maximum objective value, and to subtract the objective function from this constant to obtain the fitness function. Both of these strategies were used in the present work.

2.2 The Design Representation Scheme

In biological systems, generational memory is preserved and transferred to progeny populations in the form of chromosomes, which are a specific ordering (loci) of genes along a string-like structure. Several such chromosome strings comprise a genotype, which is the total genetic makeup for an organism. An analogous prescription that defines a complete physical system is referred to as a structure, and consists of features or descriptors that occupy designated positions on a chromosome-like string. A typical structural design problem is completely described by the design variables which must be varied to attain an optimum. In order to transform these design variables into a string-like structures for processing by genetic transform operators, a simple yet effective approach is to represent them by a fixed length binary coded number. As an example, each design variable may be represented by a p-digit binary number, and for n design variables, these n strings may be stacked end to end, to create a $p \times n$ length structure.

Just as in biological processes, the location of the 0's and 1's along the string is significant. For a 5-digit binary representation, the minimum and maximum values of the design variables (lower and upper bounds) may be represented by $\boxed{0}\boxed{0}\boxed{0}\boxed{0}\boxed{0}$ and $\boxed{1}\boxed{1}\boxed{1}\boxed{1}\boxed{1}$, respectively, with a linear scaling used to obtain design variable values in between. Two characteristics of such a representation are particularly significant and are explained as follows. First, numbers of the type $\boxed{1|*|*|*}$ and $\boxed{0|0|0|*|*}$ denote large and small values of design variables, respectively, where '$*$' is a wild-card designation admitting values of 0 or 1. That is, presence of 1's to the beginning end of the string favors larger design variable values, which in turn may be related to larger structural weight or lower constraint violations. The second feature pertains to special arrangements of portions of the string, or schema [9], which have identifiable physical characteristics. An example of this may be strings of the type $\boxed{1|1|*|*|*}$ or $\boxed{1|*|*|0|*}$, where the consequences of specific locations of 0's and 1's may be as above, with additional significance attached to the occurrence of combinations of numbers over a certain string length.

The number of fixed 0's or 1's in a string is referred to as the order of the string. Likewise, the distance between the first and last specified position on a string is the defining length. The significance of these parameters will become obvious when the transform operators of genetic search are discussed. It suffices to say here that a coding where low order, short defining length schemata model the pertinent features of the problem, are preferred to obtain the best performance from the genetic search. Furthermore, since we look for similarity templates along strings to identify favorable physical characteristics of the structure (presence of 1 at the beginning of string means lower constraint violation), it is logical to select the smallest number of characters that can suitably represent the problem. A 0-1 binary representation is considered more effective than a nonbinary representation based on a 30 character alphabet.

For a m-digit string, $2m$ schemata are possible. If the population consists of n strings, then depending on the uniqueness of the population, the total number of schemata possible are between $2m$ and $n \cdot 2m$. It is possible to show [11] that in each generation of evolution requiring n function evaluations, the number of schemata actually processed is of the order n^3. This large amount of processing ability was described by Holland as implicit parallelism, and reflects the power of genetic search.

2.3 Transform Operators in Genetic Search

Various simulations of genetic evolution and adaptation are conceivable. The three principal components of this process are reproduction, crossover, and mutation. Let us consider a structural design process in which an initial population of designs has been created, the designs have been represented by artificial chromosome strings, and a fitness function has been defined. The simulation of genetic evolution described here is somewhat contrived in that the population size is maintained at a fixed level, and two mating pairs produce two progenies but are themselves eliminated. The process of reproduction is one which simply biases the search towards producing more fit members in the population and eliminating the less fit ones. One simplistic approach of selecting members for reproduction is by constructing a biased roulette wheel, with each member occupying an area on the wheel in proportion to its fitness function. Spins of such a wheel may be simulated to select from the initial pool, a pool of the same size but with higher average fitness. Note that this process does not change or create new schema. It simply makes available for the crossover and mutation transforms, the most favorable schema from the initialized population.

The actual generation of new schema is done in the crossover transform, which is akin to transfer of genetic material in biological reproduction processes facilitated by DNA strings. Crossover entails selecting a pair of mating strings, selecting sites on these strings for exchanging schema information, and simply swapping the string lengths between the chosen sites among the mating pair. As an example, consider string lengths of 20, for which crossover sites were selected at random and indicated by an underline. The two parent strings cross to produce two child strings as follows.

$$\text{Parent1} = \boxed{1|1|0|1|1|0|0|1|1|1|1|0|0|1|1|0|0|0|0|1}$$

$$\text{Parent2} = \boxed{0|0|1|1|0|0|1|0|1|0|1|0|1|1|1|0|0|0|1|1}$$

$$\text{Child1} = \boxed{1|1|0|1|0|0|1|0|1|0|1|0|1|1|1|0|0|0|0|1}$$

$$\text{Child2} = \boxed{0|0|1|1|1|0|0|1|1|1|1|0|0|1|1|0|0|0|1|1}$$

As is obvious from this exercise, new schema have been created in the crossover operation. Furthermore, the wisdom of using a coding scheme which would produce schema of short defining lengths is underscored. Schemas with long defining lengths are more likely to be disrupted by the crossover operation, resulting in an instability or introduction of an oscillation in the genetic search. Invoking of the crossover transformation is done on the basis of probabilistic transition rules. A probability of crossover p_c is defined, and is used to determine if crossover should be implemented. In the present work, this was achieved by a simulation of a biased coin toss. Mutation is the third step in this genetic refinement process, and is one that safeguards the process from a complete premature loss of valuable genetic material during the reproduction and crossover. This step corresponds to selecting few members of the population, determining at random a location on the strings, and switching the 0 or 1 at that location. This step is also invoked on the basis of a biased coin toss, but with a probability of mutation p_m. DeJong has conducted numerical experiments for determining nominal values of p_c and p_m [10], and this issue was also examined in the context of structural design problems.

The steps described above are repeated for successive generations of the population, until no further improvement in the fitness is attainable. The member in the final generation with the highest level of fitness is considered as the optimum design.

3 Scaling in Genetic Search

Since most structural design problems are initiated by seeding a population of designs at random, and since the population size is finite and must be kept small, special attention must be given to the definition of the fitness function. It is not rare to start the process with a few members of the population with a very high fitness in relation to others. If raw values of fitness are used in determining the reproductive pool, the more fit members would quickly dominate the population pool and result in premature convergence. Hence, the raw fitnesses must be scaled to prevent this from occurring. This is also of concern when the population is largely converged, and where average population fitness is close to the maximum fitness. In such a situation, genetic search exhibits a phenomenon similar to random walk, and fitnesses must be scaled to accentuate the differences between competing members of the population. Various forms of scaling including linear, sigma truncation, and power law scaling have been proposed [12]. A simple linear scaling was implemented in the present work, and was based on the premise that the scaled average fitness was kept equal to the average raw fitness. If we define the scaled fitness f' as,

$$f' = c_1 f + c_2$$

and further assume that the scaled maximum fitness is $k \cdot f_{avg}$, where k is typically chosen to range between 1 and 2. For a value of $k = 1.5$, the constants c_1 and c_2 can be written as follows.

$$c_1 = \frac{0.5 \cdot f_{avg}}{(f_{max} - f_{avg})}$$

$$c_2 = \frac{f_{avg} (f_{max} - 1.5 \cdot f_{avg})}{(f_{max} - f_{avg})}$$

In a mature run when above transformation can produce negative scaled fitnesses, the scaling is revised by using a scaled fitness lower bound of zero. An example of using such scaled fitnesses is presented in a later section.

4 Implementation in Structural Design

The procedure described in the preceding sections was implemented as a modular program written in Fortran. The program requires, in addition to the input parameters, a subroutine to evaluate the fitness function for a given design. In the present work, the program was integrated with a finite element analysis capability EAL [13], through a series of pre- and post- processors. This section describes the use of the approach to determine the optimum solution in simple truss and frame structural systems. A closer look at structural design problems involving nonconvex and disjoint design spaces by this approach is available in [14].

4.1 Portal Frame

A single bay, single story portal frame was sized for minimum weight and for plastic collapse. The frame was loaded as shown in Fig. 1 and had a different cross section in the horizontal and vertical members, resulting in two design variables. A total of six collapse mechanisms are possible [15], resulting in six inequality constraints. The optimal weight of the structure is 1.5 units, with optimum values of each of the two design variables obtained as 0.5. This example was used to experiment with the parameters of the genetic algorithm to determine their influence on the convergence characteristics.

Two sets of minimum and maximum values for the design variables was selected. The first set allowed a variation between 0.01 and 1.0 units, and the second between 0.3 and 0.75. Population sizes of $n = 50$, 70, 90 and 120 were considered in the experiment. Each design variable was represented by a 30 character binary string. The design variables and objective function at the end of 30 generations of evolution are shown in Table 1. Only, the results of $n = 90$ and $n = 120$ are acceptably close to the optimum. However, the crossover and mutation probabilities have some influence on these results, as populations with $n = 50$ and 70 did discover designs close to the optimum in earlier stages but subsequently lost them as a result of schema disruption. Values of $p_c = 0.8$ and $p_m = 0.009$ were used in these simulations.

For a population size of $n = 90$ and $p_c = 0.8$, values of $p_m = 0.005$, 0.007, 0.015, and 0.05 were selected. Results closest to the optimum were obtained for p_m ranging from 0.009–0.012. Values of $p_m = 0.005$ and 0.007 yielded poor results even though the convergence pattern was quite stable. Higher values of p_m had a disruptive influence on the convergence behavior. The process was repeated by fixing $p_m = 0.009$, and varying p_c by assuming values of 0.5, 0.6, 0.7 and 0.8. For a value of $p_c = 0.5$, the process quickly converged to design variable values of 0.634 and 0.603 in about 12 generations, and made no significant moves from this point. Increasing p_c resulted in a more active exchange of schema at later stages in the evolution.

The effect of fitness scaling was studied with a small population size of $n = 25$, where its influence was expected to be more pronounced. With no scaling, the process converged in 6 generations to design variable values of 0.488 and 0.6146, and an objective function value of 1.597. With linear scaling, and values of $k = 1.2$ and 1.5, there was evidence of evolutionary activity upto the final generation, even though the limited population size was not expected to yield good results.

4.2 Six-Bar Truss

A six bar truss shown in Fig. 2 was sized for minimum weight, with constraints on allowable stresses in each member, and maximum permissible vertical displacements at nodes 1 and 2. The structure was subjected to two load conditions $P1$

and $P2$, as shown in the figure. The genetic algorithm was used with a fixed population size of $n = 90$, and probabilities of crossover and mutation set to $p_c = 0.8$ and $p_m = 0.009$, respectively. A comparison of the exact solution with the results obtained in 30 generations of evolution, is presented in Table 2.

4.3 Three-Bar Truss

This statically indeterminate truss, shown in Fig. 3, was chosen to demonstrate the concept of an integrated formulation in genetic search. The equations of equilibrium and those relating stress and displacements were treated as additional equality constraints in the problem. This required an addition of three member stresses and two displacements to the design variable set, bringing to seven the total number of variables for the problem. The representation of equality constraints as two equivalent inequalities resulted in a total of thirteen constraints for the problem. As above, the parameters of the population were selected as $n = 90$, $p_c = 0.8$ and $p_m = 0.009$. The population was allowed to evolve over 30 generations. A comparison of the results with the exact solution is shown in Table 3.

4.4 Twentyfive-Bar Truss

This space truss is shown in Fig. 4 and is sized for minimum weight and constraints on allowable stresses and displacements. The material properties and load conditions were selected as in [15]. The purpose of this example was to exercise the finite element based design capability used in conjunction with the genetic search methods. Parameters for the genetic search were selected as $n = 70$, $p_c = 0.8$, $p_m = 0.009$, and the genetic reproduction allowed to proceed for 20 generations. The final results and their comparisons with exact solutions are shown in Table 4.

5 Discussions and Closing Remarks

The use of genetic algorithms in optimum structural synthesis has been explored with a class of representative problems. Some broad conclusions may be drawn on the suitability of these methods in structural design. First, no guarantee exists that the method will converge to an optimum solution. This is in contrast to gradient based methods which guarantee convergence to the nearest stationary point. However, genetic search methods are good at identifying regions in which the global optimum may exist. A refined solution of the optimization problem is sometimes difficult to obtain, especially if the minimum to maximum variations of the design variables are large. This may be the case if no a *priori* information about the optimum is known. An approach that works well to circumvent the problem is to reduce the range of variation after a few generations of evolution, and when the design variables appear to oscillate in a narrower range.

648

In constrained optimization problems, the inclusion of constraints into the objective function by the penalty function approach has to be handled carefully. In the present work the penalty multiplier was selected so that the constraints contributed a term to the penalty function which was of the same order of magnitude as the objective function. The penalty multiplier was adjusted with the generation count, as the more violated designs were eliminated. Since the starting population is created by random assignment, the range of constraint violations is large in the first few generations. Hence, special attention must be directed in the selection of penalty parameters to account for constraint violations.

The problem parameters of genetic search were varied to examine their influence on the convergence pattern. The results presented for the portal frame problem cannot be considered general, as other numerical evidence indicates that the convergence behavior is problem dependent. However, some general remarks can be made that are qualitatively accurate. Probabilities of crossover of 0.7 to 0.8 were the most effective in obtaining good convergence behavior. Very low values of p_c are undesirable as they limit the exchange of schema between fit members of the population, and hence also the rate at which fitness of the population is enhanced. The results of varying the mutation rate were as expected. Low values of mutation rate are undesirable as early loss of genetic information or schema from the strings sometimes yields poor results. On the other hand, very high values of mutation rate introduce an instability in the convergence pattern. The population size was not of significance in the relatively small problems attempted here. However, scaling of fitness is most significant in small population sizes, where a lack of scaling results in a domination by the initially fit members.

The principal drawback of the approach continues to be the large number of function evaluations necessary to obtain the optimum. Numerical evidence in a class of problems with known nonconvexities in the design space indicates that the region of global optimum is identified within the first few generations. A hybrid approach has been implemented, wherein a feasible directions search is initiated with the most fit design after a few iterations of genetic search. It is worthwhile to note that genetic search may perform quite well with only the information that a design is feasible or infeasible. Consequently, approximate methods of constraint evaluation may be quite useful in this approach. Another possibility, and one currently under investigation, is the implementation of neural computing to determine the feasibility of design. At the present, however, genetic search is most useful for problems with nonconvex design spaces.

References

[1] Rao, S. S., (1978), *Optimization: Theory and Applications*, pp. 251–257, Wiley Eastern Limited, New Delhi.

[2] Fox, R. L. and Kapoor, M. P., (1970), 'Structural Optimization in the Dynamic Response Regime,' *AIAA Journal* **8**, 10, 1798–1804.

[3] Gwin, L. B. and Taylor, R. F., (1973), 'A General Method for Flutter Optimization,' *AIAA Journal* **11**, 12, 1613–1617, (December).

[4] Hajela, P. and Lamb, A., (1986), 'Automated Structural Synthesis for Nondeterministic Loads,' *Computer Methods in Applied Mechanics and Engineering* **57**.

[5] Clarkson, J., (1965), *The Elastic Analysis of Flat Grillages*, Cambridge University Press, Cambridge.

[6] Johnson, E. H., (1976), 'Optimization of Structures Undergoing Harmonic or Stochastic Excitation,' SUDAAR No. 501, Stanford University.

[7] Millls-Curran, W. C., (1983), 'Optimization of Structures Subjected to Periodic Loads,' Ph.D. Thesis, University of California, Los Angeles.

[8] Holland, J. H., (1975), *Adaptation in Natural and Artificial Systems*, The University of Michigan Press, Ann Arbor.

[9] Goldberg, D. E., (1989), *Genetic Algorithms in Search, Optimization, and Machine Learning*, Addison Wesley, Reading.

[10] DeJong, K. A., (1975), 'Analysis of the Behavior of a Class of Genetic Adaptive Systems,' Ph.D. Thesis, University of Michigan, Ann Arbor.

[11] Goldberg, D. E., (1985), 'Optimal Initial Population Size for Binary-Coded Genetic Algorithms,' TCGA Report No. 85001, November.

[12] Forrest, S., (1985), 'Documentation for PRISONERS DILEMMA and NORMS Programs that Use the Genetic Algorithm,' Unpublished Manuscript, University of Michigan, Ann Arbor.

[13] Whetstone, D., (1977), *SPAR—Reference Manual*, NASA CR-145098-1, February.

[14] Hajela, P., (1989), 'Genetic Search—An Approach to the Nonconvex Optimization Problem,' *Proceedings of the 30th AIAA/ASME/ASCE/AHS/ASC SDM Conference*, Mobile, Alabama, April.

[15] Haftka, R. T. and Kamat, M. P., (1985), *Elements of Structural Optimization*, Martinus Nijhoff Publishers, Dordrecht.

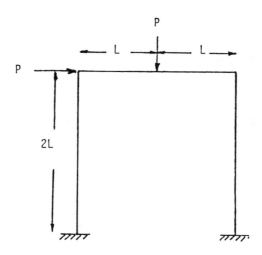

Figure 1. *Portal frame sized for plastic collapse.*

Table 1. *Summary of optimum designs for the portal frame with different population sizes. Asterisk indicates an infeasible design.*

n	$x_{\min} = 0.3$ $x_{\max} = 0.75$			$x_{\min} = 0.01$ $x_{\max} = 1.0$		
	x_1	x_2	OBJ	x_1	x_2	OBJ
120	0.582	0.485	1.649	0.502	0.566	1.571
90	0.495	0.537	1.623*	0.524	0.509	1.550
70	0.730	0.384	1.845	0.488	0.949	2.430*
50	0.581	0.726	1.888	0.664	0.430	1.758

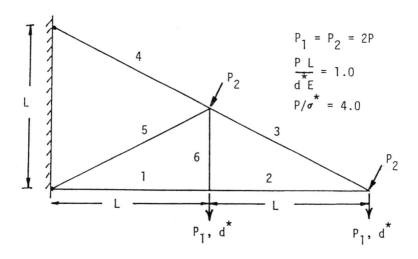

Figure 2. *Geometry and loading for the six bar planar truss.*

Table 2. *Comparison of NLP and GA results for the six bar truss.*

Solution Technique	x_1	x_2	x_3	x_4	x_5	x_6	OBJ
NLP	40.25	40.08	44.72	63.35	10.17	8.08	433.13
GA	40.35	39.65	40.88	63.51	10.81	7.97	432.68

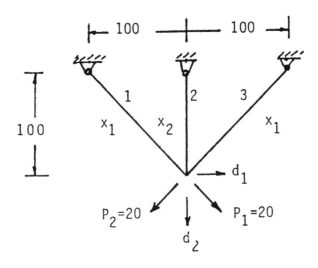

Figure 3. *Geometry and loading for the three bar planar truss.*

Table 3. *Results for the three bar truss problem obtained by a direct NLP based solution, and from a solution of the integrated problem formulation.*

Solution	Areas		Stress Variables			Displacements	
	x_1	x_2	s_1	s_2	s_3	d_1	d_2
Direct Approach	0.788	0.410	20.0	14.62	−5.38	0.084	0.048
Integrated Approach GA	0.833	0.498	19.74	14.84	−5.89	0.083	0.044

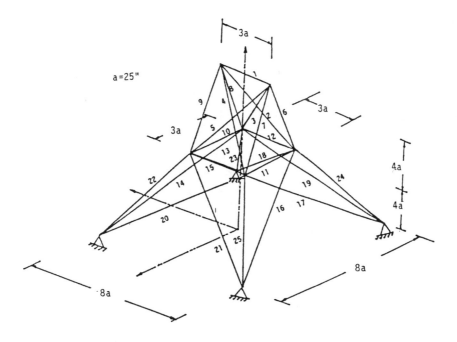

Figure 4. *Twentyfive bar planar truss (Ref. [15]).*

Table 4. *Summary of results for the twentyfive bar truss.*

Members in design variable group	Area (in^2)	
	Genetic Algorithm	Nonlinear Programming
1	0.0742	0.01
2	2.2661	1.9870
3	2.4674	2.9910
4	0.0191	0.0100
5	0.0746	0.0120
6	0.5843	0.6830
7	1.4890	1.6790
8	2.9123	2.6640
Final Weight (lbs)	546.49	545.22